ANNUAL REVIEW OF
PLANT PHYSIOLOGY

EDITORIAL COMMITTEE (1983)

ANNUAL REVIEW OF PLANT PHYSIOLOGY

VOLUME 34, 1983

WINSLOW R. BRIGGS, *Editor*
Carnegie Institution of Washington, Stanford, California

RUSSELL L. JONES, *Associate Editor*
University of California, Berkeley

VIRGINIA WALBOT, *Associate Editor*
Stanford University

ANNUAL REVIEWS INC.   4139 EL CAMINO WAY   PALO ALTO, CALIFORNIA 94306 USA

ANNUAL REVIEWS INC.
Palo Alto, California, USA

COPYRIGHT © 1983 BY ANNUAL REVIEWS INC., PALO ALTO, CALIFORNIA, USA. ALL RIGHTS RESERVED. The appearance of the code at the bottom of the first page of an article in this serial indicates the copyright owner's consent that copies of the article may be made for personal or internal use, or for the personal or internal use of specific clients. This consent is given on the condition, however, that the copier pay the stated per-copy fee of $2.00 per article through the Copyright Clearance Center, Inc. (21 Congress Street, Salem, MA 01970) for copying beyond that permitted by Sections 107 or 108 of the US Copyright Law. The per-copy fee of $2.00 per article also applies to the copying, under the stated conditions, of articles published in any *Annual Review* serial before January 1, 1978. Individual readers, and nonprofit libraries acting for them, are permitted to make a single copy of an article without charge for use in research or teaching. This consent does not extend to other kinds of copying, such as copying for general distribution, for advertising or promotional purposes, for creating new collective works, or for resale. For such uses, written permission is required. Write to Permissions Dept., Annual Reviews Inc., 4139 El Camino Way, Palo Alto, CA 94306 USA.

International Standard Serial Number: 0066-4294
International Standard Book Number: 0-8243-0634-1
Library of Congress Catalog Card Number: A-51-1660

Annual Review and publication titles are registered trademarks of Annual Reviews Inc.

Annual Reviews Inc. and the Editors of its publications assume no responsibility for the statements expressed by the contributors to this *Review*.

PRINTED AND BOUND IN THE UNITED STATES OF AMERICA

PREFACE

The successful publication of the *Annual Review of Plant Physiology* requires three different kinds of individuals. First, there are the editorial committee members, five in number, who meet annually to plan topics and authors for the volume to appear approximately two years later. Second, there are the scientific editors—one editor and two associate editors—who also participate in the annual meeting and in addition read the manuscripts for scientific content and clarity and do the sorts of things a reviewer might do for a manuscript submitted to a scientific journal for possible publication. Finally, and absolutely central to the whole process, there is the production editor—Jean Heavener for the *Annual Review of Plant Physiology*. She reads every manuscript, does detailed copy editing, sees to it that the house rules are followed in editorial matters, queries editors on apparent inconsistencies, corrects everybody's grammar (including that of editors, on occasion), prepares manuscripts for the compositor, and throughout all of this activity keeps the volume precisely on schedule. When articles return to editors for final checking, they normally are in very fine condition, as a consequence—not too different from what you will see in this volume. Without her efforts, all of the noble thoughts and actions of the rest of us would come to a sorry end.

Editorial committee members serve five-year terms that are staggered so that one member rotates off each year. Maarten Chrispeels completed his term at the end of 1982, and we thank him for his five years of effective service. We welcome Deborah Delmer, who will replace him. We also wish to thank Joseph Berry and William Taylor, who served as guest members of the committee at its 1982 meeting. Finally, Paul Green has resigned after 10 years as associate editor. His contributions over these years have been major, and we shall miss him very much. We gratefully acknowledge his important association with this series. We welcome Virginia Walbot, who has accepted the appointment as Paul's replacement, and has been busily editing manuscripts for the current volume.

Winslow R. Briggs
Editor

Annual Review of Plant Physiology
Volume 34, 1983

CONTENTS

(*continued*) vii

ARTICLES IN OTHER *ANNUAL REVIEWS* OF INTEREST TO PLANT PHYSIOLOGISTS

From the *Annual Review of Biophysics and Bioengineering,* Volume 12 (1983):
  Coupling of Proton Flux to the Hydrolysis and Synthesis of ATP, *Jui H. Wang*
  Intracellular Measurements of Ion Activities, *Roger Y. Tsien*
  Mechanisms of Assembly and Disassembly of Microtubules, *John J. Correia and Robley C. Williams, Jr.*
  Protein and Nucleic Acid Sequence Database Systems, *B. C. Orcutt, D. G. George, and M. O. Dayhoff*

From the *Annual Review of Biochemistry,* Volume 52 (1983):
  Comparative Biochemistry of Photosynthetic Light-Harvesting Systems, *A. N. Glazer*
  Proton ATPases: Structure and Mechanism, *L. Mario Amzel and Peter L. Pedersen*
  Mechanism of Free Energy Coupling in Active Transport, *Charles Tanford*
  DNA Methylation and Gene Activity, *Walter Doerfler*
  Ribulose 1,5-Bisphosphate Carboxylase-Oxygenase, *Henry M. Miziorko and George H. Lorimer*
  Fatty Acid Synthesis and its Regulation, *Salih J. Wakil, James K. Stoops, and Vasudev C. Joshi*
  The Pathway of Eukaryotic mRNA Formation, *Joseph R. Nevins*

From the *Annual Review of Ecology and Systematics,* Volume 13 (1982):
  The Molecular Clock Hypothesis: Biochemical Evolution, Genetic Differentiation and Systematics, *John P. Thorpe*
  Interactions Between Bacteria and Algae in Aquatic Ecosystems, *Jonathan J. Cole*
  Sex Change in Plants and Animals, *David Policansky*

From the *Annual Review of Genetics,* Volume 16 (1982):
  Developmental Genetics of Maize, *John D. Scandalios*
  T-DNA of the Agrobacterium TI and RI Plasmids, *Michael Webster Bevan and Mary-Dell Chilton*
  The Genetics of Development in Dictyostelium Discoideum, *Susan S. Godfrey and Maurice Sussman*

From the *Annual Review of Microbiology,* Volume 36 (1982):
  Phycobilisomes: Structure and Dynamics, *A. N. Glazer*
  The Plant Pathogenic Corynebacteria, *Anne K. Vidaver*

Ann. Rev. Plant Physiol. 1983. 34:1–19

ASPIRATIONS, REALITY, AND CIRCUMSTANCES: The Devious Trail of a Roaming Plant Physiologist

Pei-sung Tang

Institute of Botany, Academia Sinica, Peking, China

CONTENTS

OF EMPERORS, REBELLIONS, INVASIONS, AND INDEMNITIES (Personal and historical background)

I was born in the county of Kishui (now Xishui) in Hupeh province, China, on November 12, 1903. Times were hard for the Chinese people in that period under the decaying rule of the Manchu (Ching) dynasty. During the few decades prior to the turn of the present century, the doors of the then self-imposed "Central Kingdom" had repeatedly been pried open by Western invaders, and the nation under Manchurian rule was in danger of, if not actually being, dismembered by the preying powers. Faced with the constant threat of invasion from without and the rising wrath and discontent of the impoverished populace from within, the tottering Manchurian rulers promised political reform by taking steps toward constitutional monarchy.

1

0066-4294/83/0601-0001$02.00

Socially they were forced to introduce modern (European) education by sending students abroad for new knowledge. In short, I was born at a time when the old (imperial rule) was giving way to the new, and when East and West clashed and also mingled, in a land of turmoil and contradiction.

I came from a family of classical Chinese scholars. My father, Tang Hua-lung, received most of his training in Chinese classics from private tutors. Upon passing the imperial (civil) examination, he was appointed a junior clerk in the Board of Punishments (equivalent to the ministry of justice). Because he believed that the acquisition of new (Western) knowledge was essential in effecting political reform of Manchurian rule, he asked to be sent to Japan for further studies in law and political science (3).

When the Manchurian rulers, in an attempt to quell popular discontent generated by the Boxer Uprising, took their first steps toward constitutional reform in 1907, they established the system of government by election. Soon after my father's return from Japan he was elected to the newly formed Hupeh provincial assembly and became its speaker in 1910.

On October 10, 1911, revolution against the Ching (Manchurian) dynasty broke out. My father supported the revolution and functioned as the civil administrator in the provisional military government of the revolutionaries. I was then 7 years old and can still recall vividly the noise of fighting all through the night of that day which quickly led to the end of centuries of feudal rule and the beginning of the modern era in China. As an irony of fate, the last child-emperor, the overthrow of whose dynasty my father and I witnessed in 1911, worked for a short period of time in the Botanical Garden of the Botany Institute of Academia Sinica in the 1960s as a caretaker of the gardens. I was then a director of the institute and hence his boss! Acually, of course, he was "employed" there to provide him with a quiet place to write his memoirs.

Because of my father's connections, I was sent to Tokyo for grammar school education. Had it not been for World War I, and also my mother's death in 1915, I would have remained there and been educated as a law and political science student. I returned to Peking before the war's end and took the entrance examination to enter the then much coveted Tsing Hua College, actually a preparatory school for teenagers who would be sent to study in the United States after eight years of training. They were then sent to enter American universities as juniors, all on government scholarships, the so-called "Boxer Indemnity Scholarships." This is a part of the "Indemnity Fund" extorted by the eight invading powers in 1900 to "punish" the Manchurian government on the pretext of "property damage" to their nationals on Chinese soil due to the Boxer Uprising. The indemnity fund came from heavy taxation on the then 450 million Chinese people, some 90 percent of whom were peasants. The few candidates (ca 50 per year) were selected from hundreds of applicants on open examination, and the competition was keen. I was among the five students from Hupeh province who qualified for the 1917 examination. The remainder came from quotas assigned to each province proportional to the amount of tax levied on them. The great majority of these students had very good school records and excellent grades. I had a good father. He was by then Minister of Education in the Central Government. My father was assassinated by his political adversaries in 1918, a year after I entered Tsing Hua College, and I was entirely supported by that scholarship which came through taxation. I remained at the college until graduation in 1925, when I left for America. To this day I have a strong feeling of deep indebtedness to my country and my people who nurtured and nourished me.

IN QUEST OF KNOWLEDGE

I went to the University of Minnesota because it had one of the best agricultural colleges known to us at that time. But I soon transferred to the School of Science, Letters and Arts because I discovered what I really needed was a solid general foundation in science and liberal arts before specialization. I filled my credit hours with a full load of the general and intermediate courses in physics, chemistry, biochemistry, and botany, graduating with an AB degree magna cum laude, majoring in botany and minoring in chemistry and physics. My first plant physiology instructor was Rodney Harvey, who translated Maximov's textbook into English; the second instructor, also at Minnesota, was George O. Burr, who later almost discovered the C_4 (or Hatch and Slack) pathway of carbon metabolism in photosynthesis. But the most unforgettable classes I attended then were the lucid and inspiring lectures on physical chemistry by Professor Frank McDougall. His lectures sparked my interest not only in thermodynamics, but through it, the energetics of living matter in general.

Johns Hopkins Days

On recommendation from my major adviser, and because of the reputation of Johns Hopkins, I went to study with Burton E. Livingston in the fall of 1928. His doctoral thesis on osmotic pressure and his treatment of the "three-salt mineral nutrient solution" in terms of physical chemistry interested me. But by then Livingston's interests had shifted to the more ecological aspects of plant physiology which did not appeal to me. Against his wish to work on physiological ecology problems, I started to perform some exploratory experiments on photosynthesis and seed respiration. Our relationship became somewhat strained. Maybe because I was the first and perhaps the only Chinese student he ever had, and perhaps also because I was sent to him by his good friend, William S. Cooper, he did not quite have the heart to dismiss me as he did some of my contemporaries. Behind his back we students called him "Uncle Burt," and on occasions, "the old crank." Finally we compromised on doing a PhD thesis on "the influence of temperature and aeration on the germination of wheat seeds under water." This research topic itself is meaningless if not absurd. Who would germinate wheat seeds under water, even for "pure" research? And what is the meaning of the redundant effort of bubbling air through the water to overcome anaerobiosis? Livingston's aim was to obtain some experimental data to refute Blackman's "theory of limiting factors." My aim was to finish getting my degree. By 1930 I had the experiments finished and written up. Neither of us paid much attention to the physiological meaning of this research, but Livingston was very impressed with the graphical analysis of the data, a method which I learned in the physical chemistry course at Minnesota. He was pleased and elaborated my thesis into a long paper which appeared in 1931 (9). I was also very pleased, not so much because he had conferred on me the PhD degree, but also because I learned much from his writing abilities. I admit that whatever scientific writing ability I have acquired is due in large part to Livingston's legacy. More importantly, the idea of multifunctional relations in physiological processes which was employed in the analysis of the data was the germinal beginning of my later concept of multiple pathways in respiratory metabolism and their relations to other physiological functions.

It was also at Johns Hopkins that I had the good fortune of befriending Robert Marshall, who shared the laboratory with me. I do not quite remember whether he finally got his degree after I left Hopkins, but I do well remember that it was he who introduced me to the writings and views of H. L. Mencken, Norman Thomas, and Oliver Wendell Holmes. It was he, too, who introduced me to the plays of Eugene O'Neil and the superb performances of Lynn Fontanne and Alfred Lunt, among many others. The biology seminars organized by H. S. Jennings and his student Tracey Sonneborn opened my eyes to the then rising science of genetics. It was Jennings who greeted and congratulated me upon hearing that I was awarded the same fellowship he was awarded some years earlier when he was at Harvard. These friends and many others of that period taught me aspects of American life other than that found on the campus. In my days at Johns Hopkins the laboratories were modest and austere, but exchange of ideas was rich and fruitful.

I left Johns Hopkins in the summer of 1930. Little did I know that not so long afterwards, when Livingston retired, that same small and austere laboratory would become the very famous McCullum-Pratt Institute of Biological Science. Tony San Pietro, when I met him at the Columbus meeting in 1979, told me that no less than eight brilliant stars in biochemistry were crowded in that very building where I once worked. Much excellent research on enzymology flowed out from that small red brick building, including the now classical volumes of *Methods of Enzymology* edited by Colowick and Kaplan.

When I returned to visit the United States in 1979, I revisited Johns Hopkins—after a lapse of exactly half a century. That small laboratory was still there, intact, but with its greenhouse abandoned. When I palmed and turned the ever-so-familiar brass doorknob and shouldered open the laboratory door, nobody greeted or stopped me. Rows of animal cages took the place of laboratory benches, and the bleak walls had been enameled white. The once austere laboratory where we sweated in the summer has been turned into a deluxe air-conditioned animal house! And the basement of Gilman Hall, where we had seminars, is annexed to a much bigger biological laboratory where molecular biology thrives. Such is the march of time for Biology!

Woods Hole Summers and the MBL

My early interest in the physical chemical aspects of living phenomena naturally led me to seek for an apprenticeship with W. J. V. Osterhout. His earlier work on the use of relatively simple plant parts or cells for respiratory and photosynthetic studies impressed me. I wrote to inquire if I might come to work with him at Harvard. He wrote back that he had already left for the Rockefeller Institute, but I might contact his successor, W. J. Crozier. I wrote to Crozier and informed him of my acquaintance with the Arrhenius equation and chemical kinetics in general. I got an unexpectedly quick response and was offered the Parker Fellowship from Harvard University to work in his recently established Laboratory of General Physiology.

In the meantime, the praise from my colleagues at Johns Hopkins for the Marine Biological Laboratory at Woods Hole intrigued me. I applied for and was granted a space there for the course in general physiology for the summer of 1930, before I was to report for work at Harvard.

Those three months at Woods Hole, followed by a second visit in the summer of 1931, were memorable ones—both because of the course which I took the first summer and the numerous great biologists whom I had the good fortune to meet or to study with. The two summers I spent with the biologists there, especially with cellular and general physiologists, set the theme for my life work. Among the many friends who had a direct influence on me are: Ralph S. Lillie and Ralph Gerald of Chicago, Leanor Michaelis and W. J. V. Osterhout of the Rockefeller Institute. I learned in the general physiology course the then "modern" method of Warburg microrespirometry, the Van Slyke and Haldane gas analysis machines, and oxidation-reduction titrometry, among other things. Very important were the lectures given with the course and the evening lectures and discussion afterwards. Among the instructors of the general physiology course, Ralph Gerald influenced me the most. His brilliant lectures and witty remarks impressed me. He told us about A. V. Hill's research on heat production in muscle, electric action potentials in nerves, and especially the work of Otto Warburg on the rapid rise in respiration of sea urchin eggs upon fertilization, as well as his use of the light-reversible property of CO inhibition on cell respiration to obtain the action spectrum of the *Atmungsferment.* These and Otto Meyerhof's attempt to measure "energy of organization" enthralled me. I decided then and there to take up energetics of cellular and plant respiration and of photosynthesis as my life work. Fortified with such "modern" techniques, and more importantly with such lines of thought, I was well prepared to do some serious work as a "post doc," as it is termed now, at Harvard.

Besides the many friends I made at Woods Hole and the knowledge and skill in general and cell physiology which I learned from them, I also made some contribution of my own to them and to all my colleagues at the old MBL. I entertained them with almost daily after-lunch tennis exhibitions between myself and Hallowell Davis of Harvard on the court next to the mess hall. These matches were remembered by my fellow students of those days, such as Emil L. Smith and David Green, when I met the former in China in 1976 and the latter during my visit to the United States in 1979. My other contribution was the publication of three little papers: The rate of oxygen-consumption by *Asterias* eggs before and after fertilization (11); the oxygen-consumption curve of unfertilized sea urchin eggs (10); and the oxygen-consumption curve of fertilized *Arbacia* eggs (12). These three papers have been generously quoted by Joseph Needham in his book on *Chemical Embryology* (5), and by Rashevsky in his book on *Mathematical Biophysics* (8).

Harvard Days

I went straight to Harvard from Woods Hole in the fall of 1930 and reported for work at the Laboratory of General Physiology, perhaps the first institution in the country to bear that modern name. I must confess my spirit was a little dampened when I was ushered to the basement of the age-worn Peabody Museum, world famous for the glass flowers it housed, but incompatible to the name of a modern laboratory which it sheltered. I went up the squeaky stairs to a large room on the second floor where Crozier shared his office with his secretary, separated only with a placard on a wooden stand which read: "Please do not disturb" between their desks. After a brief introduction, I was returned to his second in command, Albert Navez, in the basement. While going down the stairs, I ran into Bob Emerson and had a brief exchange of greetings. Bob had recently returned from Warburg's laboratory in Germany, and was in a rush to gather his things for departure to Pasadena to join T. H. Morgan at Caltech.

Navez, who first came to Harvard as a fellow on the Belgium Relief Program, arranged for me to share his dark and crowded basement laboratory, together with two or three others, including Trevor Robinson, who was then Navez's graduate student. Navez took me on a tour of the other basement laboratories and introduced me to all the junior members and graduate students. In touring the laboratory rooms, one outstanding feature came to my attention. That was the presence of at least one or two thermostat tanks of all sorts and sizes in each and every room: the hallmark of Crozier's trade. After a few days of brief visits with the colleagues at the laboratory to acquaint myself with them and their work, I quickly settled down to work.

Since I knew exactly what I wanted to do and had my project planned ahead, I had no difficulty getting my work started. I took over the then newly designed Warburg respirometers that Bob Emerson left behind, grabbed the then perhaps most versatile refrigerated thermostat tank to go with the set, and started experiments on the temperature characteristics of seed respiration during germination. I worked very hard and very efficiently. Not infrequently I had to make continuous observations at 15 minute intervals for 24 or even 48 hours at a stretch. On those occasions I took a camp cot to the laboratory for momentary naps between temperature adaptations. I made plans for the next series of experiments before the first ones ended. In this way my research program went off with the efficiency of a one-man production line at an average output of one paper published every three months. This pace of research activities was kept up essentially throughout my period of studies in Crozier's laboratory with a slight letup after the first year due to my marriage and having a family, but the general tempo was essentially the same. Both Crozier and I were pleased during the first two years, but for different reasons. He was pleased by the additional evidence for the universality of his temperature characteristic idea, while I was mainly concerned with repetitive rappings at the door of Science's opportunity.

This happy coexistence did not last long, however. As with many of my more independent colleagues in the laboratory at that time, I began to deviate from the boss's assembly line and set up a little side business of my own. The goods from this unauthorized sideline are: the publication in 1932 of the now well-documented report on the discovery of cytochrome oxidase in plants (13), and a thorough survey of the then existing literature on the relation of oxygen tension and rate of oxygen consumption in cells and tissues (14). I summarized the findings with an equation describing the hyperbolic relation between the two. That equation had been used and mentioned in some biochemical and physiological textbooks at that time, e.g. in C. Ladd Prossors's *Comparative Animal Physiology* (7). Similarity between this and the equation of Michaelis was pointed out in that paper, but the true meaning of the constant was not understood until years later. The constant of that equation has since been used to express the affinity between oxygen and cytochrome oxidase in the cell. In other words, the constant is actually the Michaelis constant for oxygen affinity of the substrate cytochrome oxidase in vivo.

With these two side products from the benches of the temperature characteristic shop, together with some half a dozen of the "standard ware" type, I edged into the threshold of science, at the expense of my relations with Crozier. So, in the summer of 1933, with the shocks of the stock market crash still reverberating, and when my first wife lost her eyesight, I accepted an invitation from Wuhan University to return to China.

On hearing of the difficult position I was in, and my desire to leave the United States for home, my very close friend of Johns Hopkins days, Bob Marshall, made a special trip from New York to Boston to persuade me to remain in the States and continue with my scientific research. He promised to provide a private endowment at a New York university where I would be accommodated with full research facilities. He came from a very wealthy family of lawyers in New York City. Knowing his intellectual inclinations in Hopkins days, I knew his offer was sincere, and not out of philanthropy. But with my personal misfortune and my self-imposed commitment to my country, I regretfully declined.

In spite of all that I have said about my "grind mill" life in Cambridge, it was certainly not "all work and no play." The social life among many of the friends at Harvard, the numerous spontaneously organized and unscheduled discussions and seminars, were not only stimulating, they somehow banded us "dissenters" into close unity. One of these is the *"Chlorella* club" that Stacy French mentioned in his account of Harvard days in Volume 30 of the *Annual Review of Plant Physiology*. (4). I can never forget the friends I made in Cambridge (Mass); the vacation at the summer house of Walter B. Cannon and his family in the New Hampshire hills; Bill Arnold of the Emerson-Arnold photosynthetic unit fame, who constantly harassed me with such "nonsensical" (then, but not now) questions such as whether carotenoids play a part in photosynthesis; B. Fred Skinner, who introduced me to the philosophies of Ralph Waldo Emerson (who by the way was an ancestor of Bob Emerson whose respirometers I took over at Harvard), and especially of his hero, Henry David Thoreau; and finally my very close and lifelong friend and colleague, Stacy French, of derivative spectroscopy and of the French press fame, a sequoia among the few old giants of American photosynthesis.

Unlike at Johns Hopkins, I have been in touch with friends at Harvard off and on through the decades, first through Stacy French and B. F. Skinner, and more recently through Lawrence Bogorad. I came to know Lawrence during his visit to China in 1978 when he led a group of ten botanists sent over by the Botanical Society of America, sponsored by the head of its China Relations Committee, Peter Raven. Lawrence learned of my desire to revisit my old laboratory, room 208 of the Biological Laboratories on Divinity Avenue. So when we paid a return visit to the USA, he very thoughtfully arranged a grand "homecoming" for me. After a tour of the various departments at the Biological Laboratories, I was first taken to the section to which the glass flowers were moved and housed in modern display cases. To further link me with the olden days, I was taken to visit the octogenarian Ralph Wetmore. He was professor of botany when I was busily grinding out my temperature characteristic papers in his neighboring General Physiology wing. Then Lawrence took me to his own office and I immediately recognized it to be the expanded and refurnished office that Crozier once occupied. The climax came when he took me to the laboratory next to his office: room 208 where I worked half a century ago! The door opened. Standing before me was its present occupant, Daniel Branton, a young scientist in his 40s. My mind immediately shifted back 40 years. I was of that age, too, 40 years ago, and working in the same room! But the similarity ends there. The change has been enormous. Whereas it was a sparsely equipped large room with a barren bench on which stood only a set of Warburg respirometers, it has now been completely renovated and sectioned, filled to every corner with modern sophisticated instruments for molecular biology and biochemistry. Dan politely gave an account of the brilliant work he was doing, but most of his account escaped my attention. I caught little of what he told me, not only because of my inadequate knowledge of modern biology, but more because my mind was occupied by something much deeper than the subject matter itself. What I was thinking was between Dan and me and how much water had passed under the biological bridge! And more importantly, how wide is the gap between this science in my country and that of the scientifically more advanced countries!

To top the "homecoming," Lawrence kindly arranged for a reunion between me and many of my old friends at the Harvard Club. Among them were Stacy French, who came specially for the occasion from his trip to the Caribbean, B. F. Skinner, John and Wilma Fairbank, Ralph Wetmore, and Martin Gibbs. And to wind up the occasion, they arranged for a dinner at the Pier Four Restaurant. They told me they chose this place because of its food and its name. But I doubly appreciated this location for another reason. Nearly half a century before it was from these wharfs that I embarked and left America for home.

CHINA RETURNED: THE CALM BEFORE THE STORM

I accepted the invitation of and chose to remain at the National Wuhan University because they were good enough to offer me time and funds for research, a privilege rare in Chinese universities at that time. They allotted $2000 (US) to set up a small laboratory for cell and plant physiology. I spent the first two years designing and building the laboratory and purchasing the equipment and chemicals needed for studies in cellular respiration and photosynthesis. The first of a series of seven papers on the kinetics of cell respiration appeared in 1936 (15), intermingled with another series of short papers on practical aspects of biochemistry—a survey of the iodine contents of seaweed off the South China Coast in collaboration with the phycologist C. K. Tseng.

The happy and productive years at Wuhan University did not last long. At the time when the experiments for the seventh paper of the respiration series were in progress, the Japanese army invaded North China—soon the entire country—the beginning of eight years of holocaust and devastation, the effects of which were felt years afterwards. My feelings at that time were expressed in the paragraph which ended the last paper (16) published together with my close friend and colleague Dr. Mao-i Woo, thus:

> It was the original plan of this series of studies to investigate the respiration of unicellular organisms in simple physical chemical systems such as those used in the present account so that we might obtain some knowledge of the physical chemistry of cell respiration like that of the kinetics of reaction in simple chemical and in enzyme systems. Circumstances, however, do not permit the completion of this plan. It is our earnest hope that investigations along this line of thought may be continued by more fortunate workers who are at the present enjoying the blessings of tranquility.

Nevertheless, I did not give up. I turned my laboratory into manufacturing active carbon for gas masks. When Nanking fell and Wuhan was in immediate danger, I dismantled my laboratory as well as the active carbon furnace in the factory and took up a new task: that of establishing a medical school in the more interior city of Kweiyang.

THE WAR YEARS; THE ROAMING BEGAN

At the time when Wuhan (Wuchang then) city was crowded with refugees and soldiers retreating from all the Japanese occupied areas, I accepted an emergency appointment on a five-men committee to establish a medical school in the interior city of Kweiyang for training physicians to supplement the needs of the army. The first batch of students were those who fled to Wuhan from the medical schools of the occupied areas. Since I was the only nonmedical member of the staff, I was entrusted with the admission and transportation of the first batch of some 40 students to the remote city of Kweiyang. We spent 16 days on a steamer overcrowded with refugees of all sorts and went up through the treacherous rapids of the Yangtze River to Chungking. From there we had a further two days' journey by bus over the mountainous staircase highway to Kweiyang, passing many overturned vehicles down the steep cliffs below the narrow paths.

I remained with the Kweiyang Medical School for six months. After fulfilling my assignment of organizing the preclinical teacher staff and after designing and building the teaching laboratories, I moved farther to the interior stronghold of Kunming, Yunnan province, in answer to the call of my alma mater, which had long since attained the full status of a renowned university.

While I was a founding member of the medical school in Kweiyang, I was concurrently a staff member of the Chinese Red Cross (war area) Medical Relief Corps, headquartered in the same city. I was called back to that organization on numerous missions during the years I was in Kunming as an adviser on nutrition problems of the soldiers at the front. In fact, a part of the research of my plant physiology laboratory at Tsing Hua University in Kunming was on the Red Cross Medical Relief Corps program, the soybean milk nutrition project.

Kunming Days: Tapuchi, a Sanctuary for Wartime Chinese Plant Physiology

I spent practically the entire period of the war (1938–1946) in Kunming. Those were eventful and of course difficult years. During the eight years in Kunming, my laboratory was bombed out three times, moved to four different locations, and rebuilt as many times. The last location, where we worked the longest, was in the small village of Tapuchi, about 10 kilometers in the city's northern suburb, where we set up adobe laboratories. This village of not more than a score of dwellings was made widely known during and after the war by the Needhams' account of its scientific "immigrants" in their book *Science Outpost* (6).

During the years at Tapuchi, my laboratory served as a sanctuary as well as an assembly post for youthful and dedicated physiologists of all kinds: microbial, cellular, plant, and animal. That was the major goal which I set myself to attain when I accepted the call to establish a laboratory of plant physiology (actually general physiology) at the Institute of Agricultural Research of Tsing Hua University soon after the war broke out. On the aims of my laboratory, Joseph Needham in his introduction to my book of essays

(17) had this to say: "Yet at Tapuchi. . . . Tang Pei-sung created laboratories of general physiology, not ill equipped, though constructed of mud bricks and wood work; and above all knew how to surround himself with eager young scientific workers in the true atmosphere."

No less than 40 scientists passed through the doors of that humble laboratory during the eight difficult years in Kunming (Tapuchi). Almost all of them were later sent to work or study in America or in Europe. Many of them attained success in their own respective fields but never returned to China. But many of them did return and are the backbone of Chinese plant physiology today. Some of these are: H. C. Yin, director of the Institute of Plant Physiology, Academia Sinica; C. H. Lou, Professor and vice president, Peking Agricultural University; S. W. Loo, also of the Institute of Plant Physiology; T. S. Tsao, Professor at the Peking University; Y. L. Hsieh, Professor at Futan University; B. L. Cheng and T. Y. Hsieh, Professors at the Shantung College of Oceanology.

The research projects of my laboratory had been diverse, but the main theme was very clear: to do what we could to serve the nation at war, to utilize what knowledge we had to improve agricultural production, to exploit the abundant and varied plant resources in the rich subtropical regions of southwest China, and finally of course to train a young crop of scientists.

In war work, besides what we did for the Chinese Red Cross Medical Relief Corps, I served for a short period as a technical adviser with the rank of colonel to the then only armed division of the Chinese army which thwarted a Japanese offensive to the southwest from occupied Indo-China. I was trying to persuade them to use castor bean oil as a substitute for hard-to-get mineral oil lubricant.

In the exploitation of natural resources, we made a general survey of the vegetable oil resource in the southwest provinces and experimented on castor oil, sumac wax, and laurel oil for civil as well as for military use.

In the field of agriculture, we worked extensively both on the application of plant hormones for plant growth and the use of colchicine to induce polyploidy. We even worked on the physiology of the silkworm.

Looking back, those were difficult, eventful, but rewarding years; rewarding because of the gratifying feeling that I have done something, however insignificant, for my country in time of her great misfortune, and also because to some extent it was adventuresome. In performing the many varied missions during these years, I traveled all over the southwestern part of the country by bus, pushing the bus sometimes. On occasions when it was dark and no lights were available, I held two torches in my hands and ran before the bus so we could find our way to the city for the night. The athletic prowess of my college days won its belated applause.

As the war dragged on, I began to write a series of essays on my scientific activities through the 15 years since I first started my work at Johns Hopkins in the 1930s. These essays were meant to be both an account of what I had done during those years and a collecting of my scientific line of thought, a philosophical look at living phenomena in general. When Joseph and Dorothy Needham came to visit me again in Tapuchi at war's end in 1945, I showed them these essays. At their suggestion and persuasion, I agreed to have the essays published in London. Joseph wrote a generous introduction to the book and christened it *Green Thraldom* (17) from the title of the first essay dealing with my views on the conversion of solar energy through photosynthesis. This book marks the maturation of my views on the bioenergetic and metabolic concept of life phenomena, a concept which I had been closely attached to all through my later research. The theme sounded in the opening paragraphs of the book:

> Life connotes activity, and activity implies the expenditure of energy. In man, the energy for his activities is derived from the food and fuel he consumes, and these are manufactured for him, directly or indirectly, by the green plant. Primitive man, living in the forest, feeding on its fruits and deriving warmth from burning its timber, was a parasite. When gradually he learned the usefulness of certain kinds of plants and animals for food and fuel, and singled them out for cultivation, propagation and protection, he initiated the art of agriculture and thereby passed from a state of parasitism into a state of symbiosis. . . .
>
> Civilized man, long having abandoned his worship of fire and sun, is scarcely further advanced than his ancestors, since the fundamental relation between the sun, the green plant and himself is still shrouded in mystery. Not until the time comes when he can reproduce the apparently simple process of photosynthesis in his laboratory, independent of the green plant, can man claim to be free from the vestiges of his ancestral worship of sun and fire. When that day comes the energy awaiting his disposal is enormous.

This theme is continued throughout the book and the crescendo is reached in a passage in its final chapter where my metabolic views on living organisms are emphasized thus:

> A living organism is a state of aggregation of matter which in a limited range of physical environment can utilize relatively unorganized matter: a part of the material is used for the maintenance of itself and for the reproduction of its kind, while the other part is converted into energy necessary for the performance of these processes.

This is my metabolic concept of living matter, or my concept of bioenergetics. It has been the main theme on which my later work was based. We shall come back to this later. As an aside, I am grateful that the above passages were quoted and used as epigrams to the appropriate chapters in two of Bladergroen's books (1, 2).

INTERMEZZO

With the completion of these essays, my scientific activities of the prewar and the war years came to an end. At the close of the eight-year Sino-Japanese war in 1945, I, like all the scientists and intellectuals of my country, had to start from scratch again, richer with experience gained through the years, but impoverished by the soaring inflation. I left my mud-brick laboratory at Tapuchi in the summer of 1946 and returned with the university to its devastated home in Peking to build laboratories again.

We expanded the Institute of Agricultural Research into the College of Agriculture in the university. It was my goal to make the college both a research institution and a teaching school to train students of such quality that they could do advanced teaching and research in other colleges and in the experiment stations, as well as doing biological research. This was the influence of the Johns Hopkins system. We had only a dozen or so senior members on the faculty in the beginning, but these were selected from institutions all over the land on merit of their research and teaching achievements.

After a year's preparation, we began to admit students in 1947, and again in 1948. The dream and aspiration of an advanced school of agriculture, somewhat similar to the schools of biological sciences in some Western countries of recent years, did not last long. Just after the first batch of students entered the respective departments of their choice after finishing two years of training in the college of arts and science, Peking was under siege in the fall of 1948 by the People's Liberation Army, and activities of Tsing Hua as well as other universities were suspended. Very soon Peking was liberated, followed by the rapid collapse of the entire old regime, and New China was founded by the Chinese Communist Party. This change was much more profound than the 1911 revolution which overthrew the Manchurian dynasty. It was the dawning of a new era in an ancient country comparable only to the "October Revolution" of the USSR which Sidney and Beatrice Webb described as "A New Civilization" (21).

NEW CHINA

Plant physiology, like all other branches of science, grew rapidly in New China, but not without its ups and downs. Since I have given an account of the general state of this science in China at the 1979 Columbus meeting of the ASPP (18) and more recently at the Thirteenth International Botanical Congress (19), I shall only give a very sketchy account of some of my personal experiences during this period.

Soon after the founding of New China, my college of agriculture was merged with two others to form the Peking Agricultural University. I was one of three deputies on the governing board. The chief of the board was also the chief proponent for "New Biology" of the Lysenko version. He turned the university into a stronghold for propaganda of Lysenkoism. I was open minded and willing to learn anything new, but when scientific argumentation degenerated into idealogical polemics, I supported the then underdogs who adhered to facts and theories of modern genetics, however "decadent." I left the Agricultural University for the Institute of Plant Physiology, Academia Sinica, in the fall of 1952, and I have been with the Academy ever since, with a concurrent post as professor of plant physiology at Peking University from then to the present. Although my position was changed to director of the Institute of Botany in 1977, I have continued with teaching and research in plant physiology throughout the 32 years since the founding of New China in 1949. In spite of frequent disturbances and interruptions mentioned in my earlier account (18), I have managed to continue with my research in both bioenergetics and basic problems in relation to practical needs of the country, e.g. on biological nitrogen fixation and on bioenergetics of solar energy conversion.

Soon after the founding of the People's Republic of China, fundamental changes in the social and political structure of the nation took place—a complete overhauling, so to speak. It was not merely a change in the form of government, but in ideology and in all systems, including the educational system. For these changes we naturally sought the assistance of the Soviet Union. During the early years of the newly established republic, literally thousands of advisers were sent to assist us in every phase of our national reconstruction, including, of course, those who came to renovate the educational system down to the actual texts to be used for teaching the courses.

The old system of university teaching was abandoned, and new ones were introduced. Plant physiology, for example, was being taught in agricultural "universities" which had been detached from the professional colleges of the original universities. The original universities still maintained their biology departments in which there was a special faculty for plant physiology! This duplication was caused by the segregation of the professional colleges from their main universities. The regular universities retained their departments of biology with faculties of plant physiology, animal physiology, biochemistry, zoology, botany and some others. In this way, when a student graduated, he was trained as a "specialist" in his own narrow field, say of plant physiology. Even in this field he could be further "specialized" in such topics as growth, mineral nutrition, or photosynthesis. This system of teaching plant physiology was then new to us. In addition, there was a lack of teachers for the "advanced" courses. There was also the very difficult problem of obtaining the "progressive" texts even for the course of introductory plant physiology.

Faced with these difficulties, the teachers of plant physiology as well as those for the other fields were rather at a loss of what to do. Fortunately, the government foresaw this need and invited "specialists" in the various fields as advisers to teach us how to handle the situation. I was at the time fortunate enough to be a professor of plant physiology at the Peking Agricultural University which was specially favored because it was entrusted with the task of indoctrinating the "Progressive Russian New Biology" to the country. Because of this we were privileged in two aspects: first, we could receive outlines and syllabi for the biology courses from Russia; and secondly, we had a strong staff of translators to render the materials into Chinese. In spite of these advantages, it took a long time to have all the materials translated, checked, and tried out before they could be distributed to the plant physiologists at large who were anxiously awaiting intellectual food to feed the hungry students. Unable to resist their cry for help, I, with the collaboration of my junior colleagues of the plant physiology faculty, mimeographed hand-printed copies of parts of the translated outlines and syllabi in the form of "Plant Physiology Newsletters" (from 1951 on) as quickly and as frequently as they were available. These newsletters were so well received that after the first four mimeographed issues (1951–1952) and beginning from the fifth issue in 1953 they had to be changed to regular printed pamphlets. Finally, in 1955, beginning with the twenty-first issue, they were published as a regular journal which now bears the respectable name *Plant Physiology Communications,* with a wide circulation within the country.

The availability of the translated outlines of Russian plant physiology relieved the anxiety of the country's plant physiologists only for awhile, for we still needed the all-important specific materials to fill in the outlines. There was further confusion because of the then denunciation of the backward, "decadent" old biology and the indoctrination of the "Progressive New Biology" of Lysenko vintage. To help us solve this very important problem, our Russian friends came to our rescue again. They sent in the nick of time a couple of very nice young damsels to teach us how to teach the "progressive" plant physiology as it was done in Russia. They were very dedicated and pleasant young teachers who had perhaps just received their "Kandidat" from some teacher's college back home. We discussed in detail the syllabus and outlines of the introductory course. At first they were a little apprehensive about whether we were ready to accept the "new" plant physiology in place of the old. They were soon relieved and perhaps even pleased to find that I, among a few other old souls, not only could accept the material and the contents, but embraced them wholeheartedly! In fact,

I could even supply the necessary experimental evidence or factual and theoretical documentation at the right places. And, in a few instances where the translation appeared to be inexact, I supplied my alternative interpretations. The training class went on very smoothly and came to a successful conclusion, with the teachers marveling at my quick grasp of the "new progressive" plant physiology. Little did they know that out of scientific courtesy and of bourgeois etiquette, I refrained from telling my teachers that the syllabus was that of Maximov's *Text Book of Plant Physiology* with which I had been teaching my students for at least a decade. Only it was the English rendition by Rodney Harvey that I used, instead of the Chinese translation which my Russian teachers brought to us. Science, like the sun, begets its light and warmth to all people on Earth, irrespective of country, race, or belief.

To this day I still have fond memories of that training class and of the many accomplished Russian plant physiologists whom they sent us later, among them B. A. Rubin of Moscow University and P. A. Henkel of the Timiriazev Agricultural College. And it was my pleasure indeed to have the chance to serve on the same committee with such an eminent Russian scientist as M. Kh. Chailakhyan at the recent Thirteenth International Botanical Congress in Sydney.

With the initial worries over, plant physiology, together with other similar courses, began to be taught throughout the country. But there soon arose another difficulty: the unification of the proper interpretation of many of the points raised when the syllabus was taught in the various colleges. In addition, there were the contents of the laboratory courses, especially of the General Laboratory (the "Gross Prakticum" in the old European countries). Then there was the problem for the training of teachers for the advanced courses. I was then with Peking University as the head of its plant physiology faculty. Under pressure from colleagues of other colleges and universities, I sponsored and organized the first training course for plant physiology teachers of the entire country in the summer of 1956. With the collaboration of C. H. Lou of the Peking Agricultural University and C. Tsui of Nankai University and our younger colleagues at these and some other universities, we gave a comprehensive program in plant physiology in which laboratory experiments were offered, together with lectures and seminars on advanced topics. Discussions on points needing clarification or unification were conducted. A total of 130 plant physiologists from every institution in the country attended, with about a dozen or so of the senior members acting as lecturers. Fortified with what they learned from this course, all, especially the younger teachers, went back to their teaching posts with confidence. From this group that received their training from us emerged the new crop of younger generations of plant physiologists in New China.

This training course, the initiation of the journal *Plant Physiology Communications,* and the sanctuary for Chinese plant physiologists at Tapuchi which I established during the war years are perhaps the only three pebbles which I gathered on my wandering trail as a roaming plant physiologist. During the some 30 years in New China, I had on the whole enjoyed stability in my academic life as compared with the turbulent days in old China. I have essentially been left to do research on basic problems. But with a nation of nearly one billion people, over 80 percent of whom are agrarian, aspiring to lift itself from austerity, one cannot shut one's self in an ivory tower. All through the past three decades there had been many occasions when we had to pack our equipment and move to the communes or to the factories to work with the peasants and the factory hands. I call these the "Open-door Laboratory" days. Many of the things we did may appear unorthodox, but they had their educational value to me and to my younger colleagues. They enriched me both professionally and morally through contact with life's realities. In these practical activities I have profitted as much as the children in some industrially more advanced countries who had to be taken to the zoological garden or the botanical garden to learn that chickens, pigs, monkeys, or cabbages really existed as living things, and are much more sophisticated and different from what they appear on the dinner table or in the picture books.

Aside from such interruptions, I was left fairly alone to do my own research. Since a general account of what my colleagues and I did were given in my recent talk (18), and since a summary of my research on plant respiratory metabolism has just appeared (20), I shall not dwell on specific items. But I would like to recapitulate the main theme of thought underlying these research efforts to close my account on the eve of my exit from the stage of plant physiology. I shall quote from selected passages from that article (20):

When I resumed my research on plant respiration in the early fifties. . . . I initiated the series on the theme of treating it as a physiological function which furnishes energy and material for the performance of the plant as a living organism. In other words, the theme deals with the functional and regulatory aspects of respiratory metabolism in higher plants: integrating metabolic changes (material and energetic) with the other physiological functions in the living plant.

This concept, formulated in the early fifties, was presented in 1965. . . . After an interruption of another decade, . . . it can now be summarized as follows:

"Respiratory metabolism is the process whereby a part of the material stored in the plant (organism) is converted into biological work (function) for maintaining its state of being alive, while another part of the same material is converted into substances of higher degrees of orderliness (negative entropy) in the form of structure and organization. Within limits imposed by the genetic potential, these processes are regulated by internal and external factors."

This last paragraph summarizes the entire line of thought which has been the central theme of my research on bioenergetic aspects of living organisms since the early 1930s, gathering form in my book of essays in 1949 (17) and leading to the present mature form. And in concluding that article (20), which summarized my 50 years of research in this field, I made a final assessment of my accomplishments:

> ... the decrease in entropy is the unique function of metabolism and of life. This is perhaps the inner sanctum of biology and which remained stubbornly invulnerable to assaults up to the present ... This was the aim which I set myself to attain when I performed my first experiments on oxygen-consumption by starfish eggs before and after fertilization as a young and innocent novice, inspired by the elegant works of Otto Meyerhof and of Otto Warburg. After almost half a century, I have not even approached the fringe of the fortress, but neither has any one else, I think.

These, in summary, are my aspirations and the realities. I published much, but contributed little; I labored hard, but I accomplished even less. Looking back over the years, I have left many things undone which I should have done, and I have done many things which I should have refrained from doing. Such are circumstances and reality!

Literature Cited

1. Bladergroen, W. 1960. *Problems in Photosynthesis,* p. vii. Springfield, Ill: Thomas
2. Bladergroen, W. 1955. *Einführung in die Energetik und Biologischer Vorlängen,* p. 330. Basel: Wepf
3. Boorman, H. L., Edward, R. C., eds. 1967–71. *Bibliographical Dictionary of Republic of China.* 4 vols. New York: Columbia Univ. Press
4. French, C. S. 1979. Fifty years of photosynthesis. *Ann. Rev. Plant Physiol.* 30:1–26
5. Needham, J. 1931. *Chemical Embryology.* Cambridge: Cambridge Univ. Press
6. Needham, J., Needham, D. M. 1948. *Science Outpost.* London: Pilot Press
7. Prosser, C. L., Brown, F. A. 1961. *Comparative Animal Physiology,* p. 167. London: Saunders. 2nd ed.
8. Raschevsky, N. 1938. *Mathematical Biophysics.* Chicago: Univ. Chicago Press. 1st ed.
9. Tang, P-S. 1931. An experimental study of germination of wheat seeds under water. *Plant Physiol.* 6:203–48
10. Tang, P-S. 1931. The O_2-consumption curve of unfertilized *Arbacia* eggs. *Biol. Bull.* 60:242–44
11. Tang, P-S. 1931. The rate of O_2-consumption by *Asterias* eggs before and after fertilization. *Biol. Bull.* 61:468–71
12. Tang, P-S. 1932. The O_2 tension-O_2 consumption curve of fertilized *Arbacia* eggs. *J. Cell. Comp. Physiol.* 1:503–13

13. Tang, P-S. 1932. The effects of CO and light on the oxygen consumption and on CO_2 production by germinating seeds of *Lupinus albus. J. Gen. Physiol.* 16: 65–73
14. Tang, P-S. 1933. On the rate of oxygen consumption by tissues and lower organisms as a function of oxygen tension. *Q. Rev. Biol.* 8:260–74
15. Tang, P-S. 1936. Studies on the kinetics of cell respiration I. The rate of oxygen consumption by *Saccharomyces* wanching as a function of pH. *J. Cell. Comp. Physiol.* 8:109–15
16. Tang, P-S., Wu, M. 1938. Studies on the kinetics of cell respiration VII. Respiration of *Saccharomyces* wanching in acetate, lactate and pyruvate buffer solutions. *J. Cell. Comp. Physiol.* 11:495–502
17. Tang, P-S. 1949. *Green Thraldom.* London: Allen & Unwin
18. Tang, P-S. 1980. Fifty years of plant physiology in China: A prelude to the new long march. *Bioscience* 30:524–28
19. Tang, P-S. 1981. Aspects of botany in China. *Search* 10:344–49
20. Tang, P-S. 1981. Regulation and control of multiple pathways of respiratory metabolism in relation to other physiological functions in higher plants: Recollections and reflections on fifty years of research in plant respiration. *Am. J. Bot.* 68:443–48
21. Webb, S., Webb, B. 1945. *Soviet Communism: A New Civilization?* London: Longman's. 3rd ed.

Ann. Rev. Plant Physiol. 1983. 34:21–45
Copyright © by Annual Reviews Inc. All rights reserved

PHOTOSYNTHETIC REACTION CENTERS

Richard J. Cogdell

Department of Botany, University of Glasgow, Glasgow G12 8QQ, United Kingdom

CONTENTS

INTRODUCTION

The majority of the light-absorbing pigments in photosynthesis simply serve as light-harvesters. They absorb the incident solar radiation and transfer that absorbed energy to a few specialized sites for "trapping." These traps are called reaction centers. When the reaction center is electronically excited, either directly or, more usually, indirectly by energy transfer from the antenna system, it catalyzes the primary photochemical reaction. As a result of the primary photochemical reaction, solar energy is transduced into useful chemical energy. The aim of this contribution is to review what is presently known about the structure and function of reaction centers, both in oxygen-evolving photosynthetic organisms and in the anaerobic photosynthetic bacteria.

21

0066-4294/83/0610-0021$02.00

Photosynthesis is one of the most basic biological processes in the biosphere, and an understanding of the details of the primary photochemical reaction is therefore of fundamental importance in its own right. There is, however, another long-term, more practical goal to this area of research. Solar energy is an abundant renewable energy source. Reaction centers are rather efficient solar energy converters, and it is hoped that a complete description of their structure and function will provide some useful clues on how to construct more efficient solar cells.

A General Model of Reaction Center Function

$$P(I_1—I_n)X \overset{h\nu}{\rightarrow} P^* (I_1—I_n)X \rightarrow P^+(I_1—I_n)^-X \rightarrow P^+ (I_1—I_n)X^-$$

When a reaction center is excited, the primary electron donor P is raised to its first excited singlet state P^*. Then, usually in less than a few nanoseconds (122, 130, 146, 148, 151), a series of electron transfer steps through intermediates $(I_1—I_n)$ takes place, which results in the oxidation of P to P^+ and the reduction of an electron acceptor X to X^-. The state X^- is usually stable for microseconds or longer (56, 120, 155) and used to be called the primary electron acceptor. Recently, however, several intermediate electron carriers [collectively called I (7, 121, 141)] have been shown to function between P and X and so the term primary electron acceptor is no longer useful. X is now usually called the first stable electron acceptor. This series of electron acceptors is now recognized to be an important feature of reaction center function. It effectively prevents the energy-wasting back reactions which would otherwise return the electron to P^+ and release the energy of the absorbed photon as heat.

In the past few years, preparations of isolated reaction centers have been obtained from a variety of photosynthetic bacteria, for example *Rhodopseudomonas sphaeroides* (14, 64), *Rps. viridis* (124, 161), *Rps. gelatinosa* (13), *Rps. capsulata* (106, 125), *Rhodospirillum rubrum* (107), and *Chromatium vinosum* (87). The availability of these purified reaction centers, devoid of antenna bacteriochlorophyll, has allowed both their structure and function to be studied in great detail. Unfortunately, the same is not true of oxygen-evolving photosynthetic organisms where only enriched reaction center particles are currently available (9, 119, 162). However, it will be clear from the preceding sections that the insights gained from the studies on the bacterial reaction centers have greatly stimulated and aided research into the nature of the primary reactions of photosystems I and II.

ANAEROBIC PHOTOSYNTHETIC BACTERIA

The best characterized of the bacterial reaction centers is the one from *Rps. sphaeroides* (42, 49, 167). Most of this section will therefore concentrate upon the *Rps. sphaeroides* reaction center to illustrate the general features of a typical bacterial reaction center. Only occasional mention will be made of preparations from other species where the differences are particularly informative.

Reaction Center Structure

Reaction centers from *Rps. sphaeroides* are hydrophobic, integral-membrane, pigment-protein complexes. They consist of one each of three polypeptides [with apparent molecular weights of 21 KD, 24 KD, and 28 KD, estimated from their mobilities on SDS polyacrylamide gels (117), though these values are probably underestimated by about 30% (42, 131, 167)]; 4 molecules of bacteriochlorophyll *a;* two molecules of bacteriopheophytin *a;* 1-2 molecules of ubiquinone (UQ_{10}); one atom of ferrous iron; and one molecule of carotenoid (17, 64, 65, 157). The carotenoid is lacking in preparations from carotenoidless strains (14); however, carotenoidless reaction centers retain the specific carotenoid binding site as evidenced by reconstitution studies (1). When bacteriopheophytin was first detected in reaction center preparations, there was a question as to whether it might not be just an artifact of the preparative procedure. Bacteriopheophytin and pheophytin are usually assumed to be the degradation products of bacteriochlorophyll and chlorophyll respectively. The best evidence that bacteriopheophytin is a true member of the set of reaction center pigments comes from studies on the reaction centers from *R. rubrum* (172). In *R. rubrum,* Walter (172) has shown that the reaction center bacteriopheophytin "head group" is esterified to phytol, while the reaction center bacteriochlorophyll "head group" is esterified geranyl-geraniol. The reaction center bacteriopheophytin is therefore being specifically biosynthesized and is not just a simple breakdown product of the bacteriochlorophyll.

The individual reaction center polypeptides from *Rps. sphaeroides* have been isolated and purified, and their amino acid compositions have been determined (154). All three polypeptides have a high proportion of nonpolar amino acids. Sequence studies have begun on these subunits but as yet they have not been very successful. So far only the partial N-terminal sequence of the 24 KD molecular weight subunit has been reported: Ala-Leu-Leu-X-Phe-Glu-Arg-Lys-Tyr-Arg-Val-Pro-Gly-Gly-Thr-Leu-Val-Gly-Gly-Asn-Leu-Phe-Asp-Phe- (131). Efforts are now continuing in this area, making use of more modern molecular biological technology. The genes coding for the synthesis of the reaction center polypeptides are now being cloned, and hopefully the amino acid sequences will soon be determined by way of DNA sequencing (J. P. Thornber, personal communication).

The reaction center pigments are noncovalently bound to the two lowest molecular weight subunits (17, 117). A pigmented dimer of these two subunits still shows photochemical activity, although it is much less stable when the 28 KD molecular weight subunit has been removed (117). Within the dimer there is no information about which pigments bind to which polypeptide, except for one of the two ubiquinone molecules. Okamura et al (114) developed a method for the reversible removal and replacement of the reaction center quinone. They depleted reaction centers of ubiquinone and then reconstituted them with a photoaffinity analog, ^{3}H-2-azidoanthraquinone. The reconstituted reaction centers were photochemically active. Irradiation with UV light covalently attached the quinone to the 24 KD molecular weight polypeptide (94, 95), and it was concluded that the quinone binding site was on this polypeptide.

Very recently there has been an extremely exciting development in this area with the report that reaction centers from *Rps. viridis* have been successfully crystalized (D. Oesterhelt and H. Michel, personal communication). It appears that the crystals are large enough for X-ray crystallography, and there is now the real prospect of a three-dimensional structure of a reaction center being available in the next few years.

It has been predicted from biophysical studies on chromatophores that reaction centers from *Rps. sphaeroides* lie perpendicular to the plane of the membrane, such that the primary photochemical reaction takes place across the membrane (61, 62). The membrane topology of these reaction centers has been investigated biochemically, using surface-specific labeling (3, 47, 173), interaction with ferritin labeled antibodies (129, 168), and accessibility to mild proteolytic digestion (3, 52). In the most comprehensive topographical study so far, Bachmann et al (3) were able to show in *Rps. sphaeroides* that all three reaction center subunits were asymmetrically oriented across the photosynthetic membrane.

In addition to the reaction center components mentioned above, some reaction centers, such as those from *Chr. vinosum* or *Rps. viridis* (87, 124, 161), contain tightly bound c-type cytochromes. These cytochromes function (on the microsecond time scale or less) as rapidly reacting secondary electron donors.

The Identity of the Primary Electron Donor P

Figure 1 shows the absorption spectrum of a reaction center preparation from *Rps. sphaeroides* R26. The absorption bands near 870, 800, and 600 nm are due to bacteriochlorophyll, those near 760 and 530 are due to bacteriopheophytin, while the band near 365 nm is due to both bacteriochlorophyll and bacteriopheophytin. Upon illumination the absorption band near 870 nm bleaches almost completely, the band near 600 nm bleaches rather incompletely, the bands at 800 and 365 nm shift to the blue, and a new absorption band appears at 1260 nm (12). These absorption changes largely reflect the oxidation of P → P$^+$.

Most of the available experimental evidence suggests that P is a "special pair" of bacteriochlorophyll molecules, and that in the state P^+ the unpaired electron is shared (121). The best evidence for this view comes from the ESR and the ENDOR spectra of P^+ (39, 40, 88, 100, 109–112). Both P^+ and the bacteriochlorophyll cation are free radicals, and they both exhibit strong ESR signals at $g = 2.0025 \pm 0.0001$ (88, 100, 112). However, in the case of P^+ the linewidth ($9.4 \pm 0.2G$) is narrower by approximately a factor of $\sqrt{2}$ than for the monomeric bacteriochlorophyll cation ($12.0 \pm 0.2G$) (88, 112). The same relationship between the linewidths holds when $^2H\ P^+$ is compared with ^2H bacteriochlorophyll$^+$ (88, 109, 112). When an unpaired electron is delocalized over more than one molecule, the linewidth of the ESR signal will narrow. If that delocalization is symmetrical, then the narrowing will be proportional to $\sqrt{N}$, where N is the number of molecules which share the unpaired spin. Clearly, in the case of *Rps. sphaeroides* reaction centers $N = 2$, and this model predicts that P is a bacteriochlorophyll dimer (88, 112). The ENDOR spectrum of P^+ shows splittings half as large as those of bacteriochlorophyll$^+$ (39, 40, 108, 110, 111), and this again can be explained by assuming that P is a dimer. On the other hand, in *Rps. viridis* reaction centers, which contain bacteriochlorophyll *b* rather than bacteriochlorophyll *a,* neither the ESR linewidths nor the ENDOR splittings of P^+ are decreased by as much as they are in other species (36, 161). However, it has been suggested that this may not necessarily exclude a dimer if that dimer were to be asymmetric (22). A full discussion of the relative merits of a monomer versus a dimer is beyond the

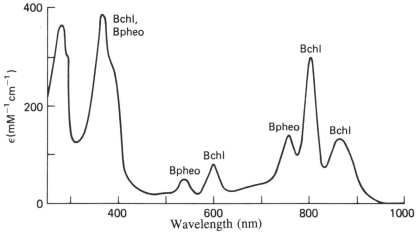

Figure 1   The absorption spectrum of a preparation of reaction centres from *Rps. sphaeroides,* R26 (carotenoidless). Data taken from ref. 12 and reproduced with the kind permission of the author.

scope of this review, and the reader is directed to a recent article by Parson (121) where this issue is considered in great detail.

Various structural models have been proposed to explain how the pair of bacteriochlorophyll molecules in P may interact (45, 68, 156). It will be interesting, when a full, high-resolution, three-dimensional structure of the reaction center is available, to see how close these informed theoretical guesses actually are.

Redox titrations of the P/P^+ couple indicate a one-electron transition with a midpoint potential (E_m) of +450 mV (26, 60). The E_m is independent of pH. A significantly lower E_m than this for the P/P^+ couple seems to be a unique feature of the green sulfur photosynthetic bacteria. For example, the E_m of the primary donor in reaction centers from *Chlorobium limicola* is + 250 mV (128).

The Stable Electron Acceptor X

For awhile there was considerable argument as to the identity of X (123). This was mainly because the ESR spectrum of X^- was dramatically different from the narrow spectrum near $g = 2.00$ that would be expected for the anionic ubisemiquinone (27, 38, 43, 86, 89). However, it is now generally accepted that in most reaction centers X is a quinone (7, 15, 114). In *Rps. sphaeroides* it is ubiquinone, while in *Chr. vinosum,* for example, it is probably menaquinone (113). The exceptions to this, are again the green sulfur photosynthetic bacteria (see below).

Initially both the iron atom in the reaction centers from *Rps. sphaeroides* and the ubiquinone were considered as possible candidates for the role of stable electron acceptor. The evidence in favor of ubiquinone and against the iron can be summarized as follows:

1. Removal of the iron atom, or its replacement with other metal ions such as manganese, did not prevent the primary photochemistry (41).
2. Mossbauer spectroscopy on ^{57}Fe-enriched reaction centers revealed that, irrespective of whether X was oxidized or reduced, the iron atom remained in the high spin ferrous state (23).
3. In contrast, removal of the last molecule of ubiquinone from the reaction centers completely blocked the primary photochemical reaction. Normal photochemistry was then quantitatively recovered when one molecule of ubiquinone was reconstituted with each depleted reaction center (15, 114).

The anomalous ESR spectrum from X^- appears to be the result of a strong magnetic interaction between the iron atom and the reduced quinone, which causes the unusually broad ESR signal with a principal g factor near 1.8 rather than 2.00 (114). *Rps. sphaeroides* reaction centers contain two bound quinones. Recognizing this, and because it is now clear that X is ubiquinone, most workers in the field currently refer to these quinones as Q_A and Q_B (174–176). Q_A now corresponds to what used to be called X. In chromatophores from *Rps. sphaeroides,* Q_A has an equilibrium midpoint potential at pH 7.0 of –20 mV (26, 60, 62). This is an $n = 1$ redox reaction and is pH dependent (-60mV/pH unit) until above its pK_a at pH 9.8 (126, 127). However, in the course of the normal reaction center photochemistry, the electron only resides upon Q_A for a few microseconds (127). During this time there is no detectable proton uptake, and this had led to the idea that the true operating E_m of Q_A is the value it assumes at pH values above the pK_a, i.e. –180 mV (126, 127). In isolated reaction centers from *Rps. sphaeroides,* Q_A behaves rather anomalously, giving a pH-independent E_m of –50 mV (28, 178). The reason for this remains to be clarified.

Most photosynthetic bacteria generate reduced pyridine nucleotides (required for CO_2 fixation) by way of energy-linked reversed electron transport (80). However, it was noted several years ago (31, 66) that green sulfur photosynthetic bacteria were able to reduce NAD to $NADH_2$ by direct electron transfer in an uncoupler-insensitive reaction. This difference could only be rationalized if the green sulfur photosynthetic bacteria contained a stable electron acceptor of rather low redox potential. The midpoint of the stable electron acceptor in *Chlorobium* has indeed now been shown to be low enough to account for the direct reduction of NAD, $\sim$ –550 mV (128). The ESR spectrum of the reduced form of this acceptor suggests that it is an iron sulfur protein (63, 128).

The Kinetics of Primary Photochemical Reaction

The kinetic details of the primary reaction have been resolved by a combination of nanosecond (10^{-9} sec) and picosecond (10^{-12} sec) absorption spectroscopy (69, 122, 130). The time interval of one picosecond is hard to grasp; however, its extreme brevity can be put into perspective by the realization that light can only travel 0.3 mm in one picosecond.

When *Rps. sphaeroides* reaction centers are excited by a 10 ps mode-locked laser pulse, the absorption changes which herald the formation of P^+ are complete within the period of the flash (105, 122, 130). Superimposed upon these there are additional changes which reflect the reduction of various intermediate electron acceptors, I (69, 122, 130). These additional changes include a broad absorbance increase around 670 nm and bleachings centered at 545, 760, and 800 nm (69, 130). Subsequently, Q_A is reduced by I^- and these additional, transient absorption changes disappear (122, 130). Since the absorbance changes at 545 and 760 nm are due to bacteriopheophytin and those at 800 nm are due to bacteriochlorophyll, it seems clear that I involves both those pigments. The absorbance increase at 670 nm could arise from either bacteriochlorophyll or bacteriopheophytin, since this spectral feature is characteristic of the anionic radicals of both of these molecules (34–36).

The most probable reaction sequence is as follows (2, 57, 69, 122, 130, 138, 149, 150). In less than 1 ps electron transfer occurs from P^* to one of the two bacteriochlorophyll (B) molecules, which absorb at 800 nm. The electron then moves in 4 – 10 ps from B to reduce one of the two bacteriopheophytins (H). Subsequently, in about 150 – 250 ps, Q_A is reduced by H^-.

$$PBHQ_A \xrightarrow{h\nu} P^*BHQ_A \xrightarrow{1ps} P^+B^-HQ_A \xrightarrow{4-10\ ps} P^+BH^-Q_A \xrightarrow{150\text{-}250\ ps} P^+BHQ_A^-$$

The initial events in this reaction sequence have been most clearly resolved when the reaction centers were excited with subpicosecond pulses (0.5 ps in duration) of 615 nm light, obtained from a passively mode-locked continuous-wave dye laser (57). Under these conditions absorption changes consistent with the formation of P^+B^- are complete in less than 1 ps. The bleaching at 545 nm (that is, the reduction of H) lags behind these initial changes and grows in gradually during the next 4 – 10 ps.

It is extremely difficult to be certain exactly how many of the reaction center pigments are involved as intermediate electron carriers between P and Q_A. This problem arises because of two inherent properties of the system. First, under normal functional conditions, I^- only lasts for a few hundred picoseconds, and this makes its detailed characterization very difficult. However, this problem can be overcome and reaction centers can be trapped in the state I^- (115, 116, 145, 150, 158, 165). This is accomplished by prolonged illumination of reduced reaction centers (i.e. in the state Q_A^-) in the presence of a suitable electron donor for P^+ such as reduced cytochrome c. Eventually all the reaction centers end up in the state $PI^-Q_A^-$ [or $I^-Q^=$ (see 115, 116)], and I^- persists long enough for it to be studied optically and by ESR. Unfortunately, this is where the second difficulty arises. The reaction center pigments are packed into a small space and form a set of closely interacting molecules. The presence of a charge on one molecule can cause large spectral shifts and changes in the magnetic properties of the others. For example, the presence of the charge on P in the state P^+ causes a red shift in the absorption spectrum of the reaction center carotenoid (18, 54). Because of these problems it is unclear whether or not B and H are monomeric bacteriochlorophyll and bacteriopheophytin respectively or whether the remaining bacteriochlorophyll and bacteriopheophytin are also involved. In reaction centers from both *Rps. sphaeroides* and

Rps. viridis it is possible to photoreduce both the reaction center bacterio-pheophytins (145, 163), but it is not clear whether these are just side reactions. For a full description of this topic, the recent review of Parson (121) should be consulted.

If Q_A is extracted from the reaction centers, or if it has been chemically reduced, then excitation results in the formation of I^- and it now persists for between 5 – 10 ns (70, 122, 152). Under these conditions P^+I^- decays by a variety of routes, several of which yield triplet states (27, 122, 152). Depending upon the temperature and the charge on Q_A, three possible triplet states have been proposed (152): ^3P, $^3(P^+B^-)$ and $^3(P^+H^-)$. In reaction centers from carotenoid-containing strains, these triplets decay rapidly to yield carotenoid triplets (16). The primary photochemical reaction in *Rps. sphaeroides* reaction centers has a quantum yield of about 100% (177), and these triplet states only occur when the normal photochemical route has been blocked. However, even though they are generated by side reactions, their detailed study has yielded a great deal of information upon the ener-getics of the primary reactions and the way in which the pigments interact together within the reaction center [for example see (152)].

OXYGEN-EVOLVING PHOTOSYNTHETIC ORGANISMS

The pattern of photosynthesis in oxygen-evolving organisms is more com-plex than that of the anaerobic photosynthetic bacteria in that they have two types of reaction center (photosystem I and photosystem II) rather than one. It may be as a result of this that it has not yet proved possible to isolate reaction centers from plants in the same way as has been achieved with the bacteria. However, it is questionable whether reaction centers in the bac-terial sense, of discrete, antenna-free, pigment-protein complexes, do exist in plants. So far only enriched preparations of the two photosystems are available, in which 20 or more antenna chlorophyll molecules are still present per "reactive center." It may well be in plants that the polypeptides which are associated with the "core" of photosystems I and II normally bind both the reaction center pigments and a set of antenna pigments as well. Thus, what in structural terms is meant by a reaction center in plants is still very much an open question.

There is, however, another basic problem in this area that also serves to confuse. Since there is no one, universally recognized, well-characterized method of preparing photosystem I and II particles, almost every labora-tory working in this area uses both different preparative methods and different biological material. Each method has its good and its bad points, and each method produces photoactive particles with subtly differing properties. This makes detailed comparisons extremely difficult, and this area of ambiguity needs to be borne carefully in mind.

However, in spite of these problems, a general picture of the structure of the two photosystems is now beginning to emerge. The best model would appear to be a central "core" region, presumably where the primary photochemical reactions take place, surrounded by a variety of specific antenna pigment-protein complexes. Each photosystem, then, has its own complement of these antenna complexes, which may be best thought of as "dedicated" to that photosystem (9, 162, 164).

The Composition of Photosystem I

The primary electron donor of photosystem I is called P700, after the wavelength of the maximum bleaching in the red when photosystem I is oxidized (84). The chlorophyll a P700 pigment-protein complex has two major polypeptides. Their molecular weights depend upon the biological material; for example, 68 KD and 66 KD in peas (104), 64 KD and 49 KD in *Chlamydomonas reinhardi* (144), and 52 KD and 46 KD in *Anabena flos-aquae* (75). It has been generally assumed that these two polypeptides form the "core" of photosystem I. However, a typical chlorophyll a P700 complex also contains a range of additional polypeptides. A good example of this added complexity comes from a study by Mullet et al (104). These workers prepared a photosystem I particle from pea chloroplasts following solubilization with triton X100. The preparation contained 110 chl a molecules per P700 and had at least 11 polypeptides (68, 66, 24.5, 24, 27.5, 21, 17, 16.5, 11.5, 11, and 10.5 KD respectively). The chlorophyll a content could be reduced by treatment with a higher concentration of triton to yield a particle with 65 chl a molecules per P700. This removal of chlorophyll a is accompanied by the loss of the 24.5, 24, and 22.5 KD polypeptides. It has been proposed that the remaining lower molecular weight polypeptides represent further antenna complexes and some of the iron sulfur electron acceptor proteins.

Detailed models of P700 have been proposed by Bengis & Nelson (5, 6) and Thornber et al (159, 160). Bengis & Nelson (5, 6) have proposed that there is one P700 per pair of the two higher molecular weight polypeptides, while Thornber et al (159, 160) suggest a model with two trimers, one containing two of the larger subunits and one of the smaller subunits and one containing three of the larger subunits. In the Thornber model only the smaller subunits have P700. However, both these models must be viewed as tentative at present. The major red absorption band of the P700 preparations is usually between 675 and 677 nm (119, 162). In those preparations with 40 or less chlorophyll a molecules per P700 chlorophyll b is usually lacking, though in those with greater than 100 chlorophyll a molecules per

P700, a little chlorophyll b is sometimes present (9, 119, 162). Most preparations usually contain at least one molecule of β-carotene per P700 (9, 119, 162). It has also been reported that some photochemically active P700 preparations lack pheophytin a (159).

Several P700 preparations contain cytochromes f and b_6 (119), and these maybe responsible for some of the lower molecular weight polypeptides seen in those preparations.

The Composition of Photosystem II

Photosystem II preparations are less well characterized than those of photosystem I. Undoubtedly, one of the main reasons for this is the difficulty in assaying directly the photochemical activity of photosystem II when it has been solubilized. The primary electron donor of photosystem II is P680 (24, 25). Its lifetime, when oxidized, is rather short (25, 97) and this, coupled with the general lability of photosystem II activity, has made its direct measurement very difficult. There is also a considerable problem of making measurements at 680 nm because of fluorescence artifacts. The indirect assay of measuring electron flow from diphenylcarbazide to dichlorophenolindophenol is good for following where PSII activity is, but a direct assay of P680 is really needed if vital data such as numbers of chlorophyll molecules per P680 is to be determined.

Two of the most commonly used photosystem II preparations are the TSF-2a particle of Vernon et al (170) and the FII preparation of Satoh (139, 140). Of these the Satoh preparation is probably the best characterized. It has three major proteinaceous components with apparent molecular weights of 43, 27, and 10 KD (140). It has been suggested that the 43 and 27 KD components represent the "core" of photosystem II, while the lower molecular weight polypeptide may be the apoprotein of cytochrome b_{559} (140).

Another approach to the problem of defining the polypeptide composition of photosystem II has involved the use of mutants lacking photosystem II activity. In this method the polypeptide composition of the photosynthetic membranes of the wild type and mutant strains are compared and the missing polypeptides are correlated with the loss of photosystem II. Two studies will serve to illustrate this approach.

A mutant of *Chlamydomonas reinhardtii*, F34, is deficient in photosystem II reaction centers. When compared with the wild-type it is lacking three major polypeptides of 50 KD, 47 KD, and 3 KD (6a). Metz & Miles (103) have analyzed a high flourescence nuclear mutant of maize which appears to lack photosystem II. They showed that six polypeptides were lacking and of those they also suggested that two with apparent molecular weights of 49 and 45 KD represented the "core" of photosystem II. So far the evidence from these genetic experiments has served nicely to confirm the conclusions drawn from the analysis of the isolated PSII preparations.

Photosystem II preparations normally show a maximum absorbance in the red between 670 – 675 nm. They contain chlorophyll a, a variable amount of chlorophyll b (it can be as low as 1 chlorophyll b to 30 chlorophyll a molecules) and usually some β-carotene (119). Plastoquinone is also usually present but in variable amounts, as is cytochrome b_{559} (119). The ratio of P680 per chlorophyll present is rather unreliable, but has been variously estimated to be anywhere from 1 : 20 to 1 : 120 depending upon the preparation (119).

The Structure and Localization of Photosystems I and II

Apart from the data upon the composition of the two photosystems described above, there is really very little known about the structure of the green plant and algal reaction centers. There is even a lack of data upon the amino acid composition of the "core" polypeptides. There is, however, some indirect evidence for the membrane location of the two photosystems.

Charge separation at both photosystem I and photosystem II generates the 515 nm shift (67), and this has lead to the suggestion that, just as with the reaction centers from *Rps. sphaeroides,* both photosystems lie across the plane of the membrane, such that their primary reactions take place across the membrane dielectric. Similar conclusions have been reached by comparing the interaction of the two photosystems with penetrating and nonpenetrating redox dyes (166).

The Primary Electron Donors of Photosystems I and II

The primary electron donor of photosystem I, P700, was first described by Kok (84). Oxidation of P700 to P700$^+$ results in absorbance decreases at 430, 680, and 700 nm (24, 84) and a broad absorbance increase at about 815 nm (59, 99). The increase at 815 nm is typical of chlorophyll cations (10). P700$^+$ is a free radical and shows a strong ESR signal, centered at g $=$ 2.0025 (68, 111, 112). The linewidth of this signal (7.2 G) is narrower than that of monomeric chlorophyll a^+ (9.3 G) by $\sqrt{2}$ and has lead to the suggestion that P700 is a chlorophyll a dimer (68, 99, 112). The dimer hypothesis is supported by the ENDOR spectra of P700$^+$ and the monomeric chlorophyll cation (110, 111), where with P700$^+$ several proton hyperfine splittings are found to be half those of chlorophyll a^+. There is some question as to the exact redox potential of P700 (82, 90, 97); however, most determinations yield values between $+400$ and $+500$ mV. The P700/P700$^+$ couple is an $n = 1$ redox reaction and is pH independent (90). It seems, therefore, that in many ways P700 is quite similar to the primary electron donor of most photosynthetic bacteria.

P680 (the primary donor of photosystem I), on the other hand, is completely different. P680 was first described by Doring et al (24, 25) and called by them chlorophyll a_{II}. Photoxidation of P680 causes bleaching at 430 and 680 nm and a broad increase at about 825 nm (53, 96, 99, 169). There is, however, some variation in the published difference spectra for P680$^+$. This probably arises from a combination of the difficulty in obtaining accurate spectra for a transient lasting < 10 μ sec (97) and a problem in correctly allowing for other interfering absorbance changes (97). A transient ESR signal for P680$^+$ has been described (92, 169, 171). It is centered at g = 2.002 and has a linewidth of between 7 – 9 G (21, 92). For several years P680 was also viewed to be a chlorophyll a dimer. However, this has recently been questioned (21). It has now been suggested that P680 could be a monomer and that narrowing of the ESR signal could be due to the interaction of the chlorophyll with the protein. ENDOR data on P680 seems to be lacking. If P680 is a monomer, this would solve an energetic problem. The redox potential of P680 has not yet been determined, but it must be higher than +820 mV since it can oxidize water. The redox potential of monomeric chlorophyll a in vitro can be as high as +900 mV, whereas the midpoint potentials of chlorophyll aggregates (such as dimers) are generally much lower (21).

The Reducing Side of Photosystems I and II

The reducing side of photosystem I has been investigated by a combination of fast optical spectroscopy (44, 71, 72, 98, 141–143, 146–148) and slower low temperature ESR spectroscopy (8, 32, 33, 48, 55, 90, 91, 93, 101, 102). These two approaches have yielded a great deal of information, but it is still difficult at present to reconcile all the available data and produce a fully consistent picture.

Hiyama & Ke (56) investigated the primary reactions in photosystem I particles and discovered a broad absorbance decrease centered at 430 nm. They called this change P430 and suggested that it represented the stable electron acceptor of photosystem I. Subsequently, in order to take account of the ESR experiments, it was suggested that P430 was an iron sulfur protein(s) (71). At room temperature and at moderate redox potentials laser flash excitation of photosystem I particles generates P700$^+$ P430$^-$. This state then decays by a back reaction, with a half-time of ~ 45 msec (56, 71). More recently, Sauer et al (142) have shown that if photosystem I particles were poised at low redox potentials where P430 is chemically reduced [the E_m of P430 is –470 to – 500 mV (71)], then P700 is still oxidized by a laser flash but it now decays with a halftime of ~ 250 μ sec. This result leads to the postulation of another electron acceptor closer to P700 than P430. Sauer et al (142, 143) called this acceptor A_2. If A_2 was now reduced prior to excitation (142, 143), or if it was removed by the addition of SDS (98, 143), then the laser flash still oxidized P700 but now the back reaction was even faster (t½ ~ 7μ sec). Continuing with the same logic, yet another electron acceptor closer still to P700 was postulated. This acceptor was called A_1 (142, 143).

At about the same time that P430 was first being described, Malkin & Bearden (91) were able to observe an ESR signal from the supposed stable electron acceptor of photosystem I. They found that if chloroplasts were illuminated at 77°K, then cooled down to 25°K, a stable free radical was generated whose ESR spectrum had resonances at g values of 1.86 and 2.05. They called this signal "bound ferredoxin" because of its similarities to soluble ferredoxin. This bound ferredoxin appeared to have just the properties that would be predicted for the stable electron acceptor. However, upon more detailed characterization of this signal it has become clear that the bound ferredoxin does in fact consist of two apparently separate iron sulfur centers (32, 33, 55, 90, 93). These two centers have been called A and B and they are most probably Fe 4 S4 clusters (30, 132). It appears that the optical signal P430 is equivalent to both these centers.

When chloroplasts or photosystem I particles are illuminated at low temperatures (77°K), the ESR signal from $P700^+$ is largely irreversible (4, 91, 93). Yet if the "bound ferredoxin" is chemically reduced prior to freezing, then the $P700^+$ ESR signal is still produced upon illumination, but now it is almost completely reversible (101). Under these experimental conditions a new electron acceptor signal can be detected with g values at 1.78, 1.88, and 2.08. This signal has been called "X," and it may be yet another iron sulfur center (101).

Nanosecond and picosecond optical spectroscopy have now also been applied to photosystem I preparations (44, 72, 142, 146–148). Following excitation from a mode-locked dye laser, P700 is oxidized within 10 ps. In addition, an extra transient species is also generated. This transient is formed as P700 is oxidized and, in photochemically "open" photosystem I preparations, decays in a few hundred picoseconds (148). If, however, normal photochemistry is blocked by chemically reducing P430 prior to excitation, then the transient shows a biphasic decay with half-times of ~ 10 ns and ~ 3 μ sec (146, 148). Just as with state I in bacterial reaction centers, this transient state can be "trapped" by prolonged illumination under reducing conditions. Making use of this method, it has been possible to obtain good optical ESR and ENDOR spectra of this state (37). Although it has not been proved, it has been suggested that this state is an anion radical of monomeric chlorophyll a (37).

The redox potentials of center A, center B, and X have been determined by several groups [for example see Malkin (90)]. In each case they have n values of 1, and the most frequently quoted values for the E_ms are -550 mV for center A, -594 mV for center B, and -730 mV for X. If the anionic radical of monomeric chlorophyll a is the reductant for X, then presumably it must have a redox potential which is more negative than -730 mV. This is not unreasonable since in organic solution the redox potential of the Chl$a$/Chla^- couple can be as low as -880 mV (37, 46).

How do the results of the optical studies relate to those obtained from the ESR experiments, and what is the sequence of the various electron acceptors described above? There are no really firm answers to these questions, but if the acceptors revealed by the optical studies are assumed to be equivalent to those seen by ESR, then two possible schemes can be drawn.

(a)   P700 $\longrightarrow$ Chl$a$ $\longrightarrow$ X $\longrightarrow$ P430

$\qquad\qquad\quad$ (A$_1$) $\qquad\quad$ (A$_2$) $\qquad\qquad$ (Centers A and B)

(b)   P700 $\longrightarrow$ Chl$a$ $\longrightarrow$ X $\Big\langle\!\!\begin{array}{l} \text{Center A} \\ \text{Center B} \end{array}$

$\qquad\qquad\quad$ (A$_1$) $\qquad\quad$ (A$_2$) $\qquad$ (P430)

Further work is required to see whether either of these models or perhaps a different one is correct.

The stable electron acceptor of photosystem II was originally called Q (Q for quencher) (29). Its presence was inferred from studies on the variable fluorescence emitted from chloroplasts at room temperature (29, 85). When Q is reduced the level of fluorescence is high, and when Q is oxidized the level of fluorescence is low (11, 20, 31, 74). Subsequently, an absorbance change at 320 nm (X320) was discovered, and its properties were found to correlate closely to those predicted for Q (155). The difference spectrum of oxidized-reduced X320 was very similar to that of the anionic semiquinone of plastoquinone (155). This then suggests that the acceptor side of photosystem II may be similar to the acceptor side of the bacterial reaction centers, with both using quinones as the stable electron acceptor.

Recently a more complicated picture of Q has emerged. Careful redox titrations of Q have shown that there are in fact two quenchers, Q$_L$ and Q$_H$ (11, 20, 51, 58, 74). Q$_H$ has a midpoint potential of -45 mV at pH 7.8 while Q$_L$ has a midpoint potential of -247 mV at the same pH (58). Both these redox reactions have n values of 1 and show some pH dependence (58). It is not yet clear how X320 is related to the presence of two Qs. There have been some reports that there are two components in the kinetics of reoxidation of X320 which may correspond to the two Qs (50, 153); however, until the redox potential dependence of X320 has been determined, this must remain an open question.

An absorbance change at 550 nm, called C550, has also been shown to correlate closely with the expected behavior of Q (79, 81). C550 titrates with the same redox potential as Q_H (79) (its redox potential dependance in the region of Q_L has not been determined). However, extraction of β-carotene from photosystem II particles results in the loss of C550 (19, 83), and it seems likely that C550 represents an electrochromic shift of the β-carotene in response to the negative charge on Q, rather than to a component which is directly oxidized and reduced. If this is correct, then the C550 change is rather similar to the reaction center carotenoid band shift seen in *Rps. sphaeroides* (18, 54).

The similarity of the acceptor side of photosystem II to the acceptor side of the bacterial reaction centers stimulated the search for a transient intermediate electron acceptor in photosystem II analogous to I in the bacteria. Klimov et al (78), using the same experimental approach which lead to the "trapping" of I⁻ (illumination at low redox potential), were able to record some low potential absorbance changes in DT-20 photosystem II particles. The difference spectrum of these changes suggested that they were due to the reduction of pheophytin *a*. The ESR and ENDOR spectra of this state have been recorded and are consistent with the suggestion that it is the anionic radical of monomeric pheophytin *a* (37, 76). If photosystem II preparations are poised at low redox potential and excited by a 30 ps laser flash, the state $P680^+$ pheoa^- is formed rapidly and then decays with a lifetime of $\sim$ 4 ns (73, 77). Some of this decay yields triplet states which have been detected both optically and by ESR (76–78, 134, 135–137, 141, 151). Redox titration of the extent of the light-induced triplet state, monitored by ESR, has allowed the midpoint potential of pheophytin *a* to be estimated (135). The E_m determined in this way, –604 mV, compares well with the value determined by direct titration of the optical changes of –610 mV (37, 78). Very recently there has been a report suggesting that there could even be another intermediary electron acceptor between P680 and pheophytin *a* (133). Illumination of photosystem II particles in which pheophytin *a* is already reduced causes a new ESR signal to accumulate. This may be due to chlorophyll *a*⁻. If this proves to be correct, then the similarity between the reducing side of photosystem II and the bacterial reaction centers will be even more striking.

It is, as yet, difficult to write out a model for the reactions on the reducing side of photosystem II with the status of the two Qs, Q_L and Q_H, being entirely unclear. There is even the possibility of heterogeneous populations of photosystems II, some using Q_L and others Q_H. Any definite conclusions in this area must await further experimentation.

FINAL REMARKS

In preparing this review I have been very conscious that several areas of reaction center research have not been covered [for example, indirect structural analysis using polarized light techniques (12), fluorescence (85), and other types of bacterial reaction centers (49, 118, 119, 158)]. This was a deliberate decision in order to have the space to concentrate on trying to present a unified view of reaction center function and to try to emphasize the basic similarities in the mechanisms of the primary reactions of both the plant and the bacterial reaction centers.

As to the future, this area is in need of advances on the biochemical front. The sophisticated biophysical analyses of reaction center function have now reached the point where detailed structural information is urgently required. In the plant and algal systems, this unfortunately seems a long way off. However, in the bacteria, with the prospect of a three-dimensional structure of the *Rps. viridis* reaction center on the horizon, this information may soon be available. If this proves to be correct, then the time between now and the next review of this topic should prove to be extremely exciting.

ACKNOWLEDGMENTS

I would like to thank all the workers in this area who sent me numerous reprints and preprints, and also Drs. Michael Hipkins and Gordon Lindsay for helpful criticism. Thanks also to Irene Durant for typing the manuscript.

Literature Cited

1. Agalidis, I., Lutz, M., Reiss-Husson, F. 1980. Binding of carotenoids to reaction centres from *Rps. sphaeroides* R26. *Biochim. Biophys. Acta* 589:264–74
2. Akhmanov, S. A., Borisov, A. Yu., Danielius, R. V., Gadonas, R. A., Kozlowski, V. S., et al. 1980. One and two photon picosecond processes of electron transfer among the porphyrin molecules in bacterial reaction centres. *FEBS Lett.* 114:149–52
3. Bachmann, R. C., Gillies, K., Takemoto, J. Y. 1981. Membrane topography of the phosynthetic reaction centre polypeptides of *Rps. sphaeroides. Biochemistry* 20:4590–96

4. Bearden, A. J., Malkin, R. 1976. Correlation of reaction centre chlorophyll (P700) oxidation and bound iron-sulphur protein photoreduction in chloroplast photosystem I at low temperatures. *Biochim. Biophys. Acta* 430:538–47
5. Bengis, C., Nelson, N. 1975. Purification and properties of the photosystem I reaction centre from chloroplasts. *J. Biol. Chem.* 250:2783–88
6. Bengis, C., Nelson, N. 1977. The subunit structure of choloroplast photosystem I reaction centre. *J. Biol. Chem.* 252:4564–69

6a. Bennoun, P., Wollman, F. A., Diner, B. A. 1981. Thylakoid polypeptides associated with photosystem II centres in *Chlamydomonas reinhardii:* comparison of system II mutants and particles. *Abstr. 5th Int. Congr. Photosynth.,* p. 55

7. Blankenship, R. E., Parson, W. W. 1978. The photochemical electron transfer reactions of photosynthetic bacteria and plants. *Ann. Rev. Biochem.* 47:635–53

8. Blankenship, R. E., McGuire, A., Sauer, K. 1975. Chemically induced dynamic electron polarisation in chloroplasts at room temperature: evidence for triplet state participation in photosynthesis. *Proc. Natl. Acad. Sci. USA* 72:4943–47

9. Boardman, N. K., Anderson, J. M., Goodchild, D. J. 1978. Chlorophyll-protein complexes and structure of mature developing chloroplasts. *Curr. Top. Bioenerg.* 8:35–109

10. Borg, D. C., Fajer, J., Felton, R. H., Dolphin, D. 1970. The π cation radical of chlorophyll *a. Proc. Natl. Acad. Sci. USA* 67:813–20

11. Butler, W. L. 1973. Primary photochemistry of photosystem II in photosynthesis. *Acc. Chem. Res.* 6:177–83

12. Clayton, R. K. 1980. *Photosynthesis: Physical Mechanisms and Chemical Patterns.* IUPAB, Biophys. Ser. Cambridge: Cambridge Univ. Press

13. Clayton, R. K., Clayton, B. J. 1978. Properties of photochemical reaction centres purified from *Rps. gelatinosa. Biochim. Biophys. Acta* 501:470–77

14. Clayton, R. K., Wang, R. T. 1971. Photochemical reaction centres from *Rps. sphaeroides. Methods Enzymol.* 23:696–704

15. Cogdell, R. J., Brune, D. C., Clayton, R. K. 1974. Effects of extraction and replacement of ubiquinone upon the photochemical activity of reaction centres and chromatophores from *Rps. sphaeroides. FEBS Lett.* 45:344–47

16. Cogdell, R. J., Monger, T. G., Parson, W. W. 1975. Carotenoid triplet states in reaction centres from *Rps. sphaeroides* and *R. rubrum. Biochim. Biophys. Acta* 408:189–99

17. Cogdell, R. J., Parson, W. W., Kerr, M. A. 1976. The type, amount and location of carotenoids within reaction centres of *Rps. sphaeroides. Biochim. Biophys. Acta* 430:83–93

18. Cogdell, R. J., Celis, S., Celis, H., Crofts, A. R. 1977. Reaction centre carotenoid band shifts. *FEBS Lett.* 80: 190–94

19. Cox, R. P., Bendall, D. S. 1974. The function of plastoquinone and β-carotene in photosystem II of chloroplasts. *Biochim. Biophys. Acta* 347:49–59

20. Cramer, W. A., Butler, W. L. 1969. Potentiometric titration of the fluorescence yield of spinach chloroplasts. *Biochim. Biophys. Acta* 172:503–10

21. Davis, M. S., Forman, A., Fajer, J. 1979. Ligated chlorophyll cation radicals: Their function in photosystem II of plant photosynthesis. *Proc. Natl. Acad. Sci. USA* 76:4170–74

22. David, M. S., Forman, A., Hanson, L. K., Thornber, J. P., Fajer, J. 1979. Anion and cation radicals of bacteriochlorophyll and bacteriopheophytin *b.* Their role in the primary charge separation of *Rps. viridis. J. Phys. Chem.* 83:3325–32

23. Debrunner, P. G., Schulz, C. E., Feher, G., Okamura, M. Y. 1975. A Mossbauer study of reaction centers from *Rps. sphaeroides. Biophys. Soc. Abstr.* 15:226

24. Doring, G., Bailey, J. L., Kreutz, W., Wiekand, J., Witt, H. T. 1968. Some new results in photosynthesis. *Naturwissenschaften* 55:219–20

25. Doring, G., Renger, G., Vater, J., Witt, H. T. 1969. Properties of the photoreactive chlorophyll a_{II} in photosynthesis. *Z. Naturforsch. Teil B* 24:1139–43

26. Dutton, P. L., Jackson, J. B. 1972. Thermodynamic and kinetic characterization of electron transport components *in situ* in *Rps. sphaeroides* and *R. rubrum. Eur. J. Biochem.* 30:495–10

27. Dutton, P. L., Leigh, J. S., Reed, D. W. 1973. Primary events in the photosynthetic reaction centres from *Rps. sphaeroides,* strain R26. Triplet and oxidised states of bacteriochlorophyll and the identification of the primary electron acceptor. *Biochim. Biophys. Acta* 292:654–64

28. Dutton, P. L., Leigh, J. S., Wraight, C. A. 1973. Direct measurement of the midpotential of the primary acceptor in *Rps. sphaeroides in situ* and in the isolated state: some relationships with pH and o-phenanthroline. *FEBS Lett.* 36:169–73

29. Duysens, L. N. M., Sweers, H. E. 1963. Mechanisms of two photochemical reactions in algae as studied by means of fluorescence. In *Microalgae and Photosynthetic Bacteria,* ed. Jpn. Soc. Plant Physiol., pp. 353–72. Tokyo: Univ. Tokyo Press

30. Evans, E. H., Rush, J. D., Johnson, C. E., Evans, M. C. W. 1979. Mossbauer

spectra of photosystem 1 reaction centres from the blue-green alga *Chlorogloea fritschii. Biochem. J.* 182:861–65

31. Evans, M. C. W. 1969. Ferredoxin: NAD reductase and the photoreduction of NAD by *Chlorobium thiosulpatophilum.* In *Progress in Photosynthesis Research,* ed. H. Metzner, 3:1474–74. Tubingen: Laupp

32. Evans, M. C. W., Reeves, S. G., Cammack, R. 1974. Determination of the oxidation-reduction potential of the bound iron-sulphur proteins of the primary electron acceptor complex of photosystem I in spinach chloroplasts. *FEBS Lett.* 49:111–14

33. Evans, M. C. W., Sihra, C. K., Cammack, R. 1976. The properties of the primary electron acceptor in the photosystem I reaction centre of spinach chloroplasts and its interaction with P700 and the bound ferredoxin in various oxidation-reduction states. *Biochem. J.* 158:71–77

34. Fajer, J., Brune, D. C., Davis, M. S., Forman, A., Spaulding, L. D. 1975. Primary charge separation in bacterial photosynthesis: oxidised chlorophylls and reduced pheophytin. *Proc. Natl. Acad. Sci. USA* 72:4956–60

35. Fajer, J., Davis, M. S., Brune, D. C., Forman, A., Thornber, J. P. 1978. Optical and paramagnetic identification of a primary electron acceptor in bacterial photosynthesis. *J. Am. Chem. Soc.* 100:1918–20

36. Fajer, J., Davis, M. S., Brune, D. C., Spaulding, L. D., Borg, D. C., Forman, A. 1977. Chlorophyll radicals and primary events. *Brookhaven Symp. Biol.* 28:74–103

37. Fajer, J., Davis, M. S., Forman, A., Klimov, V. V., Dolan, E., Ke, B. 1980. Primary electron acceptors in plant photosynthesis. *J. Am. Chem. Soc.* 102:7143–45

38. Feher, G. 1971. Some chemical and physical properties of a bacterial reaction centre particle and its primary photochemical reactants. *Photochem. Photobiol.* 14:373–87

39. Feher, G., Hoff, A. J., Isaacson, R. A., Ackerson, L. C. 1975. ENDOR experiments on chlorophyll and bacteriochlorophyll *in vitro* and in the photosynthetic unit. *Ann. NY Acad. Sci.* 244:239–59

40. Feher, G., Hoff, A. J., Isaacson, R. A., McElroy, J. D. 1973. Investigation of the electronic structure of the primary electron donor in bacterial photosynthesis by the ENDOR technique. *Biophys. J. Abstr.* 13:61

41. Feher, G., Isaacson, R. A., McElroy, J. D., Ackerson, L. C., Okamura, M. Y. 1974. On the question of the primary acceptor in bacterial photosynthesis: Manganese substituting for iron in reaction centres of *Rps. sphaeroides* R26. *Biochim. Biophys. Acta* 368:135–39

42. Feher, G. Okamura, M. Y. 1978. Chemical composition and properties of reaction centers. In *The Photosynthetic Bacteria,* ed. R. K. Clayton, W. R. Sistrom, pp. 349–86. New York/London: Plenum

43. Feher, G., Okamura, M. Y., McElroy, J. D. 1972. Identification of an electron acceptor in reaction centres of *Rps. sphaeroides* by EPR spectroscopy. *Biochim. Biophys. Acta* 267:222–26

44. Fenton, J. M., Pellin, M. J., Govindjee, Kaufmann, K. J. 1979. Primary photochemistry of the reaction centre of photosystem I. *FEBS Lett.* 100:1–4

45. Fong, F. K. 1974. Molecular basis for the photosynthesis primary process. *Proc. Natl. Acad. Sci. USA* 71:3692–95

46. Forman, A., Davis, M. S., Fujita, I., Hanson, L. K., Smith, K. M., Fajer, J. 1982. Mechanisms of energy transduction in plant photosynthesis: ESR, ENDOR and MOs of the primary acceptors. *Isr. J. Chem.* In press

47. Francis, G. A., Richards, W. R. 1980. Localisation of photosynthetic membrane components in *Rps. sphaeroides* by a radioactive labelling procedure. *Biochemistry* 19:5104–11

48. Frank, H. A., McLean, M. B., Sauer, K. 1979. Triplet states in photosystem I of spinach chloroplasts and subchloroplast particles. *Proc. Natl. Acad. Sci. USA* 76:5124–28

49. Gingras, G. 1978. A comparative review of photochemical reaction center preparations from photosynthetic bacteria. See Ref. 42, pp. 119–31

50. Glaser, M., Wolff, Ch., Buchwald, H-E., Witt, H. T. 1974. On the photoactive chlorophyll reaction in system II of photosynthesis. Detection of a fast and large component. *FEBS Lett.* 42:81–85

51. Golbeck, J. H., Kok, B. 1979. Redox titration of electron acceptor Q and the plastoquinone pool in photosystem II. *Biochim. Biophys. Acta* 547:347–60

52. Hall, R. L., Doorley, P. F., Niederman, R. A. 1978. Trans-membrane localisation of reaction centre proteins in *Rps. sphaeroides* chromatophores. *Photochem. Photobiol.* 28:273–76

53. Haveman, J., Mathis, P. 1976. Flash-induced absorption changes of the primary donor of photosystem II at 820 nm in chloroplasts inhibited by low pH or Tris-treatment. *Biochim. Biophys. Acta* 440:346–55

54. Heathcote, P., Vermeglio, A., Clayton, R. K. 1977. The carotenoid band shift in reaction centres from *Rps. sphaeroides. Biochim. Biophys. Acta* 461: 358–64

55. Heathcote, P., Williams-Smith, D. L., Sihara, C. K., Evans, M. C. W. 1978. The role of the membrane-bound iron-sulphur centres A and B in the photosystem I reaction centre of spinach chloroplasts. *Biochim. Biophys. Acta* 503:333–42

56. Hiyama, T., Ke, B. 1971. A further study of P430: A possible primary electron acceptor of photosystem I. *Arch. Biochem. Biophys.* 147:99–108

57. Holten, D., Hoganson, C., Windsor, M. W., Schenck, C. C., Parson, W. W., et al. 1980. Subpicosecond and picosecond studies of electron transfer intermediates in *Rps. sphaeroides* reaction centres. *Biochim. Biophys. Acta* 592: 461–77

58. Horton, P., Croze, E. 1979. Characterisation of two quenchers of chlorophyll fluorescence with different mid-point oxidation-reduction potentials in chloroplasts. *Biochim. Biophys. Acta* 545: 188–201

59. Inoue, Y., Ogawa, T., Shibata, K. 1973. Light-induced spectral changes of P700 in the 800 nm region in *Anacystis* and spinach lamellae. *Biochim. Biophys. Acta* 305:483–87

60. Jackson, J. B., Cogdell, R. J., Crofts, A. R. 1973. Some effects of o-phenanthroline on electron transport in chromatophores from photosynthetic bacteria. *Biochim. Biophys. Acta* 290:218–25

61. Jackson, J. B., Crofts, A. R. 1971. The kinetics of light-induced carotenoid changes in *Rps. sphaeroides* and their relation to electric field generation across the chromatophore membrane. *Eur. J. Biochem.* 18:120–30

62. Jackson, J. B., Dutton, P. L. 1973. The kinetic redox potentiometric resolution of the carotenoid shifts in *Rps. sphaeroides* chromatophores: Their relationship to electric transport and energy coupling. *Biochim. Biophys. Acta* 325:102–13

63. Jennings, V. V., Evans, M. C. W. 1977. The irreversible photoreduction of a low potential component at low temperatures in a preparation of the green photosynthetic bacterium *Chlorobium thiosulphatophilium. FEBS Lett.* 75: 33–36

64. Jolchine, G., Reiss-Husson, F. 1974. Comparative studies on two reaction centre preparations from *Rps. sphaeroides* Y. *FEBS Lett.* 40:5–8

65. Jolchine, G., Reiss-Husson, F. 1975. Studies on pigments and lipids in *Rps. sphaeroides* Y reaction centres. *FEBS Lett.* 52:33–36

66. Jones, O. T. G., Whale, F. R. 1970. The oxidation and reduction of pyridine nucleotides in *Rps. sphaeroides* and *Chl. limicola f thiosulfatophilum. Arch. Mikrobiol.* 72:48–59

67. Junge, W. 1977. Membrane potentials in photosynthesis. *Ann. Rev. Plant Physiol.* 28:503–36

68. Katz, J. J., Norris, J. R. 1973. Chlorophyll and light energy transduction in photosynthesis. *Curr. Top. Bioenerg.* 5:41–75

69. Kaufmann, K. J., Dutton, P. L., Netzel, T. L., Leigh, J. S., Rentzepis, P. M. 1975. Picosecond kinetics of events leading to reaction center bacteriochlorophyll oxidation. *Science* 188: 1301–4

70. Kaufmann, K. J., Petty, K. M., Dutton, P. L., Rentzepis, P. M. 1976. Picosecond kinetics in reaction centers of *Rps. sphaeroides* and the effects of ubiquinone extraction and reconstitution. *Biochem. Biophys. Res. Commun.* 70: 839–45

71. Ke, B. 1973. The primary electron acceptor of photosystem I. *Biochim. Biophys. Acta* 301:1–33

72. Ke, B., Demeter, S., Zamaraev, K. I., Khairutdinov, R. F. 1979. Charge recombination in photosystem I at low temperatures. Kinetics of electron tunneling. *Biochim. Biophys. Acta* 545: 265–84

73. Ke, B., Dolan, E. 1980. Flash induced charge separation and dark recombination in a photosystem II subchloroplast particle. *Biochim. Biophys. Acta* 590: 401–6

74. Ke, B., Hawkridge, F. M., Sahu, S. 1976. Redox titration of fluorescence yield of photosystem II. *Proc. Natl. Acad. Sci. USA* 73:2211–15

75. Klein, S. M., Vernon, L. P. 1977. Composition of a photosystem I chlorophyll protein complex from *Anabaena flosaquae. Biochim. Biophys. Acta* 459: 364–75

76. Klimov, V. V., Dolan, E., Ke, B. 1980. EPR properties of an intermediary electron acceptor (pheophytin) in photosys-

tem II reaction centres at cryogenic temperatures. *FEBS Lett.* 112:97–100

77. Klimov, V. V., Ke, B., Dolan, E. 1980. Effect of photoreduction of the photosystem II intermediary electron acceptor (pheophytin) on triplet state of carotenoids. *FEBS Lett.* 118:123–26

78. Klimov, V. V., Klevanik, A. V., Shuvalov, V. A., Krasnovsky, A. V. 1977. Reduction of pheophytin in the primary light reaction of photosystem II. *FEBS Lett.* 82:183–86

79. Knaff, D. B. 1975. The effect of o-phenanthroline on the midpoint potential of the primary electron acceptor of photosystem II. *Biochim. Biophys. Acta* 376:583–87

80. Knaff, D. B. 1978. Reducing potentials and the pathway of NAD^+ reduction. See Ref. 42, pp. 629–40

81. Knaff, D. B., Arnon, D. I. 1969. Spectral evidence for a new photoreactive component of the oxygen-evolving system in photosynthesis. *Proc. Natl. Acad. Sci. USA* 63:963–69

82. Knaff, D. B., Malkin, R. 1973. The oxidation-reduction potentials of electron carriers in chloroplast photosystem I fragments. *Arch. Biochem. Biophys.* 159:555–62

83. Knaff, D. B., Malkin, R., Myron, J. C., Stoller, M. 1977. The role of plastoquinone and β-carotene in the primary reaction of plant photosystem II. *Biochim. Biophys. Acta* 459:402–11

84. Kok, B. 1961. Partial purification and determination of oxidation-reduction potential of the photosynthetic chlorophyll complex absorbing at 700 nm. *Biochim. Biophys. Acta* 48:527–33

85. Lavorel, J. 1975. Luminescence. In *Bioenergetics of Photosynthesis*, ed. Govindjee, pp. 223–317. New York: Academic

86. Leigh, J. S., Dutton, P. L. 1972. The primary electron acceptor in photosynthesis. *Biochem. Biophys. Res. Commun.* 46:414–21

87. Lin, L., Thornber, J. P. 1975. Isolation and partial characterisation of the photochemical reaction center of *Chr. vinosum* (strain D). *Photochem. Photobiol.* 22:37–40

88. Loach, P. A., Androes, G. M., Maksim, A. F., Calvin, M. 1963. Variation in electron paramagnetic resonance signals of photosynthetic systems with the redox level of their environment. *Photochem. Photobiol.* 2:443–54

89. Loach, P. A., Hall, R. L. 1972. The question of the primary electron acceptor in bacterial photosynthesis. *Proc. Natl. Acad. Sci. USA* 69:786–90

90. Malkin, R. 1982. Redox properties and functional aspects of electron carriers in chloroplast photosynthesis. In *Topics in Photosynthesis*, ed. J. Barber, 4:1–28. Amsterdam/New York/Oxford: Elsevier Biomed.

91. Malkin, R., Bearden, A. J. 1971. Primary reactions of photosynthesis: Photoreduction of a bound chloroplast ferredoxin at low temperature as detected by EPR spectroscopy. *Proc. Natl. Acad. Sci. USA* 68:16–19

92. Malkin, R., Bearden, A. J. 1975. Laser-flash-activated electron paramagnetic resonance studies of primary photochemical reactions in chloroplasts. *Biochim. Biophys. Acta* 396:250–59

93. Malkin, R., Bearden, A. J. 1978. Membrane-bound iron-sulphur centres in photosynthetic systems. *Biochim. Biophys. Acta* 505:147–81

94. Marinetti, T. D., Okamura, M. Y., Feher, G. 1979. Localisation of the primary quinone binding site in reaction centres from *Rps. sphaeroides* R26 by photoaffinity labelling. *Biochemistry* 18:3126–33

95. Marinetti, T. D., Okamura, M. Y., Feher, G. 1979. Photoaffinity labelling of the quinone binding site in bacterial reaction centres of *Rps. sphaeroides*. *Biophys. J.* 25:204a

96. Mathis, P., Haveman, J., Yates, M. 1976. The reaction centre of Photosystem II. *Brookhaven Symp. Biol.* 28:267–77

97. Mathis, P., Paillotin, G. 1981. Primary processes of photosynthesis. In *The Biochemistry of Plants: Photosynthesis*, ed. M. D. Hatch, N. K. Boardman, 8:98–161. London/New York: Academic

98. Mathis, P., Sauer, K., Remy, R. 1978. Rapidly reversible flash-induced electron transfer in a P700 chlorophyll-protein complex isolated with SDS. *FEBS Lett.* 88:275–78

99. Mathis, P., Vermeglio, A. 1975. Chlorophyll radical cation in photosystem II of chloroplasts. Millisecond decay at low temperatures. *Biochim. Biophys. Acta* 369:371–81

100. McElroy, J. D., Feher, G., Mauzerall, D. C. 1972. Characterisation of primary reactants in bacterial photosynthesis I. Comparison of the light-induced EPR signal (g=2.0026) with that of a bacteriochlorophyll radical. *Biochim. Biophys. Acta* 267:363–74

101. McIntosh, A. R., Bolton, J. R. 1976. Electron spin resonance spectrum of

species 'X' which may function as the primary electron acceptor in photosystem I of green plant photosynthesis. *Biochim. Biophys. Acta* 430:555–59

102. McLean, M. B., Sauer, K. 1982. The dependence of reaction centre and antenna triplets on the redox state of photosystem I. *Biochim. Biophys. Acta* 679:384–92

103. Metz, J., Miles, D. 1982. Composition of PSII as determined by analysis of a high fluorescent nuclear mutant of maize. *Abstr. Ann. Meet. Am. Soc. Plant Physiol.* No. 153

104. Mullet, J. E., Burke, J. J., Arntzen, C. T. 1980. Chlorophyll proteins of photosystem I. *Plant Physiol.* 65:814–22

105. Netzel, T. L., Rentzepis, P. M., Leigh, J. S. 1973. Picosecond kinetics of reaction centre bacteriochlorophyll excitation. *Science* 182:238–41

106. Nieth, K. F., Drews, G., Feick, R. 1975. Photochemical reaction centres from *Rps. capsulata. Arch. Microbiol.* 105: 43–45

107. Noel, H. M., van der Rest, M., Gingras, G. 1972. Isolation and partial characterisation of a P_{870} reaction centre complex from wild-type *R. rubrum. Biochim. Biophys. Acta* 275:219–30

108. Norris, J. R., Druyan, M. E., Katz, J. J. 1973. Electron nuclear double resonance of bacteriochlorophyll free radical *in vitro* and *in vivo. J. Am. Chem. Soc.* 95:1680–82

109. Norris, J. R., Katz, J. J. 1978. Oxidised bacteriochlorophyll as a photoproduct. See Ref. 42, pp. 397–418

110. Norris, J. R., Scheer, H., Druyan, M. E., Katz, J. J. 1974. An electron nuclear double resonance of bacteriochlorophyll free radical *in vitro* and *in vivo. Proc. Natl. Acad. Sci. USA* 71:4897–4900

111. Norris, J. R., Scheer, H., Katz, J. J. 1975. Models for antenna and reaction centre chlorophylls. *Ann. NY Acad. Sci.* 244:261–80

112. Norris, J. R., Uphaus, R. A., Crespi, H. L., Katz, J. J. 1971. Electron spin resonance of chlorophyll and the origin of signal 1 in photosynthesis. *Proc. Natl. Acad. Sci. USA* 68:625–28

113. Okamura, M. Y., Ackerson, L. C., Isaacson, R. A., Parson, W. W., Feher, G. 1976. The primary electron acceptor in *Chr. vinosum* (strain D). *Biophys. Soc. Abstr.* 16:67

114. Okamura, M. Y., Isaacson, R. A., Feher, G. 1975. Primary acceptor in bacterial photosynthesis: Obligatory role of ubiquinone in photoactive reaction centre of *Rps. sphaeroides. Proc. Natl. Acad. Sci. USA* 72:3491–95

115. Okamura, M. Y., Isaacson, R. A., Feher, G. 1977. On the trapping of the transient acceptor in reaction centres of *Rps. sphaeroides* R26. *Biophys. J. Abstr.* 17:149

116. Okamura, M. Y., Isaacson, R. A., Feher, G. 1979. Spectroscopic and kinetic properties of the transient intermediate acceptor in reaction centres of *Rps. sphaeroides. Biochim. Biophys. Acta* 546:394–417

117. Okamura, M. Y., Steiner, L. A., Feher, G. 1974. Characterisation of reaction centres from photosynthetic bacteria I. Subunit structure of the protein mediating the primary photochemistry in *Rps. sphaeroides* R26. *Biochemistry* 13:1394–1402

118. Olson, J. M. 1980. Chlorophyll organisation in green photosynthetic bacteria. *Biochim. Biophys. Acta* 594:33–51

119. Olson, J. M., Thornber, J. P. 1979. Photosynthetic reaction centres. In *Membrane Proteins in Energy Transduction,* ed. R. A. Capaldi, pp. 279–340. New York/Basle: Decker

120. Parson, W. W. 1969. The reaction between primary and secondary electron acceptors in bacterial photosynthesis. *Biochim. Biophys. Acta* 189:384–96

121. Parson, W. W. 1982. Photosynthetic bacterial reaction centres: Interactions among the bacteriochlorophylls and bacteriopheophytins. *Ann. Rev. Biophys. Bioeng.* 11:57–80

122. Parson, W. W., Clayton, R. K., Cogdell, R. J. 1975. Excited states of photosynthetic reaction centres at low redox potentials. *Biochim. Biophys. Acta* 387:265–78

123. Parson, W. W., Cogdell, R. J. 1975. The primary photochemical reaction of bacterial photosynthesis. *Biochim. Biophys. Acta* 416:105–49

124. Peucheu, N. L., Kerber, N. L., Garcia, A. 1976. Isolation and purification of reaction centres from *Rps. viridis* NHTC 133 by means of LDAO. *Arch. Microbiol.* 109:301–5

125. Prince, R. C., Crofts, A. R. 1973. Photochemical reaction centres from *Rps. capsulata* Ala Pho[+]. *FEBS Lett.* 35:213–16

126. Prince, R. C., Dutton, P. L. 1976. The primary acceptor of bacterial photosynthesis: its operating midpoint potential. *Arch. Biochem. Biophys.* 172:329–34

127. Prince, R. C., Dutton, P. L. 1978. Protonation and the reducing potential

of the primary electron acceptor. See Ref. 42, pp. 439–53

128. Prince, R. C., Olson, J. M. 1976. Some thermodynamic and kinetic properties of the primary photochemical reactions in a complex from a green photosynthetic bacterium. *Biochim. Biophys. Acta* 423:357–62

129. Reed, D. W., Raveed, D., Reporter, M. 1975. Localisation of photosynthetic reaction centres by antibody binding to chromatophore membranes from *Rps. sphaeroides* strain R26. *Biochim. Biophys. Acta* 387:368–78

130. Rockley, M. G., Windsor, M. W., Cogdell, R. J., Parson, W. W. 1975. Picosecond detection of an intermediate in the photochemical reaction of bacterial photosynthesis. *Proc. Natl. Acad. Sci. USA* 72:2251–55

131. Rosen, D., Okamura, M. Y., Feher, G., Steiner, L. A., Walker, J. E. 1977. Separation and N-terminal sequence analysis of the subunits of the reaction centre protein from *Rps. sphaeroides* R26. *Biophys. J.* 17:67a Abstr. W-PM-F15

132. Rush, J. D., Johnson, C. E., Evans, E. H., Evans, M. C. W. 1980. Identification of membrane bound ferredoxin-like centres in photosystem I of blue-green algae by Mossbauer and EPR spectroscopy *J. Physiol.* (Paris) 41:481–82

133. Rutherford, A. W. 1982. EPR evidence for an acceptor functioning in photosystem II when the pheophytin acceptor is reduced. *Biochem. Biophys. Res. Commun.* In press

134. Rutherford, A. W., Mullet, J. E. 1981. Reaction centre triplet states in photosystem I and photosystem II. *Biochim. Biophys. Acta* 635:225–35

135. Rutherford, A. W., Mullet, J. E., Crofts, A. R. 1981. Measurement of the midpoint potential of the pheophytin acceptor of photosystem II. *FEBS Lett.* 123:235–37

136. Rutherford, A. W., Paterson, D. R., Mullet, J. E. 1981. A light-induced spin-polarised triplet detected by EPR in photosystem II reaction centres. *Biochim. Biophys. Acta* 635:205–14

137. Rutherford, A. W., Paterson, D. R., Mullet, J. E. 1981. A light-induced triplet state in PSII reaction centres detected by EPR. *Abstr. 5th Int. Congr. Photosynth.* p. 487

138. Rutherford, A. W., Thurnauer, M. C. 1982. EPR evidence for an acceptor in purple photosynthetic bacteria which undergoes photoreduction when the intermediate bacteriopheophytin acceptor is reduced. *FEBS Lett.* In press

139. Satoh, K. 1979. Polypeptide composition of the purified photosystem II pigment-protein complex from spinach. *Biochim. Biophys. Acta* 546:84–92

140. Satoh, K., Butler, W. L. 1978. Low temperature spectral properties of subchloroplast factors purified from spinach. *Plant Physiol.* 61:373–79

141. Sauer, K. 1981. Charge separation in the light reactions of photosynthesis. *Abstr. 5th Int. Congr. Photosynth.* p. 502

142. Sauer, K., Mathis, P., Acker, S., Van Best, J. A. 1978. Electron acceptors associated with P700 in triton solubilised photosystem I particles from spinach chloroplasts. *Biochim. Biophys. Acta* 503:120–34

143. Sauer, K., Mathis, P., Acker, S., Van Best, J. A. 1979. Absorption changes of P700 reversible in milliseconds at low temperature in triton-solubilised photosystem I particles. *Biochim. Biophys. Acta* 545:466–72

144. Schantz, R., Barn-Nun, S., Ohad, I. 1977. Preparation of antibodies against specific chloroplast membrane polypeptides associated with the formation of photosystems I and II in *Chlamydomonas reinhardi* y-1. *Plant Physiol.* 59:167–72

145. Schenck, C. C., Parson, W. W., Holten, D., Windsor, M. W. 1981. Transient states in reaction centres containing reduced bacteriopheophytin. *Biochim. Biophys. Acta* 635:383–92

146. Shuvalov, V. A., Dolan, E., Ke, B. 1979. Spectral and kinetic evidence for two early electron acceptors in photosystem I. *Proc. Natl. Acad. Sci. USA* 76:770–73

147. Shuvalov, V. A., Ke, B., Dolan, E. 1979. Kinetic and spectral properties of the intermediary electron acceptor A_1 in photosystem I. Subnanosecond spectroscopy. *FEBS Lett.* 100:5–8

148. Shuvalov, V. A., Klevanik, A. V., Sharkov, A. V., Kryukov, P. G., Ke, B. 1979. Picosecond spectroscopy of photosystem I reaction centres. *FEBS Lett.* 107:313–16

149. Shuvalov, V. A., Klevanik, A. V., Sharkov, A. V., Matveetz, Ju. A., Krukov, P. G. 1978. Picosecond detection of Bchl-800 as an intermediate electron carrier between selectively excited P_{870} and bacteriopheophytin in *R. rubrum* reaction centres. *FEBS Lett.* 91:135–39

150. Shuvalov, V. A., Klimov, V. V. 1976. The primary photoreaction in the complex cytochrome P_{890}—P (bacteriopheophytin)$_{760}$ of *Chr. minutissimum* at low redox

potentials. *Biochim. Biophys. Acta* 440:587–99

151. Shuvalov, V. A., Klimov, V. V., Dolan, E., Parson, W. W., Ke, B. 1980. Nanosecond fluorescence and absorbance changes in photosystem II at low redox potential. *FEBS Lett.* 118:279–82

152. Shuvalov, V. A., Parson, W. W. 1981. Energies and kinetics of radical pairs involving bacteriochlorophyll and bacteriopheophytin in bacterial reaction centres. *Proc. Natl. Acad. Sci. USA* 78:957–61

153. Siggel, U., Khanna, R., Renger, G., Govindjee. 1977. Investigation of the absorption changes of the plastoquinone system in broken chloroplasts. The effect of bicarbonate-depletion. *Biochim. Biophys. Acta* 462:196–207

154. Steiner, A., Okamura, M. Y., Lopes, A. D., Moskowitz, E., Feher, G. 1974. Characterisation of reaction centres from photosynthetic bacteria. II Amino acid composition of the reaction centre protein and its subunits in *Rps. sphaeroides* R26. *Biochemistry* 13:1403–10

155. Stiehl, H. H., Witt, H. T. 1969. Quantitative treatment of the function of plastoquinone in photosynthesis. *Z. Naturforsch. Teil B* 24:1588–99

156. Strouse, C. E. 1974. The crystal and molecular structure of ethyl chlorophyllide *a*. 2H₂O and its relationship to the structure and aggregation of chlorophyll *a*. *Proc. Natl. Acad. Sci. USA* 71:325–28

157. Straley, S. C., Parson, W. W., Mauzerall, D. C., Clayton, R. K. 1973. Pigment content and molar extinction coefficients of photochemical reaction centres from *Rps. sphaeroides. Biochim. Biophys. Acta* 305:597–609

158. Swarthoff, T., Amesz, J. 1979. Photochemically active pigment-protein complexes from the green photosynthetic bacterium *Prosthecochloris aestuarii. Biochim. Biophys. Acta* 548:427–32

159. Thornber, J. P., Alberte, R. S., Hunter, F. A., Shiozawa, J. A., Kan, K-S. 1977. The organisation of chlorophyll in the plant photosynthetic unit. *Brookhaven Symp. Biol.* 28:132–48

160. Thornber, J. P., Barber, J. 1979. Photosynthetic pigments and models for their organisation *in vivo*. In *Photosynthesis in Relation to Model Systems*, ed. J. Barber, pp. 1–44. Amsterdam/New York/Oxford: Elsevier Biomed.

161. Thornber, J. P., Dutton, P. L., Fajer, J., Forman, A., Holten, D., et al. 1977. Isolated photochemical reaction centres from Bchl*b*-containing organisms.

Proc. 4th Int. Congr. Photosynth., ed. D. O. Hall, J. Coombs, T. W. Goodwin, pp. 55–70. London: Biochem. Soc.

162. Thornber, J. P., Markwell, J. P., Reinman, S. 1979. Plant chlorophyll-protein complexes: recent advances. *Photochem. Photobiol.* 29:1205–16

163. Thornber, J. P., Seftor, R. E. B., Cogdell, R. J. 1981. Intermediary electron carriers in the primary photochemical event of *Rps. viridis. FEBS Lett.* 134:235–39

164. Tiede, D. M., Prince, R. C., Dutton, P. L. 1976. EPR and optical spectroscopic properties of the electron carrier intermediate between the reaction centre bacteriochlorophylls and the primary acceptor in *Chr. vinosum. Biochim. Biophys. Acta* 449:447–67

165. Tiede, D. M., Prince, R. C., Reed, G. H., Dutton, P. L. 1976. EPR properties of the electron carrier intermediate between the primary reaction centre bacteriochlorophylls and the primary acceptor in *Chr. vinosum. FEBS Lett.* 65:301–4

166. Trebst, A. 1974. Energy conservation in photosynthetic electron transport of chloroplasts. *Ann. Rev. Plant Physiol.* 25:423–58

167. Vadeboncoeur, C., Marnet-Bratley, M., Gingras, G. 1979. Photoreaction centre of photosynthetic bacteria. 2. Size and quaternary structure of the photoreaction centres from *R. rubrum* strain G9 and *Rps. sphaeroides* strain 2.4.1. *Biochemistry* 18:4308–14

168. Valkirs, G., Rosen, D., Tokuyasu, K. T., Feher, G. 1976. Localisation of reaction centre protein in chromatophores from *Rps. sphaeroides* by ferritin labeling. *Biophys. Soc. Abstr.* 16:223

169. Van Gorkham, H. J., Pulles, M. P. J., Wessels, J. S. C. 1975. Light-induced changes of absorbance and electron spin resonance in small photosystem II particles. *Biochim. Biophys. Acta* 408:331–39

170. Vernon, L. P., Shaw, E. R., Ogawa, T., Raveed, D. 1971. Structure of photosystem I and photosystem II of plant chloroplasts. *Photochem. Photobiol.* 14:343–57

171. Visser, J. W. M. 1975. *Photosynthetic reactions at low temperatures*. PhD thesis. Univ. Leiden, The Netherlands

172. Walter, E. 1978. *Die Chemische Natur der Pigmente aus Photosynthetischen Reaktionszentrensen von Rhodospirillum rubrum G-9⁺*. PhD thesis. Fed. Inst. Technol., Zurich, Switzerland

173. Webster, G. D., Cogdell, R. J., Lindsay, J. G. 1980. Localisation of the reaction centre subunits in the intracytoplasmic membrane of *Rps. sphaeroides* and *Rps. capsulata. Biochem. Soc. Trans.* 8: 184–85
174. Wraight, C. A. 1977. Electron acceptors of photosynthetic bacterial reaction centres. Direct observation of oscillatory behaviour suggesting two closely equivalent ubiquinones. *Biochim. Biophys. Acta* 459:525–31
175. Wraight, C. A. 1978. Iron-quinone interactions in the acceptor region of bacterial photosynthetic reaction centres.

FEBS Lett. 93:283–88
176. Wraight, C. A. 1979. The role of quinones in bacterial photosynthesis. *Photochem. Photobiol.* 30:767–76
177. Wraight, C. A., Clayton, R. K. 1974. The absolute quantum efficiency of bacteriochlorophyll photo-oxidation in reaction centres of *Rps. sphaeroides. Biochim. Biophys. Acta* 333:246–60
178. Wraight, C. A., Stein, R. R. 1980. Redox equilibrium in the acceptor quinone complex of isolated reaction centres and the mode of action of o-phenanthroline. *FEBS Lett.* 113:73–77

Ann. Rev. Plant Physiol. 1983. 34:47–70

ARABINOGALACTAN-PROTEINS:
Structure, Biosynthesis, and Function

Geoffrey B. Fincher and Bruce A. Stone

Department of Biochemistry, La Trobe University, Bundoora, Vic. 3083, Australia

Adrienne E. Clarke

Plant Cell Biology Research Center, School of Botany, University of Melbourne, Parkville, Vic. 3052, Australia

CONTENTS

0066-4294/83/0601-0047$02.00

INTRODUCTION

Arabinogalactan-proteins are found in most higher plants and in many of their secretions. They are a group of macromolecules characterized by a high proportion of carbohydrate in which galactose and arabinose are the predominant monosaccharides; there is also a low proportion of protein, typically containing high levels of hydroxyproline. The nature of the carbohydrate-protein linkage is known in only a few cases; most arabinogalactan-proteins examined have low levels of protein whose association with carbohydrate survives the isolation procedure and is thus presumed to be covalent. In this discussion we refer to this group of proteoglycans[1] as arabinogalactan-proteins (AGPs), even though the proportion of protein to carbohydrate and the nature of the linkage between them is not known in all cases. In many cases, the arabinogalactan moiety has been the focus of analytical work, and there is no information regarding associated protein; these arabinogalactans are referred to as AGs. This subject was reviewed in detail in 1979 (35); in the present review we have given particular attention to work which has appeared since 1978.

A second group of glycoconjugates, the cell wall glycoproteins[2] and the lectins of the Solanaceae, also contain carbohydrate (arabinose and galactose as the major monosaccharides) covalently bound to protein. These glycoproteins differ from the AGPs in the proportions and organization of the monosaccharides as well as the nature of the carbohydrate-protein linkages. We briefly review their major features and occurrence to allow comparison with the AGPs; extensive reviews of these glycoproteins are available (76, 78, 79).

DISTRIBUTION AND STRUCTURE OF ARABINOGALACTAN-PROTEINS AND ARABINOGALACTANS

Distribution

OCCURRENCE IN PLANTS   AGPs and AGs are found in flowering plants from every taxonomic group tested. AG-based gums are found in representatives of 14 orders of angiosperms and are specially abundant in species from the Combretaceae (Myrtales), Rutales, Rosales, Fabales, and Proteales. Three orders of gymnosperms—Cycadales, Coniferae, and Gnetales—also have species producing AG-based gums (103). In some species, the exudates contain mixtures of AG-based gums and gums of the rhamno-

[1]Defined as polysaccharide covalently attached to protein.
[2]Defined as protein possessing covalently attached mono- or oligosaccharide units.

galacturonan (pectin) or xylan types. The occurrence of AGPs and AGs in lower plants has not been explored systematically; to date the moss *Fontinalis anti-pyretica* (52) is the only reported source. A component of the mucilage surrounding individual cells of the colonial green alga *Eudorina californica* has the general characteristics of an AGP (114).

TISSUE DISTRIBUTION   In higher plants, AGPs occur in leaves, stems, roots, floral parts, seeds, and in large quantities in the trunks of some angio- and gymnosperms. Cultured cells derived from embryo, endosperm, root, and leaf tissue continue to produce and secrete AGPs into the medium. Some specific cell types are able to produce copious quantities of AG and AGPs, for example, stylar canal cells and secretory cells producing gum exudates (see 35).

CELLULAR LOCALIZATION OF ARABINOGALACTAN-PROTEINS   Attempts to localize AGPs in their tissues of origin are hampered by their extreme solubility; being essentially polysaccharide in nature they are not fixed by commonly used aldehyde fixatives. Cetyl pyridinium chloride (1%) effectively fixes some AGPs (37), presumably those with a high uronic acid content. The β-glycosyl artificial carbohydrate antigens (see p. 57) specifically precipitate AGPs from solution and can be used as cytochemical reagents. Material which stains with the β-glycosyl antigen is associated with granules in the peripheral cytoplasm of cultured *Lolium multiflorum* endosperm cells, the aleurone layer of cereal seeds at the cytoplasm-wall interface (10), with vesicles in the intercellular spaces of cotyledon parenchyma cells of *Zantedeschia aethiopica* and *Alocasia macrorrhizos,* in secretory ducts of *Hedera helix* leaves (36), and with pistils (stigma and style) of both mono- and dicots (55, 62). Material with similar staining properties is found in the cytoplasm of pollen grains (35) and associated with the surfaces of plant protoplasts derived from callus cells of both mono- and dicots (82). Membrane-associated 1,3;1,6-β-galactans structurally analogous to AGPs are present in *Lolium multiflorum* cultured endosperm cells (72, 85, 98) and *Prunus avium* styles (86), and an AGP is released from a crude membrane fraction of *Phaseolus vulgaris* hypocotyls by sonication (119).

The cavity of the gum-resin duct of *Commiphora mukul* develops schizogenously, and both gum and resin are secreted simultaneously (100). In *Opuntia ficus-indica,* mucilage is present only in mucilage cells, probably as its calcium and magnesium salt (116).

Immunochemical localization using antisera raised to an isolated AGP is theoretically possible; however, polyvalent antisera are directed to both terminal Ara and Gal residues and cross-react with other glycoconjugates containing these monosaccharides in the appropriate configuration (56).

FITC (fluorescein isothiocyanate)-labeled Gal-specific lectins and the 1,6-β-galactosyl specific myeloma protein J539 are useful probes (54, 98; A. Gell, personal communication). No studies on the localization of AGPs at the ultrastructural level have been published.

Structures of AGs and AGPs

COMPOSITION OF AGs AND AGPs   The protein content is usually between 2 and 10%, but is as high as 59% in *Acacia hebeclada* gum (8). The major monosaccharides are D-galactopyranose and L-arabinofuranose; the proportions (Gal:Ara) vary between 10:90 and 85:15 with most samples containing more Gal than Ara. Other monosaccharides which can be present are L-rhamnopyranose (up to 11%), D-mannopyranose (up to 16%), D-xylopyranose (up to 7%), D-glucopyranose (up to 4%), D-glucuronic acid and its 4-0-methyl derivative (up to 28%), and D-galacturonic acid and its 4-0-methyl derivative (up to 26%). In most cases, however, Gal and Ara predominate. Details of the composition of AGPs and AGs reported up to 1978 are given in the earlier review (35), and those reported since 1978 are given in Table 1.

GENERAL FEATURES   The AGs are a family of structurally related polysaccharides with a branched β-galactopyranose framework having predominantly 1,3-linkages with varying amounts of 1,6-linkages. The molecule is often depicted as a linear backbone with short side branches (Figure 1). The galactosyl branches of the framework may be substituted by Ara*f* and Ara*p* residues and other less abundant monosaccharides which are often in terminal positions. These features are shown in Figure 1, which represents a portion of the *Lolium multiflorum* AGP. The more complex AG from *Acacia senegal* gum is depicted in Figure 2.

SPECIAL FEATURES OF THE GALACTAN FRAMEWORK   There are several structural possibilities for the galactan framework including a comb-like organization or a branch-on-branch structure like amylopectin. Often the data available do not allow a distinction to be made, but Churms, Stephen, and their associates interpret their data in favor of a comb-like organization: first a homologous series of products is formed on partial acid hydrolysis of some AGs (33). Second, *Acacia* spp (25, 27, 30, 32) and *Prosopis spp* (31) exudate gums and the AGs from larch (26) and *Brassica campestris* (34) produce a series of low molecular weight monodisperse products after oxidation by periodate followed by borohydride reduction and controlled acid hydrolysis (Smith degradation). This suggests that the framework of the polysaccharides is composed of blocks of 1,3-linked

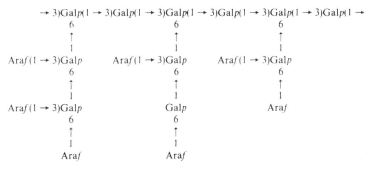

Figure 1   Tentative structure of ryegrass (*Lolium multiflorum*) endosperm arabinogalactan-protein (10).

Gal*p* residues, usually comprising about 12, 24, or 36 monosaccharides, separated at regular intervals by sugar residues that are oxidized by periodate. The identity of the periodate-vulnerable sugars is not established, but Ara*f* (125) or unsubstituted 1,6-linked Gal*p* residues (58), which have been reported in the main chain of some AGs, are candidates.

NATURE OF THE SUBSTITUENTS TO THE GALACTAN FRAMEWORK
Uronic acids, either D-Gal*p*A, D-Glc*p*A, or their 4-0-Me derivatives, are common nonreducing terminal substituents of the galactan; Man*p*A has also been reported (119). Some larch AGs are exceptional in having no uronic acid components. The AG-based exudate gums often have high proportions (up to 28%) of uronic acids; in tissue AGs and AGPs, the uronic acid component may account for up to 43% of the monosaccharides, although in most cases the proportion is lower. Of the neutral sugar substituents, L-Ara*f* is the most abundant; some residues are attached singly to Gal residues and others may be present in short chains. The terminal Ara residues occur in the pyranose or α-furanose form (cf both α- and β- linked Ara-*f* residues in cell wall glycoprotein; see Table 2). Other substituents are D-Gal *p* in both α- and β-forms, L-Rha*p*, D-Man*p*, and less often, D-Glc*p*, L-Fuc, and D-Xyl [see Tables 1 and 2, and Table 3 in (35)]. Structural heterogenity of fractions from single preparations of AGs and AGPs and between preparations from different plant parts is due to differences in the nature, proportions, and linkage of peripheral substituents [for references see (35) and Table 1].

PROTEIN CORE   Many AGs are associated with protein rich in Hyp, Ser, Ala, and Gly. The glycosylated protein component is resistant to proteolysis (66), presumably because of its substitution with bulky carbohydrate

Table 1 Chemical composition and physical properties of AG and AGPs from plant tissues, tissue cultures, and exudate gums[a]

Source	Monosaccharides (molar ratios or percentage of composition)								Protein (%)	Specific rotation $[\alpha]_D$	10^{-5} MW	References
	D-Galp	L-Araf	D-GlcpA	D-GalpA	D-Manp	L-Rhap	D-Xylp	D-Glcp				
Plant Tissues												
Nicotiana tabacum (tobacco) leaves	1.29	1.0	(10.5%)[c]			0.15			9.5	−37°	2.2	2
Phaseolus vulgaris (bean) hypocotyl organelles	61	22	14 includes ManpA		tr	tr		tr	10		1.4	118, 119
Gossypium arboreum (cotton) seed hairs	1.2	1	0.2 (8%)			0.1	tr	tr	2.5	−18°	20	
Populus alba (poplar) cambium	10	4	1			1					2	101
Phaseolus mungo (black gram) seeds	20	30		3		5					1.4	107
Carica papaya (papaya) latex PP-I	31.9	unidentified pentose 25.0		11.5		unidentified deoxyhexose 9.3		2.8				22
PP-II	27.9	23.2		14.7		11.7		1.0				

Opuntia ficus-indica (cactus) stems	0.7		1.0				0.2	9.9	+59°	92
Anacardium occidentale (cashew nut) shell	6.9	1.2 / Arap 1.0		1.0					+21°	16
Tissue-cultured Cells Nicotiana tabacum	68 / 36.2	30 / 40.0	10	+		+ / 0.8		5.5	−28°	2.24	64 / 3
Trunk Exudate Gums[b] Spondias dulcis	19.8	48.5			20				−4°		13, 14
Aegle marmelos (bael)	9	1	3			3			−23°		83
Prosopis chilensis	39	47	12			2			+67°	4.5	31
Prosopis glandulosa	34	55	9			2			+65°	4.0	31
Chlorisia speciosa	8	1	3		1	2	tr		+18°	1.05 (80%) / 0.40 (20%)	45

This gum has a glucuronomannan backbone substituted by arabino-3,6-oligogalactosyl substituents (see 35)

[a]See (35) for information prior to 1978.

[b]Further analytical data for gums from several *Acacia* spp. [8, 8a, 9, 28, 29] and *Grevillea* spp. [9a] are available.

[c]Entries centered between columns headed D-GlcpA and D-GalpA indicate that the analysis is for both acids rather than one.

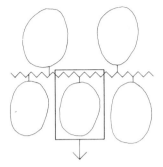

"Wattle blossom" — type structure of AGP
∧∧∧ = polypeptide backbone
○ = AG substituent
In an AGP there may be 25 Hyp residues
each of which may bear an AG substituent;
the molecule as a whole is spheroidal.

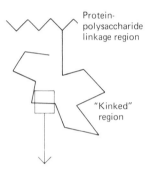

Protein-polysaccharide linkage region

AG substituent showing blocks of 1,3-linked
galactan backbone interrupted by periodate-
susceptible residues (kinked region).
(? Ara or 6-substituted Gal).
There may be 12 residues in each stretch of
the backbone of the galactan framework
between periodate-suscepible linkages and
perhaps 10 or more galactan stretches per
substituent.

"Kinked" region

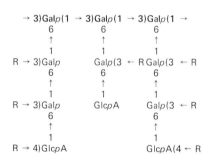

→ 3)Galp(1 → 3)Galp(1 → 3)Galp(1 →   Galactan backbone.
     6   6   6
     ↑   ↑   ↑
     1   1   1
R → 3)Galp   Galp(3 ← R Galp(3 ← R
     6   6   6   Portion of AG substituent showing 1,3-linked
     ↑   ↑   ↑   Gal backbone and galactosyl side branches
     1   1   1   with saccharide substituents. (Data from
R → 3)Galp   GlcpA   Galp(3 ← R   *Acacia senegal* gum analysis). The galactan
     6   6   backbone may assume a helical conformation.
     ↑   ↑
     1   1
R → 4)GlcpA   GlcpA(4 ← R

R = Rhap(1, Araf(1, Galp(1 → 3)Araf(1, Araf(1 → 3)Araf(1.

Figure 2   Hypothetical structure of an arabinogalactan-protein.

groups. Rigorous examination of the point of attachment of the AG to the
protein portion of the molecule has been made only in one case, the AG-
peptide from wheat endosperm (87, 104), in which Hyp is linked in an
alkali-stable glycosidic linkage to the β-D-Galp residue of the reducing
terminus of the AG. In rice bran proteoglycan (122), α-Araf is linked to

Table 2 Comparative properties of galactose- and arabinose-containing proteoglycans and glycoproteins

	Proteoglycans (AGPS)	Glycoproteins	
Sources	Plant tissues, tissue-cultured cells and media, exudate gums	Cell walls (higher plant)	Lectins from Solanaceae
Carbohydrate	β-D-Galp backbone entirely or chiefly 1,3-linked bearing 1,6-linked Galp oligosaccharaide substituents. Terminal α-L-Araf units mostly monomeric; other sugars D-GlcpA, D-GalpA, L-Rhap, D-Manp, D-Glcp, L-Arap	α-D-Galp (monomeric) β-L-Araf mono-, di-, tri-, and tetramers Terminal α-L-Araf on tetramers Other sugars?	α-D-Galp (mono- and dimeric) β-L-Araf tri- and tetramers Terminal α-L-Araf on tetramers
Protein	Ranges 2–59% (most 2–10%); rich in L-Hyp, L-Ala, and L-Ser, low ½ Cys; one or more cross-linked polypeptide chains (?)	40–50% Polypeptide rich in Hyp and Ser, domains of poly-Hyp	~50% Polypeptide rich in Hyp. Ser, ½ Cys
Carbohydrate-protein linkage(s)	β-D-Galp-Hyp [87, 104] (alkali-stable) Galp-Ser/Thr [61, 66, 118] (alkali-labile) α-L-Araf-Hyp [122] (alkali-stable)	α-D-Galp-Ser [23, 80, 90] (alkali-labile) β-L-Araf-Hyp [77] (alkali-stable)	α-D-Galp-Ser [12] (alkali-labile) β-L-Araf-Hyp [12] (alkali-stable)
Solubility			
Water	Water-soluble, removal of terminal Araf reduces solubility Forms insoluble complexes with β-glycosyl Yariv antigens	Insoluble in water, soluble in sodium chlorite-acetic acid (may be chemically modified during extraction).	Water-soluble
TCA	Soluble?	Soluble?	Soluble?
Moleular weight	Range 10^4–10^6 (mostly 10^5)	?	3×10^5 [12]

Hyp; in *Cannabis sativa* leaves (60, 61, 66), both a Galp-Ser linkage and an alkali-resistant carbohydrate-protein linkage are present, while in *Phaseolus vulgaris* AGP, carbohydrate is attached to Ser, Thr, and Hyp (118). Whether these linkages are involved in joining the AG (polysaccharide) to a polypeptide or whether they reflect the attachment of mono- or oligosaccharide substituents is not established.

It is not clear if one or more polypeptides are present in the AGP molecule. Three N-terminal amino acids, Ser, Gly, and Ala, in molar proportions 1:2:1 are found in *Acer pseudoplatanus* AGP (66), consistent with the presence of four chains cross-linked in some way, each bearing several AG molecules at Hyp sites. Attempts to sequence the peptides by the Edman procedure are thwarted by glycosylation of Hyp at residue 3 of the chain and by cross-linking structures further into the peptide chain. The

N-terminal amino acids in AGPs from *Ipomea batatas, Daucus carota, Pisum sativum, Acacia elata, Pinus taeda,* and *Pseudotuga menziesii* are the same as in *A. pseudoplatanus,* supporting the idea that the N-terminal amino acids of the AGP polypeptides are evolutionarily conserved (66).

Many analyses of AGPs show low levels of hexosamines and sometimes an unusual amino acid with chromatographic characteristics of ornithine (35, 119).

ISOLATION OF AGPs AND AGs   AGPs and AGs apart from those associated with membranes are water soluble. This sets them apart from the cell wall (Ara- and Gal-containing) glycoproteins, but not the lectin (Ara, Gal-containing) glycoproteins of the Solanaceae (Table 2). They are therefore readily extracted from plant tissues and also occur as soluble components in tissue culture filtrates. A convenient initial fractionation of extracts is treatment to saturation with $(NH_4)_2SO_4$, which does not usually precipitate AGPs (50, 86). Ion-exchange and affinity chromatography can be used to isolate the AGPs, and precipitation by complexing with the β-glucosyl (Yariv) antigen is a specific (and diagnostic) method of recovering AGPs from tissue extracts (35).

PHYSICOCHEMICAL PROPERTIES OF AGs AND AGPs   The physical properties of AGPs will depend on the structures of both the polypeptide and the substituent AG polysaccharides. The lack of precise information regarding the carbohydrate-protein linkage and organization of these molecules makes it difficult to predict their physical behavior. The polypeptide cores vary in molecular size and may consist of a single chain or several cross-linked chains (66). These differences would lead to different degrees of flexibility of the core. The polypeptides are characteristically rich in Hyp, but there is no evidence for the polyproline II type helices found in Ara- and Gal-rich glycoproteins and lectins (6, 44, 120). In the polysaccharide, sequences of 1,3-β-Gal residues are likely to exist as open helices as do the 1,3-β-glucans and 1,3-β-xylans (15). However, this framework may in some cases be interrupted by other linkages (e.g. 1,6-) or monomer types. Intra- and intermolecular associations between 1,3-β-galactan portions of AGPs could be stabilized by triple helix formation as has been found for 1,3-β-glucans and 1,3-β-xylans. By analogy with C(O)6 substituted 1,3-β-glucans which can be induced to form triple helices (15), C(O)6-linked side branches would project from the outer regions of 1,3-β-galactan helices. The extent of these interactions and associations will also be influenced by the relative amount and distribution of neutral and uronic acid substituents. While the helical conformation of 1,3-β-galactan chains would lead to a relatively extended chain over short distances, the presence of bulky side chains, the flexibility of the helix itself, and the flexibility of the protein core would lead to an overall spheroidal shape. The low viscosity of larch AG (1), wheat AG-peptide [(η) 0.108dl/g; (49)] and *Acacia* gums [(η_{sp}) 4.80 ml/g; (108)] in solution is consistent with a spheroidal shape for the molecule. The AGP can be envisaged to have a "wattle blossom" appearance with clusters of ovoid or spheroidal AGs anchored to a polypeptide core (Figure 1).

COMPLEXING OF Agps WITH ARTIFICIAL CARBOHYDRATE ANTIGENS, LECTINS, AND ANTISERA

AGP-Artificial Carbohydrate Antigen Interactions

Artificial carbohydrate antigens (Figure 3) prepared by coupling diazotized 4-amino phenyl glycosides to phloroglucinol were first prepared by Yariv in 1962 (124) as reagents to detect antibodies to glycosides. In 1967, Yariv and co-workers made the chance observation that the β-glucosyl artificial carbohydrate antigen precipitated AGPs from bean and maize extracts (123). Jermyn & Yeow (67) extended these observations to show the same phenomenon of AGP precipitation from a wide range of plant extracts. The precise nature of the interaction between the β-glucosyl artificial carbohydrate antigen and AGPs is still not established. Artificial antigens must have a β-D-glycopyranose residue and the D-gluco-configuration at C(O)2; the azo and glycosyloxyl groups must be oriented 1 : 4 to the phenyl ring (Figure 3) to be effective in precipitating AGPs. The nature of the domain of the

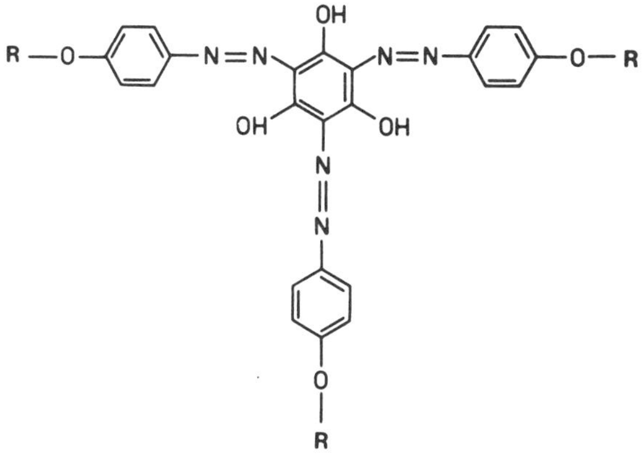

Figure 3   Structure of the β-glucosyl artificial carbohydrate antigen (R = glucosyl residue).

AGP which interacts with the artificial antigen is not clear. It probably depends on the overall physical and chemical properties of the AGP rather than a specific binding site because removal of Ara$f$ groups does not abolish complexing (3, 54) but progressive acid hydrolysis or dithionite reduction causes loss of complexing ability (57). Neither an AG-Hyp complex derived from tobacco AGP by alkaline treatment (3) nor the AG peptide from wheat endosperm (MW 2 X 10^4) form complexes (49), but the larch AG (0.1% protein) does.

AGP-Lectin Interactions

Lectins which bind terminal β-galactosyl residues, for example tridacnin, peanut agglutinin, and the *Ricinus communis* (RCA$_{120}$) lectins would be expected to bind AGPs. Whether in practice a particular AGP is bound by a particular lectin probably depends on the number and accessibility of terminal β-galactosyl residues as well as the sequence of subterminal saccharides (55). Tridacnin and myeloma protein J539 affinity chromatography are effective means of isolating AGPs (35).

AGP-Antisera Interactions

The AGP of *Gladiolus gandavensis* style is an effective antigen in rabbits. The resultant antiserum is directed to the carbohydrate portion of the AGP, with terminal β-D-Gal and α-L-Ara residues being the dominant antigenic determinants (56). The myeloma protein J539, which is an IgA specific for 1,6-β-linked galactosyl oligosaccharides (53), also binds many AGPs (12a, 55).

CELL WALL GLYCOPROTEINS AND LECTINS OF THE SOLANACEAE

The cell wall glycoproteins and chitobiose-specific lectins of the Solanaceae are characterized by Hyp-rich proteins covalently associated with carbohydrate, rich in Ara and Gal. The fundamental difference between these and the AGPs lies in the carbohydrate moiety: in the cell wall glycoproteins it consists of monosaccharides and relatively short, oligosaccharide chains; in the AGPs it is polymeric and accounts for a much higher proportion of the molecule. The properties of these cell wall glycoproteins and lectins will now be reviewed briefly.

Isolation

Structures of the cell wall glycoproteins as they occur in vivo are not established, as samples are always isolated under degradative conditions.

Alkali (81) and hydrazine (59) can be used to extract glycopeptide fragments of the glycoprotein; sodium chlorite-acetic acid treatment solubilizes cell wall glycoprotein but is also likely to cause partial oxidation or hydrolysis (90, 99). Cross-linking of polypeptides through ether linkages occurs between two tyrosine residues to form isodityrosine bridges; these bridges are vulnerable to acid-chlorite and are believed to be responsible for the insolubility of the Hyp-rich cell wall glycoprotein (51a). The salt-soluble cell wall glycoprotein "precursors" reported from carrots are not sufficiently characterized to distinguish them from AGPs (19, 106). In contrast to higher plant cell wall glycoproteins, lectins from the Solanaceae are water soluble (6, 44, 73), and wall glycoproteins from the unicellular green alga *Chlamydomonas reinhardtii* can be extracted with chaotropic agents (21).

Composition

Primary plant cell walls generally contain 5–10% protein (40) as glycoprotein. Of the total glycoprotein, about 40–60% is carbohydrate with Gal and Ara as the major monosaccharides, and Xyl, Rha, Glc, and uronic acids as minor monosaccharides. In dicots, Hyp accounts for 20–25% of amino acids in the protein (40), although higher levels do occur (99, 106). The Hyp content of monocot cell walls is often much lower (40) or absent (11, 48, 84). In addition to Hyp, cell wall glycoproteins are characterized by high levels of Ser. The Ala content is usually lower than that in AGPs. After the Hyp-rich cell wall glycoprotein is extracted from parenchyma cell walls of runner beans (*Phaseolus coccineus*) with sodium chlorite/acetic acid, a second, distinct, Hyp-poor glycoprotein can be extracted from the wall residue with alkali (99). A similar Hyp-poor glycoprotein is present in *P. vulgaris* cell walls (17, 18).

The soluble lectins from potato (*Solanum tuberosum*) (6), tomato (*Lycopersicon esculentum*) (89), and thorn apple (*Datura stramonium*) (44) are glycoproteins containing 40–50% carbohydrate, with Ara:Gal ratios of between 6:1 and 12:1 and protein rich in Hyp and Ser. A striking difference between the amino acid composition of the cell wall glycoproteins and the lectins is the high concentration of half-cysteine in the lectins. However, half-cysteine would be lost in many of the procedures used to extract cell wall glycoproteins. Ara- and Gal-containing glycoproteins account for the bulk of the cell wall of *Chlamydomonas reinhardii;* the composition resembles that of higher plant cell wall glycoprotein but the mannose content is higher (21).

Structure

In the glycoproteins of higher plant cell walls, carbohydrate is linked to protein through Ara-O-Hyp (77) and Gal-O-Ser (23, 80). The carbohydrate consists of oligomers of Ara$f$ residues of DP 1–4; the average DP varies with the source (81). Hyp is usually glycosylated, although unsubstituted Hyp residues occur in cell wall glycoproteins from monocots (81); the degree of glycosylation changes during development (74).

The tri- and tetra-arabinosides from cell wall glycoprotein of *Nicotiana tabacum* suspension cultures defined by ^{13}C-NMR (4) are:

$$\beta\text{-L-Ara}f(1\rightarrow2)\text{-}\beta\text{-L-Ara}f(1\rightarrow2)\text{-}\beta\text{-L-Ara}f(1\rightarrow4) \overset{|}{\underset{|}{\text{—Hyp}}}$$

and

$$\alpha\text{-L-Ara}f(1\rightarrow3)\text{-}\beta\text{-L-Ara}f(1\rightarrow2)\text{-}\beta\text{-L-Ara}f(1\rightarrow2)\text{-}\beta\text{-L-Ara}f(1\rightarrow4) \overset{|}{\underset{|}{\text{—Hyp}}}$$

and are identical with the Hyp tri- and tetra-arabinosides from potato lectin (12). Single α-D-Gal$p$ residues in higher plant cell wall glycoproteins are linked to Ser residues through O-glycosidic linkages (23, 80, 90). The disaccharide 3-0-β-D-galactopyranosyl-D-galactose is linked to Ser residues in the thorn apple but not the potato lectin (12). Hyp and its associated carbohydrate are asymmetrically distributed along the polypeptide of the cell wall glycoprotein (59). Sequencing studies indicate regions of poly-Hyp flanked by Ser residues (76), and circular dichroism data demonstrate regions of polyproline II-type conformation (120). Strong hydrogen bonding of the oligo-arabinosyl side chains with the polypeptide backbone reinforces the left-handed helices of the polyproline II conformation (120), which occur as rigid rod-like domains in the glycoprotein. Similar protein conformations are suggested for the cell wall glycoprotein from *Chlamydomonas reinhardii* (63) and for the potato and thorn apple lectins (6, 44). A cystine-rich region of the thorn apple polypeptide, which probably contains the carbohydrate binding sites, may be located adjacent to the glycosylated regions (44). Thus, in each case, the molecule is divided into distinct domains. Whether the glycosylated sequences of the glycoprotein are distributed evenly along the peptide chain or are restricted to a particular region is not established. The possible functions of cell wall glycoproteins have been discussed (43, 77, 79).

BIOSYNTHESIS OF AGPs

Little information is available regarding the biosynthesis of AGPs. On the other hand, considerable effort has been directed to understanding the synthesis of cell wall glycoprotein, but in many cases the results cannot be interpreted solely in terms of cell wall glycoprotein biosynthesis. For example, incorporation of ^{14}C- or ^{3}H-Ara or the appearance of labeled Hyp after providing tissues or extracts with ^{14}C- or ^{3}H-Pro may be due to synthesis of AGP, cell wall glycoprotein, or both (96), and rigorous product analysis is seldom reported.

Peptidyl-Proline Hydroxylation

As in animal systems, Hyp in plant proteins is synthesized by post-translational modification of peptidyl Pro (93, 94, 97), and the process requires molecular oxygen (75), Fe^{2+}, α-oxoglutarate, and an appropriate peptidyl Pro substrate (97, 113). Examination of the hydroxylase in vitro is difficult because defined substrates are not readily available. A partially purified hydroxylase from carrot parenchyma catalyzes hydroxylation of prolyl residues both in protocollagen and in a carrot protein fraction synthesized in vivo in the presence of α,α'-dipyridyl to inhibit hydroxylation (97). Most hydroxylase was in the 40,000 g supernatant. In extracts from suspension-cultured cells of *Vinca rosea* almost 90% of detectable hydroxylase is associated with membrane fractions (113). Poly-L-Pro is the preferred substrate, and the enzyme requires an intact poly-L-Pro II helix (112); the protocollagen analog (Pro-Pro-Gly)$_n$ is a relatively poor substrate. *Vinca rosea* cells in suspension culture also secrete an AGP (109, 110), and it is not clear whether the hydroxylase is involved in AGP or cell wall glycoprotein biosynthesis or both. It is possible that the same enzyme hydroxylates peptidyl proline residues in precursor proteins of both classes of macromolecule.

Suspension-cultured endosperm cells of *Lolium multiflorum* are a convenient system for studying AGP biosynthesis specifically, because they lack significant levels of cell wall glycoprotein (11) but secrete AGP into the medium (10). In these cells, a membrane-bound hydroxylase hydroxylates poly-L-Pro in vitro (39).

In vitro Synthesis of the Protein Precursor of the Cell Wall Glycoprotein

RNA for the precursor polypeptide of the cell wall glycoprotein can be translated in vitro (102, 105). Hyp-rich sequences are translated from cytosine-rich mRNA which can be isolated from carrots by affinity chromatography on oligo (dG)-cellulose (105). This technique provides an approach for isolating the gene for the cell wall glycoprotein.

Biosynthesis of AG Polysaccharide

An unfractionated membrane preparation from suspension-cultured *Lolium multiflorum* endosperm cells, incubated with UDP-^{14}C-Gal in the presence of ADP-ribose, cayalyzes incorporation of ^{14}C into 66% ethanol-insoluble products (apparent MW $>$ 60,000). ADP-Ribose is added to inhibit a membrane bound UDP-Gal 4-epimerase and thus prevent incorporation of ^{14}C into β-glucans (85). The radioactivity is associated with unidentified glycoconjugates which have 1,6-β-Gal$p$ residues, because it is retained by a myeloma protein J539-Sepharose affinity column (85).

Biosynthesis of the AG portion of AGPs might proceed by polymerization of large oligosaccharide subunits. This would be consistent with the organization of the AG backbone into blocks of 1,3-linked-β-Gal residues separated at regular intervals by periodate-susceptible residues (Figure 1). Whether these oligosaccharides are assembled on lipids of the dolichol phosphate or pyrophosphate type (47) is not known. Radioactivity from UDP-^{14}C-Gal is incorporated into a $CHCl_3$:MeOH-soluble fraction by membrane preparations from *Lolium multiflorum;* a small proportion of this material behaves as dolichol pyrophosphate-sugars during DEAE-cellulose chromatography (85). Transfer of this radioactivity into AGP has not been demonstrated.

No attempts have been made to study the transferases for Ara and other terminal residues involved in AGP biosynthesis, although a particulate fraction from *Acer pseudoplatanus* suspension cultures incorporates Ara from UDP-L-Ara into products which might include AGPs (68).

Subcellular Location of AGP Biosynthesis

Some clues to the subcellular aspects of AGP biosynthesis might be gleaned from parallel studies of cell wall glycoprotein biosynthesis. The Golgi apparatus is implicated as a site of cell wall glycoprotein synthesis on the grounds that Pro → Hyp conversion and the enzyme which catalyses the transfer of Ara from UDP-Ara to cell wall glycoprotein co-sediment with latent IDPase in a fraction containing intact dictyosomes (51b). Golgi-associated arabinosyl transferases are known in other systems (71, 91), and the prolyl hydroxylase from ryegrass endosperm cells is also probably Golgi-associated (39). However, Dashek (41) concludes that most Hyp-rich protein is transferred via smooth membranous components in suspension cultures of *Acer pseudoplatanus.* When cells are incubated with ^{14}C-Pro, the label appears in both the 5% TCA-soluble and TCA-insoluble cytoplasmic protein (24, 42). In elongating pollen tubes of *Lilium longiflorum,* the cytoplasmic TCA-insoluble ^{14}C-Hyp is incorporated into the cell wall (42). The TCA-soluble material may be a cell wall glycoprotein precursor, but may also be nascent AGP en route to the extracellular medium. TCA or salt solubility are insufficient criteria for distinguishing the two classes of glycoconjugates (Table 2). Weinecke et al (121a) conclude that in carrot (*Daucus carota*) discs, synthesis of both AGP and cell wall glycoprotein is initiated in the endoplasmic reticulum, and that both are transported via the Golgi apparatus to the plasma membrane.

Secretion of an AG from suspension-cultured cells in *Acer pseudoplatanus* is limited by the rate of fusion of dictyosome-derived vesicles with the plasma membrane and is potentiated by cations (88). Secretion of Hyprich proteins in carrots is not obligatorily coupled to protein synthesis but requires metabolic energy (46). During AGP and cell wall biosynthesis, the ferrous ion chelator α,α'-dipyridyl has no apparent effect on the total amount of peptide-bound radioactivity secreted from cells provided with ^{14}C- or ^3H-Pro, even though the inhibitor completely blocks formation of Hyp from Pro (94, 96, 102).

Turnover of AGPs

In pulse-chase experiments with *Vinca rosea* suspension cultures, ^{14}C-Ara and -Gal, believed to originate from AGP, were lost from cell walls (111). This may be turnover of cell wall bound AGP, but may also reflect transitory extracellular AGP in the cell wall region during secretion (51). A range of hydrolytic enzymes capable of degrading AGPs is present in plant tissues suggesting that in vivo, hydrolysis of AGPs is possible. In addition, an effective mechanism for scavenging Hyp operates in a number of systems (38, 93, 95, 121). Free Hyp is rapidly taken up by endosperm cells of *Lolium multiflorum* and converted to Pro which is incorporated into protein, hydroxylated and secreted as AGP (95). In this process Pro formed from Hyp appears to have preferential access to the incorporating system over free Pro. Conversion of Hyp to Pro in carrots occurs through the intermediate 4,5-dehydroproline; a high proportion of free Hyp is eventually converted to peptidyl-Pro and peptidyl-Hyp via this route (121).

FUNCTION

Possible functions of AGPs can be inferred from their physical properties and their cellular localization. All statements regarding function must be regarded as speculative.

Potential for Specific Interactions

AGPs have potential for two major types of interaction: macromolecule-macromolecule or macromolecule-small ligand interactions. If portions of the galactan framework adopt either a helical or triple helical conformation with the substituents oriented to the outside of the helix, several surfaces would be available for cooperative interactions: the regular areas of helix face, the helical grooves of the triple helix, and the substituent chains. Macromolecules such as antisera and lectins (Gal-specific) bind to AGPs by specific interactions with particular components of the AG chains; although lectins specific for the other saccharide components (Ara, Rha) are not known, they may exist and could be involved in specific recognition reactions. Other macromolecules such as polysaccharides and proteins may interact nonspecifically with the AGPs causing changes in the viscosity and gel-forming potential of AGP solutions. Small ligands such as flavonol glycosides (65) and the artificial β-glucosyl antigens (67), which may aggregate to a relatively large complex, also bind AGPs. It seems likely that the AGPs could interact with other polysaccharides such as pectins in the middle lamella, and it is possible that they may also make contact with membrane-bound and wall-associated lectins (70). As AGPs are present at the plasma membrane, they may be receptors of external signals such as the low molecular weight β-glucan elicitors of the phytoalexin response (5).

Informational Potential

It is possible that the outer chains are the functional regions of the molecules with the protein and β-galactan backbone serving only to present information inherent in the variety of terminal substituents. There is some evidence that terminal disaccharides of plant gums have taxonomic significance (7); an extension of this is that these groups may be implicated in the expression of identity of individual plant tissues or cell types, just as the identity of animal cells is expressed by cell surface determinants such as the blood group and transplantation antigens.

Water-Holding Capacity

Suggestions that the water-holding capacity of AGPs may be important in wound healing, frost hardiness, and drought resistance (35, 79) are not strongly supported by the available data. Also, the water-holding capacity of mucilage (possibly containing AGP) from the cactus *Opuntia ficus-indica* (0.75g/g) is considered insufficient to explain drought resistance (117). Although AGPs, particularly the AG gums, are secreted by wounded tissues of some plants, just how they might contribute to wound healing is obscure.

Analogy with Animal Glycosaminoglycans

The AGPs have many properties analogous to those of the animal proteoglycans, the glycosaminoglycans (GAGs). Like AGPs, GAGs are found in highest concentration in extracellular spaces; they are stored in subcellular organelles of specialized cells and are secreted by many cultured mammalian cells into the medium. Some GAGs (e.g. heparin sulfate) are associated with animal cell surfaces either embedded in the plasma membrane via hydrophobic regions or held via specific binding sites (115). AGPs are also located at the plasma membrane, and there are glycolipids containing Hyp, Gal, and Ara in tobacco leaves (69). The GAGs may be involved in cell-cell adhesion (115); perhaps the AGPs play a similar role in cell-cell adhesion. The high AGP content of suspension cultures is consistent with the medium being an "extended middle lamella" (79) or extracellular matrix.

Another suggested function for GAGs is in controlling morphogenesis: the idea is that each tissue has a characteristic complement of sulfated GAGs whose composition is altered during morphogenesis. Perhaps AGPs play an analogous role. For both groups of proteoglycans, sound structural information is available; but in neither case is the function established.

CONCLUSIONS

There is a wealth of information on the structure of AGPs. The early literature was dominated by the gum chemists reporting analyses of the gums of commerce; gum chemistry still makes a significant input to the literature which is now augmented with analyses of AGPs from many plant tissues and exudates. Often these AGPs have "turned up" during the course of research designed around seemingly unrelated topics. Cell wall glycoproteins are structurally quite distinct from the AGPs, and this must be recognized in biosynthetic studies where labeled precursors may ultimately be incorporated into either the AGP, the glycoprotein, or both. The function of AGPs remains obscure. As they are so widely distributed and as their structure is so conserved, it seems unlikely that their function is trivial. No one has reported any attempts to establish their function by direct experimentation—indeed, just how it could be done is not obvious; one way might be to find an AGP-deficient mutant or selectively block in vivo formation of AGP. Until such definitive experiments are performed, the function of AGP will remain speculative. The information available on the localization of AGPs is also rather limited; however, they are clearly often present in the extracellular milieu, and their function, like that of the glycosaminoglycans of animal tissues, may well be in cell-cell adhesion, communication, or morphogenesis.

ACKNOWLEDGMENTS

We thank our colleagues in Melbourne, particularly Dr. Michael Jermyn, and graduate students for many fruitful discussions; we are also grateful to many colleagues who made preprints available to us. We also thank Ms. Anne Pottage, who prepared Figure 2, and Miss Genevieve Loy for expert typing and editorial assistance.

We thank the Australian Research Grants Scheme whose support over many years has made possible our research cited in this manuscript.

Literature Cited

1. Adams, M. F., Ettling, B. V. 1973. Larch arabinogalactan. In *Industrial Gums,* ed. R. L. Whistler, pp. 415–72. New York: Academic. 2nd ed.
2. Akiyama, Y., Eda, S., Kato, K. 1982. An arabinogalactan-protein from the leaves of *Nicotiana tabacum. Agric. Biol. Chem.* 46:1395–97
3. Akiyama, K., Kato, K. 1981. An extracellular arabinogalactan-protein from *Nicotiana tabacum. Phytochemistry* 20:2507–10
4. Akiyama, Y., Mori, M., Kato, K. 1980. ^{13}C-NMR analysis of hydroxyproline arabinosides from *Nicotiana tabacum. Agric. Biol. Chem.* 44:2487–89
5. Albersheim, P., Valent, B. S. 1978. Host-pathogen interactions in plants. *J. Cell Biol.* 78:627–43
6. Allen, A. K., Desai, N. N., Neuberger, A., Creeth, J. M. 1978. Properties of potato lectin and the nature of its glycoprotein linkages. *Biochem. J.* 171: 665–74
7. Anderson, D. M. W., Dea, I. C. M. 1969. Chemotaxonomic aspects of the chemistry of *Acacia* gum exudates. *Phytochemistry* 8:167–76
8. Anderson, D. M. W., Farquhar, J. G. K. 1979. The composition of eight *Acacia* gum exudates from the series Gummiferae and Vulgares. *Phytochemistry* 18:609–10
8a. Anderson, D. M. W., Farquhar, J. G. K., Gill, M. C. L. 1980. Chemotaxonomic aspects of some *Acacia* gum exudates from series Juliflorae. *Bot. J. Linn. Soc.* 80:79–84
9. Anderson, D. M. W., Pinto, G. 1980. Variations in the composition and properties of the gum exuded by *Acacia karroo* Hayne in different African locations. *Bot. J. Linn. Soc.* 80:85–89
9a. Anderson, D. M. W., de Pinto, G. L. 1982. Gum exudates from the genus *Grevillea* (Proteaceae). *Carbohydr. Polym.* 2:19–24
10. Anderson, R. L., Clarke, A. E., Jermyn, M. A., Knox, R. B., Stone, B. A. 1977. A carbohydrate-binding arabinogalactan-protein from liquid suspension cultures of *Lolium multiflorum* endosperm. *Aust. J. Plant Physiol.* 4:143–58
11. Anderson, R. L., Stone, B. A. 1978. Studies on *Lolium multiflorum* endosperm in tissue culture III. Structural

studies on the cell walls. *Aust. J. Biol. Sci.* 31:573–86
12. Ashford, D., Desai, N. N., Allen, A. K., Neuberger, A., O'Neill, M. A., Selvendran, R. R. 1982. Structural studies of the carbohydrate moieties of lectins from potato (*Solanum tuberosum*) tubers and thorn apple (*Datura stramonium*) seeds. *Biochem. J.* 201:199–208
12a. Baldo, B. A., Neukom, H., Stone, B. A., Uhlenbruck, G. 1978. Reaction of some invertebrate and plant agglutinins and a mouse myeloma anti-galactan protein with an arabinogalactan from wheat. *Aust. J. Biol. Sci.* 31:149–60
13. Basu, S. 1980. Some structural studies on degraded *Spondias dulcis* gum. *Carbohydr. Res.* 81:200–1
14. Basu, S., Rao, C. V. N. 1981. Structural investigations on degraded *Spondias dulcis* gum. *Carbohydr. Res.* 94:215–24
15. Bluhm, T. L., Sarko, A. 1977. Conformational studies of polysaccharide multiple helices. *Carbohydr. Res.* 54: 125–38
16. Bose, S., Soni, P. L. 1974. Structure of the shell polysaccharide from cashew nuts (*Anacardium occidentale*). *Indian J. Chem.* 12:680–83
17. Brown, R. G., Kimmins, W. C. 1977. Glycoproteins. *International Review of Biochemistry, Plant Biochemistry II,* ed. D. H. Northcote, 13:183–209
18. Brown, R. G., Kimmins, W. C. 1979. Linkage analysis of hydroxyproline-poor glycoprotein from *Phaseolus vulgaris. Plant Physiol.* 63:557–61
19. Brysk, M. M., Chrispeels, M. J. 1972. Isolation and partial characterization of a hydroxyproline-rich cell wall glycoprotein and its cytoplasmic precursor. *Biochim. Biophys. Acta* 257:421–32
20. Buchala, A. J., Meier, H. 1981. An arabinogalactan from the fibres of cotton (*Gossypium arboreum* L.). *Carbohydr. Res.* 89:137–43
21. Catt, J. W., Hills, G. J., Roberts, K. 1976. A structural glycoprotein, containing hydroxyproline isolated from the cell wall of *Chlamydomonas reinhardii. Planta* 131:165–71
22. Chandrasekaran, E. V., BeMiller, J. N., Lee, S.-C. D. 1978. Isolation, partial characterization, and biological proper-

ties of polysaccharides from crude papain. *Carbohydr. Res.* 60:105–15

23. Cho, Y. P., Chrispeels, M. J. 1976. Serine-O-galactosyl linkages in glycopeptides from carrot cell walls. *Phytochemistry* 15:165–69

24. Chrispeels, M. J. 1969. Synthesis and secretion of hydroxyproline containing macromolecules in carrots. I. Kinetic analysis. *Plant Physiol.* 44:1187–93

25. Churms, S. C., Merrifield, E. H., Miller, C. L., Stephen, A. M. 1979. Some new aspects of the molecular structure of the polysaccharide gum from *Acacia saligna* (syn. *cyanophylla*). *S. Afr. J. Chem.* 32:103–6

26. Churms, S. C., Merrifield, E. H., Stephen, A. M. 1978. Regularity within the molecular structure of arabinogalactan from Western larch (*Larix occidentalis*). *Carbohydr. Res.* 64:C1–C2

27. Churms, S. C., Merrifield, E. H., Stephen, A. M. 1978. A comparative examination of two polysaccharide components from the gum of *Acacia mabellae*. *Carbohydr. Res.* 63:337–41

28. Churms, S. C., Merrifield, E. H., Stephen, A. M. 1980. Alkaline degradation of methylated *Acacia* gum polysaccharides from the series Phyllodineae. *S. Afr. J. Chem.* 33:39–40

29. Churms, S. C., Merrifield, E. H., Stephen, A. M. 1981. Chemical and chemotaxonomic aspects of the polysaccharide gum from *Acacia longifolia*. *S. Afr. J. Chem.* 34:8–11

30. Churms, S. C., Merrifield, E. H., Stephen, A. M. 1981. Comparative examination of the polysaccharide gums from *Acacia implexa* and *Acacia cyclops*. *S. Afr. J. Chem.* 34:68–71

31. Churms, S. C., Merrifield, E. H., Stephen, A. M. 1981. Smith degradation of gum exudates from some *Prosopis* species. *Carbohydr. Res.* 90:261–67

32. Churms, S. C., Merrifield, E. H., Stephen, A. M., Stephen, E. W. 1978. Evidence for sub-unit structure in the polysaccharide gum from *Acacia mearnsii*. *S. Afr. J. Chem.* 31:115–16

33. Churms, S. C., Stephen, A. M. 1972. Molecular-weight distribution of hydrolysis products from the gum of *Acacia elata* A. Cunn. *Carbohydr. Res.* 21:91–98

34. Churms, S. C., Stephen, A. M., Siddiqui, I. R. 1981. Evidence for repeating sub-units in the molecular structure of the acidic arabinogalactan from rapeseed (*Brassica campestris*). *Carbohydr. Res.* 94:119–22

35. Clarke, A. E., Anderson, R. L., Stone, B. A. 1979. Form and function of arabinogalactans and arabinogalactanproteins. *Phytochemistry* 18:521–40

36. Clarke, A. E., Gleeson, P. A., Jermyn, M. A., Knox, R. B. 1978. Characterization and localization of β-lectins in lower and higher plants. *Aust. J. Plant Physiol.* 5:707–22

37. Clarke, A. E., Knox, R. B., Jermyn, M. A. 1975. Localization of lectins in legume cotyledons. *J. Cell Sci.* 19: 157–67

38. Cleland, R., Olson, A. C. 1967. Metabolism of free hydroxyproline in *Avena* coleoptiles. *Biochemistry* 6:32–36

39. Cohen, P. B., Schibeci, A., Fincher, G. B. 1983. Biosynthesis of arabinogalactan-protein in *Lolium multiflorum* (ryegrass) endosperm cells. III. In vitro hydroxylation of peptidyl proline. Submitted for publication

40. Darvill, A., McNeil, M., Albersheim, P., Delmer, D. P. 1980. The primary cell walls of flowering plants. In *Biochemistry of Plants*, ed. N. E. Tolbert 1:91–162

41. Dashek, W. V. 1970. Synthesis and transport of hydroxyproline-rich components in suspension cultures of sycamore maple cells. *Plant Physiol.* 46:831–38

42. Dashek, W. V., Harwood, H. I. 1974. Proline, hydroxyproline, and lily pollen tube elongation. *Ann. Bot.* 38:947–59

43. Delmer, D. P., Lamport, D. T. A. 1977. The origin and significance of plant glycoproteins. In *Cell Wall Biochemistry*, ed. B. Solheim, I. Raa, pp. 85–104. Universitats-forlaget

44. Desai, N. N., Allen, A. K., Neuberger, A. 1981. Some properties of the lectin from *Datura stramonium* (thorn apple) and the nature of its glycoprotein linkages. *Biochem. J.* 197:345–53

45. Di Fabio, J. L., Dutton, G. G. S., Moyna, P. 1982. The structure of *Chorisia speciosa* gum. *Carbohydr. Res.* 99:41–50

46. Doerschug, M. R., Chrispeels, M. J. 1970. Synthesis and secretion of hydroxyproline-containing macromolecules in carrots. III. Metabolic requirements for secretion. *Plant Physiol.* 46:363–66

47. Elbein, A. D. 1979. The role of lipid-linked saccharides in the biosynthesis of complex carbohydrates. *Ann. Rev. Plant Physiol.* 30:239–72

48. Fincher, G. B. 1975. Morphology and chemical composition of barley endo-

sperm cell walls. *J. Inst. Brew.* 81:116–22

49. Fincher, G. B., Sawyer, W. H., Stone, B. A. 1974. Chemical and physical properties of an arabinogalactan-peptide from wheat endosperm. *Biochem. J.* 139:535–45

50. Fincher, G. B., Stone, B. A. 1974. A water-soluble arabinogalactan-peptide from wheat endosperm. *Aust. J. Biol. Sci.* 27:117–32

51. Fincher, G. B., Stone, B. A. 1981. Metabolism of noncellulosic polysaccharides. In *Encyclopedia of Plant Physiology* (New ser.). *Plant Carbohydrates II: Extracellular Carbohydrates*, ed. W. Tanner, F. A. Loewus, 13B:68–132. Berlin: Springer

51a. Fry, S. C. 1982. Isodityrosine, a new cross-linking amino acid from plant cell wall glycoprotein. *Biochem. J.* 204:449–55

51b. Gardiner, M., Chrispeels, M. J. 1975. Involvement of the golgi apparatus in the synthesis and secretion of hydroxyproline-rich cell wall glycoproteins. *Plant Physiol.* 55:536–41

52. Geddes, D. S., Wilkie, K. C. B. 1971. Hemicelluloses from the stem tissues of the aquatic moss *Fontinalis antipyretica.* *Carbohydr. Res.* 18:333–35

53. Glaudemans, C. P. J. 1975. The interaction of homogeneous, murine myeloma immunoglobulins with polysaccharide antigens. *Adv. Carbohydr. Chem. Biochem.* 31:313–46

54. Gleeson, P. A., Clarke, A. E. 1979. Structural studies on the major component of *Gladiolus* style mucilage, an arabinogalactan-protein. *Biochem. J.* 181:607–21

55. Gleeson, P. A., Clarke, A. E. 1980. Arabinogalactans of sexual and somatic tissues of *Gladiolus* and *Lilium.* *Phytochemistry* 19:1777–82

56. Gleeson, P. A., Clarke, A. E. 1980. Antigenic determinants of a plant proteoglycan, the *Gladiolus* style arabinogalactan-protein. *Biochem. J.* 191:437–47

57. Gleeson, P. A., Jermyn, M. A. 1979. Alteration of the composition of β-lectins caused by chemical and enzymic attack. *Aust. J. Plant Physiol.* 6:25–38

58. Haq, S., Adams, G. A. 1961. Structure of an arabinogalactan from Tamarack (*Larix laricina*). *Can. J. Chem.* 39: 1563–73

59. Heath, M. F., Northcote, D. H. 1971. Glycoprotein of the wall of sycamore tissue-culture cells. *Biochem. J.* 125:953–61

60. Hillestead, A., Wold, J. K. 1977. Water-soluble glycoproteins from *Cannabis sativa* (South Africa). *Phytochemistry* 16:1947–51

61. Hillestead, A., Wold, J. K., Engen, T. 1977. Water-soluble glycoproteins from *Cannabis sativa* (Thailand). *Phytochemistry* 16:1953–56

62. Hoggart, R. M., Clarke, A. E. 1983. Arabinogalactans in stylar tissues of mono- and dicotyledons. *Phytochemistry.* Submitted for publication

63. Homer, R. B., Roberts, K. 1979. Glycoprotein conformation in plant cell walls. *Planta* 146:217–22

64. Hori, H., Takeuchi, Y., Fujui, T. 1980. Structure of an arabinogalactan of extracellular hydroxyproline-rich glycoprotein in suspension-cultured tobacco cells. *Phytochemistry* 19:2755–56

65. Jermyn, M. A. 1978. Isolation from the flowers of *Dryandra praemorsa* of a flavonol glycoside that reacts with β-lectins. *Aust. J. Plant Physiol.* 5:563–71

66. Jermyn, M. A. 1980. The nature of the protein moiety of the arabinogalactan-proteins. *AGP News* 3:26–31

67. Jermyn, M. A., Yeow, Y. M. 1975. A class of lectins present in the tissues of seed plants. *Aust. J. Plant Physiol.* 2:501–31

68. Karr, A. L. Jr. 1972. Isolation of an enzyme system which will catalyse the glycosylation of extensin. *Plant Physiol.* 50:275–82

69. Kaul, K., Lester, R. L. 1978. Isolation of six novel phosphoinositol-containing sphingolipids from tobacco leaves. *Biochemistry* 17:3569–75

70. Kauss, H. 1980. Lectins and their physiological role in slime moulds and in higher plants. See Ref. 51, pp. 627–57

71. Kawasaki, S. 1981. Synthesis of arabinose-containing cell wall precursors in suspension-cultured tobacco cells. I. Intracellular site of synthesis and transport. *Plant Cell Physiol.* 22:431–42

72. Keller, F., Stone, B. A. 1978. Preparation of *Lolium* protoplasts and their purification using an antigalactan-Sepharose conjugate. *Z. Pflanzenphysiol.* 87:167–72

73. Kilpatrick, D. C. 1980. Purification and some properties of a lectin from the fruit juice of the tomato (*Lycopersicon esculentum*). *Biochem. J.* 185:269–72

74. Klis, F. M., Eeltink, H. 1979. Changing arabinosylation patterns of wall-bound hydroxyproline in bean cell cultures. *Planta* 144:479–84

75. Lamport, D. T. A. 1963. Oxygen fixation into hydroxyproline of plant cell

wall protein. *J. Biol. Chem.* 238:1438–40

76. Lamport, D. T. A. 1973. The role of hydroxyproline-rich proteins in the extracellular matrix of plants. In *Macromolecules Regulating Growth and Development.* 30th Symp. Soc. Dev. Biol., pp. 113–30

77. Lamport, D. T. A. 1977. Structure, biosynthesis and significance of cell wall glycoproteins. *Recent Adv. Phytochem.* 11:79–115

78. Lamport, D. T. A. 1980. Structure and function of plant glycoproteins. In *Biochemistry of Plants,* ed. J. Preiss, 3:501–41

79. Lamport, D. T. A., Catt, J. W. 1981. Glycoproteins and enzymes of the cell wall. See Ref. 51, pp. 133–64

80. Lamport, D. T. A., Katona, L., Roerig, S. 1973. Galactosylserine in extensin. *Biochem. J.* 133:125–31

81. Lamport, D. T. A., Miller, D. H. 1971. Hydroxyproline arabinosides in the plant kingdom. *Plant Physiol.* 48: 454–56

82. Larkin, P. J. 1977. Plant protoplast agglutination and membrane-bound β-lectins. *J. Cell Sci.* 26:31–46

83. Mandal, P. K., Mukherjee, A. K. 1980. Structural investigations on bael exudate gum. *Carbohydr. Res.* 84:147–59

84. Mares, D. J., Stone, B. A. 1973. Studies on wheat endosperm. II. Properties of the wall components and studies on their organization in the wall. *Aust. J. Biol. Sci.* 26:813–30

85. Mascara, T., Fincher, G. B. 1982. Biosynthesis of arabinogalactan-protein in *Lolium multiflorum* (ryegrass) endosperm cells. II. In vitro incorporation of galactosyl residues from UDP-galactose into polymeric products. *Aust. J. Plant Physiol.* 9:31–45

86. Mau, S-L., Raff, J. W., Clarke, A. E. 1982. Isolation and partial characterization of components of *Prunus avium* L. styles, including an S-allele-associated antigenic glycoprotein. *Planta.* In press

87. McNamara, M. K., Stone, B. A. 1981. Isolation, characterization and chemical synthesis of a galactosyl-hydroxyproline linkage compound from wheat endosperm arabinogalactan-peptide. *Lebensm.-Wiss.-Technol.* 14:182–87

88. Morris, M. R., Northcote, D. H. 1977. Influence of cations at the plasma membrane in controlling polysaccharide secretion from sycamore suspension cells. *Biochem. J.* 166:603–18

89. Nachbar, M. S., Oppenheim, J. D., Thomas, J. O. 1980. Lectins in the U.S. diet. Isolation and characterization of a lectin from the tomato (*Lycopersicon esculentum*). *J. Biol. Chem.* 255:2056–61

90. O'Neill, M. A., Selvendran, R. R. 1980. Glycoproteins from the cell wall of *Phaseolus coccineus.* *Biochem. J.* 187:53–63

91. Owens, R. J., Northcote, D. H. 1981. The location of arabinosyl:hydroxyproline transferase in the membrane system of potato tissue culture cells. *Biochem. J.* 195:661–67

92. Paulsen, B. S., Lund, P. S. 1979. Water-soluble polysaccharides of *Opuntia ficus-indica* cv "Burbank's Spineless". *Phytochemistry* 18:569–71

93. Pollard, J. K., Steward, F. C. 1959. The use of ^{14}C-proline by growing cells: its conversion to protein and to hydroxyproline. *J. Exp. Bot.* 10:17–32

94. Pollard, P. C., Fincher, G. B. 1981. Biosynthesis of arabinogalactan-protein in *Lolium multiflorum* (ryegrass) endosperm cells. I. Hydroxylation of peptidyl proline. *Aust. J. Plant Physiol.* 8:121–32

95. Pollard, P. C., Way, P. W., Fincher, G. B. 1981. Uptake and metabolism of hydroxyproline in endosperm cells of *Lolium multiflorum* (ryegrass). *Aust. J. Plant Physiol.* 8:535–46

96. Pope, D. G. 1977. Relationships between hydroxyproline-containing proteins secreted into the cell wall and medium by suspension-cultured *Acer pseudoplatanus* cells. *Plant Physiol.* 59:894–900

97. Sadava, D., Chrispeels, M. J. 1971. Hydroxyproline biosynthesis in plant cells; peptidyl proline hydroxylase from carrot disks. *Biochim. Biophys. Acta* 227:278–87

98. Schibeci, A., Fincher, G. B., Stone, B. A., Wardrop, A. B. 1982. Isolation of plasma membranes from protoplasts of *Lolium multiflorum* (ryegrass) endosperm cells. *Biochem. J.* 205:511–19

99. Selvendran, R. R. 1975. Cell wall glycoproteins and polysaccharides of parenchyma of *Phaseolus coccineus.* *Phytochemistry* 14:2175–80

100. Setia, R. C., Parthasarathy, M. V., Shah, J. J. 1977. Development, histochemistry and ultrastructure of gumresin ducts in *Commiphora mukul* Engl. *Ann. Bot.* 41:999–1004

101. Simson, B. W., Timell, T. E. 1978. Polysaccharides in cambial tissues of *Populus tremuloides* and *Tilia americana.* III. Isolation and constitution of an arabinogalactan. *Cellul. Chem. Technol.* 12:63–77

102. Smith, M. A. 1981. Characterization of carrot cell wall protein. I. Effect of α,α'-dipyridyl on cell wall protein synthesis and secretion in incubated carrot discs. *Plant Physiol.* 68:956–63

103. Stephen, A. M. 1980. Plant carbohydrates. In *Encyclopedia of Plant Physiology* (New ser.). *Secondary Plant Products,* ed. E. A. Bell, B. V. Charlwood, 8:555–84. Berlin: Springer

104. Strahm, A., Amado, R., Neukom, H. 1981. Hydroxyproline-galactoside as a protein-polysaccharide linkage in a water soluble arabinogalactan-peptide from wheat endosperm. *Phytochemistry* 20:1061–63

105. Stuart, D. A., Mozer, T. J., Varner, J. E. 1982. Cytosine-rich messenger RNA from carrot root discs. *Biochem. Biophys. Res. Commun.* 105:582–88

106. Stuart, D. A., Varner, J. E. 1980. Purification and characterization of a salt-extractable hydroxyproline-rich glycoprotein from aerated carrot discs. *Plant Physiol.* 66:787–92

107. Susheelamma, N. S., Rao, M. V. L. 1978. Isolation and characterization of arabinogalactan from black gram (*Phaseolus mungo*). *J. Agric. Food Chem.* 26:1434–37

108. Swenson, H. A., Kaustinen, H. M., Kaustinen, O. A., Thompson, N. S. 1968. Structure of gum arabic and its configuration in solution. *J. Polym. Sci.* 6:1593–1606

109. Takeuchi, Y., Komamine, A. 1978. Composition of the cell wall formed by protoplasts isolated from cell suspension cultures of *Vinca rosea. Planta* 140:227–32

110. Takeuchi, Y., Komamine, A. 1980. Turnover of cell wall polysaccharides of a *Vinca rosea* suspension culture. III. Turnover of arabinogalactan. *Physiol. Plant* 50:113–41

111. Takeuchi, Y., Komamine, A., Saito, T., Watanabe, K., Morikawa, N. 1980. Turnover of cell wall polysaccharides of a *Vinca rosea* suspension culture. II. Radio gas chromatographical analyses. *Physiol. Plant* 48:536–41

112. Tanaka, M., Sato, K., Uchida, T. 1981. Plant prolyl hydroxylase recognizes poly(L-proline) II helix. *J. Biol. Chem.* 256:11397–11400

113. Tanaka, M., Shibata, H., Uchida, T. 1980. A new prolyl hydroxylase acting on poly-L-proline, from suspension cultured cells of *Vinca rosea. Biochim. Biophys. Acta* 616:188–98

114. Tautvydas, K. J. 1978. Isolation and characterization of an extracellular hydroxyproline-rich glycoprotein and a mannose-rich polysaccharide from *Eudorina californica* (Shaw). *Planta* 130:213–20

115. Toole, B. P. 1981. Glycosaminoglycans in morphogenesis. In *Cell Biology of the Extracellular Matrix,* ed. E. Hay. New York: Plenum

116. Trachtenberg, S., Fahn, A. 1981. The mucilage cells of *Opuntia ficus-indica* (L.) Mill.—Development, ultrastructure and mucilage secretion. *Bot. Gaz.* 142:206–13

117. Trachtenberg, S., Mayer, A. M. 1981. Composition and properties of *Opuntia ficus-indica* mucilage. *Phytochemistry* 20:2665–68

118. van Holst, G.-J., Klis, F. M. 1981. Hydroxyproline glycosides in secretory arabinogalactan-protein of *Phaseolus vulgaris* L. *Plant Physiol.* 68:979–80

119. van Holst, G.-J., Klis, F. M., de Wildt, P. J. M., Hazenberg, C. A. M., Buijs, J., Stegwee, D. 1981. Arabinogalactan protein from a crude cell organelle fraction of *Phaseolus vulgaris* L. *Plant Physiol.* 68:910–13

120. van Holst, G.-J., Varner, J. E. 1983. Polyproline II conformation in hydroxyproline-rich cell wall glycoproteins from carrot root. *Biochem. J.* Submitted for publication

121. Varner, J. E. 1980. The direct conversion of hydroxyproline to proline. *Biochem. Biophys. Res. Commun.* 96:692–96

122. Yamagishi, T., Matsuda, K., Watanabe, T. 1976. Characterization of the fragments obtained by enzymic and alkaline degradation of rice-bran proteoglycans. *Carbohydr. Res.* 50:63–74

123. Yariv, J., Lis, H., Katchalski, E. 1967. Precipitation of arabic acid and some seed polysaccharides by glycosylphenylazo dyes. *Biochem. J.* 105:1C–2C

124. Yariv, J., Rapport, M. M., Graf, L. 1962. The interaction of glycosides and saccharides with antibody to the corresponding phenylazo glycosides. *Biochem. J.* 85:383–88

125. Young, R. A., Sarkanen, K. V. 1977. The use of alkaline degradation for structural characterization of branched-chain polysaccharides. *Carbohydr. Res.* 59:193–201

Ann. Rev. Plant Physiol. 1983. 34:71–104
Copyright © 1983 by Annual Reviews Inc. All rights reserved

PHOTOSYNTHETIC ASSIMILATION OF EXOGENOUS HCO₃ BY AQUATIC PLANTS

William J. Lucas

Department of Botany, University of California, Davis, California 95616

CONTENTS

INTRODUCTION

The topic of photosynthetic assimilation of exogenous HCO_3^- by aquatic plants (algae and aquatic angiosperms) has been, and still is, controversial. In the present review, our attention will be confined to freshwater species. Much of the early work in this field focused on the influence of inorganic carbon availability on organismal growth (85–88). These studies revealed that many aquatic species can grow very effectively under culture conditions where the available inorganic carbon exists almost entirely as HCO_3^-. This led to the conclusion that these species could utilize the exoge-

71

0066-4294/83/0601-0071$02.00

nous inorganic carbon of both CO_2 and HCO_3^- for photosynthesis. Prior reviews have dealt with the problems of interpreting such experiments (99, 120).

Raven (99) developed a more rigorous form of analysis to ascertain whether a species could utilize exogenous HCO_3^-. This approach consisted of comparing photosynthetic carbon fixation at high and low external pH values. If the species is *unable* to utilize HCO_3^- (i.e. transport it into the cell), fixation is predicted to vary in proportion to the external H^+ concentration, because this parameter determines the concentration of CO_2 within the medium (sealed systems are generally used). An excellent example of such a response is seen with high CO_2-grown cells of *Scenedesmus obliquus* (25). However, following adaptation to normal air levels of CO_2, *S. obliquus* can utilize exogenous HCO_3^-. Numerous species appear to be capable of higher levels of fixation, at alkaline pH, than would be predicted by the presumed concentration of CO_2 present under these conditions (10, 14, 23, 29, 51, 59, 80, 110). These data have been rather widely accepted as demonstrating that certain aquatic plants have the ability to transport HCO_3^- from the external medium into the cytoplasm.

A further difficulty in interpreting results pertaining to HCO_3^- assimilation is caused by the fact that HCO_3^- may (can) act as a reservoir for extracellular CO_2 (see Figure 1). In recent years, several investigators have addressed this issue by developing theoretical analyses of the rate(s) at which CO_2 could be produced from HCO_3^-. Lucas (60) was able to show that the maximum rate at which CO_2 could be supplied, from the equilibrium reactions involving HCO_3^-, was insufficient to account for ^{14}C-fixation in *Chara corallina* as measured in alkaline media. Similar analyses have indicated that *Microcystis aeruginosa, Ceratium hirudinella* (121), *Chlamydomonas reinhardtii* (55), *Egeria densa* (14), *Scenedesmus obliquus* (96) and *Coccochloris peniocystis* (80) can assimilate exogenous inorganic carbon at rates in excess of those possible by HCO_3^- to CO_2 conversion (cf 34, 35, 95). Thus, there is a growing body of evidence which supports the contention that numerous aquatic species can utilize both CO_2 and HCO_3^- for photosynthesis. It should be stressed, however, that CO_2 is the chemical species utilized by the ribulose bisphosphate carboxylase (15).

STRATEGIES USED FOR HCO_3^- ACQUISITION

In a morphological sense, the photosynthetic tissues which utilize exogenous HCO_3^- are extremely variable, ranging from single cells (6, 10, 24, 28, 50, 79) to the submerged leaves of aquatic angiosperms (14, 38, 94, 119). Within this range of cellular complexity, it is possible to elucidate certain basic strategies that appear to have been adopted to enable the tissue to acquire exogenous HCO_3^-. At this point it is important to stress that follow-

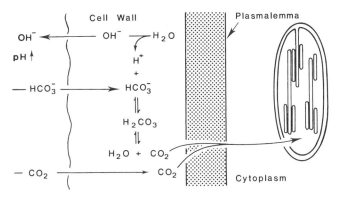

Figure 1 Diffusion of inorganic carbon (CO_2 and HCO_3^-) from the medium up to the plasmalemma. Conversion of HCO_3^- to CO_2 at the outer surface of the membrane enhances the flux into the cell.

ing HCO_3^- transport into the cell, CO_2 is fixed via the photosynthetic reactions and, as a result of this removal of CO_2, alkalinity (OH⁻) is produced in near stoichiometric quantities. To control the cytoplasmic pH during this assimilation process, this alkalinity must be disposed of via OH⁻ efflux to the bathing medium. It is the operation and spatial location of the HCO_3^- and OH⁻ transport systems that will now be examined.

Polarized Transport Across Aquatic Leaves

Arens, an early pioneer in the field of HCO_3^- assimilation, was one of the first to develop the concept of a spatial separation of physiological events associated with HCO_3^- assimilation. Arens talked of the submerged leaves of *Potamogeton* and *Elodea* as being polarized (3, 4), in that only the lower surface appeared to be able to transport exogenous HCO_3^- while the upper surface became very alkaline, presumably due to OH⁻ excretion. Extensive research on the species *Potamogeton lucens* by Steemann Nielsen (119; see also 120), Helder & co-workers (38–42), and others (75, 92–94) has confirmed Aren's hypothesis. Similar examples of spatially separate (polar) HCO_3^- and OH⁻ transporting systems have been demonstrated in *Elodea* (5, 102) and *Myriophyllum* (119). However, in a recent report, Helder et al (41) showed that the upper leaf surface of *P. lucens* could be "induced" to take up HCO_3^-. Whether this involved true induction of transport capacity at the upper surface or apoplastic transfer of HCO_3^- to the lower leaf surface remains to be elucidated.

During assimilation of HCO_3^-, polar cation transport also occurs across the leaves of *P. lucens,* i.e. cations are transported from the lower to the upper leaf surface (3, 39, 40, 119). This phenemenon is light-mediated and exhibits a specific requirement for the presence of HCO_3^- at the lower leaf surface (39). It has also been found that during this process an electrical potential gradient of approximately 15 to 20 mV is established across the leaf (38, 40, 42, 94). A potential of this magnitude could only be developed if at least one of the transported species is moved via an active electrogenic mechanism. (Electrogenic transport involves the movement of unbalanced charge.) This polar transport system is illustrated schematically in Figure 2A.

Numerous hypotheses have been presented regarding the nature and interrelationship(s) between HCO_3^-, OH^-, and polar cation fluxes. Serious

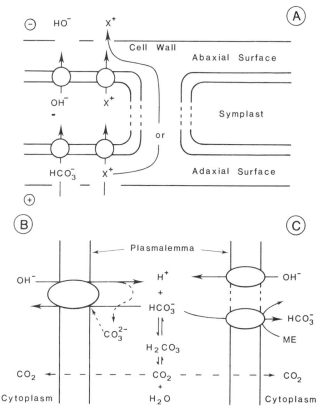

Figure 2   Possible mechanisms by which exogenous HCO_3^- may be transported into the cytoplasm. A: Polar transport situation observed in *Potamogeton* and *Elodea;* X^+ represents cations that are transported either through the apoplast or symplast. B: Mitchell-type antiporter; extracellular production of CO_3^{2-} may inhibit HCO_3^- transport via competitive binding to the HCO_3^--binding site. C: Electrogenic HCO_3^- and OH^- transport systems that are spatially separated along the plasmalemma; ME represents the investment of metabolic energy.

disagreement exists, however, between the various proposals. Steemann Nielsen (120) proposed that HCO$_3^-$ entry was passive, with the active process being the excretion of OH$^-$ from the upper leaf surface. The polar cation fluxes were thought to be passive (4, 119). Lowenhaupt (57, 58), and later Helder & Boerma (39), proposed that it was the polar cation transport process that was active. Helder (38) later suggested that the polar fluxes may result from either a cation pump or via some form of coupling to the transport of HCO$_3^-$ and OH$^-$ ions.

Since an electric potential is established across the leaf during HCO$_3^-$ assimilation, this potential could explain in part the polar cation flux. A completely apoplastic pathway for the cations could be tested using HCO$_3^-$ solutions containing large cations that could not be transported across the plasmalemma. Further developments concerning the energetics of HCO$_3^-$ and OH$^-$ transport will be dealt with in later sections.

HCO$_3^-$/OH$^-$ Antiport?

In contrast to the polar transport situation found in *Potamogeton, Elodea,* and *Myriophyllum,* it has been suggested that in *Chara* (74) and *Scenedesmus* (24, 28) the HCO$_3^-$ and OH$^-$ ions may move on a "Mitchell-type" antiporter. In this situation, HCO$_3^-$ and OH$^-$ transport would occur at the same location on the plasmalemma and involve a single transport entity (see Figure 2B). Provided HCO$_3^-$ influx and OH$^-$ efflux were balanced, such a transport system would be electrically neutral, thereby eliminating any effect of the membrane potential on either flux. If this were the case, photosynthesis could develop a sufficient HCO$_3^-$ concentration gradient to facilitate the influx of HCO$_3^-$ and efflux of OH$^-$.

During photosynthesis, when exogenous HCO$_3^-$ is utilized, alternating acid and alkaline bands (zones) develop along the internodal cells of *Chara corallina* (74, 111). Diffusion analysis of the alkaline bands indicated an equivalence between OH$^-$ efflux from these regions and the rate of H^{14}CO$_3^-$ fixation by the internodal cells (59), a result consistent with the antiporter hypothesis. However, Lucas (62) later invalidated this hypothesis by showing that when the solution bathing a particular alkaline band (region of OH$^-$ efflux) was HCO$_3^-$-free, OH$^-$ efflux continued. This result demonstrated categorically that the OH$^-$ efflux process could function in the absence of exogenous HCO$_3^-$ at the actual efflux site. Thus, in *Chara,* the transport of HCO$_3^-$ and OH$^-$ appears to be independent insofar as the two processes can be spatially separated (62, 72). This conclusion has been verified by results obtained using a range of techniques (21, 65, 69, 72, 128), and will be discussed further in a later section.

High CO$_2$-grown (2%) cells of *Scenedesmus obliquus* cannot utilize exogenous HCO$_3^-$; however, after 2–3 hr adaptation to 0.03% CO$_2$, cells appear to be able to utilize the exogenous inorganic carbon in proportion to the levels of CO_2 and HCO_3^- present in the bathing medium (26, 28). Findenegg has shown that during this period of adaptation, cells develop carbonic anhydrase activity (23, 25), and it has been suggested that this enzyme may act as a permease for HCO_3^- at the plasmalemma of *Scenedesmus* (24). However, carbonic anhydrase is essential for HCO_3^- assimilation at low pH (5.7–7.0) but not at pH 9–9.5 (26–28). This finding is inconsistent with a direct role of carbonic anhydrase in the transport of HCO_3^- across the plasmalemma of *Scenedesmus* (27).

Nonetheless, in assaying carbonic anhydrase activity, which was identified as being within the cytoplasm of *S. obliquus,* Findenegg obtained evidence consistent with there being extremely large plasmalemma fluxes (200 nmol $cm^{-2}s^{-1}$) of HCO_3^- and OH^- (28). These high rates were obtained at 3°C and were independent of metabolism. *Scenedesmus* may, therefore, possess a HCO_3^-/OH^- antiport system which is not metabolically driven at low pH, but becomes energy dependent under high pH conditions (28). This hypothesis should be tested by utilizing microprobe techniques, as has been done for *Chara* (71, 72, 128). If the transport of HCO_3^- is electrogenic, it may be possible to detect local electric fields within the region where this ion is being transported. Although these cells are small (diameter approximately 10–15 μm), it should be possible to use extracellular voltage electrodes (tip diameter 0.5 μm) and/or the vibrating probe technique (46, 72) to confirm or refute the above hypothesis.

Electrogenic Transport of HCO_3^- and OH^-

RELATIONSHIP BETWEEN LIGHT, MEMBRANE POTENTIAL, AND HCO_3^- TRANSPORT   The cell membrane potential of many green plants appears to be sensitive to light [for a general review see (43)]. This phenomenon has been studied in only a limited number of aquatic plant species, with most research having been done on the giant cells of the Characeae. Much of the early work is quite confusing, the light-response being obviously complex, with more than one transport system (both active and passive) influencing the direction in which the potential changes upon cell illumination. In some cases, a light-induced hyperpolarization (transient or stable) of the membrane potential was observed (82, 83, 123), while other workers reported a depolarization (2). However, in *Chara* and *Nitella,* it has been shown that the electrical changes produced by light have an action spectrum similar to that of photosynthesis (123, 125). Andrianov and co-workers (1) also found a close correlation between the shape of the light-induced change in membrane potential and the induction of photosynthesis. The involvement of photosynthesis has also been shown by the sensitivity of the plasmalemma response to photosynthetic inhibitors (103, 104, 123, 124). Excellent confirmation of the coupling between changes in membrane potential and photosynthesis was obtained by Andrianov et al (1). These workers were able to demonstrate that when the rhizoids of *Nitella flexilis* were free of chlorophyll, no light-induced changes occurred in the membrane potential. However, as the rhizoids developed chloroplasts, the membrane potential became sensitive to light.

The nature of the coupling between the chloroplasts and the plasmalemma is still speculative. An electrogenic role for HCO_3^- (see Figure 2C) was first proposed by Hope (45), who showed that the membrane potential of *Chara* became hyperpolarized upon the addition of HCO_3^-, and this response was light-sensitive, as was the decrease in the membrane resistance, which occurred upon addition of HCO_3^-. Spanswick (114) refuted this role of HCO_3^-; he claimed that the electrical potential change was independent of the chemical nature of the buffer employed, i.e. HCO_3^- produced a hyperpolarization via its buffering action. However, the conditions used by Spanswick were not HCO_3^--free and, therefore, his results did not eliminate a specific role for HCO_3^-. Reports of the influence of HCO_3^- on the membrane potential have been presented by other workers (16, 65, 83, 103, 123).

Saito & Senda (103), using *Nitella flexilis* and *Nitella axilliformis,* found that addition of $NaHCO_3$ to the medium caused a small hyperpolarization (approximately 20 mV) when added in the dark, and illumination then caused a further stable hyperpolarization of about 40 mV. This response is similar to that reported by Hope (45). However, Nishizaki's (83) results with *Chara* were quite different from those obtained by Hope. He found that illumination in the presence of 0.2 mM $NaHCO_3$ caused a slight hyperpolarization, followed by a rapid depolarization; in the dark, the potential again hyperpolarized. This result is quite similar to that obtained on submerged leaves of *Potamogeton schweinfurthii* (16), and is also consistent with the recent results obtained by Lucas (65).

Much of the early work relied almost entirely on electrophysiological measurements to elucidate the influence of HCO_3^- on the electrical properties of the plasmalemma. As illustrated above, many of these results were either inconsistent or contradictory, and this is probably attributable to the complexity of the transport processes which reside in the plant plasma membrane (see 76, 91, 117 for reviews of this literature). In addition, the ability of a particular species to assimilate HCO_3^- can vary quite dramatically as a function of season (47, 60), age (81, 88, 89), and storage after tissue excision (W. J. Lucas, unpublished results), and this variation most surely contributed toward impeding progress in this area. Thus, until the mid-1970s, the hypothesis that HCO_3^- was being transported across the plasmalemma of photosynthesizing (aquatic) cells in an electrogenic manner received only equivocal support.

EXTRACELLULAR pH AND ELECTRICAL MEASUREMENTS   Following the demonstration that *Chara corallina* could assimilate exogenous HCO_3^- at quite high rates (60) via transport processes that were not coupled in the form of an antiporter (62), considerable attention was focused on this system to further elucidate its characteristics. In studies where both the pH and the external electric potential were measured at the cell surface, an excellent correlation was observed between alkaline band pH centers and the negative electric potential maxima (21, 128). In the regions between the alkaline bands, where HCO_3^- transport was thought to occur, positive external electric potentials were observed. These results clearly demonstrated that electric currents circulate between the alkaline and "acid" zones of the *Chara* internodal cell. In general, these currents were dependent upon cell illumination and the presence, within the bathing medium, of HCO_3^-. These findings offered strong experimental support for the hypothesis that HCO_3^- transport (in this system) was electrogenic and that the alkaline bands were being formed by electrogenic efflux of OH^- (or H^+ influx).

Using their theoretical numerical analysis model, Ferrier & Lucas (21) were able to use their experimental values of pH and electric potential to compute the magnitude of the putative HCO_3^- fluxes. As a result of this analysis, they suggested that along the length of a particular *Chara* cell, the actual HCO_3^- fluxes may vary quite significantly. This prediction was later confirmed by Lucas & Nuccitelli (72) when they used the Jaffe-Nuccitelli vibrating probe (46) to further investigate these extracellular currents. This technique, in which a small (10 μm) platinum spherical probe is vibrated over a known distance (usually 10 to 20 μm) at a particular frequency (300 to 500Hz), enables an accurate measurement of local voltage gradients that are established near the cell wall. The local voltage gradient can be converted to local current density or ionic flux (both in the direction of vibration). Using this technique, these authors found that extracellular bands of positive inward current (OH^- efflux) 1–3 mm wide were separated by wider bands of outward current (HCO_3^- influx) along the length of the cell. The measured peaks of inward current ranged from 20 to 60 μA cm^{-2}, which would correspond to an ionic flux across the cell surface of 270–800 pmol cm^{-2}s^{-1}. The peaks of outward current (HCO_3^- influx) ranged from 10 to 30 μA cm$^-$, or 140–400 pmol cm^{-2}s^{-1}. The inward current bands matched the regions of surface alkalinity very well. The HCO_3^-–associated current declined when the exogenous HCO_3^- level was reduced; a commensurate readjustment in the strength and pattern of the OH^- current was always observed under these conditions.

The greatly improved spatial and temporal resolution provided by this vibrating probe technique enabled Lucas & Nuccitelli (72) to detect details of transport activity within single HCO_3^- or OH^- bands. Bicarbonate-associated currents were much more stable than those measured in the alkaline bands, and the HCO_3^- currents were surprisingly close to those predicted by Ferrier & Lucas (21). A further point of importance was that no strong correlation was found between the absolute pH values along the interalkaline band regions and the HCO_3^- current strengths measured in these locations. This was true even for regions where the surface pH was depressed below the background value, a phenomenon which was often used as representing a site of strong HCO_3^- transport (72). We will return to this point in a later section.

It will be of considerable interest to see what details this technique can elucidate when it is applied to the polar HCO_3^- transport situation of *Potamogeton* and *Elodea* leaves.

HCO_3^- TRANSPORT IN UNICELLULAR SYSTEMS   A considerable amount of elegant work has been published during the past few years which shows that many unicellular algal species and certain blue-green algae (Cyanobacteria) can assimilate exogenous HCO_3^- (7, 8, 10, 23, 48, 50, 79). General aspects of nutrient transport (including HCO_3^-) in the microalgae have been reviewed recently (100). Some of the most outstanding work has been done on the Cyanobacteria, where, although their smallness of size precludes the attainment of spatial details possible with the Characeae, the simplified cytoplasmic phase (no cellular compartmentation) has enabled the measurement of HCO_3^- accumulation ratios, internal pH, and membrane potentials under HCO_3^--assimilating conditions (50, 51).

Badger et al (7) used the silicone oil centrifugation technique to separate low CO_2-grown cells of *Anabaena variabilis* and *Chlamydomonas reinhardtii* rapidly from a $H^{14}CO_3^-$-incubation medium. This separation could be achieved in 2 sec or less (7). Analysis of the acid-labile and -stable components within these two cell types revealed that inorganic carbon concentrations within the cytoplasm were as much as 1000 times that of the incubation medium; i.e. the values were far in excess of those possible by passive equilibration across the plasmalemma. Similar results have now been obtained on a variety of microalgal species, including *Coccochloris peniocystis* (79, 80), *Chlorella emersonii* (10), *Dunaliella salina* (130), and *Spirulina platensis* (48). Using ^{14}C-dimethyloxazolidinedione, Badger et al (7, 8) were able to measure the cytoplasmic pH, and this parameter enabled a calculation of the extent to which internal CO_2 was raised above the level of the bathing medium. In *C. reinhardtii,* low CO_2-grown cells were able to concentrate CO_2 up to 40-fold in relation to the external medium (8). Similar or somewhat smaller CO_2 concentration ratios have been reported for *Anabaena* (50, 78), *Coccochloris* (79, 80), *Chlorella* (10), and *Dunaliella*

(130). In the above-mentioned studies, various metabolic inhibitors were shown to reduce severely the accumulation of inorganic carbon.

The ability of certain low CO_2-grown microalgae to concentrate CO_2 within the cell has been generally attributed to an increased capacity by these cells to utilize exogenous HCO_3^-. This conclusion is consistent with the analysis of photosynthesis with respect to the theoretical rate of CO_2 production from HCO_3^-. Talling (121) showed that both *Microcystis aeruginosa* and *Ceratium hirudinella* have photosynthetic rates, at pH values of 10 or greater, that are in excess of the rates that could be supported by the spontaneous dehydration of HCO_3^- within the medium. Similarly, Miller & Colman (80) have shown, in an elegant study on *Coccochloris peniocystis,* that the rate of CO_2 fixation at alkaline pH is as much as 50-fold the maximum rate of CO_2 production from HCO_3^- dehydration.

The observed accumulation ratios in the microalgae could be caused by facilitated diffusion of HCO_3^- across the plasmalemma. However, as pointed out by Kaplan et al (50), this would require that the membrane potential be positive; the value for *Anabaena* would have to be approximately $+150$ mV (inside positive). In a recent study (51) it was shown that during HCO_3^- assimilation, the membrane potential of *Anabaena* was negative, having a value of approximately -100 mV. Negative membrane potentials are a common feature of plant cells (43), and thus it is probably safe to conclude that in many microalgae there is an active HCO_3^- transport system.

The relationship between O_2 evolution and OH^- efflux has been studied in *Anabaena* (49) and *Coccochloris* (80). In both cases a close equivalance was observed between O_2 evolution and OH^- efflux. In the *Anabaena* system, however, Kaplan (49) found that addition of $NaHCO_3$ to cells which had previously depleted the inorganic carbon in the medium resulted in an almost immediate evolution of O_2, whereas a lag of approximately 20 to 40 sec was observed before OH^- efflux reached its steady state. This temporal separation in O_2 and OH^- efflux suggests that the transport of HCO_3^- and OH^- may not be coupled in a compulsory manner (49). Thus, HCO_3^- transport in this system may also be electrogenic in nature. Since no inner membranes enclose the photosynthetic system in the Cyanobacteria, the location of such an electrogenic HCO_3^- transport system would be at the plasmalemma. Beardall & Raven (10) have argued, albeit not all that convincingly, that in *Chlorella* and *Chlamydomonas* the site of HCO_3^- transport may be the chloroplast envelope rather than the plasmalemma.

In a recent study, Kaplan and co-workers (51) investigated the influence of HCO_3^- transport on the membrane potential of *A. variabilis.* To do this they used the lipid-soluble cation tetraphenyl phosphonium (TPP^+). They found that when 1.0 mM $NaHCO_3$ (pH 8.0) was added to cells bathed in a CO_2-depleted medium, the membrane potential immediately hyperpolarized, with a maximum value being established after 90 to 120 sec. Approximately the same time was required for similarly treated cells of *A. variabilis* to reach a steady state with respect to accumulation of inorganic carbon (50). An important point to note concerning this work was that the membrane potential hyperpolarized quite rapidly and then remained in this state for up to 20 min. The correlation between inorganic carbon uptake and the development of a stable hyperpolarized membrane potential is certainly consistent with the involvement (operation) of an electrogenic HCO_3^- transport system.

The question of monovalent cation coupling to the movement of HCO_3^- has also been examined in *A. variabilis* (51). Addition of $NaHCO_3$ to CO_2-depleted cells resulted in an immediate uptake of K^+ from the bathing medium which ceased after 2 min, and, shortly thereafter, K^+ began to be released from the cells. Approximately 9 min after the addition of $NaHCO_3$ there was no further net flux of K^+, but O_2 evolution continued in a linear manner. These results could be explicable in terms of a symport of K^+ with HCO_3^-, followed by a coupled release of K^+ and OH^-. If this were the case, photosynthesis in *A. variabilis* should be influenced by the presence or absence of K^+ in the bathing medium. Kaplan and co-workers (50) have demonstrated that only under low levels of exogenous HCO_3^- is photosynthesis limited by the transport capacity of the HCO_3^- accumulating system. Thus, the critical K^+ concentration experiments were performed under low levels of exogenous HCO_3^-. From these experiments it was clear that photosynthetic assimilation of exogenous HCO_3^- was independent of external K^+ concentration (51). A similar independence on the external monovalent cation concentration was observed in *Chara* (64). In his studies on *Chara*, Lucas used choline bicarbonate so that there would be no transportable cation available. It would have been of interest to see the effect of choline bicarbonate on the *Anabaena* system.

In view of the observed insensitivity to K^+ availability, and since the membrane potential appeared to hyperpolarize and remain in this state, it is intriguing to speculate as to the reasons for the observed K^+ fluxes. Uptake of K^+, upon addition of $NaHCO_3$, could be due to the passive movement of this ion down its electrochemical potential gradient. However, K^+ efflux cannot be explained on this basis, since the membrane potential did not decline and thus a net K^+ loss from the cells should not have occurred. It would appear that events occurring across the plasmalemma are more complex than can be resolved using a single probe like TPP^+.

Any model proposed to explain HCO_3^- uptake and accumulation in the microalgae must take account of the experimental findings outlined above. Beardall & Raven (10) favor the operation of a primary, electrogenic HCO_3^- transport system; however, they speculated that this pump may be present in the chloroplast envelope of eukaryotes. Miller & Colman (80) discuss the possibility that electroneutrality could be maintained by an equivalent uptake of H^+ with HCO_3^-, the protons coming from the dissociation of H_2O; the resultant OH^- ions would cause an alkalinization of the medium. They rejected this model on the basis that photosynthesis in *Coccochloris* was as rapid at pH 10 as it was at pH 8; i.e. a 100-fold reduction in H^+ concentration had no observable effect on HCO_3^- transport. Since the cytoplasmic pH in the Cyanobacteria appears to be between 7 and 8 (6, 49), Miller and Colman rationalized that electroneutrality could more reasonably be maintained via the operation of an OH^- efflux system. (The concentration of OH^- ions available to the cytoplasmic face of the plasmalemma would be approximately 1.0 μM compared with a H^+ concentration of 0.1 nM at the outer surface.) Based on the available evidence, these workers could not distinguish between a tightly coupled obligate exchange of OH^- for HCO_3^- or separate electrogenic transport components.

A HCO_3^-/OH^- antiport system could not explain the hyperpolarizing response observed in *Anabaena* (51). Such a transport system is also at variance with the lag between onset of O_2 evolution and OH^- efflux that was observed when HCO_3^- was introduced to CO_2-depleted cells of *Anabaena* (49). Similarly, unless such an antiporter had a variable stoichiometry, it could not account for the change in the O_2 to OH^- ratio that Kaplan (49) observed when cells of *Anabaena* were exposed to pH values above 9.5.

Two possibilities remain: first, a $HCO_3^--H^+$ cotransport system driven by the H^+ electrochemical potential gradient, which could be maintained by a plasmalemma-bound H^+-translocating ATPase; and second, a primary, active, electrogenic HCO_3^- transport system. Beardall & Raven (10) regarded $HCO_3^--H^+$ transport as unlikely because, on energetic grounds, 2 H^+ per HCO_3^- may be required to account for movement against the measured inorganic carbon accumulation ratios. Such a transport system would tend to depolarize the membrane potential upon addition of $NaHCO_3$. However, as pointed out by Kaplan et al (51), a $HCO_3^--2H^+$ cotransport system could be operative and still be consistent with the observed membrane potential hyperpolarization, provided the H^+ current through the putative ATPase is sufficiently stimulated by the addition of HCO_3^-. In other words, the H^+-translocating ATPase would have to be stimulated, by addition of $NaHCO_3$, in advance of the activation of the putative $HCO_3^--2H^+$ transport system. In both models the efflux of OH^- could occur via a passive OH^- uniport system (51). (This aspect of HCO_3^- assimilation will be discussed in a separate section on OH^- transport.) Experimental evidence presently available does not permit us to select between these two possibilities, but some workers favor the electrogenic HCO_3^- transport hypothesis (51).

A NEW HYPOTHESIS/CONTROVERSY FOR CARBON ACQUISITION

Following the pioneering work of Kitasato on the involvement of an electrogenic H^+ pump in the plasmalemma of *Nitella* (52), innumerable plant systems have been examined with respect to the operation of such an efflux system. The existence of this H^+ pump is now widely accepted (91, 117).

Aquatic Angiosperms

The operation of this pump recently has become entwined in a novel way with photosynthetic utilization of exogenous HCO_3^-. Prins and co-workers (92–94) have suggested that the cells of the lower surface of *Elodea* and *Potamogeton* leaves contain H^+ pumps that act to acidify the wall. It has been claimed that the cells of the lower epidermis are transfer cells and that the wall ingrowths act to retain the protons within the wall. As a consequence of H^+ pump activity, HCO_3^- would be converted to CO_2. It is suggested that the increased plasmalemma surface area of the transfer cells allows for an efficient "capture" of the CO_2 molecules so produced. This hypothesis suggests that a HCO_3^- transport system does not exist; carbon transfer occurs via diffusion of CO_2. A model of this system is presented in Figure 3. According to this model, H^+ efflux at the lower epidermis is balanced by OH^- efflux (or H^+ influx) across the plasmalemma of the upper epidermal cells. These fluxes would form a current of negative charges, moving from the lower to the upper leaf surface, which would create an electric potential difference across the leaf (see earlier section and Figure 2A). The current would have to be balanced by a polar cation flux (Na^+ or K^+) and the cations could move either apoplastically or through the symplasm.

Using miniature pH electrodes placed on both leaf surfaces, Prins and co-workers (92–94) have shown that upon illumination, the lower surface of both *Elodea canadensis* and *Potamogeton lucens* (previously established to be HCO_3^- assimilators) became acidified. Acidification could have been due to H^+ escape from the transfer cell wall labyrinth or, alternatively, HCO_3^- transport into the cytoplasm could have raised the internal level of CO_2 above that of the bathing medium. In this situation, back diffusion of CO_2 would cause an acidification near the leaf surface.

This second possibility was investigated by examining the pH changes at the leaf surface in the "absence" of exogenous inorganic carbon (94). The results were most intriguing. Following illumination, the pH value at both surfaces remained rather stationary until, after approximately 20 min, the upper surface began a gradual alkalinization and the lower surface acidified. In the absence of exogenous buffer the pH at the lower surface approached values as low as pH 4.5. The changes at both surfaces were more or less in

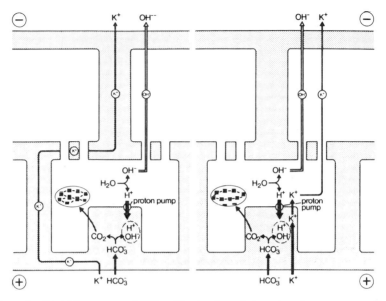

Figure 3   Model for exogenous HCO_3^- utilization through the action of a proton pump. Two possible pathways are given for K^+ transport: in the left half through the cell wall and in the right half via the symplast (from Ref. 94).

phase. After a further 30–40 min the pH values at both leaf surfaces returned toward the dark-adapted value, and then 2 hr later the entire cycle started again. When the leaf was kept in continuous light these oscillations in pH continued until the light was turned off (94). These data indicate that the acidification and alkalinization phenomena are coupled, but it is not possible to determine from the pH data alone whether the associated H^+ and OH^- fluxes are stoichiometric. It would be difficult to analyze these fluxes using diffusion analysis (see 59), but the vibrating probe technique (72) may be useful to further study this system.

At present the available data from *Potamogeton* and *Elodea* are consistent with the hypothesis as outlined in Figure 3. Whether other species have also adopted this strategy remains to be shown.

Characean Cells

It has long been known that *Nitella* and *Chara* develop regions of acidity and alkalinity along their internodal cells (74, 111, 118). The acid bands are easily detectable when the *Chara* cells are bathed in media low in total inorganic carbon (74). In analyzing the kinetics of $H^{14}CO_3^-$ fixation obtained on *Chara corallina,* Walker and co-workers stressed the influence that the unstirred layer near the cell wall would have on the supply of inorganic carbon to these cells (113, 126, 129). As a result of their evaluation of the data available for Characean cells, Walker and co-workers (129) put forward an hypothesis not unlike that of Prins et al (92–94). They suggested that the unstirred layer, in conjunction with the plasmalemma elaboration made possible by the charasome (see 30, 31, 70), would permit a H$^+$ efflux system to generate H$_2$CO$_3$ at a sufficient rate to account for the observed ^{14}C-fixation data (60). They stress that a specific HCO$_3^-$ transport system may not be necessary; the H$_2$CO$_3$ produced in the unstirred layer-cell wall-periplasmic space could diffuse across the plasmalemma, being driven by its concentration gradient.

Ferrier & Lucas (69) were also aware that during HCO$_3^-$ assimilation in *C. corallina*, the interalkaline regions appeared as sinks for OH$^-$. They suggested that these "sinks," detected using numerical analysis of diffusion profiles for OH$^-$ (H$^+$), could be caused by OH$^-$ influx, H$^+$ efflux, or CO$_2$ diffusion (leakage) from the cytoplasm following HCO$_3^-$ transport into the cell. Ferrier also raised the question of whether the action of a H$^+$ efflux pump could generate molecular CO$_2$ at a rate sufficient to explain photosynthesis in *Chara* (20). Figure 4 represents the basic details of the new

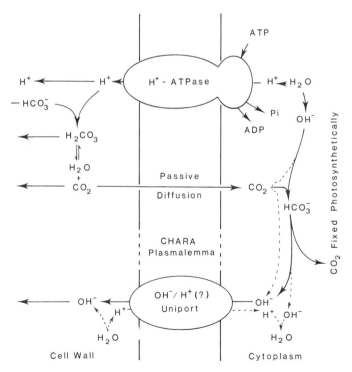

Figure 4 Model for extracellular CO$_2$-production in *Chara* (from Ref. 62; see text for details).

hypothesis by which *Chara* may acquire inorganic carbon for photosynthesis.

The theoretical models developed to examine the H^+ efflux-extracellular CO_2 production hypothesis predict that the pH at or near the cell surface should be at least as low as pH 6.0 in order to achieve the experimentally observed rates of ^{14}C-fixation. The conditions used in the models were those which Lucas used to achieve near maximum rates of fixation; i.e. 1.0 mM $NaHCO_3$, pH 8–9. An examination of the pH profiles obtained on *Chara* under these conditions indicates that the measured pH values in the HCO_3^--acquiring regions of the cell never approach this value of pH 6. In most of the profiles published by Lucas, the pH value in the interalkaline regions was above that of the background bathing medium, but below that which would be established by diffusion of OH^- along the cell from the neighboring alkaline bands (69).

Walker et al (129) explained that the discrepancy between the measured pH values and the predicted value of 6.0 might be attributed to the physical dimensions of the miniature pH electrodes used in these experiments; it was argued that the pH-sensitive area of these electrodes was as large as the unstirred layer itself. In a recent study on the *Chara* system, a comparison was made between the pH profiles obtained using miniature and recessed-tip pH microelectrodes (tip diameter approximately 0.5 μm). A dramatic difference was observed between the pH profiles measured by the two electrode systems; in the alkaline regions the profiles were quite similar, but in the HCO_3^--transporting regions, only the recessed-tip pH microelectrode detected acidification of the medium at the cell surface (71).

Although the pH dropped below that of the background medium (pH 8.2), it did not approach the theoretical values generated by the models. Even as close as 2 μm from the wall the pH was generally between 7.2 and 7.6. Thus, although Walker et al (129) were correct in their assessment that the miniature pH electrode was unable to measure accurately the pH within the unstirred layer, the values obtained using the appropriate technique did not provide supporting evidence for their hypothesis.

Further Tests of the CO_2-Production Hypothesis

The CO_2-production model proposed that the outward current measured in the acid regions (72) was due to (or generated by) electrogenic H^+ efflux. If this hypothesis were correct, one should find a correlation between the electrogenic fluxes (currents) and the level of cell surface acidification measured in these "acid" regions; similar correlations should exist from cell to cell. However, as mentioned previously, no such correlation has been observed, either with the external electric potential work (21) or with the more recent combined recessed-tip pH microelectrode-vibrating probe studies (71).

Since the charasome has been implicated in the acquisition of HCO_3^- in *Chara* (126), and transfer cells may be involved in *Elodea* and *Potamogeton* (94), it is possible that these structures mask the true events that are occurring within the vicinity of the wall. The spatial distribution of charasomes along the cell has been examined with respect to the position of the alkaline and HCO_3^--acquiring (acid) regions (30). Charasomes were always found in the "acid" regions, but were fewer in number or absent in the areas engaged in OH^- efflux. The observed distribution is consistent with their involvement in some way with HCO_3^- acquisition. However, since *Nitella flexilis* exhibited the same acid and alkaline banding pattern but did not possess the charasome structure along its plasmalemma, Franceschi & Lucas (30) concluded that the charasome must be involved in the transport of some other species, perhaps Cl^- (32).

Doubt has also been cast on the involvement in HCO_3^- acquisition of the transfer cell wall ingrowth in *Elodea* and *Potamogeton.* Prins et al (94) have recently reported that these "membrane foldings are not always present" in the leaves of species which are still capable of HCO_3^- utilization. They suggest that the wall ingrowth may not be essential for extracellular CO_2 production, but may increase the efficiency of the process. Further experiments should, therefore, be conducted on species which can assimilate exogenous HCO_3^- but which do not possess complex plasmalemma structures. In these species the predicted correlations between electrogenic currents and surface acidification may be found. (Experiments of this nature are presently being conducted in our laboratory.)

Walker et al (129) found that their theoretical model was relatively insensitive to the presence of the cell wall. Thus, almost all of their computations and predictions ignored the influence of the wall. It is possible that the wall may develop specific properties in response to localized membrane transport function. This view is supported by the recent report that the cell wall of *Anabaena variabilis* undergoes ultrastructural modification when cells are grown under low CO_2 culture conditions (78). There are other reports of changes in the cytoplasmic ultrastructure that occur during adaptation to HCO_3^- utilization (54). It would seem that further studies of this nature may provide valuable insight into the mechanism of cellular utilization of exogenous HCO_3^-.

UNSTIRRED LAYER MANIPULATION   Walker et al (129) suggested that extracellular production of CO_2 would be highly efficient "because the diffusion resistance of the unstirred layer retains the CO_2 that is produced near the cell surface, and the H^+ that flow out from the cell." An explicit prediction of their model is, therefore, that a reduction in the unstirred layer should cause a decline in the rate of photosynthesis. This prediction was tested by measuring ^{14}C-fixation in *Chara* as a function of the rate of solution flow (71). Again, contradictory results were obtained; photosynthesis was found to increase, not decrease, as the unstirred layer was reduced in thickness. Although this result appears to be incompatible with the extracellular CO_2-production hypothesis for *Chara*, it should be stressed that interpretation of these data is complex. Flowing the solutions past the cell surface would sweep away the OH^- ions that are released across the plasmalemma within the alkaline bands. This would contain the alkaline band-associated pH increase which would otherwise occur along the cell wall. A larger membrane surface area would therefore be available to engage in HCO_3^- acquisition, which may offset a reduction in efficiency due to the sweeping away of H^+ and CO_2 (71).

BUFFER EXPERIMENTS   A further prediction of the extracellular CO_2-production hypothesis is that the addition of buffer to the bathing medium should reduce the rate of photosynthesis. The reduction would occur because the buffer would compete with HCO_3^- for the transported H^+, thereby reducing the level of CO_2 available. It has been suggested that the well-documented inhibitory effect of high pH on ^{14}C-assimilation (60, 99, 101) could be caused by OH^- ions competing with HCO_3^-/CO_3^{2-} for transported H^+ (129).

In *C. corallina*, $H^{14}CO_3^-$ assimilation is inhibited in what appears to be a competitive manner by the presence of certain amino-type buffers (63). Lucas interpreted the inhibition as being caused by the formation of carbamate, which then competed with HCO_3^- at the binding site on the transport system. Walker and co-workers (129) interpreted the inhibition as resulting from a nonspecific buffer effect in which competition for H^+ was occurring. However, it should be stressed that other workers have reported inhibitory buffer effects (49, 94), and in at least one case there is clear evidence for competitive inhibition. Photosynthesis in *A. variabilis* was drastically reduced in the presence of 10mM 3-[cyclohexylamino]-1-propanesulfonic acid (CAPS); the V_{max} was unchanged but the K_m was raised by a factor of 100 (49). Since the pK_a of this buffer is 10.4, and these experiments were conducted in the presence of 50 mM Hepes, pH 8.0, this effect must be explained by the chemical nature of the buffer rather than via competition for H^+.

In a recent study on *Chara*, the effect of artificial buffer on the pH and extracellular current profiles and $H^{14}CO_3^-$ fixation was investigated (71). In the presence of 5 mM phosphate buffer (or Hepes), the pH profiles were significantly reduced, with the acid regions being more affected than the alkaline bands. This level of buffer almost completely dissipated the radial pH gradients within the "acid" regions of the cell surface. Raising the pH value at the cell surface in this way would reduce the CO_2 concentration and, according to the CO_2-production hypothesis, photosynthesis should therefore be dramatically reduced. Photosynthesis was reduced, but only by about 30% (71). When the extracellular currents were measured under the same conditions, they were found to be completely unaffected by the presence of the buffer. (This result will be discussed in the next section.)

Collectively, these results indicate that in *Chara* the extracellular production of CO_2 is unlikely to account for the ability of these cells to assimilate exogenous HCO_3^-.

HCO_3^--H^+ COTRANSPORT OR HCO_3^- UNIPORT?

Plasmalemma transport of HCO_3^- via either electrogenic uniport or a HCO_3^--H^+ contransport mode could also result in acidification of the external medium near the *Chara* cell wall. These systems were modeled by Walker et al (129), and reasonable agreement has been found between their predicted radial pH profile for the HCO_3^--H^+ cotransport model and the experimental results obtained using the recessed-tip pH microelectrode system (71). However, since there may well be insufficient resolution in the theoretical model to distinguish between these two transport modes, the observed correlation should be viewed only as offering support for the HCO_3^--H^+ cotransport hypothesis and not as diagnostic proof.

At pH 8–9, a HCO_3^- uniport system should not be influenced by the presence of exogenous buffer. The partial sensitivity to phosphate or Hepes buffer that was observed in *C. corallina* (71) is consistent, therefore, with the operation of a HCO_3^--H^+ cotransport system in this species. Although data on the extent of inhibition were not presented, Prins et al (94) reported that photosynthesis in *E. densa* was affected by artificial buffers (phosphate, Taps, and Tris). This effect was claimed to support the hypothesis that an extracellular CO_2-generating mechanism operates in this species; however, it is also consistent with transport via HCO_3^--H^+ cotransport.

If we follow this line of reasoning for the *A. variabilis* data, the apparent insensitivity to 50 mM Hepes would offer support for the hypothesis that inorganic carbon in *Anabaena* is accumulated via a primary electrogenic HCO_3^- transport system. Further support for this conclusion may be obtained by applying CAPS buffer to cells which have been preincubated in TPP⁺ (pH 8) but starved of inorganic carbon. Application of HCO_3^- should cause one of two possible responses: a hyperpolarization of the membrane potential, which would be consistent with a HCO_3^--H^+ system, or little change in the potential, which would indicate the operation of a HCO_3^- uniport that has been competitively blocked by CAPS.

Electrophysiological Evidence

In a recent study, the effects of light/dark and HCO_3^- were reexamined on the *Chara* system (65). It was found that when an internodal cell was assimilating exogenous HCO_3^- (pH 8.2), the membrane potential was extremely sensitive to changes in the level of exogenous HCO_3^-. Removal of HCO_3^- resulted in a rapid 50 to 70 mV depolarization of the potential; returning HCO_3^- to the bathing medium elicited an equivalent hyperpolarization. Since the pH of the bathing medium was constant during these experiments, the results invalidate Spanswick's (114) claim that HCO_3^- does not have a direct effect on the electrical properties of the Characean plasmalemma.

Using light/dark transitions as a probe, Lucas (65) found that in the presence of exogenous HCO_3^-, the potential rapidly hyperpolarized when the *Chara* cell was given a brief dark treatment. In the light the resting (stable) membrane potential was approximately −240 mV, whereas after the cell had been in the dark for 3–6 min, the potential was −330 to −350 mV. Upon reilluminating the cell, the potential further hyperpolarized transiently (−375 mV), and then rapidly depolarized back toward the light-adapted value. Analysis of electrophysiological and vibrating probe experiments conducted on *Chara* cells indicated an excellent level of correlation between the light- and dark-induced changes in extracellular currents and membrane potential. The presence of a plasmalemma-bound electrogenic HCO_3^- influx system in association with an electrogenic OH^- efflux process could explain the observed electrical responses.

In the presence of phosphate buffer (1 mM) and the absence of HCO_3^-, however, the membrane potential still hyperpolarized when cells were given a brief dark pulse. In addition, the extracellular currents that are associated with HCO_3^- assimilation in *C. corallina* (71, 72, 128) could also be maintained in the absence of HCO_3^-, provided the bathing medium contained 1.0 mM phosphate buffer. These results demonstrate conclusively that the hyperpolarizing response cannot be explained on the basis of a HCO_3^- uniport system. The simplest explanation is that the hyperpolarizing response is caused by an electrogenic H^+-translocating ATPase. Experimental evidence outlined in the above sections indicates that the extracellular CO_2-production model is almost certainly invalid *for this species*, and thus it would appear that HCO_3^- is transported via a HCO_3^--H^+ cotransport system. Lucas (65) suggested that this system may be electrically neutral, but the possibility of a HCO_3^--$2H^+$ cotransport mechanism cannot be discounted. If such a system exists, the transport of HCO_3^- would be electrogenic, but in this situation the membrane might respond by depolarizing. The actual change in membrane potential upon activating the HCO_3^- assimilating system would depend upon the relative time-dependent change in conductance of each of the transport elements involved—namely, H^+ efflux, $HCO_3^- $-$2H^+$ cotransport, and OH^- efflux.

The finding that the extracellular currents in the HCO_3^- transporting region of the cell membrane can be maintained in the absence of exogenous HCO_3^- is of paramount importance. It indicates that in *Chara* the loop between HCO_3^- assimilation and OH^- efflux can be opened. A summary of this system is now possible and is presented in model form in Figure 5. Under normal HCO_3^--assimilating conditions, the H^+-translocating ATPase of the *Chara* plasmalemma provides a supply of protons to the outer surface of this membrane. These protons are used to drive the $HCO_3^- $-$H^+$ cotransport system; the reentering protons are again cycled through the ATPase, thereby forming a H^+ current loop. Some of the H^+ will also react with HCO_3^- to produce H_2CO_3 and thereby CO_2, which will diffuse both into the cytoplasm and the bathing medium. Photosynthetic

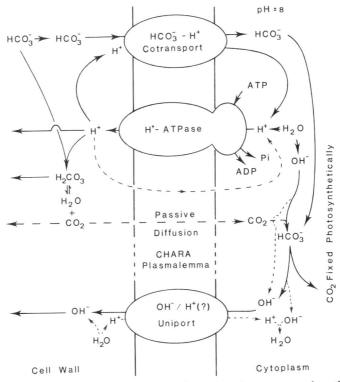

Figure 5   Schematic representation of the plasmalemma-bound transport complexes that may enable internodal cells of *Chara* to acquire exogenous HCO_3^- for photosynthesis (from Ref. 62; see text for details).

CO_2 fixation produces OH^- by causing HCO_3^- to dissociate and the OH^--transporting regions of the plasmalemma serve to dispose of this alkalinity.

Substitution of phosphate (HPO_4^{2-}) for HCO_3^- (65, 71) provided an apoplastic buffer that could penetrate to the plasmalemma, be converted to $H_2PO_4^-$ (pK_a 7.2), and then diffuse back into the bathing medium where it would dissociate to HPO_4^{2-}. Thus, phosphate (and Hepes) can provide a H^+-shuttle which is sensed by the vibrating probe as an electrogenic current. Earlier theoretical treatments have addressed this role for exogenously applied buffers (19), but the implications have been largely ignored. Clearly, since many recent studies have indicated a central role for protons in energizing substrate cotransport systems (9, 33, 53, 77, 105), the standard approach of buffering a tissue to maintain a desired pH value will have to be reassessed. It may even be necessary to develop a new series of buffers which are larger, or at least less mobile in nature, so that their ability to transfer H^+ away from the site of transport would be reduced, yet the ability to buffer the system would be maintained.

Transport Stoichiometry of the H^+-ATPase

The thermodynamics of the *Chara* H^+-translocating ATPase have been discussed by various workers (115, 127). Walker & Smith (127) showed that, based on the operation of a $2H^+$-translocating ATPase (i.e. $2H^+$ per ATP hydrolyzed), the maximum membrane potential that could be established in *Chara* was −268 mV (pH 8). However, Lucas has now measured membrane potentials in the vicinity of −350 mV; correcting for the positive tonoplast potential would raise this value to a maximum of approximately −390 mV. Lucas (65) concluded, therefore, that the H^+-translocating ATPase that operates during HCO_3^- assimilation must be a $1H^+$ per ATP hydrolyzed. This system would be far from thermodynamic equilibrium and must therefore be rate-controlled by other parameters.

The question now to be addressed is whether the plasmalemma of *Chara*, and other plants, has a single ATPase that can change (switch) its H^+ translocating stoichiometry under different physiological conditions, or whether there are at least two different ATPases having $2H^+:1ATP$ and $1H^+:1ATP$ transport stoichiometries. This will form an extremely valuable avenue for future research in terms of providing basic information on the control processes which regulate membrane transport. The technique presently being pioneered by Hansen and co-workers (36, 37) may be particularly valuable in helping to elucidate these details.

ELECTROGENIC OH^- TRANSPORT

The main OH^- transport systems that have been studied with respect to HCO_3^- assimilation are those of *Chara corallina, Potamogeton lucens,*

Elodea canadensis, and Anabaena variabilis. Although other systems have been examined, the main features can be demonstrated by drawing from the information available for the above species. Obviously, if HCO_3^- transport occurs via either a uniport, in cotransport with H^+, or by extracellular CO_2 production, alkalinity will be produced within the cytoplasm and OH^- ions will need to be excreted both to maintain cytoplasmic pH and electrical balance across the plasmalemma. The area of cytoplasmic pH regulation has been reviewed recently (112).

Spatial and Temporal Regulation

The reason for spatially separating the OH^- efflux region from that of HCO_3^- uptake is all too apparent for the systems that appear to utilize H^+ efflux to energize the transport of HCO_3^-. From the macroalgae and the aquatic angiosperms it can be seen that spatial separation was achieved by regulating the location of transport processes either along the plasmalemma of each single cell, as in *C. corallina* (62, 72), or on the plasmalemma of cells forming the opposite surface of the leaf, as in *P. lucens* and *E. canadensis* (38, 41, 42, 92–94). Whether the microalgae and Cyanobacteria follow the strategy adopted in the Characean cells remains to be elucidated, but if HCO_3^- enters via a primary, active uniport mechanism, the necessity for separation would not be as important. However, OH^- efflux produces CO_3^{2-} near the plasmalemma, and if this form of inorganic carbon competitively inhibits the uptake of HCO_3^-, then spatial separation of HCO_3^- uptake and CO_3^{2-} production may be of value. Considering the size of the Cyanobacteria, a spatial separation of these sites may have been of little value. In this regard, Kaplan (49) considers that in *A. variabilis*, the inhibition of photosynthesis at high pH is attributable to a rise in cytoplasmic pH rather than an effect of carbonate.

It should also be stressed that not all HCO_3^- assimilators have separated the HCO_3^- and OH^- transport sites; for example, *Vallisneria* appears to transport both ions across the upper and lower leaf surface (93). Whether different cells are involved in these processes is not yet known.

The question of what controls the spatial positioning of either transport system is intriguing. In the aquatic angiosperms it may be a matter of cellular development in which tissue (cell) specialization is expressed. However, it should be recalled that recent results indicate that the upper leaf surface of *P. lucens* may be induced to accumulate exogenous HCO_3^- (41). The situation in the Characeae appears to be more complex. Lucas & Dainty (67) have investigated the question of whether the discrete location of the OH^- bands was caused by a restricted distribution of the OH^- transport entities along the *Chara* plasmalemma. When cytoplasmic streaming was reduced (using cytochalasin B) below a threshold value, the normal pattern of OH^- efflux disintegrated, and numerous small discs of

OH⁻ efflux activity appeared to operate over the entire cell surface. The rate of collapse and subsequent establishment of this new mode of transport was too rapid to have been caused by lateral migration of membrane proteins. This process was reversible; removal of the cytochalasin B caused a restoration of cyclosis and subsequently the control OH⁻ efflux pattern was reestablished.

It would seem, therefore, that the OH⁻ transporters are distributed over the entire surface of the *Chara* plasmalemma, but that a regulating mechanism operates to control which of the transporters actually function. This concept of spatial control of OH⁻ efflux is also supported by the transport profiles computed using numerical analysis of the OH⁻ diffusion profiles generated by these alkaline bands (69). For those who are skeptical of the numerical analysis technique, further conformation of the OH⁻ transport profiles along the plasmalemma surface was obtained using the vibrating probe technique (72). In the central region of the alkaline band, OH⁻ efflux is uniform (in a time-averaged sense), and the width of this region depends upon the HCO_3^- concentration present, the light intensity, and the status of the band (see below). At the margins of this central region the rate of OH⁻ transport declines to zero quite rapidly. This is the region, along the cell surface, where the transition occurs from OH⁻ efflux to HCO_3^- influx. Under constant conditions, where photosynthesis is stable over many hours, the position of these transition zones is extremely stable.

In experiments on tonoplast-free *C. corallina* cells, OH⁻ efflux did not occur in its normal pattern (73). Preparation of these cells resulted in the destruction of certain organelles within the streaming cytoplasm. Lucas & Shimmen (73) demonstrated that OH⁻ efflux patterns equivalent to those obtained with perfused cells were established on the centripetal segment of centrifuged cells. In these centrifugation experiments very low gravitational forces (X180) were used, and only heavy organelles like the nuclei were displaced to the centrifugal end. Ligation of the cell at its midpoint prevented these organelles from being redistributed throughout the length of the cell. The cell segment at the centrifugal end developed normal alkaline bands. Upon redistribution of the larger cytoplasmic organelles, normal pH profiles (bands) were established along the entire cell length. These results are consistent with the hypothesis that an organelle within the streaming phase of the cytoplasm is ultimately responsible for the spatial location and control over OH⁻ transport. The identity of this putative organelle and the nature of the control signal(s) is an area worthy of future research. On a speculative note, the nucleus has been implicated and a protein has been suggested as being part of the control signal (73).

In terms of temporal regulation, *Chara, Potamogeton, Elodea,* and *Anabaena* appear to have evolved certain common attributes. Following tissue illumination, there is always a lag prior to activation of the OH$^-$ efflux process (49, 61, 93). In *Chara,* a precise relationship was found between light intensity and the lag period. This suggested that it is necessary to establish a threshold condition within the cytoplasm before OH$^-$ transport can be activated. A preliminary model to explain this phenomenon has been presented in which the threshold condition is dealt with in terms of changes in cytoplasmic pH (61).

The steady state pH value at the upper leaf surface of *P. lucens* declines as lower light intensities are employed. This suggests a simple relationship between photosynthesis, production of alkalinity, and competition for OH$^-$ by the transporters at the upper leaf surface. Such a system was not found in *Chara;* here a complex hierarchy of OH$^-$ transport exists (61). The OH$^-$ efflux bands which develop along a particular cell are not equivalent in status. The terms *primary, sub-primary,* and *subsidiary* bands were established by Lucas (61) to explain his findings that as the light intensity was reduced, the number of alkaline bands diminished, with the remaining bands being somewhat unaffected in their rates of OH$^-$ transport. At very low light intensities, only one OH$^-$ band would become activated, and this was given the status of the *primary* band. Although this hierarchical status has been clearly established at the phenomenological level, its explanation at the molecular level awaits further work.

Further insight into the complexity of the control or regulatory processes operating on the OH$^-$ transport system of *Chara* was obtained in studies using the vibrating probe technique. In their initial study, Lucas & Nuccitelli (72) found that the presence of the vibrating probe caused a perturbation of the OH$^-$ transport system. The perturbation was usually observed as a reduction in the number of operational OH$^-$ efflux bands, with an approximate doubling of the transport rate through the one or two remaining bands. This suggested that cytoplasmic control over OH$^-$ efflux at each of the functional bands modulates the transport rates at values well below their potential maxima. Whether this form of rate control is a part of the hierarchial system in these cells must await further investigation.

Another feature concerning the temporal relationship between HCO$_3^-$ and OH$^-$ transport was also discovered in experiments using the vibrating probe. The extracellular currents in the HCO$_3^-$ transport regions were very steady with time, and synchronized with respect to the cellular response to cell darkening or illumination (65). This was not true for the extracellular currents in the alkaline bands; here the currents often exhibited quite marked, but low-frequency, fluctuations. Fluctuations of a similar form were observed during external electric potential measurements (21). A low-frequency noise component has also been reported from experiments on the *Chara* membrane potential (22, 97, 98). Ogata & Kishimoto (84) have reported small oscillations in the membrane potential of *N. pulchella,* and the period of these oscillations was highly correlated with the rate of cytoplasmic streaming. Since a close relationship between the presence of a stable OH^- efflux pattern and the existence of a critical rate of streaming has been found (67), it is conceivable that these phenomena reflect the operation of a feedback process which controls HCO_3^- and OH^- transport synchronization (see 36).

Active or Passive Process?

Extensive research has shown that OH^- ions accumulate at the cell surface during photosynthetic assimilation of HCO_3^-, but it should be realized that external alkalinity could be generated by at least two different mechanisms, i.e. OH^- efflux or H^+ influx. This issue has been addressed by several workers, but little agreement has been achieved (12, 13, 64, 68). Lucas and co-workers have argued from experimental and theoretical grounds that in *Chara* the rate of H^+ production from the dissociation of H_2O within the cell wall is much too slow to account for the measured rates of OH^- efflux. However, they pointed out that if H_2O was the initial species involved in binding to the transport entity, then the theoretical computations would no longer hold (see also 109). Based on the response of the *Chara* membrane potential to high pH (9.0 to 12.0), it has been proposed that the membrane changes its state under these conditions (pH > 10.5), becoming very permeable, in a passive manner, to H^+ (12). Bisson & Walker (12) point out that their results could be discussed equally well in terms of a pH-induced change in OH^- permeability. Miller & Colman (80) have also concluded that for *Coccohloris* the relatively higher concentration of OH^- (in the cytoplasm) compared with H^+ (in the cell wall) makes the OH^- the more likely candidate responsible for charge and pH balance.

Since the exact nature of the transported species remains somewhat equivocal, both possibilities were included in Figures 4 and 5. The following section will focus on the question of passive OH^- efflux; the arguments could equally well have been applied to passive H^+ influx. To establish that OH^- efflux is *not* a passive process, it is necessary to show that OH^- ions move against their electrochemical potential gradient. This requires the measurement of both the activity and electric gradients that are "sensed" by these ions. In many situations these parameters have not been accurately determined, and so definitive conclusions cannot be drawn.

In the Cyanobacteria, the cytoplasmic pH has been measured, but under alkaline conditions the values must be considered to be only approximate. Also, only few measurements of the membrane potential are available for species which can assimilate exogenous HCO_3^-. In any event, it is possible to use a general approach to examine the nature of OH^- efflux in these cells.

Kaplan (49) and Miller & Colman (80) have shown that *Anabaena* and *Coccochloris* can assimilate exogenous HCO$_3^-$ at external pH values in excess of 10.0. Under these conditions the internal pH of these cells may be close to 8.0. Thus, a membrane potential of at least –120 mV would be needed to drive the passive efflux of OH$^-$. Potentials of approximately –100 mV were recently reported (51), and in view of the errors involved in making these measurements, it may well be that in the Cyanobacteria OH$^-$ efflux is passive.

In *Potamogeton*, pH values as high as 11.8 have been observed on the upper leaf surface (42); however, values of cytoplasmic pH under these conditions are not available. If we assume a value of 8.0, a membrane potential of approximately –240 mV would be required to drive passive OH$^-$ efflux. We are unaware of such negative values being measured across the upper leaf cells of this plant, but such values have been recorded in leaves of *Elodea* (116). In the absence of experimental data to the contrary, it would appear that OH$^-$ efflux in *Potamogeton* and *Elodea* is also passive.

For Characean cells, the membrane potential, cell surface pH, and cytoplasmic pH have been measured over a wide range of experimental conditions. Under HCO$_3^-$ assimilating conditions the central region of the alkaline band can reach a pH of 10.5, with the value at the plasmalemma being higher. The cytoplasmic pH value under these conditions would be about 8 (112, 127). For passive OH$^-$ efflux a membrane potential of –150 to –180 mV would be required; such values are now obtained routinely (65). This value is of interest with respect to recent work, where it was found that removal of exogenous HCO$_3^-$ caused the *Chara* membrane potential to depolarize quite dramatically. It was noticed that the potential never went below –175 mV (65), which may indicate passive OH$^-$ efflux. However, this conclusion should be viewed with reservation, since it has been shown that brief treatment of *Chara* cells with *N*-ethyl maleimide (0.1 mM, 1 min) caused the membrane potential to depolarize to approximately –125 mV (56), yet the alkaline bands were only marginally affected (66), the pH at the cell surface being about 10.0.

Since the plasmalemma of *Chara* is extremely heterogenous with respect to the spatial distribution of various transport processes, etc, it would seem unwise to apply a simple thermodynamic approach to these cells. Bisson & Walker (12, 13) have clearly demonstrated the complexity of this situation. However, at present it would appear that OH$^-$ efflux in Characean cells, as well as in other HCO$_3^-$-assimilating species, occurs via a passive mechanism. To change this conclusion, it will be necessary to find an experimental condition which will reduce the membrane potential without influencing OH$^-$ efflux.

BENEFITS OF BICARBONATE TRANSPORT

Any aquatic species that can fully exploit its environment with respect to exogenous inorganic carbon acquisition has the potential to establish photosynthetic dominance. A species which can utilize both exogenous CO_2 and HCO_3^- may also have available a wider range of ecological habitats. Certainly for some of the microalgae and Cyanobacteria, in which the internal levels of inorganic carbon can be raised by as much as 1000 over that of the medium, the advantage must be considerable. An outcome of inorganic carbon utilization is an alkalinization of the medium, and this must also aid in conferring species dominance to HCO_3^- assimilators; this would be especially so in bodies of water which undergo rather slow mixing. However, since metabolic energy must be expended to achieve this level of carbon accumulation, there is an additional cost associated with this aspect of photosynthetic carbon production. A vigorous analysis of the cost : benefit ratio should now be possible on some species, because many of the details relating to transport mechanisms appear to have been resolved.

In recent years considerable attention has been focused on the question of whether all aquatic plants utilize the C_3 pathway of photosynthesis (17, 18, 44, 90). One of the main issues is whether algae undergo photorespiration, since many species exhibit the C_4 characteristic of extremely low photosynthetic compensation points (see 11). Considerable controversy developed over this issue (see 106–108), and the literature pertaining to the subject is quite extensive and warrants a review in its own right. It is now generally believed that algae do possess the photorespiration pathway (122). This being the case, there is a further advantage for a species which has an effective CO_2-concentrating mechanism, because this mechanism can serve to reduce the effect of O_2 on photosynthesis in an analogous way to that seen in C_4 plants. At present there is little information available on whether the macroalgae (like *Chara*) and aquatic angiosperms (like *Potamogeton* and *Elodea*) use a carbon-concentrating mechanism to modulate the influence of photorespiration.

CONCLUDING REMARKS

Over the past few years considerable progress has been made toward elucidating the mechanism(s) by which aquatic plants acquire exogenous HCO_3^- for photosynthetic assimilation. Most species studied fall into the category of electrogenic HCO_3^- transport or $HCO_3^-H^+$ cotransporting systems, with a few examples of extracellular CO_2-production and HCO_3^-/OH^- antiport. Thus, it appears that various species may have evolved different mechanism by which they energize the transport of HCO_3^- across the plasmalemma. Many of the aquatic angiosperm, algal and cyanobacterial systems which have been established as "HCO_3^- assimilators" have not been investigated to ascertain the specific effect of buffers on photosynthesis. This will form an important area for future research, the result of which may test the hypothesis that all microalgae and cyanobacteria possess a similar transport mechanism (8). Similarly, the effect of CAPS buffer may also prove to be informative with respect to identifying "classes" of HCO_3^- transport-binding sites.

Finally, although the phenomenon of spatially separate transport sites for HCO_3^- and OH⁻ has now been well documented, little is known about the biophysical and biochemical mechanisms utilized by the tissue to maintain electrical and pH balance. Parallel studies on intact and isolated membrane systems may enable the resolution of some of these details.

ACKNOWLEDGMENTS

We thank the National Science Foundation for its continuing support of research in our laboratory through Grants PCM 78-10474, PCM 80-03133, and PCM 81-17721. Our special thanks go to Roswitha Burchardt for technical assistance and Barbara Shaneyfelt for typing the manuscript.

Literature Cited

1. Andrianov, V. K., Bulychev, A. A., Kurella, G. A. 1970. Connection between photo-induced changes in the resting potential and the presence of the chloroplasts in *Nitella* cells. *Biophysics* 15:199–200
2. Andrianov, V. K., Kurella, G. A., Litvin, F. F. 1968. Influence of light on the electrical activity of *Nitella* cells. In *Transport and Distribution of Matter in Cells of Higher Plants*, ed. K. Mothes, E. Müller, A. Nelles, D. Neuman, pp. 187–96. Berlin: Akademie-Verlag
3. Arens, K. 1933. Physiologisch polarisierter Massenaustausch und Photosynthese bei submersen Wasserpflanzen I. *Planta* 20:621–58
4. Arens, K. 1936. Physiologisch polarisierter Massenaustausch und Photosynthese bei submersen Wasserpflanzen. II. Die Ca(HCO₃)₂-assimilation. *Jahrb. Wiss. Bot.* 83:513–60
5. Arens, K. 1938. Manganablagerungen bei Wasserpflanzen als Folge des physiologisch polarisierten Massenaustausches. *Protoplasma* 30:104–29
6. Badger, M. R., Kaplan, A., Berry, J. A. 1977. The internal CO₂ pool of *Chlamydomonas reinhardtii:* response to external CO₂. *Carnegie Inst. Washington Yearb.* 76:362–66
7. Badger, M. R., Kaplan, A., Berry, J. A. 1978. A mechanism for concentrating CO₂ in *Chlamydomonas reinhardtii* and *Anabaena variabilis* and its role in photosynthetic CO₂ fixation. *Carnegie Inst. Washington Yearb.* 77:251–61
8. Badger, M. R., Kaplan, A., Berry, J. A. 1980. Internal inorganic carbon pool of *Chlamydomonas reinhardtii:* Evidence for a carbon dioxide-concentrating mechanism. *Plant Physiol.* 66:407–13
9. Baker, D. A. 1978. Proton co-transport of organic solutes by plant cells. *New Phytol.* 81:485–97
10. Beardall, J., Raven, J. A. 1981. Transport of inorganic carbon and the 'CO₂ concentrating mechanism' in *Chlorella emersonii* (Chlorophyceae). *J. Phycol.* 17:134–41
11. Birmingham, B. C., Colman, B. 1979. Measurement of carbon dioxide compensation points of freshwater algae. *Plant Physiol.* 64:892–95
12. Bisson, M. A., Walker, N. A. 1980. The *Chara* plasmalemma at high pH. Electrical measurements show rapid specific passive uniport of H⁺ or OH⁻. *J. Membr. Biol.* 56:1–7

100   LUCAS

13. Bisson, M. A., Walker, N. A. 1982. Control of passive permeability in the *Chara* plasmalemma. *J. Exp. Bot.* 33:520–32

14. Browse, J. A., Dromgoole, F. I., Brown, J. M. A. 1979. Photosynthesis in the aquatic macrophyte *Egeria densa*. III. Gas exchange studies. *Aust. J. Plant Physiol.* 6:499–512

15. Cooper, T. G., Filmer, D., Wishnick, M., Lane, M. D. 1969. *J. Biol. Chem.* 244:1081–83

16. Denny, P., Weeks, D. C. 1970. Effects of light and bicarbonate on membrane potential in *Potamogeton schweinfurthii* (Benn.). *Ann. Bot.* 34:483–96

17. Döhler, G. 1974. C_4-pathway of photosynthesis in the blue-green alga *Anacystis nidulans*. *Planta* 118:259–69

18. Döhler, G. 1974. Effect of growth conditions on the photosynthetic pathway of *Chlorella vulgaris*. *Z. Pflanzenphysiol.* 71:144–53

19. Engasser, J., Horvath, C. 1974. Dynamic role of buffers: Facilitated transport of protons, weak acids and bases. *Physiol. Chem. Phys.* 6:541–43

20. Ferrier, J. M. 1980. Apparent bicarbonate uptake and possible plasmalemma proton efflux in *Chara corallina*. *Plant Physiol.* 66:1198–99

21. Ferrier, J. M., Lucas, W. J. 1979. Plasmalemma transport of OH^- in *Chara corallina*. II. Further analysis of the diffusion system associated with OH^- efflux. *J. Exp. Bot.* 30:705–18

22. Ferrier, J. M., Morvan, C., Lucas, W. J., Dainty, J. 1979. Plasmalemma voltage noise in *Chara corallina*. *Plant Physiol.* 63:709–14

23. Findenegg, G. R. 1974. Relations between carbonic anhydrase activity and uptake of HCO_3^- and CL^- in photosynthesis by *Scenedesmus obliquus*. *Planta* 116:123–31

24. Findenegg, G. R. 1974. Carbonic anhydrase and the driving force of light-dependent uptake Cl^- and HCO_3^- by *Scenedesmus*. In *Membrane Transport in Plants*, ed. U. Zimmerman, J. Dainty, pp. 192–96. Berlin/Heidelberg/New York: Springer, 473 pp.

25. Findenegg, G. R. 1976. Correlation between accessibility of carbonic anhydrase for external substrate and regulation of photosynthetic use of CO^2 and HCO_3^- by *Scenedesmus obliquus*. *Z. Pflanzenphysiol.* 79:428–37

26. Findenegg, G. R. 1976. Zur Frage eines HCO_3^-/OH^--Austausches im Plasmalemma von *Scenedesmus obliquus*. *Ber. Dtsch. Bot. Ges.* 89:277-84

27. Findenegg, G. R. 1977. Estimation of bicarbonate fluxes in *Scenedesmus obliquus*. In *Ionic Exchange in Plants*, eds. M. Thellier, A. Monnier, M. Demanty, J. Dainty, pp. 275–81. Paris: Univ. Rouen; CNRS Publ. No. 258

28. Findenegg, G. R. 1979. Inorganic carbon transport in microalgae. I: Location of carbonic anhydrase and HCO_3^- / OH^- exchange. *Plant Sci. Lett.* 17:101–8

29. Findenegg, G. R. 1980. Inorganic carbon transport in microalgae. II: Uptake of HCO_3^- ions during photosynthesis of five microalgal species. *Plant Sci. Lett.* 18:289–97

30. Franceschi, V. R., Lucas, W. J. 1980. Structure and possible function(s) of charasomes; complex plasmalemma-cell wall elaborations present in some Characean species. *Protoplasma* 104:253–71

31. Franceschi, V. R., Lucas, W. J. 1981. The charasome periplasmic space. *Protoplasma* 107:269–84

32. Franceschi, V. R., Lucas, W. J. 1982. The relationship of the charasome to chloride uptake in *Chara corallina*: Physiological and histochemical investigations. *Planta* 154:525–37

33. Giaquinta, R. T. 1980. Mechanism and control of phloem loading of sucrose. *Ber. Dtsch. Bot. Ges.* 93:187–201

34. Goldman, J. C., Oswald, W. J., Jenkins, D. 1974. The kinetics of inorganic carbon limited algal growth. *J. Water Pollut. Control Fed.* 46:554–74

35. Goldman, J. C., Porcella, D. B., Middlebrooks, E. J., Toerien, D. F. 1972. The effects of carbon on algal growth— its relationship to eutrophication. *Water Res.* 6:637–79

36. Hansen, U-P. 1978. Do light-induced changes in the membrane potential of *Nitella* reflect the feed-back regulation of a cytoplasmic parameter? *J. Membr. Biol.* 41:197–24

37. Hansen, U-P., Keunecke, P. 1977. The parallel pathways of the action of light on membrane potential in *Nitella*. See Ref. 27, pp. 333–40

38. Helder, R. J. 1975. Polar potassium transport and electrical potential difference across the leaf of *Potamogeton lucens* L. *Proc. Kon. Ned. Akad. Wet. Ser. C* 78:189–97

39. Helder, R. J., Boerma, J. 1972. Polar transport of labelled rubidium ions across the leaf of *Potamogeton lucens*. *Acta Bot. Neerl.* 21:211–28

40. Helder, R. J., Boerma, J. 1973. Exchange and polar transport of rubidium

ions across the leaves of *Potamogeton lucens. Acta Bot. Neerl.* 22:686–93

41. Helder, R. J., Boerma, J., Zanstra, P. E. 1980. Uptake pattern of carbon dioxide and bicarbonate by leaves of *Potamogeton lucens* L. *Proc. Kon. Ned. Akad. Wet. Ser. C* 83:151–66

42. Helder, R. J., Zanstra, P. E. 1977. Changes of the pH at the upper and lower surface of bicarbonate assimilating leaves of *Potamogeton lucens* L. *Proc. Kon. Ned. Akad. Wet. Ser. C* 80:421–36

43. Higinbotham, N. 1973. Electropotentials of plant cells. *Ann. Rev. Plant Physiol.* 24:25–46

44. Hogetsu, D., Miyachi, S. 1979. Operation of the reductive pentose phosphate cycle during the induction period of photosynthesis in *Chlorella. Plant Cell Physiol.* 20:1427–32

45. Hope, A. B. 1965. Ionic relations of cells of *Chara australis. Aust. J. Biol. Sci.* 18:789–801

46. Jaffe, L., Nuccitelli, R. 1974. An ultrasensitive vibrating probe for measuring extracellular currents. *J. Cell Biol.* 63:614–28

47. Kadono, Y. 1980. Photosynthetic carbon sources in some *Potamogeton* species. *Bot. Mag.* 93:185–94

48. Kaplan, A. 1981. Photoinhibition in *Spirulina platensis:* response of photosynthesis and HCO$_3^-$ uptake capability to CO$_2$-depleted conditions. *J. Exp. Bot.* 32:669–77

49. Kaplan, A. 1981. Photosynthetic response to alkaline pH in *Anabaena variabilis. Plant Physiol.* 67:201–4

50. Kaplan, A., Badger, M. R., Berry, J. A. 1980. Photosynthesis and the intracellular inorganic carbon pool in the blue-green alga *Anabaena variabilis:* Response to external CO$_2$ concentration. *Planta* 149:219–26

51. Kaplan, A., Zenvirth, D., Reinhold, L., Berry, J. A., 1982. Involvement of a primary electrogenic pump in the mechanism for HCO$_3^-$ uptake by the cyanobacterium *Anabaena variabilis. Plant Physiol.* 69:978–82

52. Kitasato, J. 1968. The influence of H$^+$ on the membrane potential and ion fluxes of *Nitella. J. Gen. Physiol.* 52:60–87

53. Komor, E., Tanner, W. 1974. The hexose-proton cotransport system in *Chlorella:* pH-dependent change in K_m values and translocation constants of the uptake system. *J. Gen. Physiol.* 64:568–81

54. Kramer, D., Findenegg, G. R. 1978. Variations in the ultrastructure of *Scenedesmus obliquus* during adaptation to low CO$_2$ level. *Z. Pflanzenphysiol.* 89:407–10

55. Lehman, J. T. 1978. Enhanced transport of inorganic carbon into algal cells and its implications for the biological fixation of carbon. *J. Phycol.* 14:33–42

56. Lichtner, F. T., Lucas, W. J., Spanswick, R. M. 1981. Effect of sulfhydryl reagents on the biophysical properties of the plasmalemma of *Chara corallina. Plant Physiol.* 68:899–904

57. Lowenhaupt, B. 1956. The transport of calcium and other cations in submerged aquatic plants. *Biol. Rev.* 31:371–95

58. Lowenhaupt, B. 1958. Active cation transport in submerged aquatic plants. I. Effect of light upon the absorption and excretion of calcium by *Potamogeton crispus* (L.) leaves. *J. Cell Comp. Physiol.* 51:199–208

59. Lucas, W. J. 1975. Analysis of the diffusion symmetry developed by the alkaline and acid bands which form at the surface of *Chara corallina* cells. *J. Exp. Bot.* 26:271–86

60. Lucas, W. J. 1975. Photosynthetic fixation of ^{14}carbon by internodal cells of *Chara corallina. J. Exp. Bot.* 26:331–46

61. Lucas, W. J. 1975. The influce of light intensity on the activation and operation of the hydroxyl efflux system of *Chara corallina. J. Exp. Bot.* 26:347–60

62. Lucas, W. J. 1976. Plasmalemma transport of HCO$_3^-$ and OH$^-$ in *Chara corallina:* non-antiporter systems. *J. Exp. Bot.* 27:19–31

63. Lucas, W. J. 1977. Analogue inhibition of the active HCO$_3^-$ transport site in the characean plasma membrane. *J. Exp. Bot.* 28:1321–36

64. Lucas, W. J. 1979. Alkaline band formation in *Chara corallina:* Due to OH$^-$ efflux or H$^+$ influx? *Plant Physiol.* 63:248–54

65. Lucas, W. J. 1982. Mechanism of acquisition of exogenous HCO$_3^-$ by internodal cells of *Chara corallina. Planta* 156:181–92

66. Lucas, W. J., Alexander, J. M. 1980. Sulfhydryl group involvement in plasmalemma transport of HCO$_3^-$ and OH$^-$ in *Chara corallina. Plant Physiol.* 65:274–80

67. Lucas, W. J., Dainty, J. 1977. Spatial distribution of functional OH$^-$ carriers along a characean internodal cell: Determined by the effect of cytochalasin B on H^{14}CO$_3^-$ assimilation. *J. Membr. Biol.* 32:75–92

68. Lucas, W. J., Ferrier, J. M. 1980. Plasmalemma transport of OH⁻ in *Chara corallina.* III. Further studies on transport substrate and directionality. *Plant Physiol.* 65:46–50

69. Lucas, W. J., Ferrier, J. M., Dainty, J. 1977. Plasmalemma transport of OH⁻ in *Chara corallina:* Dynamics of activation and deactivation. *J. Membr. Biol.* 32:49–73

70. Lucas, W. J., Franceschi, V. R. 1981. Characean charasome-complex and plasmalemma vesicle development. *Protoplasma* 107:255–67

71. Lucas, W. J., Keifer, D. W., Sanders, D. 1983. Bicarbonate transport in *Chara corallina:* Evidence for cotransport of HCO₃⁻ with H⁺. *J. Membr. Biol.* In press

72. Lucas, W. J., Nuccitelli, R. 1980. HCO₃⁻ and OH⁻ transport across the plasmalemma of *Chara corallina:* Spatial resolution obtained using extracellular vibrating probe. *Planta* 150:120–31

73. Lucas, W. J., Shimmen, T. 1981. Intracellular perfusion and cell centrifugation studies on plasmalemma transport processes in *Chara corallina. J. Membr. Biol.* 58:227–37

74. Lucas, W. J., Smith, F. A. 1973. The formation of alkaline and acid regions at the surface of *Chara corallina* cells. *J. Exp. Bot.* 24:1–14

75. Lucas, W. J., Tyree, M. T., Petrov, A. 1978. Characterization of photosynthetic ¹⁴carbon assimilation by *Potamogeton lucens* L. *J. Exp. Bot.* 29:1409–21

76. MacRobbie, E. A. C. 1970. The active transport of ions in plant cells. *Q. Rev. Biophys.* 3:251–94

77. Malek, F., Baker, D. A. 1977. Proton cotransport of sugars in phloem loading. *Planta* 135:297–99

78. Marcus, Y., Zenvirth, D., Harel, E., Kaplan, A. 1982. Induction of HCO₃⁻ transporting capability and high photosynthetic affinity to inorganic carbon by low concentration of CO₂ in *Anabaena variabilis. Plant Physiol.* 69:1008–12

79. Miller, A. G., Colman, B. 1980. Active transport and accumulation of bicarbonate by a unicellular cyanobacterium. *J. Bacteriol.* 143:1253–59

80. Miller, A. G., Colman, B. 1980. Evidence for HCO₃⁻ transport by the blue-green alga (Cyanobacterium) *Coccochloris peniocystis. Plant Physiol.* 65:397–402

81. Mukerji, D., Glover, H. E., Morris, I. 1978. Diversity in the mechanism of

carbon dioxide fixation in *Dunaliella tertiolecta* (Chlorophyceae). *J. Phycol.* 14:137–42

82. Nagai, R., Tazawa, M. 1962. Changes in resting potential and ion absorption induced by light in a single plant cell. *Plant Cell Physiol.* 3:323–39

83. Nishizaki, Y. 1968. Light-induced changes of bioelectric potentials in *Chara. Plant Cell Physiol.* 9:377–87

84. Ogata, K., Kishimoto, U. 1976. Rhythmic change of membrane potential and cyclosis of *Nitella* internode. *Plant Cell Physiol.* 17:201–7

85. Österlind, S. 1949. Growth conditions of the algal *Scenedesmus quadricauda* with special reference to the inorganic carbon sources. *Symb. Bot. Ups.* 10:1–141

86. Österlind, S. 1950. Inorganic carbon sources of green algae. I. Growth experiments with *Scenedesmus quadricauda* and *Chlorella pyrenoidosa. Physiol. Plant.* 3:353–60

87. Österlind, S. 1950. Inorganic carbon sources of green algae. II. Carbonic anhydrase in *Scenedesmus quadricauda* and *Chlorella pyrenoidosa. Physiol. Plant* 3:430–34

88. Österlind, S. 1951. Inorganic carbon sources of green algae. III. Measurements of photosynthesis in *Scenedesmus quadricauda* and *Chlorella pyrenoidosa. Physiol. Plant.* 4:242–54

89. Österlind, S. 1951. Inorganic carbon sources of green algae. IV. Photoactivation of some factor necessary for bicarbonate assimilation. *Physiol. Plant.* 4:514–27

90. Ouellet, C., Benson, A. A. 1952. The path of carbon in photosynthesis. XIII. pH effects in ¹⁴CO₂ fixation by *Scenedesmus. J. Exp. Bot.* 3:237–45

91. Poole, R. J. 1978. Energy coupling for membrane transport. *Ann. Rev. Plant Physiol.* 29:437–60

92. Prins, H. B. A., Snel, J. F. H., Helder, R. J., Zanstra, P. E. 1979. Photosynthetic bicarbonate utilization in the aquatic angiosperms *Potamogeton* and *Elodea. Hydrobiol. Bull.* 13:106–11

93. Prins, H. B. A., Snel, J. F. H., Helder, R. J., Zanstra, P. E., 1980. Photosynthetic HCO₃⁻ utilization and OH⁻ excretion in aquatic angiosperms: Light-induced pH changes at the leaf surface. *Plant Physiol.* 66:818–22

94. Prins, H. B. A., Snel, J. F. H., Zanstra, P. E., Helder, R. J. 1982. The mechanism of bicarbonate assimilation by the polar leaves of *Potamogeton* and *Elo-*

dea. CO$_2$ concentrations at the leaf surface. *Plant Cell Environ.* 5:207–14

95. Rabinowitch, E. I. 1951. *Photosynthesis and Related Processes,* Vol. 2, Part 1. New York: Interscience. 1208 pp.

96. Radmer, R., Ollinger, O. 1980. Light-driven uptake by oxygen, carbon dioxide, and bicarbonate by the green alga *Scenedesmus. Plant Physiol.* 65:723–29

97. Rao, R. L., Pickard, W. F. 1976. The use of membrane electric noise in the study of characean electrophysiology. *J. Exp. Bot.* 27:460–72

98. Rao, R. L., Pickard, W. F. 1977. Further experiments on low frequency excess noise in the vacuolar resting potential of *Chara braunii. J. Exp. Bot.* 28:1–16

99. Raven, J. A. 1970. Exogenous inorganic carbon sources in plant photosynthesis. *Biol. Rev.* 45:167–221

100. Raven, J. A. 1980. Nutrient transport in microalgae. *Adv. Microb. Physiol.* 21: 47–226

101. Raven, J. A., Glidewell, S. M. 1978. C$_4$ characteristics of photosynthesis in the C$_3$ alga *Hydrodictyon africanum. Plant Cell Environ.* 1:185–97

102. Ruttner, F. 1947. Zur frage der karbonatassimilation der wasserpflanzen. I. Teil: Die beiden haupttypen der kohlenstoffaufnahme. *Oesterr. Bot. Z.* 94: 265–94

103. Saito, K., Senda, M. 1973. Light-dependent effect of external pH on the membrane potential in *Nitella. Plant Cell Physiol.* 14:147–56

104. Saito, K., Senda, M. 1973. The effect of external pH on the membrane potential of *Nitella* and its linkage to metabolism. *Plant Cell Physiol.* 14:1045–52

105. Sanders, D. 1980. The mechanism of Cl$^-$ transport at the plasma membrane of *Chara corallina.* I. Cotransport with H$^+$. *J. Membr. Biol.* 53:129–41

106. Shelp, B. J., Canvin, D. T. 1980. Photorespiration and oxygen inhibition of photosynthesis in *Chlorella pyrenoidosa. Plant Physiol.* 65:780–84

107. Shelp, B. J., Canvin, D. T. 1980. Utilization of exogenous inorganic carbon species in photosynthesis by *Chlorella pyrenoidosa. Plant Physiol.* 65:774–79

108. Shelp, B. J., Canvin, D. T. 1981. Photorespiration in air and high CO$_2$-grown *Chlorella pyrenoidosa. Plant Physiol.* 68:1500–3

109. Simons, R. 1979. Strong electric field effects on proton transfer between membrane-bound amines and water. *Nature* 280:824–26

110. Smith, F. A. 1968. Rates of photosynthesis in characean cells. II. Photosynthetic ^{14}CO$_2$ fixation and ^{14}C-bicarbonate uptake by Characean cells. *J. Exp. Bot.* 19:207–17

111. Smith, F. A. 1970. The mechanism of chloride transport in Characean cells. *New Phytol.* 69:903–17

112. Smith, F. A., Raven, J. A. 1979. Intracellular pH and its regulation. *Ann. Rev. Plant Physiol.* 30:289–311

113. Smith, F. A., Walker, N. A. 1980. Photosynthesis by aquatic plants: effects of unstirred layers in relation to assimilation of CO$_2$ and HCO$_3^-$ and to carbon isotope discrimination. *New Phytol.* 86:245–59

114. Spanswick, R. M. 1970. Bicarbonate pH and membrane potentials. *J. Membr. Biol.* 2:59–70

115. Spanswick, R. M. 1972. Evidence for an electrogenic ion pump in *Nitella translucens.* I. The effects of pH, K$^+$, Na$^+$, light and temperature on the membrane potential and resistance. *Biochim. Biophys. Acta* 288:73–89

116. Spanswick, R. M. 1973. Electrogenesis in photosynthetic tissues. In *Ion Transport in Plants,* ed. W. P. Anderson, pp. 113–39. London/New York: Academic. 630 pp.

117. Spanswick, R. M. 1981. Electrogenic ion pumps. *Ann. Rev. Plant Physiol.* 32:267–89

118. Spear, D. G., Barr, J. K., Barr, C. E. 1979. Localization of hydrogen ion and chloride ion fluxes in *Nitella. J. Gen. Physiol.* 54:397–414

119. Steemann Nielsen, E. 1947. Photosynthesis of aquatic plants with special reference to the carbon-sources. *Dan. Bot. Ark.* 12:1–71

120. Steemann Nielsen, E. 1960. Uptake of CO$_2$ by the plant. In *Encyclopedia of Plant Physiology, Vol. V, The Assimilation of Carbon Dioxide,* ed. W. Ruhland, pp. 70–84. Berlin: Springer. 1013 pp.

121. Talling, J. F. 1976. The depletion of carbon dioxide from lake water by phytoplankton. *J. Ecol.* 64:79–121

122. Tolbert, N. E. 1979. Glycolate metabolism by higher plants and algae. In *Encyclopedia of Plant Physiology, New Ser.* ed. M. Gibbs, E. Latzko, 6:338–52. Berlin: Springer. 578 pp.

123. Volkov, G. A. 1973. Bioelectrical response of the *Nitella flexilis* cell to illumination: A new possible state of plasmalemma in a plant cell. *Biochim. Biophys. Acta* 314:83–92

124. Volkov, G. A., Petrushenko, V. V. 1969. Studies of the resting potential of a single cell of the alga *Nitella flexilis.* IV. The correlation between the cell response to light and the photosynthesis process. *Tsitologiya* 11:1007–13

125. Walker, N. A. 1962. Effect of light on the plasmalemma of *Chara cells. Ann. Rep. Div. Plant Ind.,* CSIRO

126. Walker, N. A. 1980. The transport systems of charophyte and chlorophyte giant algae and their integration into modes of behaviour. In *Plant Membrane Transport: Current Conceptual Issues,* ed. R. M. Spanswick, W. J. Lucas, J. Dainty, pp. 287–300. Amsterdam: Elsevier/North Holland. 670 pp.

127. Walker, N. A., Smith, F. A. 1975. In-

tracellular pH in *Chara corallina* measured by DMO distribution. *Plant Sci. Lett.* 4:125–32

128. Walker, N. A., Smith, F. A. 1977. Circulating electric currents between acid and alkaline zones associated with HCO_3^- assimilation in *Chara. J. Exp. Bot.* 28:1190–1206

129. Walker, N. A., Smith, F. A., Cathers, I. R. 1980. Bicarbonate assimilation by fresh-water charophytes and higher plants: I. Membrane transport of bicarbonate ions is not proven. *J. Membr. Biol.* 57:51–58

130. Zenvirth, D., Kaplan, A. 1981. Uptake and efflux of inorganic carbon in *Dunaliella salina. Planta* 152:8–12

Ann. Rev. Plant Physiol. 1983. 34:105–36

ASPECTS OF HYDROGEN METABOLISM IN NITROGEN-FIXING LEGUMES AND OTHER PLANT-MICROBE ASSOCIATIONS

Günter Eisbrenner and Harold J. Evans

Laboratory for Nitrogen Fixation Research, Oregon State University, Corvallis, Oregon 97331

CONTENTS

INTRODUCTION

In 1956, Prof. P. W. Wilson and co-workers (164) summarized some of the early experiments which led to the conclusion that N_2 fixation and H_2

0066-4294/83/0601-0105$02.00

metabolism in some way were interrelated. As background for this review, we need to recall some of the crucial observations that provided the basis for our present understanding of this interaction. Beginning in the late 1930s, Wilson and associates showed that H_2 specifically inhibited N_2 fixation in nodulated red clover plants (182, 183) and *Azotobacter* (184). A hydrogenase system was discovered in *Azotobacter vinelandii*, the specific activity of which was markedly increased when the organism depended upon N_2 for growth (107). Hydrogenase activity also was detected in bacteroids from nodules of *Pisum sativum* (143), but for several years these observations could not be confirmed and no hydrogenase activity could be found in laboratory cultures of *Rhizobium*.

The inhibition by N_2 of the photoevolution of H_2 from *Rhodospirillum rubrum* (84) led to the discovery of N_2 fixation in this and some other photosynthetic bacteria (101). The relationship between N_2 fixation and H_2 metabolism remained obscure, however, until it was shown that N_2-fixing soybean nodules evolved H_2 (97) and that cell-free nitrogenase preparations of *Azotobacter* catalyzed an ATP and reductant-dependent reaction in which H_2 was evolved concomitantly with N_2 fixation (32, 33). Further clarification of the nitrogenase-hydrogenase relationship was provided when Dixon (56–58) showed that two strains of *R. leguminosarum* formed nodules that did not evolve appreciable H_2 and that the bacteroid forms of these strains contained an uptake hydrogenase which participated in a mechanism for recycling the H_2 produced as a by-product of the nitrogenase reaction. The potential importance of energy loss via nitrogenase-catalyzed H_2 evolution in a variety of nodulated legumes was pointed out (162), and the conditions for depression of the hydrogenase were defined in those relatively few strains of *R. japonicum* possessing a capability for hydrogenase synthesis (111, 118). This led to the discovery of chemolithotrophy in those strains of *R. japonicum* that possess the H_2-oxidizing system (95).

Concern about our dwindling supplies of fossil fuels has stimulated a renewed interest in the possibilities of increasing the use and efficiency of biological N_2 fixation for agricultural purposes, thus conserving some of the fossil fuels now used extensively for industrial N_2 fixation. As a consequence, there has been a rapid expansion of research activity concerning biologically produced H_2 as a source of fuel, factors influencing H_2 loss during N_2 fixation, the properties of hydrogenases, and the advantages of H_2 recycling to N_2-fixing organisms. Several recent reviews have been devoted to H_2 metabolism and H_2 recycling: in legume nodules (59, 75–80, 144); in blue-green algae (20, 22–26, 105); in *Azotobacter* (178, 186); and in microorganisms in general (1, 104, 150, 159, 179). In this review, we shall emphasize recent developments in our understanding of the interrelationships of H_2 metabolism and N_2 fixation in legumes and a few other plant-microbe associations. Sufficient information on H_2 metabolism in other organisms is included to provide an appropriate perspective.[1]

BIOLOGICAL SOURCES OF DIHYDROGEN

Hydrogenase-Catalyzed Reactions

Many obligate anaerobic bacteria have solved the problem of regenerating oxidized redox carriers during fermentation by transferring the excess electrons to protons resulting in H_2 evolution. Hydrogenase, the enzyme that catalyzes this reaction, has been purified and characterized from several sources, including *Clostridium pasteurianum* (see 1, 126). Reduced ferredoxin, which is formed during pyruvate cleavage in the pyruvate phosphoclastic reaction, is the carrier that transfers electrons to hydrogenase in *C. pasteurianum*. Another common soil bacterium, the facultatively anaerobic N_2-fixing *Klebsiella pneumoniae,* under anaerobic conditions also evolves H_2 via a hydrogenase-catalyzed reaction. In the absence of O_2, pyruvate: formate lyase in *K. pneumoniae* catalyzes the cleavage of pyruvate yielding acetyl CoA and formate. The latter is converted into CO_2 and H_2 by the formate-H_2-lyase complex. For every mole of glucose consumed, *K. pneumoniae* evolves 0.35 mol H_2 (170). The anaerobic and the facultatively anaerobic bacteria may be major sources of H_2 in the soil. H_2 from these organisms after diffusion to aerobic areas of soil could be taken up and utilized by soil microorganisms that possess H_2 oxidation capability (158).

By-product of N_2 Fixation

The pioneering work of Bulen and co-workers (31–33) demonstrated ATP- and reductant-dependent H_2 evolution from a reaction catalyzed by purified nitrogenase from *A. vinelandii.* Since then, ATP and $Na_2S_2O_4$-dependent H_2 evolution from cell-free nitrogenases from soybean root nodules (102), *Rhodospirillum rubrum* (35), *Anabaena cylindrica* (96), and *Alnus glutinosa* nodules (11) has been demonstrated. It seems clear that H_2 evolution is an inherent property of the nitrogenase reaction regardless of the source of the enzyme, and that both the Fe protein and MoFe protein components are essential for activity. When optimum proportions of the two nitrogenase components are utilized and an atmosphere of N_2 is provided as the reducible substrate, usually 25–30% of the nitrogenase electron flux is consumed

[1]Abbreviations used: cyt (cytochrome); CoA (coenzyme A); DBMIB (dibromothymoquinone); EDTA (ethylenediaminetetraacetic acid); Fd (ferredoxin); Fe protein (iron component of nitrogenase); Hup⁻ (hydrogen uptake negative); Hup⁺ (hydrogen uptake positive); HQNO (2-*n*-heptyl-8-hydroxyquinoline-N-oxide); MoFe protein (molybdenum iron component of nitrogenase); UQ (ubiquinone); UV (ultraviolet).

in the reduction of protons and the remainder utilized in the reduction of N_2 (32, 102). When N_2 in the nitrogenase assay mixtures is replaced by an inert gas, the entire nitrogenase electron flux is utilized for proton reduction (32). The factors affecting H_2 loss from the nitrogenase reaction are discussed later.

Although it seems clear that the H_2 evolved from aerobic N_2-fixing organisms is catalyzed by nitrogenase, a small proportion of nodulated legumes do not evolve measurable amounts of H_2 (162). It is now apparent that both symbiotic and free-living N_2-fixing organisms that evolve little or no H_2 during N_2 fixation contain an effective H_2-oxidizing system that recycles essentially all the H_2 that is produced during N_2 fixation. In three surveys of *R. japonicum* strains, 21% (41), 13% (152), and 25% (114) were reported to be H_2 uptake positive (Hup$^+$). Most of the cowpea strains of *Rhizobium* examined are Hup$^+$ (161), but no strain of *R. meliloti* or *R. trifolii* has been identified which synthesizes an effective H_2 recycling system (155). There was a claim (130a) that a mutant of *R. trifolii* expressed hydrogenase activity; however, this was proved (116a) to be a result of contamination by a Hup$^+$ strain of *R. japonicum*. The magnitude of the effect of H_2 recycling capability on H_2 loss from legumes grown under controlled conditions is illustrated by a recent tabulation (80) showing that a mean of 3.8% of the nitrogenase electron flux was lost as H_2 from nodules formed by Hup$^+$ strains of *Rhizobium*. In comparison, the mean H_2 loss from a series of legumes inoculated with known H_2-uptake negative (Hup$^-$) strains was 32% of the nitrogenase electron flux. Since most strains of *Rhizobium* lack an effective H_2 recycling system, energy loss through H_2 evolution is enormous. From measurements of rates of H_2 loss or uptake at the soil surface in fields in Germany and from Burns & Hardy's (36) estimates of global annual rates of N_2 fixation, Conrad & Seiler in 1980 (46) estimated that 2.1 to 4.4 $\times$ 10^{12} g of H_2 were evolved annually from agricultural legumes.

None or relatively little H_2 evolution was found by actinorhizal N_2-fixing symbiotic associations such as *Alnus rubra* (12, 162), *Alnus glutinosa* (11, 124, 151), *Alnus rugosa, Elaeagnus commutabi* (124), *Elaeagnus augustifoli, Ceanothus velutinus, Myrica californica, Purshia tridentata* (162), and of *Azolla caroliniana* (129, 139). Also, cultures of *Azospirillum brasiliense* do not evolve H_2 under normal growth conditions (43); however, they behave like *Azotobacter chroococcum* (169) and *Anabaena cylindrica* (28, 47), which evolve H_2 when the hydrogenase is inhibited by C_2H_2 and CO (43). The presence of an active uptake hydrogenase in some free-living and symbiotic N_2-fixing associations either prevents or diminishes the loss of H_2 into the atmosphere.

FACTORS INFLUENCING NITROGENASE-CATALYZED H_2 FORMATION

Relation to Proposed Nitrogenase Mechanisms

A discussion of the factors influencing H_2 evolution from purified nitrogenase requires a consideration of some aspects of the nitrogenase mechanism. Several recent reviews (37, 38, 72, 168, 185, 188) indicate general agreement on the sequence of electron transfer during nitrogenase catalysis; however, many details of the mechanism remain to be clarified (Figure 1). Ferredoxin and flavodoxin have been firmly established as effective electron donors for the Fe protein. In soybean nodule bacteroids, however, no conclusive evidence for a functional flavodoxin has been presented, but a ferredoxin with a midpoint redox potential of -485 mV is present at relatively high concentrations and is active as a proximal donor to purified bacteroid nitrogenase (42). The binding of MgATP to the Fe protein results in a conformational change and a lowered redox potential which is sufficient to reduce the MoFe protein. Electrons are transferred from the reduced MoFe protein to substrates such as N_2 and H^+ which become reduced to NH_3 and H_2, respectively.

Since CO has been shown to inhibit N_2 fixation, but not H_2 evolution, it has been concluded that the site for binding H^+ and N_2 are not identical (149). This view is supported by a report that EPR signals of H_2-evolving nitrogenase were altered by addition of other substrates (115). The conclusion that separate binding sites are involved in the reduction of N_2 and H^+ also is based in part on the observation that H_2 evolution from functioning nitrogenase cannot be eliminated by increasing the partial pressure of N_2 over reaction mixtures (88, 149).

The substitution of an inert gas such as Ar for N_2 over nitrogenase reactions markedly increases H_2 evolution, indicating that N_2 competes with protons for electrons at some point in the electron transport sequence or that the absence of N_2 makes available more sites for the reduction of protons. To explain H_2 evolution in the absence and presence of N_2, Mortensen (125) proposed the following two reactions:

$$E + 2e \xrightarrow[2H^+]{nATP} E^*:2H \xrightarrow{} E + H_2 \qquad \qquad 1.$$

$$E^*:2H + N_2 \rightleftharpoons E^*:N_2 + H_2 \qquad \qquad 2.$$

Reaction (1) was postulated to proceed only in the absence of N_2, whereas reaction (2) would require N_2 for H_2 evolution. Reaction (1) would account for the observed minimum of one mole of H_2 evolved per mole of N_2

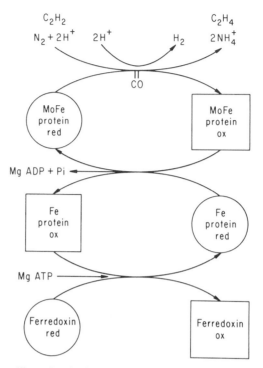

Figure 1   A scheme illustrating the electron flow from reductants to substrates of nitrogenase. As shown by the double bar, the reduction of N_2 and C_2H_2, but not H^+, is inhibited by CO.

reduced, and a combination of reactions (1) and (2) would explain H_2 evolution under conditions in which N_2 fixation was limited. If reactions (1) and (2) proceeded simultaneously, more than one mole H_2 evolved per mole of N_2 reduced would be expected.

Recently, Chatt (44, 45) has presented a scheme that provides an explanation for some of the patterns of nitrogenase-catalyzed H_2 evolution. He believes that Mo is part of the nitrogenase FeMo protein catalytic site which is located in a "pocket" in the enzyme, rendering the site inaccessible to O_2. From the known chemistry of the reaction of N_2 with metal hydrides, he postulates that reducing equivalents supplied to the nitrogenase active site generate a trihydride of bound Mo and that N_2 may displace two H atoms from the Mo trihydride, forming a bound molybdenum-N_2 complex and one mole of H_2. This reaction would account for the observed evolution of a minimum of one mole of H_2 for each mole of N_2 fixed. Furthermore, Chatt has proposed that CO also would be expected to displace two hydrogen atoms from the enzyme-bound Mo trihydride, forming a tightly bound MoCO complex which would block access to N_2. The third Mo-bound hydride ligand would remain free to react with a proton, forming H_2, a reaction that would provide a logical explanation for the CO insensitivity of the catalysis of H_2 evolution by nitrogenase.

It seems reasonable to expect that a higher oxidation state of the MoFe protein such as monohydride of bound molybdenum could provide a site for H_2 evolution but would lack a capacity for N_2 binding and subsequent reduction. This mechanism also could explain the observed minimum H_2 evolution from nitrogenase reactions provided with C_2H_2. During N_2 fixation, the MoFe protein must become reduced sufficiently to supply six electrons to N_2 for reduction to two moles of NH_3. The reduction of C_2H_2, like the reduction of two protons, requires only two electrons and, therefore, C_2H_2 could compete directly with protons for electrons from the MoFe protein. The different responses to CO of nitrogenase-catalyzed H^+ and C_2H_2 reduction (149) could be explained by considering the stereospecific reduction of C_2H_2. It is known that C_2H_2 when reduced in the presence of ^2H_2O was converted into cis-$C_2^1H_2{}^2H_2$ only (55). This can be explained by assuming that both carbon atoms of C_2H_2 receive electrons simultaneously. It seems reasonable to conclude, therefore, that C_2H_2 reduction involves two Mo-bound hydride ligands which would not be available when CO is present. Chatt's postulations are indeed interesting and should stimulate the design of experiments to test his hypothesis.

In a series of papers, Hageman and Burris (37, 89–91, 93) have presented strong evidence supporting the view that the Fe protein and MoFe protein components of nitrogenase associate and dissociate with each electron transfer from one component to the other. In fact, they refer to the Fe protein as dinitrogenase reductase and the MoFe protein as dinitrogenase, a nomenclature that is not accepted by all (168). When Hageman et al (93) utilized a nitrogenase assay mixture in which the Fe protein was limiting and the MoFe protein present in considerable excess, a lag was observed in the rate of H_2 evolution, but no lag occurred in the initial rate of ATP hydrolysis. If ATP were being consumed in the transfer of electrons from the MoFe protein to the substrate, a lag in both H_2 evolution and ATP hydrolysis would have been expected. Since this was not the case, they argue that ATP hydrolysis is limited to the transfer of electrons from the Fe protein to the MoFe protein. From these and other experiments, a model was proposed in which electrons are transferred one at a time from the Fe protein to the MoFe protein until an electron pool is accumulated that is sufficient to reduce the various substrates. Since substrates of different types require different multiples of electron pairs for reduction, it is obvious that not all substrates are necessarily reduced by the same oxidation-reduction state of the MoFe protein.

Of significant physiological implications are data showing that those variables in the nitrogenase assay such as concentration of ATP, supply of reductant, and ratio of Fe protein to MoFe protein, all of which affected the rates of MoFe protein turnover, strikingly influenced the allocation of electrons to substrates (92). For example, an adequate concentration of ATP and reductant and a high ratio of Fe protein to MoFe protein increased the turnover rate of the MoFe protein and resulted in an increase in the proportion of electrons transferred to N_2 and minimized electron transfer to H^+. In contrast, an inadequacy of reductant and MgATP and a low ratio of Fe protein to MoFe protein in assays greatly decreased the MoFe turnover rate which was associated with a marked increase in the allocation of electrons to H^+ and a minimum allocation of electrons to N_2. In agreement with these results are those of Wherland et al (180), who reported that decreasing the ratio of *Azotobacter* Fe protein to MoFe protein favored H_2 evolution over NH_3 production.

In our opinion, the experimental data of Hageman & Burris (92) and the model that they have proposed are compatible with the previously discussed scheme of Chatt (44). The accumulation of a pool of electrons in the MoFe protein which is sufficient for a maximum rate of reduction of N_2 (38) could correspond to a rate of electron flow sufficient to generate Chatt's bound Mo trihydride which is postulated to have a capacity to react with N_2. A smaller pool of electrons in the MoFe protein resulting from lower rates of electron flow might be caused by a situation in which the oxidation-reduction potential was insufficient to generate the Mo trihydride. It may be adequate, however, to generate a less reduced state of the MoFe protein capable of transferring electrons to substrates such as H^+ or C_2H_2, both of which require less than three pairs of electrons for reduction. Although there is good agreement on the general aspects of electron transfer from reductant via the nitrogenase to substrates, and of the factors which affect the allocation of electrons to substrates, the significance of a persistent complex of the two nitrogenase proteins in catalysis (62) and the question of whether any mechanistic importance can be assigned to the binding of MgATP to the MoFe protein (63), or to a complex of the two protein components (63), remains a subject of controversy (38, 168).

It has been known for many years (98) that the nitrogenase system in soybean root nodules catalyzed a reaction in which D_2 is converted to HD and that the rate of the reaction under N_2 is much more rapid than under He. The catalysis of this reaction by purified nitrogenase requires electrons and ATP and is believed to involve an interaction of D_2 with an enzyme-bound intermediate stage of the reduction of N_2 to $2NH_3$ (34). This subject has been reviewed thoroughly by Robson & Postgate (150).

Intact N_2-Fixing Organisms

A major factor determining whether N_2-fixing legume nodules evolve H_2 is the presence of an active uptake hydrogenase system which catalyzes the oxidation of the H_2 evolved during N_2 fixation. It was Dixon (56) who first observed that *Pisum sativum* inoculated with *R. leguminosarum* strain 311 evolved a negligible quantity of H_2 from nodules and that plants inoculated with all but one other strain evolved H_2 at rapid rates (58). Hydrogenase activity was demonstrated in bacteroids of strain 311, but no activity was detected in bacteroids from nodules that evolved H_2 (58). The effect of an uptake hydrogenase on the magnitude of H_2 evolution was discussed above.

Evidence that hydrogenase expression in nodules is controlled primarily by the bacterial inoculant strain has been shown by Carter et al (41), who used a Hup$^+$ *R. japonicum* strain and a Hup$^-$ strain as two inoculants for a series of soybean cultivars differing greatly in their genetic background. All cultivars nodulated with the Hup$^-$ strain evolved H_2 from nodules, whereas little or no H_2 was evolved from nodules from all cultivars inoculated with the Hup$^+$ strain. These experiments indicated no effect of the legume host on hydrogenase expression within a single species, but Dixon (58) has reported that a Hup$^+$ strain of *R. leguminosarum* expressed hydrogenase activity in nodules of *Pisum sativum* and *Vicia bengalensis* but failed to express activity in nodules of *Vicia faba*. More recently, Gibson et al (85) have reported that *Rhizobium sp.* CB756 and 32H1 produced Hup$^-$ nodules on *Vigna radiata* and Hup$^+$ nodules on *Vigna unguiculata* and *Vigna mungo*. From reciprocal grafts of several *Vigna* species, they claim that the root rather than the shoot genotype was the important factor in Hup expression. The effect of the host plant on hydrogenase expression may well be related to the type and quantity of carbon substrates translocated from leaves to nodules, because it is well known that the concentration of carbon substrates plays a major role in the derepression of hydrogenase in free-living *R. japonicum* (119). It is clear that the genetic information for hydrogenase synthesis resides in *Rhizobium* and not the plant (see below).

Since H_2 evolution from N_2-fixing symbionts is derived primarily from the nitrogenase reaction, the extent of H_2 evolution in an organism lacking an effective H_2 recycling mechanism would be expected to coincide with the activity of the N_2-fixing process. That this is indeed the case was dramatically demonstrated by Conrad & Seiler (46), who monitored the rates of H_2 production and uptake at the soil surface in a clover field during the growth season. They observed no net H_2 production during the early part of spring, a striking increase in H_2 production which peaked about June 1 when the clover was growing vigorously, and then a marked decline in H_2 production as the clover began to senesce. Practices such as removal of plant tops, placing them in the dark, and fertilization with NH_4Cl all decreased N_2 fixation and H_2 evolution. Measurements of rates of H_2 uptake and evolution in grass fields showed no trends comparable to those observed in fields of clover. From these results, it seems obvious that H_2 was produced by the Hup⁻ bacteroids of clover nodules and that the rate of H_2 production in soil containing actively growing clover was greater than the rate of H_2 consumption by indigenous H_2-oxidizing bacteria.

The extent of H_2 evolution is reported to vary during the growth cycle of N_2-fixing legumes and to be influenced by environmental conditions. According to Bethlenfalvay & Phillips (18), the percentage of the nitrogenase electron flux lost as H_2 from *Pisum sativum* nodules changed from about 60% during the early stages of growth to about 30% for the period 45 days to 68 days after planting. The *R. leguminosarum* strain 128C53, which was used as an inoculum in these experiments, is known to recycle only a portion of the H_2 evolved during N_2 fixation (154). Interpretation of these results is complicated by changes during the growth cycle of the allocation of electrons to N_2 or H^+ and activity of an inefficient uptake hydrogenase which may be sufficient to recycle all H_2 produced by the nitrogenase reaction during the latter part of the life cycle but insufficient to recycle all H_2 when N_2 fixation was proceeding rapidly. Bethlenfalvay & Phillips (19) also reported that increasing the irradiance over nodulated *Pisum sativum* from 400 to 800 $\mu E \cdot cm^{-2} \cdot sec^{-1}$ resulted in more than a tenfold increase in the rate of H_2 loss per plant. Again, it is not possible to determine whether the effect of irradiance was caused by a change in allocation of nitrogenase electrons to substrates or to a change in hydrogenase activity in *R. leguminosarum* bacteroids (strain 128C53). No effect was observed from increasing irradiance from 400 to 1000 $\mu E \cdot cm^{-2} \cdot sec^{-1}$ on the extent of H_2 loss from subterranean clover nodules (165).

Temperature and partial pressure of O_2 also have an effect on the extent of H_2 loss from legume nodules. As shown by Dart & Day (49), raising the temperature from 20° to 40°C approximately doubled the rate of H_2 evolution from nodules of *Vigna unguiculata*. Experiments with soybean nodules in our laboratory have produced results that are consistent with this observation (N. T. Jennings, unpublished 1979). Drevon et al (61) have observed that an increase in the partial pressure of O_2 from 20% to 40% over soybean nodules markedly increased the nitrogenase activity and decreased by 16% the proportion of the nitrogenase electron flux lost as H_2. This was interpreted by assuming that an increased respiratory rate at the higher O_2 partial pressure resulted in an increased turnover rate of the MoFe protein and a decreased allocation of electrons to H^+ (see above).

Relatively little is known about the effect of environmental factors on H_2 evolution from other N_2-fixing associations. According to Roelofsen &

Akkermans (151), significant increases in H_2 evolution from nodules of *Alnus glutinosa* were observed at the end of the growing season. Experiments with the *Azolla caroliniana-Anabaena azollae* symbiosis by Newton (129) have shown that the addition of nitrate to the medium resulted in up to 35% of the nitrogenase electron flux being evolved as H_2. On the basis of our discussion above, one might speculate that the presence of nitrate as an additional electron sink would result in a decreased turnover rate of the MoFe protein and an increased allocation of electrons to H^+.

HYDROGENASES

General Properties

Hydrogenases are widely distributed in bacteria and algae and are responsible for the catalysis of a reversible reaction in which H_2 yields two protons and two electrons. They are usually characterized as "uptake" or bidirectional in respect to their physiological roles in cells. However, once the membrane-bound hydrogenases of the uptake type are solubilized, the concentrations of substrates and other conditions in assays may be altered so that reversible reactions may be demonstrated. Recently the properties and physiological roles of hydrogenases have been reviewed comprehensively (1, 104, 105, 159). All of the extensively characterized hydrogenases are iron-sulfur proteins and most of them are very susceptible to O_2 inactivation.

Hydrogenase from R. japonicum Bacteroids

The most thoroughly characterized hydrogenase from a plant-bacterial N_2-fixing association is the uptake hydrogenase that occurs in the membranes of *R. japonicum* bacteroids. Arp & Burris in 1979 (6) used Triton X100 to solubilize the enzyme from *R. japonicum* strain 110 bacteroids and were successful in obtaining a highly purified preparation with an increase in specific activity of 196–fold. The preparation was estimated to be over 90% pure. The molecular weight of the iron-sulfur protein was estimated to be about 63,300. The purified enzyme readily transferred electrons to methylene blue, ferricyanide, and 2,6 dichlorophenol-indophenol but failed to catalyze the reduction of NADP, FAD, FMN, or O_2. By use of reduced methylviologen, a rate of hydrogenase-catalyzed H_2 evolution was observed that was about 2 percent of the forward rate. The purified hydrogenase catalyzed a rate of exchange between $^2H_2$ and H_2O that was about 10% of the oxidation rate (8). From isotope discrimination experiments, they concluded that dihydrogen bond breakage was not the rate-limiting step in H_2 oxidation. This finding is consistent with the report of Emerich et al (74), who measured the effect of $^2H_2$ vs $^1H_2$ on the rate of oxyhydrogen reaction in *R. japonicum* bacteroids. In a study of the mechanism of the bacteroid hydrogenase, Arp & Burris (7) have shown that CO is a competitive inhibitor of H_2. Apparently, CO is accessible to the active site of the purified enzyme, but inaccessible to the active site of the enzyme in membranes of *R. japonicum* bacteroids (153). Even though O_2 inactivates the enzyme, this gas is a reversible inhibitor of hydrogenase at low concentrations (7). A two-site "ping pong" mechanism in which H_2 is reversibly activated at one site and electron acceptors react at another site was proposed.

Suzuki & Maruyama (171) report that two hydrogenase activities measured by $^3H_2$ uptake were detected in the bacteroids from soybean and lupin nodules. The activity of one of the preparations was dependent upon ATP or an ATP-generating system while activity of the other fraction failed to respond to ATP. They claim that the ATP-dependent activity was separated from nitrogenase activity.

It seems clear that purified hydrogenases from membranes of *Azotobacter chroococcum* (176), *Alcaligenes eutrophus* (157), and *R. japonicum* bacteroids (6–8) share many common properties. To our knowledge, there are no detailed studies of properties of purified hydrogenases from N_2-fixing plant-microbe associations other than *Glycine max* nodules.

Nickel Role in H_2 Metabolism

Until recently the biological importance of Ni was not appreciated. However, Thauer et al in 1980 (175) pointed out that there is convincing evidence of its essential role in several processes of plant and animal cells and in the synthesis and function of hydrogenase in several bacteria. A specific Ni requirement for chemolithotrophic growth of cultures of *Alcaligenes eutrophus* (strains H1 and H16) was shown by Bartha & Ordal in 1956 (9) and confirmed by Repaske & Repaske (148) and Gruzinskii et al (87). Organisms of this type depend upon H_2 oxidation via a hydrogenase as a source of energy for their metabolism during autotrophic growth. More recently, Tabillion et al (172) showed a need for Ni for autotrophic growth of five strains of *Alcaligenes eutrophus,* two strains of *Xanthobacter autotrophicum,* a single strain of *Pseudomonas flava,* and two strains of *Arthobacter sp.* In experiments involving 12 strains representing 6 species, 10 exhibited a Ni requirement when cultured autotrophically, but none responded to Ni when cultured heterotrophically. By use of EDTA to complex Ni in media, Friedrich et al (81) have concluded that Ni is required for the synthesis of both the soluble and membrane-bound hydrogenases in *Alcaligenes eutrophus* and that ribulose-1,5-bisphosphate carboxylase activity in this organism was not impaired by Ni deficiency.

More recently, the addition of Ni to cultures of *Rhodopseudomonas capsulata* grown photosynthetically has been shown to increase the specific activity of hydrogenase. In tests of a series of trace metals with this bacterium, the response to Ni was highly specific and the maximum effect was obtained at about 10^{-5} M (174). An implication of a role of Ni in hydrogenase synthesis by *Azotobacter chroococcum* has been reported by Partridge & Yates (133), who cultured the organism in a purified medium containing EDTA or other chelating agents. They observed an inhibition of hydrogenase expression by chelating agents which was reversed by adding Ni alone or Ni and Fe together. The rate of hydrogenase derepression in the absence of added chelating agents also was markedly stimulated by Ni or Ni and Fe.

In recent investigations of the metabolism of methanogens (bacteria that require Ni for growth with H_2 and CO_2 as energy and carbon sources, respectively) (160), Ni was shown to be a component of factor F_{430} (53, 181) which Ellefson et al (70) recently have shown to function as an essential cofactor for methyl reductase. The yellow cofactor containing ^{63}Ni was released from the enzyme by the addition of 80% methanol. The involvement of Ni in factor F_{430} by several methanogenic bacteria also has been shown by Diekert et al (54), and convincing biosynthetic evidence was presented demonstrating that factor F_{430} consists of a Ni tetrapyrrole (52). In a continuing series of excellent papers by Thauer and associates, hydrogenase has been isolated from *M. thermoautotrophicum* and shown to contain about one gram atom of Ni per mole of enzyme (86). A homogeneous preparation of the soluble hydrogenase from *Alcaligenes eutrophus* has been analyzed by X-ray fluoresence and shown to contain two nickel gram atoms per mole of protein, but for some reason that is not apparent, the Ni content of the enzyme calculated from ^{63}Ni incorporation was about one gram atom per mole of enzyme (83).

Recently, Albracht et al (3) have shown by EPR measurements that Ni participates in a reversible oxidation-reduction process during hydrogenase catalysis. They believe that Ni is functioning at the active site of hydrogenase, and by use of Ni isotope substitution, they obtained convincing evidence that the observed EPR signals were caused by Ni and not some other EPR-active species. Using a highly purified hydrogenase from *Desulfovibrio gigas,* LeGall et al (108) have shown that oxidized hydrogenase exhibits EPR signals characteristic of Ni(III). These signals were replaced by a series of other EPR signals when the enzyme was reduced by H_2. It seems clear that Ni is involved in an oxidation reduction mechanism during hydrogenase catalysis; however, the nature of Ni binding to the hydrogenase protein remains to be determined. To date, there is no evidence to indicate that the function of Ni as a redox constituent of hydrogenase and the role of Ni as a part of the factor F_{430} in methanogens are related phenomena.

Obviously, the flurry of research activity concerning a role of Ni in hydrogenase from a variety of types of organisms raises questions about the possibility that Ni may prove to be of economic importance in production of legumes and other N_2-fixing species that utilize an uptake hydrogenase for the oxidative recovery of the H_2 that is evolved during N_2 fixation. In this regard, Robert Klucas and associates, working in our laboratory, recently have shown (unpublished results, 1982) that the derepression of hydrogenase in *R. japonicum* cells cultured on a highly purified medium was strikingly stimulated by the addition of $NiCl_2$ at 0.5 μM. Also, the hydrogenase activity of bacteroids from soybean nodules from plants supplied with Ni was at least 33% higher than the activity in bacteroids from plants lacking added Ni. In these experiments all nutrient salts were purified by use of diphenylthiocarbozone and 8-hydroxyquinoline extraction. In the soybean experiments, it was necessary to employ a "cutback" procedure (2) to deplete the Ni content of the seed before the responses to Ni addition were observed. It seems highly probable that increasing attention must be devoted to the requirement and biological role of Ni in the production of legumes and other N_2-fixing species.

Regulation

Efforts to obtain expression of hydrogenase activity in free-living cultures of *Rhizobium* were unsuccessful until it was learned that a microaerobic environment, a limited supply of carbon substrates, and a source of H_2 were necessary for expression (112, 113, 118). Lim (111) claimed that *R. japonicum* cells, induced for hydrogenase, catalyzed an exchange reaction that did not require O_2; however, the basis for this contention has been questioned (78). Also, it has been reported (132) that carbon-limited N_2-fixing cultures of *A. chroococcum* showed increased hydrogenase activity and that the activity of the enzyme was higher in cells grown with N_2 than those grown with NH_4^+ or NO_3^-. These findings are in general agreement with those obtained with *Rhizobium*. Simpson et al (167) reported that the condition needed to derepress hydrogenase synthesis in *R. japonicum* also derepressed the synthesis of ribulose-1,5-bisphosphate carboxylase, but did not influence the activity of propionyl-CoA carboxylase.

The experiments of Maier et al (119) have shown that a series of carbon substrates each at 15 mM repressed the expression of hydrogenase activity in *R. japonicum,* but none of them inhibited hydrogenase activity when added to assay mixtures. The rate of hydrogenase derepression was markedly influenced by the partial pressure of O_2 over cultures and was decreased to some extent when CO_2 was removed. Neither NO_3^- nor NH_4Cl added at 10mM to cultures affected hydrogenase expression when other conditions were optimum. More recently, Maier & Merberg (120) have described a series of mutants of *R. japonicum* that are hypersensitive to repression of H_2 oxidation capability by O_2. Neither hydrogenase activity nor growth of the mutant cells was particularly sensitive to O_2. Also, Maier

& Mutafschiev (121) have reported the reconstitution of hydrogenase activity by combining and incubating at 30° for 6 to 10 hours filtered extracts of two different Hup⁻ *R. japonicum* mutants. Although control extracts from each mutant alone showed no activity, evidence regarding possible bacterial growth in the extracts during the 10 hours of anaerobic incubation would have been desirable. The results of these experiments suggest that more than one gene may be involved in the hydrogenase system.

Lim & Shanmugam (113) report that the addition of cyclic AMP to *R. japonicum* suspensions provided with malate alleviated the inhibiting effect of malate on hydrogenase synthesis and that the response to cyclic AMP was prevented by inhibitors of protein synthesis. Lim et al (112) also reported that the addition of cyclic guanosine 3'5'-monophosphate to *R. japonicum* cultures inhibited the expression of hydrogenase and nitrogenase. These and related effects of cyclic nucleotides on enzyme expression in *R. japonicum* have been discussed by Phillips (144). From the information obtained from experiments with free-living *R. japonicum,* it seems reasonable to postulate that the partial pressure of O_2 within nodules, the quantity and type of carbon substrate provided from shoots to root nodules, and the extent of H_2 evolution from the N_2-fixing process within bacteroids are the primary factors that may influence the derepression of hydrogenase in those nodule endophytes that possess the genetic potential for hydrogenase synthesis. The cyclic nucleotides may prove to play a regulatory role as "secondary messengers" in enzyme expression by *Rhizobium.* However, this area needs further study.

Genetics

Although rapid progress is being made in understanding the genetics of *Rhizobium,* our information on the genetics of the H_2 recycling process is just beginning. It has been established that the determinants for several symbiotic genes are located on large plasmids in the fast-growing species *R. leguminosarum, R. trifolii, R. phaseoli,* and *R. meliloti* (14, 29, 51, 103). Brewin et al (30) made the important observation that the transfer by conjugal mating of a plasmid with determinants for nodulation resulted in the transfer of determinants for Hup in *R. leguminosarum.* Conjugative techniques also have been used to transfer plasmids carrying determinants for H_2 oxidation in the chemolithotrophic bacterium *Alcaligenes eutrophus* (82, 173). As discussed later, DeJong et al (50) have reported increases in N_2 fixation by *Pisum sativum* inoculated with an *R. leguminosarum* strain carrying a plasmid with determinants for Nod, Nif, and Hup. Although the work with *R. leguminosarum* indicates that determinants for Hup are plasmid-borne, an examination of *R. japonicum* strains revealed plasmids with molecular weights ranging from 49–280 $\times$ 10^6 in a series of Hup⁻ strains and no evidence of plasmids in *R. japonicum* strains with high H_2 uptake activities (40). Experiments with nonrevertible Hup⁻ mutants (122) derived from a Hup⁺ parent revealed two discernible plasmid bands but no detectable plasmids in revertible Hup⁻ mutants from the same Hup⁺ parent (40). It was proposed that the plasmids detected in the nonrevertible Hup⁻ mutants may have been derived from the breakdown of a plasmid in the Hup⁺ parent that was too large for resolution by usual procedures. Obviously, this area requires further study.

Recently, Cantrell et al (39) have described the use of a conjugative cosmid in the construction of a clone bank of DNA from a Hup⁺ strain of *R. japonicum.* Cosmids from the bank containing H_2-uptake gene(s) have been detected by conjugation into a revertible Hup⁻ mutant of *R. japonicum* and selection of colonies that were tetracycline-resistant and Hup⁺. Colonies showing complementation grew chemolithotrophically and showed a capability for H_2-dependent methylene blue reduction. When plasmid DNA was isolated from a series of *R. japonicum* transconjugants and transformed into *E. coli,* conjugation of the *E. coli* transformants into Hup⁻ *R. japonicum* mutants resulted in a Hup⁺ phenotype in all transconjugants. The transconjugants retained a capacity to nodulate soybeans and fix N_2. It is hoped that the *R. japonicum* clone bank will be useful in transferring the determinants for Hup into those *Rhizobium* species lacking an effective H_2 recycling system.

Lepo et al (110) and Maier (117) have described procedures for isolation of revertible Hup⁻ mutants of *R. japonicum.* Both Lepo et al (110) and Maier (117) have used the capability for chemoautotrophic growth as a key method in selection of Hup⁻ mutants and Hup⁺ revertants. Although the capacity to reduce triphenyltetrazolium chloride was an effective procedure for distinguishing wild-type Hup⁺ and Hup⁻ *R. japonicum* (118), unpublished results of M. A. Cantrell (1982) have shown that the capacity for reduction of this dye is not a reliable method for screening for Hup⁻ point mutants. Maier (117) has described groups of mutants, some unable to oxidize H_2, some deficient in CO_2 uptake when grown under conditions that derepress hydrogenase, some capable of oxidizing H_2 with artificial electron acceptors, but not with O_2, and some lacking a capacity for expression of ribulose-1,5-bisphosphate carboxylase but retaining H_2-oxidizing capacity. All of the mutants isolated by Lepo et al (110) were defective in H_2 uptake with either O_2 or artificial electron acceptors and showed low but measurable ribulose-1,5-bisphosphate carboxylase activities. Revertible mutants not only are valuable in investigations of the genetics of *R. japonicum,* but also are useful in the evaluation of the benefits of H_2 recycling in nodules of soybean plants.

HYDROGEN OXIDATION PATHWAYS

Participation in Part of the Respiratory Chain

The H_2 uptake mechanism in aerobic N_2-fixing bacteria has been studied intensively. The relationship of the N_2-fixing and H_2 uptake processes are illustrated in Figure 2. Since the rediscovery of hydrogenase activity in *Rhizobium* (56), it has been well established that O_2 is the final acceptor for electrons for the uptake hydrogenase system from *R. leguminosarum* (57) and *R. japonicum* (41, 161). The stoichiometry of the oxyhydrogen reaction is shown by the observation that two moles of H_2 are oxidized by one mole of O_2, yielding two moles of H_2O (57, 123). Oxygen does not react directly with the hydrogenase, but electrons from H_2 are transferred via a series of electron carriers to O_2. Inhibitor experiments by several workers (27, 64, 66, 73, 74, 137, 166, 186), and the demonstration that ATP is generated during the H_2 oxidation reaction (58, 73, 100, 141), indicates an involvement of part of the respiratory pathway in the oxyhydrogen reaction. In cyanobacteria it has been established that H_2 can be consumed not only in the oxyhydrogen reaction, but also in a light-dependent oxidation process (10, 24, 65, 69, 99, 138, 142). With *R. japonicum* bacteroids, Emerich et al (74) showed that KCN, NaN_3, NH_2OH, or Na_2S, each of which was assumed to inhibit electron transport from cytochrome c to O_2, strongly

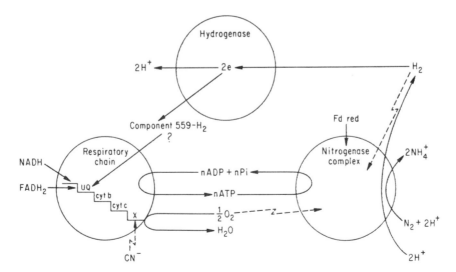

Figure 2   A generalized scheme illustrating possible interrelationships during nitrogen fixation, hydrogen uptake, and respiratory electron transport in *Rhizobium japonicum*. Inhibition is indicated by dashed lines. There are indications that component 559-H_2 may act as an electron carrier between hydrogenase and the electron transport chain (67, 68).

inhibited H_2 uptake. Furthermore, Emerich et al (74) reported that HQNO inhibited H_2 uptake to some extent but had almost no effect on O_2 uptake in the absence of H_2. In contrast, iodoacetate inhibited endogenous O_2 uptake much more than the H_2 uptake (73, 74). This was interpreted to mean a greater participation of sulfhydryl-containing dehydrogenases in the oxidation of endogenous substrates than in the oxyhydrogen reaction (73, 74). Similar inhibitory effects of iodoacetate and KCN were observed with chemolithotrophically grown R. japonicum (68).

Proximal Electron Acceptors of Hydrogenase in R. japonicum

The first acceptor for electrons from reduced uptake hydrogenase has not been identified. The purified enzyme catalyzes the reduction of the dyes methylene blue and phenazine methosulfate (6) with midpoint oxidation-reduction potentials of +11 and +80 mV, respectively. When the enzyme is solubilized and released from membranes, it fails to react with O_2 and no longer is sensitive to KCN or NaN_3 (153). With electron carriers that have a much more negative midpoint oxidation-reduction potential such as pyridine nucleotides, flavin nucleotides, and the artificial electron carrier methylviologen, the uptake hydrogenase shows either no or very poor activity (6, 153). The behavior of the uptake hydrogenase from Azotobacter chroococcum toward different acceptors is similar to that of the enzyme from bacteroids (176).

Recently, Eisbrenner et al (68) have detected a component with a difference spectrum similar to that of b-type cytochromes and have shown that it is involved in the H_2-uptake process of chemolithotrophically grown R. japonicum. Since in the presence of KCN and O_2 this component was reduced immediately upon the addition of H_2, but was not reduced with endogenous substrates or succinate under the same conditions, they (67) assumed that its oxidation-reduction potential might be lower than that of ubiquinone (48). The difference spectrum of this carrier, referred to as component 559-H_2 (67), revealed absorbance maxima for the α, β, and soret bands at 559, 532, and 428 nm, respectively. Component 559-H_2 also is present in Hup$^+$ bacteroids of R. japonicum, but at a much lower concentration than in chemolithotrophically grown cells. A positive correlation ($r = 0.98$) between the concentrations of component 559-H_2 and hydrogenase-specific activities of Rhizobium cells was observed (67). Since neither Hup$^-$ mutants nor Hup$^-$ wild-type strains of R. japonicum contained component 559-H_2, Eisbrenner & Evans (67) suggested that this component may be closely associated with hydrogenase in the H_2 uptake system. They discussed the possibility that component 559-H_2 may be a b-type cytochrome such as that found in the respiratory electron transport from ubiquinone to O_2 and that an oxidation-reduction potential more negative than that of ubiquinone might be caused by its association with the membrane-bound hydrogenase. A cytochrome of the b type, that has a lower oxidation-reduction potential than that of quinones, is known to participate in the electron transfer from hydrogenase to menaquinone in anaerobic cultures of *Escherichia coli* provided with fumarate (17; see also 1, 177). Further work is needed to characterize component 559-H_2 from *R. japonicum* and to determine its specific role in the H_2-oxidation pathway.

Involvement of Quinones in R. japonicum

There is evidence that ubiquinone is involved as a carrier in electron transfer from H_2 to O_2. Indirect evidence for this was provided by Eisbrenner & Evans (66), who showed that DBMIB, a substance known to antagonize the function of plastoquinone in chloroplasts (21) and ubiquinone in *E. coli* (145), inhibited the uptake of both H_2 and O_2 by *R. japonicum* bacteroids. Recently, O'Brian & Maier (130) also have indicated that quinones participate in the oxyhydrogen reaction in *R. japonicum*. Their conclusion was based upon an observed inhibition of H_2 oxidation after UV irradiation and then restoration of activity by adding menadione. There is no evidence, however, that menadione is a natural component of *Rhizobium*.

Involvement of Cytochrome in R. japonicum

It has been demonstrated by direct kinetic measurements that b and c-type cytochromes which catalyze electron transfer between ubiquinone and O_2 are involved in the electron transport from H_2 to O_2 (66, 68). When endogenous respiration was inhibited by iodoacetate and malonate, the addition of H_2 resulted in a marked increase in the rate of cytochrome reduction. Such high rates were never observed in suspensions unless H_2 was added. The increase in rates of cytochrome reduction resulting from the addition of H_2 could be explained only by assuming that b and c-type cytochromes with oxidation-reduction potentials more positive than ubiquinone were involved in the H_2 oxidation process. Recently, O'Brian & Maier (130) have reported the appearance of a reduced b-type cytochrome spectrum (peak at 560 nm) when they determined difference spectra of H_2-reduced membranes from free-living *R. japonicum* cells vs ascorbate-reduced membranes.

The final enzyme involved in electron transport from H_2 to O_2 appears to be a terminal oxidase that is sensitive to inhibition by KCN (153). The apparent K_m of the enzyme for O_2 is about 10 nM (71), a value that is of the same order of magnitude as some of the apparent K_m values for O_2 for respiratory terminal oxidases in bacteroids (13). It seems likely that the terminal oxidase involved in the oxidation of H_2 and other substrates are the same; however, these have not been fully characterized and further work is necessary. In chemolithotrophically cultured cells, a cytochrome of the

a-a_3 type may function as the terminal oxidase in the H_2 oxidation pathway (68). The KCN-insensitive electron pathway in $R.$ $japonicum$ supports little if any H_2 oxidation (67, 68, 73).

Nonlegume Microbe-Plant Associations

No details are known so far about the H_2 oxidation pathway in the Hup^+ nonlegume microorganism-plant associations. In the $Alnus$ $rubra$ (12, 162), $Alnus$ $glutinosa$, $Alnus$ $rhombifolia$, $Myrica$ $pensylvanica$ (12), $Azospirillum$ $brasilense$ (187), and in $Anabaena$ $azollae$ (140) the H_2 oxidation systems appear to be similar to the system in $Rhizobium$ bacteroids.

BENEFITS FROM H_2 UTILIZATION

N_2-Fixing Associations

Dixon in 1972 (58) pointed out several potential advantages of an efficient H_2 recycling system to N_2-fixing organisms: (a) The consumption of O_2 by the H_2 oxidation reaction may contribute toward the protection of the O_2 sensitive nitrogenase. (b) The oxidation and thus removal of H_2 produced via nitrogenase within the cell might prevent inhibition of nitrogenase by H_2. (c) Oxidation of H_2 may support the synthesis of ATP for use in N_2-fixation and other processes. In addition to these possibilities, it seems obvious that useful reducing power for the cell may be derived from H_2 oxidation.

These proposed advantages of H_2 recycling have been experimentally examined in several different N_2-fixing organisms. Protection of nitrogenase activity against O_2 damage has been demonstrated in $Anabaena$ $cylindrica$ (27), $A.$ $variabilis$ (23), $Azotobacter$ $chroococcum$ (178), and in $R.$ $japonicum$ bacteroids (73). Recently, Dixon et al (60) has presented a calculation based upon a mathematical model which is consistent with the conclusion that the H_2 concentration inside pea and lupin root nodules may reach concentrations sufficient to inhibit nitrogenase. They argue, therefore, that the removal of inhibitory H_2 in nodules is a major benefit of the uptake-hydrogenase system. H_2-dependent ATP synthesis has been demonstrated in $Azotobacter$ $vinelandii$ (100), $R.$ $leguminosarum$ bacteroids (57, 128), and $Anabaena$ 7120 (141). In $R.$ $japonicum$ bacteroids, Emerich et al (73) have shown that the ATP concentration was increased during H_2 oxidation. Nelson & Salminen (128) have observed that some strains of $R.$ $leguminosarum$ contain an H_2-uptake system that is coupled to ATP synthesis while other strains are not coupled. This finding may be related to the unsuccessful attempts to culture $R.$ $leguminosarum$ strains chemolithotrophically (D. W. Emerich and H. J. Evans, unpublished results, 1980). Petrosa et al (136) have reported that the addition of H_2 to carbon-starved cultures of *Azospirillum brasiliense* greatly increased rates of C_2H_2 reduction. This bacterium appears to behave like several other N_2-fixing organisms (27, 73, 178), utilizing the oxyhydrogen reaction to provide some of the energy needed for C_2H_2 reduction. So far, there is no direct evidence that electrons may be supplied directly to nitrogenase via H_2 and ferredoxin in aerobic N_2-fixing organisms.

In cyanobacteria, the transfer of electrons from H_2 to C_2H_2 was shown to be light-dependent (64, 65, 142). Since some species of *Xanthobacter* fix N_2 when grown chemolithotrophically with mineral salts, H_2, CO_2, and O_2, the electrons needed for N_2 fixation by these organisms ultimately must be supplied by H_2 (15, 16). The possibility exists, however, that electrons from H_2 are utilized in chemolithotrophic *Xanthobacter* species, first to reduce CO_2 to carbohydrates, the metabolism of which provides the reductant for N_2 fixation. Other investigations by Dixon (57), Emerich et al (74), Arima (5), Drevon et al (61), and Rainbird et al (147) have shown that rates of CO_2 evolution were substantially decreased in *Rhizobium* bacteroids or nodules formed by Hup^+ strains. This indicates that H_2-oxidizing strains of *Rhizobium* can utilize H_2 as an energy source and, as a consequence, conserve more of their carbohydrate supply than Hup^- strains (106).

The magnitude of the beneficial effects of H_2 recycling in legume nodules has been determined in several experiments, and the theoretical expectations were considered by Evans et al (75, 80) and Pate et al (134). A series of experiments have conducted in which Hup^+ and Hup^- strains of *Rhizobium* have been compared as inoculants for legumes. A majority of these (Table 1) have indicated that yield and total N content of plant materials were increased by use of Hup^+ inoculants. Since Hup^- revertible mutants have not been available for comparison until recently (110), most of the experiments have been conducted with single wild-type Hup^+ and Hup^- strains or with groups of wild-type Hup^+ and Hup^- strains. Comparison of single strains or small groups of Hup^+ and Hup^- strains, however, are open to the criticism that genetic differences other than hydrogenase may have influenced the performance. Theoretically, comparisons of Hup^- mutant strains with their Hup^+ revertants as inoculants should be ideal for the evaluation of advantages of H_2 recycling capability toward N_2 fixation and growth of legumes. However, increases in favor of the Hup^+ strains in such comparisons have ranged near 10%, a value that in some cases has been within the range of experimental error (Table 1). In the use of mutants and revertants as inoculants, consideration needs to be given to the possible deleterious effects on symbiotic performance of selection procedures and the introduction of antibiotic resistance markers.

Recently, DeJong et al (50) transferred plasmid pIJ1008, a recombinant

Table 1   Effects of Hup$^+$ and Hup$^-$ strains of *Rhizobium* on dry matter and total N content of legumes

| Types of Hup$^+$ and Hup$^-$ comparisons | Inoculants | Increase from use of Hup$^+$ strains (%)[a] | | References |
		Dry matter	Total N	
Single wild-type strains	*R. japonicum*	24	31	Schubert et al (163)
Single wild-type strains	Cowpea *Rhizobium*	11	15	Schubert et al (163)
Groups of wild-type strains	*R. japonicum*	16	26	Albrecht et al (4)
Parent and nonrevertible mutants	*R. japonicum*	32	49	Albrecht et al (4)
Groups of wild-type strains	*R. japonicum*	NS[b]	8.4	Hanus et al (94)
Parent and revertible mutants	*R. japonicum*	17	36	Lepo et al (110)
Groups of wild-type strains	Cowpea *Rhizobium*	13–56	21–46	Pahwa & Dogra (131)
Groups of wild-type strains	*R. japonicum*	NS	NS	Gibson et al (85)
Groups of wild-type strains	*R. leguminosarum*	NS	NS	Nelson & Child (127)
Single strains	Cowpea *Rhizobium*	NS[c]	NS	Rainbird et al (147)
Genetically engineered strains	*R. leguminosarum*	14–56	18–41	DeJong et al (50)
Revertants and mutants	*R. japonicum* PJ17; PH17-1	NS	NS	G. Eisbrenner et al, unpublished
Revertants and mutants	*R. japonicum* PJ18; PJ18-1	10–11	10–11	G. Eisbrenner et al, unpublished

[a] Increase in all experiments refers to whole plants except those of Hanus et al (94) where data apply to grain only.
[b] NS indicates no significant differences where observed.
[c] Plants inoculated with Hup$^+$ strain produced significantly more root weight.

of two indigenous *R. leguminosarum* plasmids, into *R. leguminosarum* strains 300 and 3622. The newly produced strains 3960 and 3963, which were Hup$^+$ and possessed increased genetic information for nitrogenase, nodulation, and other unknown functions, fixed significantly more N$_2$ and had higher dry weights than the recipient strains. DeJong et al (50) speculated that the Hup$^+$ characteristic might have caused the increased effectiveness because the transfer of a plasmid which encoded only Nif and Nod determinants, plus other genetic information into the two recipient *R. leguminosarum* strains, resulted in no increased N$_2$-fixing effectiveness.

Rainbird et al (147) have measured a striking effect of Hup$^+$ strains of cowpea *Rhizobium* on the carbon economy of *Vigna unguiculata*. Over the whole growth period, nodules formed by Hup$^+$ strains respired 10% less carbohydrate than nodules formed by Hup$^-$ strains. The conserved carbohydrates apparently were partitioned into additional root growth (147). Although the more extensive root development may not be important in the support plants in greenhouse experiments where ideal growth conditions are provided, a more extensive root system may be advantageous in field conditions (147). Gibson et al (85) found no significant effect of the addition of H$_2$ on ^{15}N$_2$ fixation or C$_2$H$_2$ reduction rates by Hup$^+$ strains of cowpea *Rhizobium*. Furthermore, they observed no significant correlation between total N or dry weight of the host plant and the relative efficiency (162) of

R. japonicum strains. The data so far presented for clover, peas, and lupin indicated that no strain contained sufficient uptake hydrogenase activity to recycle a significant amount of H_2 lost during N_2 fixation. The same is true for Hup$^+$ *R. leguminosarum* isolates tested by Nelson & Child (127), who also reported no significant difference in plant dry weight and N content of four-week-old pea plants. Recently, Nelson & Salminen (128) have identified two strains of *R. leguminosarum* (128C30 and H106) which have very high H_2 uptake rates. These strains and revertible mutants derived from them should provide good experimental inoculant strains for the evaluation of the benefits of H_2 recycling systems in *R. leguminosarum*.

From the demonstrated reproducible physiological benefits from H_2 recycling capability and the increases in dry matter yields and N contents of legumes inoculated with Hup$^+$ strains (Table 1), it seems that the weight of evidence so far indicates that efficient H_2 recycling capability is a desirable characteristic that should be present in strains of *Rhizobium* that are utilized as commercial inoculants for legumes.

Free-Living Rhizobium

Until recently, all *Rhizobium* species were considered to be heterotrophs. The demonstration (116) of a CO_2 requirement for growth of several *Rhizobium* species and more recent information on the conditions needed for derepression of the uptake hydrogenase in Hup$^+$ strains of *R. japonicum* (119, 167) provided the background for the finding that Hup$^+$ strains of *R. japonicum* are capable of growing as chemolithotrophs, using H_2 and CO_2 as sources of energy and carbon, respectively (95). Lepo et al (109) showed that at least 89% of the carbon in chemolithotrophically cultured *R. japonicum* was derived from $^{14}CO_2$ and that activity of ribulose-1,5-bisphosphate carboxylase was sufficient to account for their growth rate. The properties of a homogeneous preparation of the carboxylase from chemolithotrophically cultured *R. japonicum* were described by Purohit et al (146). By use of a programmed increase in O_2 supply, dense chemolithotrophic cultures of *R. japonicum* have been grown which tolerated up to 8% O_2 in the aeration stream (109). In the initial stages of growth, however, it was necessary to maintain O_2 partial pressures below 1%. It seems probable that continued research in this area will reveal the conditions that are necessary for chemolithotrophic cultures of Hup$^+$ strains of *Rhizobium* species other than *R. japonicum*.

The capacity for chemolithotrophic growth of Hup$^+$ cells might give them an advantage in their survival as free-living bacteria outside the roots. When Hup$^+$ strains of *R. japonicum* are provided with H_2 by anaerobic fermentative bacteria or by Hup$^-$ N_2-fixing organisms, there is no requirement for competition for organic carbon sources in the soil because carbon may be obtained through the ribulose-1,5-bisphosphate carboxylase reaction (109, 146). When *R. japonicum* is cultured under chemolithotrophic conditions, it needs to be supplied with either NO_3^- or NH_4^+ as a nitrogen source. So far, there have been no successful reports of N_2 fixation by chemolithotrophically cultured *Rhizobium japonicum*. Chemolithotrophic growth without simultaneous N_2 fixation recently has been demonstrated with cultures of *Derxia gummosa* (135, 156), *Azospirillum brasiliense*, and *A. lipoferrum* (156). It seems probable that the discovery of chemolithotrophy in *Rhizobium japonicum* has evolutionary, taxonomic and perhaps practical considerations in the biology of *Rhizobium*.

CONCLUSIONS

Under optimum conditions, about 75 percent of the electron flow through the nitrogenase reaction is utilized for the reduction of N_2, and the remainder is consumed in the reduction of protons to H_2. The apparent waste of energy through H_2 evolution can be much greater than 25 percent. The magnitude of loss is influenced greatly by those factors that affect the turnover rate of the MoFe component of nitrogenase. In N_2-fixing organisms, the extent of H_2 evolution depends to a large extent on the presence of an H_2-oxidizing system which, when present, participates in an H_2 recycling process that conserves some of the energy expended in nitrogenase-catalyzed H_2 evolution. Most of the N_2-fixing symbioses involving blue-green algae or actinomycetous endophytes possess effective H_2 recycling systems, but H_2 recycling capacity is sparsely distributed among the species of *Rhizobium*. The H_2 oxidation pathway of *Rhizobium japonicum* bacteroids includes a membrane-bound hydrogenase, cytochromes of the *b* and *c* types, and very probably ubiquinone. Nutritional and enzymological investigations with several different N_2-fixing microorganisms indicate that Ni is specifically required for hydrogenase activity and is a constituent of the enzyme.

There is convincing evidence that the capacity for H_2 oxidation in N_2-fixing microorganisms provides physiological benefits such as H_2-supported ATP synthesis, respiratory protection of nitrogenase from O_2 damage, and possibly the prevention of inhibition by H_2 of the nitrogenase in nodules. In addition, the Hup$^+$ strains of *R. japonicum* are capable of growing chemolithotrophically, a capacity that may provide survival and competitive advantages to free-living *R. japonicum* in soils. The majority of results with field and greenhouse trials in which Hup$^+$ and Hup$^-$ strains of *Rhizobium* were compared as inoculants for legumes have shown advantages from use of the Hup$^+$ strains. More definitive experiments need to be conducted in which Hup$^-$ mutants and their Hup$^+$ revertants as legume inoculants are compared. The evidence available so far indicates that the capacity for H_2 recycling is one of many desirable attributes for N_2-fixing symbiotic associations. The progress that has been made in cloning and characterizing Hup genes should be useful in the transfer of Hup to those N_2-fixing endophytes that lack efficient H_2 recycling capabilities. During the past 6 years, our appreciation of the potential significance of apparent energy waste during nitrogenase-catalyzed H_2 evolution and of the factors affecting the expression of a functional H_2-oxidizing system have increased remarkably. It is hoped that our improved understanding will contribute toward the development of more efficient nodulated legumes and other N_2-fixing associations for use in the production of food for the increasing world population.

ACKNOWLEDGMENTS

We thank Mrs. Sheri Woods-Haas for typing, Dr. M. A. Cantrell and Dr. P. J. Bottomley for reading and commenting on the manuscript. We are grateful to all our colleagues who made available to us their research data that are in press. During the time that this review was written, G. E. received a postdoctoral scholarship from the Deutsche Forschungsgemeinschaft. Research cited from our laboratory was supported by a National Science Foundation grant PCM 81–18148; a US Department of Agriculture, Science and Education Administration, CRGO grant 78–49–2411–0–1–0991-1; by a National Institute of Health Biomedical Research Support grant RR 07079; and by the Oregon Agricultural Experiment Station, from which this is Technical Paper No. 6496.

Literature Cited

1. Adams, M. W. W., Mortenson, L. E., Chen, J. -S. 1981. Hydrogenase. Biochim. Biophys. Acta 594:105–76
2. Ahmed, S., Evans, H. J. 1961. The essentiality of cobalt for soybean plants grown under symbiotic conditions. Proc. Natl. Acad. Sci. USA 47:24–36
3. Albracht, S. P. J., Graf, E. G., Thauer, R. K. 1982. The EPR properties of nickel in hydrogenase from Methanobacterium thermoautotrophicum. FEBS Lett. 140:311–13
4. Albrecht, S. L., Maier, R. J., Hanus, F. J., Russell, S. A., Emerich, D. W., Evans, H. J. 1979. Hydrogenase in R. japonicum increases nitrogen fixation by nodulated soybeans. Science 203:1255–57
5. Arima, Y. 1981. Respiration and efficiency of nitrogen fixation by nodules formed with a hydrogen uptake positive

strain of Rhizobium japonicum. Soil Sci. Plant Nutr. 27:115–20
6. Arp, D. J., Burris, R. H. 1979. Purification and properties of the particulate hydrogenase from the bacteroids of soybean root nodules. Biochim. Biophys. Acta 570:221–30
7. Arp, D. J., Burris, R. H. 1981. Kinetic mechanism of the hydrogen-oxidizing hydrogenase from soybean nodule bacteroids. Biochemistry 20:2234–40
8. Arp, D. J., Burris, R. H. 1982. Isotope exchange and discrimination by the H_2-oxidizing hydrogenase from soybean root nodules. Biochim. Biophys. Acta 700:7–15
9. Bartha, R., Ordal, E. J. 1965. Nickel-dependent chemolithotrophic growth of two Hydrogenomonas strains. J. Bacteriol. 89:1015–19
10. Benemann, J. R., Weare, N. M. 1974.

Nitrogen fixation by *Anabaena cylindrica*. III. Hydrogen-supported nitrogenase activity. *Arch. Microbiol.* 101:401-8

11. Benson, D. R., Arp, D. J., Burris, R. H. 1979. Cell-free nitrogenase and hydrogenase from actinorhizal root nodules. *Science* 205:688-89

12. Benson, D. R., Arp, D. J., Burris, R. H. 1980. Hydrogenase in actinorhizal root nodules and root nodule homogenates. *J. Bacteriol.* 142:138-44

13. Bergersen, F. J., Turner, G. L. 1980. Properties of terminal oxidase systems of bacteroids from root nodules of soybean and cowpea and of N_2-fixing bacteria grown in continuous culture. *J. Gen. Microbiol.* 118:235-52

14. Beringer, J. E. 1980. Development of *Rhizobium* genetics—4th Fleming Lecture. *J. Gen. Microbiol.* 116:1-7

15. Berndt, H., Ostwal, K. -P., Lalucat, J., Schumann, C., Mayer, F., Schlegel, H. G. 1976. Identification and physiological characterization of the nitrogen fixing bacterium *Corynebacterium autotrophicum* GZ 29. *Arch. Microbiol.* 108:17-26

16. Berndt, H., Wölfle, D. 1978. Hydrogenase: its role as electron generating enzyme in the nitrogen fixing hydrogen bacterium *Xanthobacter autotrophicus*. In *Hydrogenases, Their Catalytic Activity, Structure and Function,* ed. H. G. Schlegel, K. Schneider, pp. 327-51. Göttingen: Goltze, K. G. 453 pp.

17. Bernhard, T., Gottschalk, G. 1978. The hydrogenase of *Escherichia coli*, purification, some properties and the function of the enzyme. See Ref. 16, pp. 199-208

18. Bethlenfalvay, G. J., Phillips, D. A. 1977. Ontogenetic interactions between photosynthesis and symbiotic nitrogen fixation in legumes. *Plant Physiol.* 60:419-21

19. Bethlenfalvay, G. J., Phillips, D. A. 1977. Effect of light intensity on efficiency of carbon dioxide and nitrogen reduction in *Pisum sativum* L. *Plant Physiol.* 60:868-71

20. Bishop, N. I., Jones, L. W. 1978. Alternate fates of the photochemical reducing power generated in photosynthesis: hydrogen production and nitrogen fixation. *Curr. Top. Bioenerg.* 8:3-31

21. Böhme, H., Reimer, S., Trebst, A. 1971. The effect of dibromothymoquinone, an antagonist of plastoquinone on noncyclic and cyclic electron flow systems in isolated chloroplasts. *Z. Naturforsch. Teil B* 26:341-52

22. Bothe, H. 1982. Hydrogen production by algae. *Experientia* 38:59-64

23. Bothe, H., Distler, E., Eisbrenner, G. 1978. Hydrogen metabolism in blue-green algae. *Biochimie* 60:277-89

24. Bothe, H., Eisbrenner, G. 1978. Aspects of hydrogen metabolism in blue-green algae. See Ref. 16, pp. 353-69

25. Bothe, H., Eisbrenner, G. 1981. The hydrogenase-nitrogenase relationship in nitrogen-fixing organisms. In *Biology of Inorganic Nitrogen and Sulfur,* ed. H. Bothe, A. Trebst, pp. 141-50. Berlin: Springer. 384 pp.

26. Bothe, H., Neuer, G., Kalbe, I., Eisbrenner, G. 1980. Electron donors and hydrogenase in nitrogen-fixing microorganisms. In *Nitrogen Fixation,* ed. W. D. P. Stewart, J. R. Gallon, pp. 83-111. London: Academic. 451 pp.

27. Bothe, H. Tennigkeit, J., Eisbrenner, G. 1977. The utilization of molecular hydrogen by the blue-green alga *Anabaena cylindrica*. *Arch. Microbiol.* 114:43-49

28. Bothe, H., Tennigkeit, J., Eisbrenner, G., Yates, M. G. 1977. The hydrogenase-nitrogenase relationship in the blue-green alga *Anabaena cylindrica*. *Planta* 133:237-42

29. Brewin, N. J., Beynon, J. L., Johnston, A. W. B. 1981. The role of *Rhizobium* plasmids in host specificity. In *Genetic Engineering of Symbiotic Nitrogen Fixation and Conservation of Fixed Nitrogen,* ed. J. M. Lyons, R. C. Valentine, A. A. Phillips, A. W. Rains, R. C. Huffaker, pp. 65-78. New York: Plenum. 698 pp.

30. Brewin, N. J., Dejong, T. M., Phillips, D. A., Johnston, A. W. B. 1980. Co-transfer of determinants for hydrogenase activity and nodulation ability in *Rhizobium leguminosarum*. *Nature* 288:77-79

31. Bulen, W. A., Burns, R. C., LeComte, J. R. 1964. Nitrogen Fixation: cell-free system with extracts of *Azotobacter*. *Biochem. Biophys. Res. Commun.* 17:265-71

32. Bulen, W. A., Burns, R. C., LeComte, J. R. 1965. Nitrogen fixation: hydrosulfite as electron donor with cell-free preparations of *Azotobacter vinelandii* and *Rhodospirillum rubrum*. *Proc. Natl. Acad. Sci. USA* 53:532-39

33. Bulen, W. A., LeComte, J. R. 1966. The nitrogenase system from *Azotobacter*: two-enzyme requirement for N_2 reduction, ATP-dependent H_2 evolution and ATP hydrolysis. *Proc. Natl. Acad. Sci. USA* 56:979-86

34. Burgess, B. K., Wherland, S., Newton, W. E., Stiefel, E. I. 1981. Nitrogenase reactivity: insight into the nitrogen-fixing process through hydrogen inhibition and HD-forming reactions. *Biochemistry* 20:5140-46

35. Burns, R. C., Bulen, W. A. 1966. A procedure for the preparation of extracts from *Rhodospirillum rubrum* catalyzing N_2 reduction and ATP-dependent H_2

evolution. *Arch. Biochem. Biophys.* 113:461–63

36. Burns, R. C., Hardy, R. W. F., eds. 1975. *Nitrogen Fixation in Bacteria and Higher Plants.* Berlin: Springer. 189 pp.

37. Burris, R. H., Arp, D. J., Benson, D. R., Emerich, D. W., Hageman, R. V., et al. 1980. The biochemistry of nitrogenase. See Ref. 26, pp. 37–54

38. Burris, R. H., Arp, D. J., Hageman, R. V., Houchins, J. P., Sweet, W. J., Tso, M. -Y. 1981. Mechanism of nitrogenase action. In *Current Perspectives in Nitrogen Fixation,* ed. A. H. Gibson, W. E. Newton, pp. 56–66. New York: Elsevier/ North Holland. 534 pp.

39. Cantrell, M. A., Haugland, R. A., Evans, H. J. 1982. Construction of a *Rhizobium japonicum* gene bank and use in the isolation of a hydrogen uptake gene. Abstr. 1st Int. Symp. on "Molecular Genetics of the Bacteria-Plant Interaction," Bielefeld, West Germany

40. Cantrell, M. A., Hickok, R. E., Evans, H. J. 1982. Identification and characterization of plasmids in hydrogen uptake positive and hydrogen uptake negative strains of *Rhizobium japonicum. Arch. Microbiol.* 131:102–6

41. Carter, K. R., Jennings, N. T., Hanus, J., Evans, H. J. 1978. Hydrogen evolution and uptake by nodules of soybeans inoculated with different strains of *Rhizobium japonicum. Can. J. Microbiol.* 24:307–11

42. Carter, K. R., Rawlings, J., Orme-Johnson, W. H., Becker, R. R., Evans, H. J. 1980. Purification and characterization of a ferredoxin from *Rhizobium japonicum* bacteroids. *J. Biol. Chem.* 255:4213–23

43. Chan, Y. K., Nelson, L. M., Knowles, R. 1980. Hydrogen metabolism of *Azospirillum brasilense* in nitrogen-free medium. *Can. J. Microbiol.* 26:1126–31

44. Chatt, J. 1980. Chemistry relevant to the biological fixation of nitrogen. See Ref. 26, pp. 1–17

45. Chatt, J. 1981. Towards new catalysts for nitrogen fixation. See Ref. 38, pp. 15–21

46. Conrad, R., Seiler, W., 1980. Contribution of hydrogen production by biological nitrogen fixation to the global hydrogen budget. *J. Geophys. Res.* 85:5493–98

47. Daday, A., Platz, R. A., Smith, G. D. 1977. Anaerobic and aerobic hydrogen gas formation by the blue-green alga *Anabaena cylindrica. Appl. Environ. Microbiol.* 34:478–83

48. Daniel, R. M. 1979. The occurrence and role of ubiquinone in electron transport to oxygen and nitrate in aerobically, anaerobically and symbiotically grown *R. japonicum. J. Gen. Microbiol.* 110:333–37

49. Dart, P. J., Day, J. M. 1971. Effects of incubation, temperature and oxygen tension on nitrogenase activity of legume root nodules. *Plant Soil* 1971(SI):167–84

50. DeJong, T. M., Brewin, N. J., Johnston, A. W. B., Phillips, D. A. 1982. Improvement of symbiotic properties in *Rhizobium leguminosarum* by plasmid transfer. *J. Gen. Microbiol.* 128:1829–38

51. Denarie, J., Rosenberg, C., Boistard, P., Truchet, G., Casse-Delbart, F. 1981. Plasmid control of symbiotic properties in *Rhizobium meliloti.* See Ref. 38, pp. 137–41

52. Diekert, G., Jaenchen, R., Thauer, R. K. 1980. Biosynthetic evidence for a nickel tetrapyrole structure of factor F_{430} from *Methanobacterium theramoautotrophicum. FEBS Lett.* 119:118–20

53. Diekert, G., Klee, B., Thauer, R. K. 1980. Nickel, a component of factor F_{430} from *Methanobacterium thermoautotrophicum. Arch. Microbiol.* 124:103–6

54. Diekert, G., Konheiser, U., Piechulla, K., Thauer, R. K. 1981. Nickel requirement and factor F_{430} content of methanogenic bacteria. *J. Bacteriol.* 148:459–64

55. Dilworth, M. J. 1966. Acetylene reduction by nitrogen-fixing preparations from *Costridium pasteurianum. Biochim. Biophys. Acta* 127:285–94

56. Dixon, R. O. D. 1967. Hydrogen uptake and exchange by pea root nodules. *Ann. Bot.* 31:179–88

57. Dixon, R. O. D. 1968. Hydrogenase in pea root nodule bacteroids. *Arch. Mikrobiol.* 62:272–83

58. Dixon, R. O. D. 1972. Hydrogenase in legume root nodule bacteroids: occurrence and properties. *Arch. Mikrobiol.* 85:193–201

59. Dixon, R. O. D. 1978. Nitrogenase-hydrogenase interrelationships in rhizobia. *Biochimie* 60:233–36

60. Dixon, R. O. D., Blunden, E. A. G., Searl, J. W. 1981. Inter-cellular space and hydrogen diffusion in root nodules of pea *Pisum sativum* and lupine *Lupinus alba. Plant Sci. Lett.* 23:109–16

61. Drevon, J. J., Frazier, L., Russell, S. A., Evans, H. J. 1982. Respiratory and nitrogenase activities of soybean nodules formed by hydrogen uptake negative (Hup⁻) mutant and revertant strains of *Rhizobium japonicum* characterized by protein patterns. *Plant Physiol.* 70: 1341–46

62. Eady, R. R. 1973. Nigrogenase of *Klebsiella pneumoniae.* Interaction of the component proteins studied by ultracentrifugation. *Biochem. J.* 135: 531–35

63. Eady, R. R., Imam, S., Lowe, D. J., Miller, R. W., Smith, B. E., Thorneley, R. N. F. 1980. The molecular enzymology of nitrogenase. See Ref. 26, pp. 19–35

64. Eisbrenner, G., Bothe, H. 1979. Modes of electron transfer from molecular hydrogen in *Anabaena cylindrica*. *Arch. Microbiol.* 123:37–46

65. Eisbrenner, G., Distler, E., Floener, L., Bothe, H. 1978. The occurrence of the hydrogenase in some blue-green algae. *Arch. Microbiol.* 118:117–84

66. Eisbrenner, G., Evans, H. J. 1982. Carriers in the electron transport from molecular hydrogen to oxygen in *Rhizobium japonicum* bacteroids. *J. Bacteriol.* 149:1005–12

67. Eisbrenner, G., Evans, H. J. 1982. Spectral evidence for a component involved in hydrogen metabolism of soybean nodule bacteroids. *Plant Physiol.* 70:1667–72

68. Eisbrenner, G., Hickok, R. E., Evans, H. J. 1982. Cytochrome patterns in *Rhizobium japonicum* cells grown under chemolithotrophic conditions. *Arch. Microbiol.* 132:230–35

69. Eisbrenner, G., Roos, P., Bothe, H. 1981. The number of hydrogenases in cyanobacteria. *J. Gen. Microbiol.* 125:383–90

70. Ellefson, W. L., Whitman, W. B., Wolfe, R. S. 1982. Nickel-containing factor F_{430}:chromophore of the methylreductase of *Methanobacterium*. *Proc. Natl. Acad. Sci. USA* 79:3707–10

71. Emerich, D. W., Albrecht, S. L., Russell, S. A., Ching, T. M., Evans, H. J. 1980. Oxyleghemoglobin-mediated hydrogen oxidation by *Rhizobium japonicum* USDA 122DES bacteroids. *Plant Physiol.* 65:605–9

72. Emerich, D. W., Hageman, R. V., Burris, R. H. 1981. Interactions of dinitrogenase and dinitrogenase reductase nitrogen fixing enzymes. *Adv. Enzymol. Relat. Areas Mol. Biol.* 52:1–22

73. Emerich, D. W., Ruiz-Argüeso, T., Ching, T. M., Evans, H. J. 1979. Hydrogen-dependent nitrogenase activity and ATP formation in *R. japonicum* bacteroids. *J. Bacteriol.* 173:153–60

74. Emerich, D. W., Ruiz-Argüeso, T., Russell, S. A., Evans, H. J. 1980. Investigation of the H_2 oxidation system in *Rhizobium japonicum* 122 DES nodule bacteroids. *Plant Physiol.* 66:1061–66

75. Evans, H. J., Eisbrenner, G., Cantrell, M. A., Russell, S. A., Hanus, F. J. 1982. The present status of hydrogen recycling in legumes. *Isr. J. Bot.* 31:72–88

76. Evans, H. J., Emerich, D. W., Lepo, J. E., Maier, R. J., Carter, K. R., et al. 1980. The role of hydrogenase in nodule

bacteroids and free-living rhizobia. See Ref. 26, pp. 55–81

77. Evans, H. J., Emerich, D. W., Ruiz-Argüeso, T., Albrecht, S. L., Maier, R. J., et al. 1978. Hydrogen metabolism in legume nodules and rhizobia: some recent developments. See Ref. 16, pp. 287–306

78. Evans, H. J., Emerich, D. W., Ruiz-Argüeso, T., Maier, R. J., Albrecht, S. L. 1980. Hydrogen metabolism in the legume-*Rhizobium* symbiosis. In *Nitrogen Fixation*, ed. W. E. Newton, W. H. Orme-Johnson, 2:69–86. Baltimore: Univ. Park Press, 325 pp.

79. Evans, H. J., Lepo, J. E., Hanus, F. J., Purohit, K., Russell, S. A. 1981. Chemolithotrophy in *Rhizobium*. See Ref. 29, pp. 141–58

80. Evans, H. J., Purohit, K., Cantrell, M. A., Eisbrenner, G., Russell, S. A., et al. 1981. Hydrogen losses and hydrogenases in nitrogen-fixing organisms. See Ref. 38, pp. 84–96.

81. Friedrich, B., Heine, E., Finck, A., Friedrich, C. G. 1981. Nickel requirement for active hydrogenase formation in *Alcaligenes eutrophus*. *J. Bacteriol.* 145:1144–49

82. Friedrich, B., Hogrefe, C., Schlegel, H. G. 1981. Naturally occurring genetic transfer of hydrogen-oxidizing ability between strains of *Alcaligenes eutrophus*. *J. Bacteriol.* 147:198–205.

83. Friedrich, C. G., Schneider, K., Friedrich, B. 1982. Nickel—a component of catalytically active hydrogenase of *Alcaligenes eutrophus*. *J. Bacteriol.* 152:42–48

84. Gest, H., Kamen, M. D. 1949. Photoproduction of molecular hydrogen by *Rhodospirillum rubrum*. *Science* 109:558–59

85. Gibson, A. H., Dreyfus, B. L., Lawn, R. J., Sprent, J. I., Turner, G. L. 1981. Host and environmental factors affecting hydrogen evolution and uptake. See Ref. 38, p. 373

86. Graf, E. G., Thauer, R. K. 1981. Hydrogenase from *Methanobacterium thermoautotrophicum*, a nickel-containing enzyme. *FEBS Lett.* 136:165–69

87. Gruzinskii, I. V., Gogotov, I. N., Bechina, E. M., Semener, Y. N. 1977. Hydrogenase activity of hydrogen-oxidizing bacteria *Alcaligenes eutrophus*. *Microbiology* 46:510–15

88. Hadfield, K. L., Bulen, W. A. 1969. Adenosine triphosphate requirement of nitrogenase from *Azotobacter vinelandii*. *Biochemistry* 8:5103–8

89. Hageman, R. V., Burris, R. H. 1978. Kinetic studies on electron transfer

and interaction between nitrogenase components from *Azotobacter vinelandii. Biochemistry* 17:4117–24
90. Hageman, R. V., Burris, R. H. 1978. Nitrogenase and nitrogenase reductase associate and dissociate with each catalytic cycle. *Proc. Natl. Acad. Sci. USA* 75:2699–702
91. Hageman, R. V., Burris, R. H. 1979. Changes in the EPR signal of dinitrogenase from *Azotobacter vinelandii* during the lag period before hydrogen evolution begins. *J. Biol. Chem.* 254: 11189–92
92. Hageman, R. V., Burris, R. H. 1980. Electron allocation to alternative substrates of *Azotobacter* nitrogenase is controlled by the electron flux through dinitrogenase. *Biochim. Biophys. Acta* 591:63–75
93. Hageman, R. V., Orme-Johnson, W. H., Burris, R. H. 1980. Role of magnesium adenosine 5'-triphosphate in the hydrogen evolution reaction catalyzed by nitrogenase from *Azotobacter vinelandii. Biochemistry* 19:2333–42
94. Hanus, F. J., Albrecht, S. L., Zablotowicz, R. M., Emerich, D. W., Russell, S. A., Evans, H. J. 1981. The effect of the hydrogenase system in *Rhizobium japonicum* inocula on the nitrogen content and yield of soybean seed in field experiments. *Agron. J.* 73:368–72
95. Hanus, F. J., Maier, R. J., Evans, H. J. 1979. Autotrophic growth of H_2-uptake positive strains of *R. japonicum* in an atmosphere supplied with hydrogen gas. *Proc. Natl. Acad. Sci. USA* 76:1788–92
96. Haystead, A., Robinson, R., Stewart, W. D. P. 1970. Nitrogenase activity in extracts of heterocystous and nonheterocystous blue-green algae. *Arch. Mikrobiol.* 72:235–43
97. Hoch, G. E., Little, H. N., Burris, R. H. 1957. H_2 evolution from soybean root nodules. *Nature* 179:430–31
98. Hoch, G. E., Schneider, K. C., Burris, R. H. 1960. Hydrogen evolution and exchange, and conversion of N_2O to N_2 by soybean root nodules. *Biochim. Biophys. Acta* 37:273–79
99. Houchins, J. P., Burris, R. H. 1981. Light and dark reactions of the uptake hydrogenase in *Anabaena* 7120. *Plant Physiol.* 68:712–16
100. Hyndman, L. A., Burris, R. H., Wilson, P. W. 1953. Properties of hydrogenase from *Azotobacter vinelandii. J. Bacteriol.* 65:522–31
101. Kamen, M. D., Gest, H. 1949. Evidence for a nitrogenase system in the photosynthetic bacterium *Rhodospirillum rubrum. Science* 109:560

102. Koch, B., Evans, H. J., Russell, S. A. 1967. Properties of the nitrogenase system in cell-free extracts of bacteroids from soybean root nodules. *Proc. Natl. Acad. Sci. USA* 58:1343–50
103. Kondorosi, A., Banfalvi, Z., Sakanyan, V., Dusha, I., Kiss, A. 1981. Location of nodulation and nitrogen fixation genes on a high molecular weight plasmid of *Rhizobium meliloti*. See Ref. 38, p. 407
104. Krasna, A. I. 1979. Hydrogenase: properties and applications. *Enzyme Microbiol. Technol.* 1:165–72
105. Lambert, G. R., Smith, D. G. 1981. Hydrogen metabolism of cyanobacteria (blue-green algae). *Biol. Rev.* 56:589–660
106. Layzell, D. B., Rainbird, R. M., Atkins, C. A., Pate, J. S. 1979. Economy of photosynthate use in nitrogen-fixing legume nodules. *Plant Physiol.* 64:888–91
107. Lee, S. B., Wilson, P. W. 1943. Hydrogenase and nitrogen fixation by *Azotobacter. J. Biol. Chem.* 151:377–85
108. LeGall, J., Ljungdahl, P. O., Moura, I., Peck, H. D., Xavier, A. V., et al. 1982. The presence of redox-sensitive nickel in the periplasmic hydrogenase from *Desulvibrio gigas. Biochem. Biophys. Res. Commun.* 106:610–16
109. Lepo. J. E., Hanus, F. J., Evans, H. J. 1980. Further studies on the chemoautotrophic growth of hydrogen uptake positive strains of *R. japonicum. J. Bacteriol.* 141:664–70
110. Lepo, J. E., Hickok, R. E., Cantrell, M. A., Russell, S. A., Evans, H. J. 1981. Revertible hydrogen uptake-deficient mutants of *Rhizobium japonicum. J. Bacteriol.* 146:614–20
111. Lim, S. T. 1978. Determination of hydrogenase in free-living cultures of *Rhizobium japonicum* and energy efficiency of soybean nodules. *Plant Physiol.* 62:609–11
112. Lim, S. T., Hennecke, H., Scott, D. B. 1979. Effect of cyclic guanosine 3',5'-monophosphate on nitrogen fixation in *Rhizobium japonicum. J. Bacteriol.* 139:256–63
113. Lim, S. T., Shanmugam, K. T. 1979. Regulation of hydrogen utilization in *Rhizobium japonicum* by cyclic AMP. *Biochim. Biophys. Acta* 584:479–92
114. Lim, S. T., Uratsu, S. L., Weber, D. F., Keyser, H. H. 1981. Hydrogen uptake (hydrogenase) activity of *Rhizobium japonicum* strains forming nodules in soybean production areas of the U.S.A. See Ref. 29, pp. 131–36
115. Lowe, D. J., Eady, R. R., Thorneley, R. N. F. 1978. Electron-paramagnetic-

resonance studies on nitrogenase of *Klebsiella pneumoniae:* evidence for acetylene- and ethylene-nitrogenase transient complexes. *Biochem. J.* 173:277–90

116. Lowe, R. H., Evans, H. J. 1962. Carbon dioxide requirement for growth of legume nodule bacteria. *Soil Sci.* 94:351–56

116a. Ludwig, R. A., Raleigh, E. A., Duncan, M. J., Signer, E. R., Gibson, A. H., et al. 1979. Further examination of presumptive *Rhizobium trifolii* mutants that nodulate *Glycine max. Proc. Natl. Acad. Sci. USA* 76:3942–46

117. Maier, R. J. 1981. *Rhizobium japonicum* mutant strains unable to grow chemoautotrophically. *J. Bacteriol.* 145:533–40

118. Maier, R. J., Campbell, N. E. R., Hanus, F. J., Simpson, F. B., Russell, S. A., Evans, H. J. 1978. Expression of hydrogenase activity in free-living *Rhizobium japonicum. Proc. Natl. Acad. Sci. USA* 75:3258–62

119. Maier, R. J., Hanus, F. J., Evans, H. J. 1979. Regulation of hydrogenase in *R. japonicum. J. Bacteriol.* 137:824–29

120. Maier, R. J., Merberg, D. M. 1982. *Rhizobium japonicum* mutants that are hypersensitive to repression of H_2 uptake by oxygen. *J. Bacteriol.* 150:161–67

121. Maier, R. J., Mutaftschiev, S. 1982. Reconstitution of H_2 oxidation activity from H_2 uptake-negative mutants of *Rhizobium japonicum* bacteroids. *J. Biol. Chem.* 257:2092–96

122. Maier, R. J., Postgate, J. R., Evans, H. J. 1978. Mutants of *R. japonicum* unable to utilize hydrogen. *Nature* 276:494–95

123. McCrae, R. E., Hanus, J., Evans, H. J. 1978. Properties of the hydrogenase system in *Rhizobium japoncium* bacteroids. *Biochem. Biophys. Res. Commun.* 80:384–90

124. Moore, A. W. 1964. Note on nonleguminous nitrogen-fixing plants in Alberta. *Can. J. Bot.* 42:952–55

125. Mortenson, L. E. 1978. The role of dihydrogen and hydrogenase in nitrogen fixation. *Biochimie* 60:219–23

126. Mortenson, L. E., Chen, J.-S. 1974. Hydrogenase. In *Microbial Iron Metabolism,* ed. J. B. Nielands, pp. 231–82. London: Academic

127. Nelson, L. M., Child, J. J. 1981. Nitrogen fixation and hydrogen metabolism in *Rhizobium leguminosarum* isolates in pea *Pisum sativum* root nodules. *Can. J. Microbiol.* 27:1028–34

128. Nelson, L. M., Salminen, S. O. 1982. Uptake hydrogenase activity and ATP formation in *Rhizobium leguminosarum* bacteroids. *J. Bacteriol.* 151:989–95

129. Newton, J. W. 1976. Photoproduction of molecular hydrogen by a plant-algal symbiotic system. *Science* 191:559–60

130. O'Brian, M. R., Maier, R. J. 1982. *Electron transport components involved in hydrogenase oxidation in Rhizobium japonicum.* Abstr. Ann. Meet. Am. Soc. Microbiol., p. 153

130a. O'Gara, F., Shanmugam, K. T. 1978. Mutant strains of clover rhizobium *(Rhizobium trifolii)* that form nodules on soybean *(Glycine max). Proc. Natl. Acad. Sci. USA* 75:2343–47

131. Pahwa, K., Dogra, R. C. 1981. Hydrogen recycling system in mung bean *Vigna radiata Rhizobium* in relation to nitrogen fixation. *Arch. Microbiol.* 129:380–83

132. Partridge, C. D. P., Walker, C. C., Yates, M. G., Postgate, J. R. 1980. The relationship between hydrogenase and nitrogenase in *Azotobacter chroococcum:* effect of nitrogen sources on hydrogenase activities. *J. Gen. Microbiol.* 119:313–19

133. Partridge, C. D. P., Yates, M. G. 1982. Effect of chelating agents on hydrogenase in *Azotobacter chroococcum:* evidence that nickel is required for hydrogenase synthesis. *Biochem. J.* 204:339–44

134. Pate, J. S., Atkins, C. A., Rainbird, R. M. 1981. Theoretical and experimental costing of nitrogen fixation and related processes in nodules of legumes. See Ref. 38, pp. 105–16

135. Pedrosa, F. O., Döbereiner, J., Yates, M. G. 1980. Hydrogen-dependent growth and autotrophic carbon dioxide fixation in *Derxia. J. Gen. Microbiol.* 119:547–51

136. Pedrosa, F. O., Stephan, M., Döbereiner, J., Yates, M. G. 1982. Hydrogen-uptake hydrogenase activity in nitrogen-fixing *Azospirillum brasilense. J. Gen. Microbiol.* 128:161–66

137. Peschek, G. A. 1979. Aerobic hydrogenase activity in *Anacystis nidulans.* The oxyhydrogen reaction. *Biochim. Biophys. Acta* 548:203–15

138. Peschek, G. A. 1979. Anaerobic hydrogenase activity in *Anacystis nidulans* H_2-dependent photoreduction and related reactions. *Biochim. Biophys. Acta* 548:187–202

139. Peters, G. A., Evans, W. R., Toia, E. R. Jr. 1976. *Azolla-Anabaena azollae* relationship. IV. Photosynthetically driven, nitrogenase catalyzed H_2-production. *Plant Physiol.* 58:119–26

140. Peters, G. A., Ito, O., Tyagi, V. V. S., Kaplan, D. 1981. Physiological studies on N_2-fixing *Azolla.* See Ref. 29, pp. 343–62

141. Peterson, R. B., Burris, R. H. 1978. Hydrogen metabolism in isolated heterocysts of *Anabaena* 7120. *Arch. Microbiol.* 116:125–32

142. Peterson, R. B., Wolk, C. P. 1978. Localisation of an uptake hydrogenase in *Anabaena. Plant Physiol.* 61:688–91

143. Phelps, A. S., Wilson, P. M. 1941. Occurrence of hydrogenase in nitrogen fixing organisms. *Proc. Soc. Exp. Biol. Med.* 47:473–76

144. Phillips, D. A. 1980. Efficiency of symbiotic nitrogen fixation in legumes. *Ann. Rev. Plant Physiol.* 31:29–49

145. Poole, R. K., Haddock, B. A. 1975. Dibromothymoquinone: an inhibitor of aerobic electron transport at the level of ubiquinone in *Escherichia coli. FEBS Lett.* 52:13–16

146. Purohit, K., Becker, R. R., Evans, H. J. 1982. D-Ribulose-1,5-bisphosphate carboxylase/oxygenase from chemolithotrophically-grown *Rhizobium japonicum* and inhibition by D-4-phosphoerythronate. *Biochim. Biophys. Acta* 715:320–29

147. Rainbird, R. M., Atkins, C. A., Pate, J. S., Sandford, P. 1983. Significance of hydrogen evolution in the carbon and nitrogen economy of nodulated cowpea. *Plant Physiol.* 71:122–27

148. Repaske, R., Repaske, C. 1976. Quantitative requirements for exponential growth of *Alcaligenes eutrophus. Appl. Environ. Microbiol.* 32:585–91

149. Rivera-Ortiz, J. M., Burris, R. H. 1975. Interactions among substrates and inhibitors of nitrogenase. *J. Bacteriol.* 123:537–45

150. Robson, R. L., Postgate, J. R. 1980. Oxygen and hydrogen in biological nitrogen fixation. *Ann. Rev. Microbiol.* 34:183–207

151. Roelofsen, W., Akkermans, A. D. L. 1979. Uptake and evolution of H_2 and reduction of C_2H_2 by root nodules and nodule homogenates of *Alnus glutinosa. Plant Soil* 52:571–78

152. Ruiz-Argüeso, T., Cabrera, E., de Bertalmio, M. B. 1981. Selection of symbiotically energy efficient strains of *Rhizobium japonicum* by their ability to induce a H_2-uptake hydrogenase in the free-living state. *Arch. Microbiol.* 128:275–79

153. Ruiz-Argüeso, T., Emerich, D. W., Evans, H. J. 1979. Characteristics of the hydrogen oxidizing system in soybean nodule bacteroids. *Arch. Microbiol.* 121:199–206

154. Ruiz-Argüeso, T., Hanus, J., Evans, H. J. 1978. Hydrogen production and uptake by pea nodules as affected by strains of *Rhizobium leguminosarum. Arch. Microbiol.* 116:113–18

155. Ruiz-Argüeso, T., Maier, R. J., Evans, H. J. 1979. Hydrogen evolution from alfalfa and clover nodules and hydrogen uptake by free-living *R. Meliloti. Appl. Environ. Microbiol.* 37:582–87

156. Sampaio, M.-J. A. M., da Silva, E. M. R., Döbereiner, J., Yates, M. G., Pedrosa, F. O. 1981. Autotrophy and methylotrophy in *Derxia gummosa, Azospirillum brasilense* and *A. lipoferum.* See Ref. 38, p. 444

157. Schink, B., Schlegel, H. G. 1979. The membrane-bound hydrogenase of *Alcaligenes eutrophus. Biochim. Biophys. Acta* 567:315–24

158. Schlegel, H. G. 1974. Production, modification, and consumption of atmospheric trace gases by microorganisms. *Tellus* 26:11–20

159. Schlegel, H. G., Schneider, K. 1978. Introductory report: distribution and physiological role of hydrogenases in microorganisms. See Ref. 16, pp. 15–44

160. Schönheit, P., Moll, J., Thauer, R. K. 1979. Nickel, cobalt and molybdenum requirement for growth of *Methanobacterium thermoautotrophicum. Arch. Microbiol.* 123:105–7

161. Schubert, K. R., Engelke, J. A., Russell, S. A., Evans, H. J. 1977. Hydrogen reactions of nodulated leguminous plants I: effects of rhizobial strain and plant age. *Plant Physiol.* 60:651–54

162. Schubert, K. R., Evans, H. J. 1976. Hydrogen evolution: a major factor affecting the efficiency of nitrogen fixation in nodulated symbionts. *Proc. Natl. Acad. Sci. USA* 73:1207–11

163. Schubert, K. R., Jennings, N. T., Evans, H. J. 1978. Hydrogen reactions of nodulated leguminous plants II. Effects on dry matter accumulation and nitrogen fixation. *Plant Physiol.* 61:398–401

164. Shug, A. L., Hamilton, P. B., Wilson, P. W. 1956. Hydrogenase and nitrogen fixation. In *Inorganic Nitrogen Metabolism*, ed. W. D. McElroy, B. Glass, pp. 344–60. Baltimore: Johns Hopkins Press. 728 pp.

165. Silsbury, J. H. 1981. CO_2 exchange and dinitrogen fixation of subterranean clover in response to light level. *Plant Physiol.* 67:599–602

166. Sim, E., Vignais, P. M. 1978. Hydrogenase activity in *Paracoccus denitrificans.* Partial purification and interaction with the electron transport chain. *Biochimie* 60:307–14

167. Simpson, F. B., Maier, R. J., Evans, H. J. 1979. Hydrogen-stimulated CO_2 fixa-

tion and coordinate induction of hydrogenase and ribulose-bisphosphate carboxylase in a H_2-uptake positive strain of *Rhizobium japonicum*. *Arch. Microbiol.* 123:1–8

168. Smith, B. E., Lowe, D. J., Postgate, J. R., Richards, R. L., Thornely, R. N. F. 1981. Studies on nitrogen reduction by nitrogenase from *Klebsiella pneumoniae*. See Ref. 38, pp. 67–70

169. Smith, L. A., Hill, S., Yates, M. G. 1976. Inhibition by acetylene of conventional hydrogenase in nitrogen-fixing bacteria. *Nature* 262:209–10

170. Stanier, R. Y., Doudoroff, M., Adelberg, E. A., eds. 1970. *The Microbial World.* Englewood Cliffs: Prentice-Hall. 873 pp.

171. Suzuki, T., Maruyama, Y. 1979. Studies on hydrogenases in legume root nodule bacteroids: effect of ATP. *Agric. Biol. Chem.* 43:1833–39

172. Tabillion, R., Weber, F., Kaltwasser, H. 1980. Nickel requirement of chemolithotrophic growth in hydrogen-oxidizing bacteria. *Arch. Microbiol.* 124:131–36

173. Tait, R. C., Anderson, K., Cangelosis, G., Lim, S. T. 1981. Hydrogen uptake (Hup) plasmids: characterization of mutants and regulation of the expression of hydrogenase. See Ref. 29, pp. 131–36

174. Takakuwa, S., Wall, J. D. 1981. Enhancement of hydrogenase activity in *Rhodopseudomonas capsulata* by nickel. *FEMS Microbiol. Lett.* 12:359–64

175. Thauer, R. K., Diekert, G., Schönheit, P. 1980. Biological role of nickel. *Trends Biochem. Sci.* 5:304–6

176. Van der Werf, A. N., Yates, M. G. 1978. Hydrogenase from nitrogen-fixing *Azotobacter chroococcum*. See Ref. 16, pp. 307–26

177. Von Jagow, G., Sebald, W. 1980. *b*-Type cytochrome. *Ann. Rev. Biochem.* 49:281–314

178. Walker, C. C., Yates, M. G. 1978. The

hydrogen cycle in nitrogen-fixing *Azotobacter chroococcum*. *Biochimie.* 60:225–31

179. Weaver, P. F., Lien, S., Seibert, M. 1980. Photobiological production of hydrogen. *Sol. Energy* 24:3–45

180. Wherland, S., Burgess, B. K., Stiefel, E. I., Newton, W. E. 1981. Nitrogenase reactivity: effects of component ratio on electron flow and distribution during nitrogen fixation. *Biochemistry* 20:5132–40

181. Whitman, W. B., Wolfe, R. S. 1980. Presence of nickel in factor F_{430} from *Methanobacterium bryantii*. *Biochem. Biophys. Res. Commun.* 92:1196–1201

182. Wilson, P. W., Umbreit, W. W. 1937. Mechanism of symbiotic nitrogen fixation. III. Hydrogen as a specific inhibitor. *Arch. Mikrobiol.* 8:440–57

183. Wilson, P. W., Umbreit, W. W., Lee, S. B. 1938. Mechanism of symbiotic nitrogen fixation. IV. Specific inhibition by hydrogen. *Biochem. J.* 32:2084–95

184. Wyss, O., Wilson, P. W. 1941. Mechanism of biological nitrogen fixation: VI. Inhibition of *Azotobacter* by hydrogen. *Proc. Natl. Acad. Sci. USA* 27:162–68

185. Yates, M. G. 1980. Biochemistry of nitrogen fixation. In *Biochemistry of Plants*, ed. B. J. Miflin, 5:1–63. London: Academic. 670 pp.

186. Yates, M. G., Partridge, C. D. P., Walker, C. C., van der Werf, A. N., Campbell, F. O., Postgate, J. R. 1980. Recent research in the physiology of heterotrophic, non-symbiotic nitrogen-fixing bacteria. See Ref. 26, pp. 161–76

187. Yates, M. G., Walker, C. C., Partridge, C. D. P., Pedrosa, F. O., Stephan, M., Döbereiner, J. 1981. Hydrogen metabolism and nitrogenase activity in *Azotobacter chroococcum* and *Azospirillum brasilense*. See Ref. 38, pp. 97–100

188. Zumft, W. G. 1981. The biochemistry of dinitrogen fixation. See Ref. 25, pp. 116–30

Ann. Rev. Plant Physiol. 1983. 34:137–61

myo-INOSITOL: ITS BIOSYNTHESIS AND METABOLISM

Frank A. Loewus and Mary W. Loewus

Institute of Biological Chemistry/Program in Biochemistry and Biophysics, Washington State University, Pullman, Washington 99164-6340

CONTENTS

INTRODUCTION

An appreciation of the multifunctional role of MI[1] is slowly emerging from studies that encompass many aspects of plant physiology; notably cyclitol biosynthesis, storage of polyhydric compounds as reserves, germination of seeds, sugar transport, mineral nutrition, carbohydrate metabolism, membrane structure, cell wall formation, hormonal homeostasis, and stress

---

[1]Abbreviations used: MI, *myo*-inositol; SI, *scyllo*-inositol. Configurational prefix (D or L) is deleted from names of common naturally occurring sugars except as indicated in the text.

0066-4294/83/0601-0137$02.00

physiology. Earlier work was limited largely to the chemistry of the inositols, an effort prompted by the presence in plants of diverse cyclitols and their substituted derivatives, and to the metabolism of MI phosphates, especially phytic acid (140). Curiously, this long-standing concern with phytic acid remains a major challenge with regard to biosynthesis, deposition, and breakdown of this important phosphate reserve of seeds (23). Twice before, in 1966 and again in 1971, MI has been reviewed in this series (2, 86). The present chapter will focus on progress made since the last report, particularly those aspects which deal with the biosynthesis and metabolism of MI. Several related reports have appeared in recent years, and attention is directed to these sources for information on such aspects as nomenclature (70); chemistry (1a); related cyclitols in plants (64a, 88, 90, 94, 168); cell wall formation (87); biochemical, physiological, and nutritional properties (128a, 156, 172); and phosphate esters (21a, 23).

BIOSYNTHESIS OF MI-1-P AND MI

Evidence for cyclization of glucose to MI, a conversion anticipated by Maquenne nearly a century ago, was obtained from [1-^{14}C]glucose-labeled parsley leaves in 1962 (91) and shortly thereafter in rat testis (40). In vitro studies soon revealed that this conversion required three enzymes: hexokinase, MI-1-P synthase, and MI monophosphatase. In terms of MI biosynthesis, the first committed step is the conversion of glucose-6-P to MI-1-P; therefore, discussion will be limited to the synthase and phosphatase. Since the properties of these enzymes as derived from plant, animal, fungal, and yeast sources are similar with regard to mechanism, the following two sections will include information that was obtained from animal and microorganismal studies as well as work on plants.

1L-MI-1-P Synthase, EC 5.5.1.4

This enzyme catalyzes the reaction shown in Figure 1. It has been isolated and partially or completely purified from yeast (16, 36), fungi (43, 139, 177), several animal tissues, notably testis (104, 110, 122, 127, 135, 136), and a number of plant species (49, 96, 114, 115, 132). A mechanism has been suggested (Figure 1) which conforms to experimental observations and a chemical model of this reaction (76). Neither of the two intermediates postulated in Figure 1 has been isolated. All intermediates are tightly bound and not released until finally reduced to MI-1-P (160). The first identifiable product of the synthase is 1L-MI-1-P whether the enzyme source is animal tissue (37, 80, 81), yeast (18), or plant (103). Phosphate attached to C-6 of glucose does not migrate during formation of MI-1-P (160). Evidence for the first step is found in reduction of D-*xylo*-hex-5-ulose 6-P to glucose 6-P

myo - INOSITOL 1 - PHOSPHATE SYNTHASE (EC 5.5.1.4)

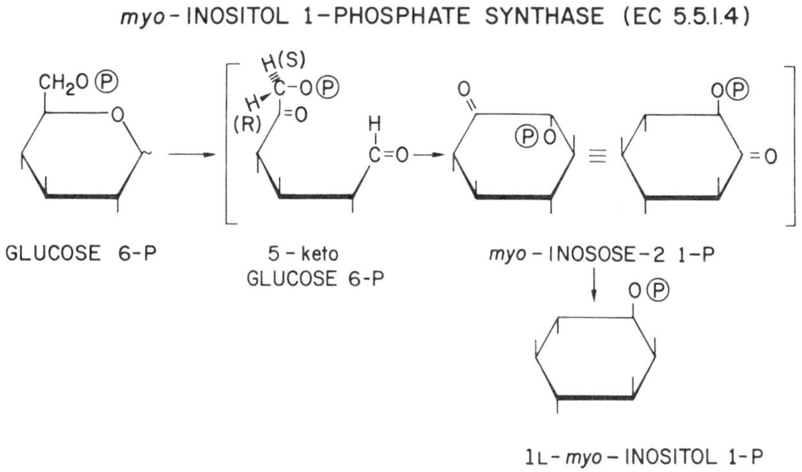

GLUCOSE 6-P   5 - keto   *myo* - INOSOSE - 2 1 - P
                      GLUCOSE 6-P

1L- *myo* - INOSITOL 1- P

Figure 1   Proposed reaction sequence of 1L-*myo*-inositol 1-phosphate synthase, EC 5.5.1.4. Bracket-enclosed compounds are enzyme-bound structures whose presence in the sequence is postulated. The second intermediate compound, *myo*-inosose-2 1-phosphate, is given in two orientations. Cofactor (not shown) for oxidation of D-glucose 6-phosphate and for reduction to *myo*-inositol 1-phosphate is NAD.

and of D-*xylo*-hex-5-ulitol 6-P to glucitol 6-P by rat testis synthase (8). Evidence for the second intermediate is found in reduction of a testis synthase-catalyzed reaction mixture with $[^3H]NaBH_4$ and isolation of tritiated MI and SI from the reaction mixture (17, 38). As noted below, the synthase requires NAD^+ for activity.

Studies with deuterated or tritiated substrate show that hydrogen at C-5 of glucose-6-P transfers to NAD^+ during the oxidation and is returned in the subsequent reduction (39, 61, 160). NADH is tightly bound during the reaction, and there is no exchange between bound nucleotide and free nucleotide in the medium. If the reaction is run in the presence of $[^3H]H_2O$ or either α- or β-$[4-^3H]NADH$, no tritium appears in the product (7, 19, 61, 95). By taking advantage of the reversibility of the first step (8), that is, reversible oxidation of glucose 6-P to D-*xylo*-5-ulose 6-P, Byun & Jenness (13) succeeded in incorporating about 7% of the tritium from [proS-4-^3H]NADH into glucose 6-P in the presence of rat mammary gland or testis synthase from which the endogenous NAD^+ had been stripped by charcoal treatment. When [proR-4-^3H]NADH was supplied, incorporation was about 1%. From these results, the authors conclude that the synthase transfers hydrogen to the proS C-4 position of NADH, corresponding to B stereospecificity (176) in the first step. That the stereospecificity of the third step, reduction of 1L-*myo*-inosose-2

1-P to MI-1-P, is the same as step 1 was shown by demonstrating that tritium from [5-^{3}H]glucose 6-P is incorporated into MI but not into NAD$^+$ and that tritium from [4-^{3}H]NAD$^+$ is not incorporated into MI. Further studies are needed to determine the mechanism of hydrogen transfer and the way in which NAD is bound at the active site. In this regard, the experimental approach used to study the mechanism of action of UDP-galactose-4-epimerase provides an excellent example (50).

With [5-^{3}H]glucose 6-P as substrate, an isotope effect at C-5 as low as 0.21 to 0.48 is obtained with synthase from plant and animal sources (95). Deuterated substrate gives an isotope effect of 0.62 (39). Byun et al (14) reported a loss of proS hydrogen from C-6 of glucose 6-P during cyclization. More recently, the proR hydrogen was reported to be the one removed (102). This conflict has yet to be resolved. Mauck et al (122) observed the uptake of deuterium from D_2O into the sixth position of glucose 6-P in the presence of bovine testis synthase, thus demonstrating reversibility in cyclization of glucose 6-P to 1L-inosose-2 1-P. There is a hydrogen effect of 0.5 during this process (14, 19). Hydrogen attached to C-1, C-2, C-3, and C-4 are minimally disturbed, if at all, by the conversion, and no isotope effects are detected at these positions (7, 95, 160). All these observations support the suggested mechanism.

The second step of the reaction, cyclization of D-*xylo*-hex-5-ulose 6-P to form L-*myo*-inosose-2 1-P, is sterically controlled by an aldol condensation, and much research has been directed toward a fuller understanding of this condensation. Discussion of results is found in several references (64, 102, 137, 157–159). Aldolases are classified as Class I when there is evidence of Schiff base formation between substrate and enzyme and as Class II if the enzyme fails to form a Schiff base and there is a divalent metal ion requirement (65). Synthase preparations from some sources (*Streptomyces griseus, Acer pseudoplatanus* cell culture, *Lemna gibba, Neurospora crassa*) are inhibited by EDTA, an indication that the enzyme behaves as a Class II aldolase, but synthase from animal tissues or *Lilium longiflorum* (lily) pollen is unaffected (158). To determine whether the latter (those not inhibited by EDTA) operate via a Schiff base, experiments were undertaken with [5-^{18}O]glucose 6 P in nonenriched medium or with unlabeled substrate in $H_2^{18}O$ using enzyme from bovine testis, rat testis, lily pollen, and *N. crassa* (158, 159). No loss of ^{18}O occurred in the conversion of [5-^{18}O]glucose to product and no uptake of ^{18}O from $H_2^{18}O$ into MI-1-P was obtained, precluding Schiff base formation during cyclization. Pittner & Hoffmann-Ostenhof (137, 138) obtained uptake of ^{18}O from $H_2^{18}O$ into MI-1-P in the presence of rat testis synthase; however, this ^{18}O was located on C-6 of MI (corresponding to C-1 of glucose). A Schiff base mechanism analogous to that found in Class I aldolase would involve an ϵ-NH$_2$ of a lysyl residue on the enzyme and C-5 of D-*xylo*-hex-5-ulose 6-P. It appears that the observation of Pittner and Hoffmann-Ostenhof involves a site of Schiff base formation whose significance is undetermined. Oxidation at C-5 is considered necessary for activation of the proton at C-6 during cyclization (75). A mechanism involving only base catalysis has been proposed (102, 157, 159).

Variable amounts of endogenous NAD^+ remain bound to the synthase during purification. Barnett & Corina (7) were the first to report that this bound NAD^+ could be removed from rat testis enzyme by activated charcoal, but synthase from *A. pseudoplatanus* and *L. gibba* were destroyed by this treatment (97, 132). Purification to homogeneity of synthase from rat liver (110) or *N. crassa* (43) gave enzyme with an almost absolute requirement for NAD^+. Ammonium ion stimulates synthase activity from plant (96), animal (110, 122), yeast (16), and fungi (43). Potassium ion stimulates but sodium ion inhibits the activity of testis synthase (110, 122). Similar studies are needed on the plant enzyme. The pH optimum ranges from 7.2 to 8.6 for different synthases.

Lower molecular weights are reported for synthase from plant as compared to animal sources (Table 1). The need for more experimental work in this area is obvious.

The synthase is inhibited by sulfhydryl reagents, heavy metals, P_i, PP_i, and substrate analogs such as 2-deoxyglucose 6-P, 6-phosphogluconate, galactose 6-P, glucuronate, mannose 6-P, 5-thioglucose, and dihydroxyacetone-P. MI-1-P, glucuronate 1-P, galacturonate, fructose 6-P, and glucosa-

Table 1 Molecular weight of MI–1–P synthase from different sources

Source	Molecular weight	Subunit weight	Number of subunits	References
Animal				
Beef testis	209,000			122
	218,000	54,500	4	135
Rat testis	210,000	68,000	3	110
	215,000	72,000	2	136
		35,000	2	
	215,000			8
Rat mammary gland	290,000			127
Fungi				
Neurospora crassa	345,000	59,000	6	43
	225,000	64,000	2	177
		50,000	2	
Plant				
Acer	150,000			96
Lemna	135,000	45,000	3	132

mine 6-P do not inhibit. In general, pentose or hexose phosphorylated at C-5 or C-6, respectively, inhibits while hexose substituted at C-1 does not (12, 16, 98, 99, 124).

The synthase is present in both cytoplasmic and chloroplastic fractions of pea leaf (68) and *Euglena gracilis* (M. W. Loewus, unpublished). In mung bean (115) and rice grain (63), synthase activity reaches a maximum about 14–16 days after flowering. Production of synthase is repressed by the presence of inositol in *A. pseudoplatanus* cell culture (98) and yeast (34). The structural gene for MI-1-P synthase in yeast has been identified as the *ino 1* locus (34–36, 117). An allele of this locus produces inactive but fully immunologically cross-reacting protein of subunit molecular weight identical to the wild type enzyme (35). A mutant of *N. crassa* has been isolated which contains an enzymatically inactive protein with properties that resemble synthase from wild type organisms (177). In animal systems, hormonal control of synthase production is indicated (60).

MI Monophosphatase, EC 3.1.3.25

This enzyme was designated 1L-MI-1-phosphatase by the Enzyme Commission (71). Hallcher & Sherman (56) suggested deletion of the chiral prefix in view of the fact that both 1D- and 1L-MI-1-P are natural substrates. Perhaps MI monophosphatase, as used here, is even better since MI-2-P is also a substrate. The enzyme has been partially purified from *Candida utilis* (16), bovine brain (56), rat mammary gland (127), and avian erythrocyte (149), and purified as judged by polyacrylamide gel electrophoresis to homogeneity from lily pollen (101). A key step in purification is heat treatment (60°-70°C, 0.25-2 h) which destroys other alkaline phosphatases and MI-1-P synthase. The molecular weight ranges from 43,000 to 58,000 depending upon the source of the enzyme and the method of determining this value. The pollen enzyme exhibits a relatively narrow specificity with K_ms for 1D- and 1L-MI-1-P and β-glycerol-P in the range 0.06 to 0.08 mm. The K_m for 1L-*chiro*-inositol 3-P is even lower, about 0.01 mm (M. W. Loewus and S. C. Gumber, unpublished observation). MI-2-P, L-α-glycerol-P, and 3'-AMP which are also hydrolyzed have K_ms nearly a magnitude larger. A summary of specificities of this enzyme from various sources has been prepared (101). Phytate is not a substrate and MI, a product of the reaction, does not inhibit. Compounds inactive as substrate or inhibitor include mannose 6-P, fructose 1-P, fructose 6-P, and *p*-nitrophenyl-P. A very weak glucose 6-P activity ($K_m = 1.7$ mm) is present in the purified enzyme.

The enzyme has a pH optimum at 8.5 and requires Mg^{2+} which is partially replaceable with Co^{2+}, Fe^{2+}, Ni^{2+}, or Mn^{2+} (101). Ca^{2+} and Mn^{2+} are competitive inhibitors and Li^+ a very potent (50% at 0.8 mm) inhibitor of the phosphatase in brain tissue (56). The pollen enzyme is also inhibited by Li$^+$ but at much higher concentration (50% at 50 mM). Phosphoglycerate is a weak competitive inhibitor of the pollen enzyme.

Unlike nonspecific alkaline phosphatases (107), MI monophosphatase does not exhibit transphosphorylase activity and the K_ms vary for different substrates, implying that release of phosphate is not the rate-limiting process (101).

It is worth noting that MI monophosphatase from avian erythrocytes and lily pollen hydrolyze MI-2-P in addition to MI-1-P and that these tissues contain MI pentakisphosphate and hexakisphosphate, respectively (72, 149). MI-2-P is the ultimate hydrolysis product of 6-phytase (23). Four isoenzymic acid phosphatases are found in rice aleurone particles (173). The major isoenzyme, F-1, is a violet protein (mol. wt. 72,000) that hydrolyzes phytic acid as well as the lower MI phosphates including MI monophosphate. It also hydrolyzes *p*-nitrophenyl-P, PP$_i$, and ATP but not α- or β-glycerol-P and glucose-6-P (174).

Free MI and MI Kinase, EC 2.7.1.64

MI-1-P synthase is probably the sole de novo synthetic route that links carbohydrate metabolism to cyclitol formation in the plant kingdom. Its product,1L-MI-1-P, must undergo dephosphorylation to provide the free MI required for such processes as the MI oxidation pathway (87), phosphoinositide biosynthesis (126), bound auxin formation (22), and the biosynthesis of sucrosyl oligosaccharides (73). Whether MI-1-P, and in particular 1L-MI-1-P, is involved in biosynthetic, isomerizing, or MI-transferring reactions is unknown, although some authors have assigned to 1L-MI-1-P the role of substrate in phytate biosynthesis (27, 125). Free MI is a ubiquitous constituent of plants. Its rate of turnover is probably quite rapid because exogenous MI is quickly translocated and converted to normal products of MI metabolism when supplied to intact plants or plant parts (77, 87, 123, 145, 152), germinated seed (10, 113, 130), germinated pollen (20, 112, 119, 146), cell cultures (5, 59, 85a, 170), plant organelles (167), and regenerating protoplasts (11, 121). Unfortunately, a careful study of the turnover of MI under conditions that allow constant monitoring of the changes within the pool of free MI has yet to be undertaken.

Plants as well as other organisms contain a soluble ATP/Mg^{2+}-dependent MI kinase (41). The product is MI-1-P and recently its chirality was shown to be 1L by comparison with DL-MI-1-P on a capillary gas chromatographic column packed with Chiro-Val, a chiral stationary phase (103). One might wonder at the biological necessity of such an enzyme which merely returns free MI to the 1L-MI-1-P pool at the expense of ATP. The answer will probably be found in the physiological relationship that links biosynthetic and storage sources of MI to metabolic sinks such as phytate deposition and other biosynthetic functions yet to be determined. In fact, it is the determination of chirality of the product of MI kinase that lends strongest support for a precursor role of 1L-MI-1-P in phytate biosynthesis. Wheat germ is a rich source of this enzyme and translocation of free MI to the kernel may represent the principal source of the carbon skeleton of phytate during grain ripening (152).

THE MI OXIDATION PATHWAY

Glucose 6-P occupies a pivotal position in the metabolism of carbohydrates in plants (3, 4, 51). Its conversion to nucleotide sugars fulfills biosynthetic requirements in the growing plant and provides sugar components which are involved in transport, storage, and development. One very important nucleotide sugar acid is UDPglucuronate. Not only does it supply glucuronosyl units during polysaccharide biosynthesis, it also functions as precursor of other glycosyl units, namely UDPgalacturonate, UDPxylose, UDParabinose, and UDPapiose. These nucleotide sugars and sugar acids, with the possible exception of UDPapiose, are virtually ubiquitous in the plant kingdom (45), and their sugar moieties are found in noncellulosic cell wall polysaccharides, glycoproteins, gums, and mucilages (162).

Two pathways provide for the conversion of glucose 6-P to UDPglucuronate in plants (Figure 2). Central to the nucleotide sugar oxidation path-

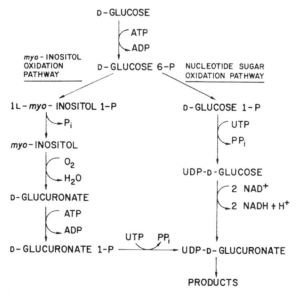

Figure 2   Alternative pathways from D-glucose 6-phosphate to UDP-D-glucuronate and products of UDP-D-glucuronate metabolism in plants.

way is UDPglucose dehydrogenase. Virtually all mechanistic information on this enzyme has been obtained with purified enzyme from microbial or mammalian sources (45). Comparative studies with plant enzyme are few (24, 25, 32). UDPglucose dehydrogenase oxidizes C-6 of UDPglucose in two discreet 2-electron NAD⁺-linked steps. Both the uridylyltransferase that forms UDPglucose and the dehydrogenase are product inhibited. Moreover, nucleotide sugar products of UDPglucuronate, notably UDP-xylose, strongly inhibit. The requirement for NAD^+ and kinetic restraints imposed by product inhibition are important considerations when invoking this pathway.

The MI oxidation pathway bypasses UDPglucose and its coupled reduction of NAD^+. This alternative route to UDPglucuronate is linked directly to glucose 6-P through MI-1-P synthase and MI monophosphatase, or indirectly by processes which translocate and/or store MI prior to MI oxidation. Conversion of free MI to UDPglucuronate involves three enzymes: MI oxygenase, glucuronokinase, and glucuronate 1-P uridylyltransferase (UDPglucuronate pyrophosphorylase).

MI Oxygenase, EC 1.13.99.1

This enzyme has been purified from oat seedlings (78), rat kidney (79), and hog kidney (141, 142). Of particular importance is the discovery that the enzyme becomes catalytically less active during purification and is reactivated by the addition of Fe^{2+} and cysteine (142). The smallest active subunit has a molecular weight of 63,000 to 65,000. The plant enzyme has a pH optimum of 7.2 while the hog kidney enzyme exhibits an optimum at pH 8 for oxygenase complexed to several proteins, including NADPH-linked glucuronate reductase, and at pH 6 for homogeneous enzyme. Four iron atoms are present in the latter, possibly bound one to each of four identical 16,000 dalton subunits. Evidence for an iron-sulfur cluster is negative. In yeast and animal tissues, the cleavage product of MI oxidation undergoes the following reactions: MI → D-glucuronate → L-gulonate → L-*xylo*-hex-3-ulosonate $\xrightarrow{-CO_2}$ L-xylulose → xylitol → D-xylulose → pentose phosphate (57) and eventually appears as hexose or products of hexose metabolism (140). In plants the MI oxidation pathway predominates with virtually all oxygenase-generated glucuronate being channeled into UDP-glucuronate and related products (86, 87, 146).

D-Glucuronokinase, EC 2.7.1.43

Phosphorylation of glucuronate is catalyzed by an ATP-dependent kinase that is found in many types of plant tissue (31) of which lily pollen remains the preferred source (82). The enzyme exhibits marked specificity for the natural substrate, glucuronate. It is inhibited by its product, α-glucuronate

1-P, and by UDPglucuronate which is the product of the subsequent step in the MI oxidation pathway. The substrate analog, β-glucuronate 1-P, also inhibits (53). Several sugars and sugar analogs including D-lyxose (a hexokinase inhibitor), 2-deoxyglucose, 6-deoxyglucose, 3-O-methyl-glucose, 5-thioglucose, and 2-deoxy-2-fluoroglucose are not effective inhibitors (31). A role as regulator of glucuronate metabolism has been proposed (82).

D-Glucuronate 1-P Uridylyltransferase, EC 2.7.7.44

This uniquely plant-derived enzyme completes the pathway from MI to UDPglucuronate. Although it was first described in 1958 (45), an appreciation of its significance emerged only after the MI oxidation pathway was formulated (92, 143). Apart from an improvement in assay procedure and the isolation of this activity from various plant tissues including lily pollen (33), very little attention has been given to this transferase as compared to its counterparts, glucose 1-P uridylyl-transferase, EC 2.7.7.9 and UDP-glucose dehydrogenase, EC 1.1.1.22 in the nucleotide sugar oxidation pathway (45).

Metabolism of UDPGlucuronate and Gluconeogenesis

Although not strictly a part of the MI oxidation pathway, the metabolism of UDPglucuronate is tightly coupled to the former process with most products of MI oxidation appearing as uronosyl and pentosyl units of polysaccharides (86–88) and products of gluconeogenesis (66, 85, 92, 93, 146–148). Appearance of free xylose among metabolic products of MI metabolism (85, 92), similarities in glucose labeling patterns from germinated lily pollen after growth in [2-^{14}C]MI or [5-^{14}C]arabinose (147), and production of labeled starch from germinated lily pollen after growth in [2-^{3}H]MI, [R5,S5-^{3}H]xylose, or [1-^{14}C]arabinose (148) are observations that support a scheme of gluconeogenesis as follows: MI → glucuronate → glucuronate 1-P → UDPglucuronate $\xrightarrow{-CO_2}$ UDPxylose → D-xylose → D-xylulose → D-xylulose 5-P → pentose phosphate intermediates and eventually, hexose phosphates. In germinated lily pollen such a scheme probably functions as conduit for pistil-derived nutrients including MI, which is eventually utilized by the elongating pollen tube for tube wall biosynthesis, and as a source of starch for energy to maintain this biosynthetic capability (93).

Functional Aspects of the MI Oxidation Pathway

There have been a number of attempts in recent years to test the functional role of the MI oxidation pathway as compared to that of the nucleotide sugar oxidation pathway. Carbon labeling studies with specifically labeled glucose fail to distinguish alternative pathways (91). Attempts to circumvent this problem led to experiments in which exogenous substrate, MI or glucose, was suppled to the tissue in the presence of [^{14}C]glucose or [2-^{3}H]MI, respectively, with the view that unlabeled substrate would dilute the corresponding endogenous pool should the labeled substrate happen to utilize that pathway. Glucose failed to perturb labeling of pectic constituents in [2-^{3}H]MI-labeled germinated lily pollen, but MI did perturb the labeling of pectic constituents in [1-^{14}C]glucose-labeled germinated lily pollen and galacturonosyl residues of pectin in [6-^{14}C]glucose-labeled corn root tips (112, 145). These findings are regarded as supportive of carbon flow from glucose to UDPglucuronate via MI.

An even more convincing demonstration that germinated lily pollen converts glucose to products of UDPglucuronate via the MI oxidation pathway makes use of the considerable hydrogen isotope effect at C-5 of glucose 6-P during MI-1-P biosynthesis (95). Ratios of ^{3}H:^{14}C in glucosyl and galacturonosyl residues of pollen tube polysaccharides were determined after labeling lily pollen with [5-^{3}H,1-^{14}C]glucose. Only in galacturonosyl residues was there an appreciable isotope effect attributable to MI-1-P synthase (100).

Other studies that have given consideration to the functional role of the MI oxidation pathway deal with the inhibitory effects of 2-*O,C*-methylene MI (21, 112, 113), xylan synthesis in differentiated xylem cells (26), utilization for cell wall polysaccharide biosynthesis of the MI component of phytate (153), and cell wall biosynthesis of carrot cell culture (5, 170).

PHYTATE METABOLISM WITH REFERENCE TO ITS MI COMPONENT

During seed development, phytate, the hexakisphosphate ester of MI, is deposited in discrete regions (called globoids) of cellular organelles which are usually referred to as protein bodies or aleurone particles (105, 120, 133). Formation of phytate and its subsequent breakdown are thought to be restricted to these subcellular regions (130, 131, 152, 175), although a recent observation reports on accumulation of phytin-containing particles in the cytoplasm prior to globoid formation (55a). This immobility of phytate, a quality similar to that found in starch or certain reserve polysaccharides, oils, and proteins of the seed, invites a consideration of processes which lead to formation and accumulation of phytate reserves as well as to the distribution of breakdown products. The subject has been reviewed (21a, 23, 89, 111, 129), and this section will merely summarize those aspects which pertain to MI metabolism.

Biosynthesis of Phytate

There is no agreement and very little experimental information concerning the first phosphorylated intermediate leading to phytate. A direct role for 1L-MI-1-P has been proposed (27, 125). Perhaps the best evidence to date is the observation that both MI-1-P synthase and MI kinase in plants produce 1L-MI-1-P (103). Others have suggested that MI-2-P is the monophosphate precursor of phytate (67, 166), drawing this conclusion from in vivo labeling studies and from identification of MI-2-P among in vitro products of the phosphorylation of MI. Such assignment should be treated with caution in view of potential acid-catalyzed migration of phosphate from C-1 or C-3 of MI to C-2 (23) or involvement of an intermediary cyclic MI-1,2-monophosphate.

A single enzyme, phosphoinositol kinase, catalyzes the stepwise ATP-dependent phosphorylation of MI monophosphate to phytate (15, 118). A phosphoprotein intermediate is involved (116). Curiously, the enzyme is isolated from germinating mung bean seed, not maturing seed, and there are no reports of this enzyme in ripening seeds or grain where phytate synthesis proceeds at a rapid rate. The stereochemistry of this phosphorylation, an interesting problem itself, has not been examined.

In vivo conversion of MI to phytate is readily observed in plants during the phytate-accumulating stage of maturation (69, 144, 152). In duckweed, phytate accumulation is initiated by withdrawal of nitrate or by addition of sucrose to the medium (154). The ease with which MI is converted to phytate in intact plants provides a convenient procedure for preparation of labeled phytate (10, 128).

Hydrolysis of Phytate (The Phytases, EC 3.1.3.X)

Hormonal control, similar to that involved in hydrolysis of starch reserves, may determine phytase formation and secretion during germination (21b, 74a). Wheat bran phytase, EC 3.1.3.26, the most actively studied enzyme of this class, is designated 6-phytase, the prefix referring to the position in phytate from which the first phosphate is removed. In subsequent steps, the first product, 1L-MI-1,2,3,4,5-pentakisphosphate, is dephosphorylated to produce MI-2-P, the ultimate product. Further hydrolysis to yield free MI requires a second enzyme, possibly a MI monophosphatase such as the one found in lily pollen (101), although this enzyme has a low affinity for MI-2-P, about one-tenth of that shown for MI-1-P (K_m =0.08 mM). A small amount of MI-1-P has been detected among the products of phytate hydrolysis by wheat bran phytase (84), leading to the suggestion that a second phytase named 2-phytase is also present. Here the ultimate product will be MI-1-P, possibly the 1L isomer, although this is unproven. Phytases from bacterial and fungal sources exhibit other specificities (23). Phytase in higher plants is associated with the phytate-rich subcellular organelle (175), but the way in which the hormonal signal is transmitted to this site and the mechanism through which it is expressed are problems that still await solutions.

A scheme encompassing the known enzymatic steps that are linked to the metabolism of MI phosphates is assembled in Figure 3. Reactions leading to formation of phytate are generally regarded as associated with seed development, although the role of the five phosphorylative steps of phosphoinositol kinase in this period of growth is undetermined as noted above. Breakdown of phytate, a process associated with seed germination, is shown as a five-step hydrolysis leading to MI-2-P in the case of 6-phytase or to 1L-MI-1-P in the case of hypothetical 2-phytase. Free MI, presented here as a single pool, may in fact represent several intra- and intercellular compartments where the concentration of MI, free or bound, is regulated by MI-producing and MI-consuming processes.

Strother (165) has proposed a homeostatic mechanism for utilization of endogenous phytate in germinating seeds. His idea is not unlike a more detailed proposal developed by Bandurski (6) for homeostatic control of IAA. In the latter study, attention is focused on the rate of hydrolysis of conjugated IAA as a determinant in tissue IAA concentration. In the case of phytate, the rate of hydrolysis may be controlled by production of phytase, by the relative activity of phytase on partially dephosphorylated forms of phytate, by inhibitors such as P_i, by activators (83), and by hor-

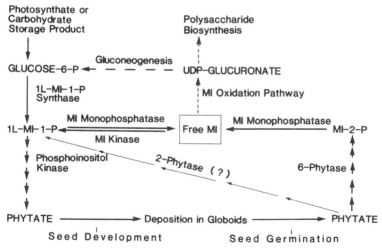

Figure 3   Schematic diagram of *myo*-inositol biosynthesis and metabolism as it is probably related to formation, accumulation, and breakdown of phytate in seed-bearing plants. Consult text for details regarding individual steps and interconversions.

mones such as gibberellic acid; each is a measure of the control that ultimately determines the concentration of P_i required for seedling development. It would be interesting to learn whether the metabolic requirement for MI in the seedling is also expressed in terms of homeostasis during phytate hydrolysis.

MI METABOLISM IN WHEAT

When [2-^{3}H]MI is supplied to germinating wheat kernels, either by imbibition or by injection into softened endosperm, the bulk of the added tritium is transferred to the seedling where it appears as uronosyl and pentosyl units of newly formed polysaccharides (113). The MI antagonist, 2-O,C-methylene MI, blocks utilization of [2-^{3}H]MI as well as that of [1-^{14}C]glucose (113). Unfortunately, such a study fails to provide information regarding endogenous sources of MI in the kernel. To obtain such information, a labeling procedure devised by Sakri & Shannon (151) was employed to inject [2-^{3}H]MI into developing wheat spikes. About 50% of the tritium was translocated to the kernels where it appeared primarily in the bran fraction (aleurone layer). Apart from that portion of the tritium that was incorporated into cell wall polysaccharide and small amounts of labeled galactinol (α-D-galactopyranosyl-(1 → 1)-MI) and MI, the major labeled constituent was phytate. When the source of label was replaced by [R-^{3}H]SI, there was no incorporation of tritium into products of the MI oxidation pathway or into phytate, even though the labeled SI was translocated to the same region of the kernel and was glycosylated to the SI analog of galactinol (SI-galactoside) (152).

Germination of mature wheat kernels which had been labeled by the procedure just described provides a means of following mobilization of endogenous MI-derived products. When this experiment was undertaken (153) it was observed that galactinol is hydrolyzed within one day of imbibition. Phytate, which is present in two forms, a water-soluble fraction and a bound form, hydrolyzes slowly over a period of many days, the water-soluble phytate in about 3 days and the bound form in about 11 days. Translocation to the new plant, presumably as MI, accounts for about 70% of the [2-^{3}H]MI-derived label in 13 days and all of this tritium appears in new cell wall polysaccharides. Arabinosyl units have over 90% of the tritium on C-5, typical of results to be expected if MI is utilized via the MI oxidation pathway (147). When wheat kernels labeled with [R-^{3}H]SI were germinated, the labeled SI-galactoside quickly hydrolyzed just as did galactinol in germinated MI-labeled grain. Virtually all of the labeled SI was translocated into the new plant where it remained as free SI, unable to function as substrate for the MI oxidation pathway yet nontoxic to the seedling. SI is also a nonmetabolizable substrate that utilizes the MI transport system in *Klebsiella aerogenes* (29).

PHYSIOLOGICAL IMPLICATIONS OF MI METABOLISM

In addition to its role as an intermediate in the MI oxidation pathway and its function as a mineral reserve in the form of phytate in dormant tissues and seeds, MI has emerged as a biologically significant molecule in many other areas of plant physiology. Its presence in bound auxin, a discovery pioneered in Bandurski's laboratory, was recently reviewed (22). The phospholipid component of plant membranes contains a relatively low percentage of phosphatidylinositol compared to other glycerolipid components (150). This quantitative viewpoint may be deceptive (62, 123a, 172). The biosynthesis of phosphoinositides was recently reviewed (126).

Two aspects of MI metabolism, possibly interrelated, with physiological implications have emerged in recent research on sugar transport and environmental stress.

Sugar Transport

Galactinol is synthesized by transfer of the galactosyl moiety of UDPgalactose to free MI (48, 134, 171, 171a). As Frydman and Neufeld first showed, MI may be replaced by SI as acceptor in this reaction, an observation that can be reproduced under physiological conditions (152). Galactinol functions as a galactosyl donor in biosynthesis of sucrosyl oligosaccharides of the raffinose series (73, 74), and this aspect has been studied in some detail in *Cucurbita pepo* leaves (52, 108). Both galactinol synthase and galactinol : raffinose galactosyltransferase activities appear to be centered in the mesophyll cells. Stachyose, the major constituent of photosynthate in *C. pepo,* probably moves symplastically from cells to minor veins (109). It is still unclear just where galactinol-linked MI fits into the sugar transport scheme unless, as suggested by Kandler, it serves only as a cofactor in the synthesis of sucrosyl oligosaccharides. In view of its other metabolic activities this appears unlikely, and new studies are needed to follow not only the galactosyl moiety of galactinol but also the MI moiety during sugar transport.

The recent discovery of a naturally occurring galactoside of pinitol (1D-2-*O*-(α-D-galactopyranosyl)-4-*O*-methyl-*chiro*-inositol) in legumes (9, 155) points toward possible involvement of this and other *O*-methyl inositols (as well as MI) in galactosyl transfer. A survey of the occurrence of these cyclitols has appeared (90).

Environmental Stress

Water-stressed plants undergo a variety of biochemical changes in metabolism (58). While studying the effects of water stress on osmotically significant solutes in tropical pasture species, Ford & Wilson (47) noted a threefold increase in pinitol in water-stressed *Macroptilium atropurpureum,* a tropical legume. Extension of this finding to several *Vigna* species revealed that all exhibited marked increases in their *O*-methyl inositol contents after water stress (46). In many *Vigna* species, the most common *O*-methyl inositol is ononitol (1D-4-*O*-methyl MI) while *O*-methyl SI is found in two species, *V. mungo* and *V. radiata* (169). In *Cajanus cajan* (pigeon pea), water-stressed leaves accumulated pinitol to 5.3% dry weight (see 46). When 3-month *Honkenya peploides* L. Ehr. (dicot) seedlings were salt-stressed (250 mm NaCl) for 3 weeks, the pinitol concentration was twice as great as that in unstressed plants (55).

Prolonged periods of light and dark do not appear to influence the pinitol concentration in clover and soybean (161). Onset of nitrogen fixation in soybean leads to brief increase in accumulation of pinitol in the nodules where it appears to be localized in infected cells but outside of bacteroids (163). Seasonally, the pinitol concentration decreases while D-*chiro*-inositol and MI concentrations remain constant as the soybean plant matures and bears fruit (164).

A possible connection between sugar transport and environmental stress is found in studies of seasonal variations in soluble carbohydrates of trees. Pinitol accumulates in needles and bark of gymnosperms during winter months (30, 42). Similar accumulations of quebrachitol and quercitol appear in angiosperms, maple and oak respectively (30). In *Populus* shoots, raffinose and stachyose account for 70% of the sugars in January but only trace amounts remain in the spring and summer. MI is low throughout the year (44). If galactosyl transfer to sucrose is galactinol-linked (73, 74), then accumulation of raffinose/stachyose in *Populus* after leaf-fall must rely on the starch-sucrose transition associated with frosthardiness and on a source of galactinol, possibly originating in the roots. The role of MI in temperature-stressed plants such as *Populus* should be studied in this regard.

CONCLUDING REMARKS

A full appreciation of the position occupied by MI in the biochemistry and physiology of plants has yet to be achieved. That one enzyme, and one alone, determines the biosynthesis of MI is a fair start. Some details regarding the mechanism of action of 1L-MI-1-P synthase are in hand, but information regarding its localization within the cell, its relationship to other aspects of carbohydrate metabolism, possible regulatory controls, and its distribution within the plant during growth and development are still needed. Similar comments may be directed at MI monophosphatase and MI kinase, two enzymes operating in opposite directions between MI monophosphate and free MI. The specificity of the former and the apparent lack of specificity of the latter as regards chirality toward MI-1-P need to be examined.

Analytical procedures are now sufficiently sensitive, specific, and rapid to explore the distribution, turnover, and transport of free MI within and among the specialized cells and tissues of the plant. An elegant demonstration of this capability as applied in animal tissue has been described (54). An isotopic dilution technique employing deuterated MI has also been reported (1). Since free MI is readily transported across the plasmalemma, work on the mechanism of membrane transport in higher plants is indicated. In this regard, SI may function as a useful nometabolizable substrate as it has in bacteria (29).

Lily pollen has proved to be a useful model system for virtually all aspects of research on the biosynthesis and metabolism of MI. It contains all enzymes of the MI oxidation pathway and probably has the full complement of enzymes associated with the biosynthesis and breakdown of phytic acid (72). There is much to be said in favor of lily pollen as a physiological model as well. It germinates under simple cultural conditions and the pollen tube, produced in less than two hours, provides a means of following cell wall formation without the elaborate preparations associated with protoplasts. In addition, it is an excellent experimental system for studying the transport of sugars across the plasmalemma (28).

Of phytate biosynthesis, deposition, and breakdown much remains to be discovered. None of the enzymes associated with these processes have been fully described. Neither the stereochemistry of the MI monophosphate precursor nor that of the intermediates leading to phytate has been established. Processes leading to accumulation of counterions within the globoid are unknown, but some progress in this direction has been made recently (106). Much research is needed to clarify the biochemistry of phytases. Is there a 2-phytase? How is hormonal control related to induction of phytase activity? Is there a homeostatic mechanism involved? What is the nature of the final product in vivo, and is there a second phosphatase required to produce free MI? These are just a few of the questions that come to mind regarding phytate.

Finally, consideration must be given to the numerous products of MI metabolism that accumulate or are stored within the plant during certain stages of development or periods of growth and dormancy. Attempts have been made to interpret the biosynthesis of these isomeric and substituted forms by interconversions based on free MI. Is it possible that the product of 1L-MI-1-P synthase is involved? The participation of compounds like pinitol, ononitol, *O*-methyl SI, quebrachitol and bornesitol in sugar transport, control of water deficit, salt tolerance, and frosthardiness needs further study. MI is a part of every plant cell and its vital roles are bound up in the answers that must still be sought.

ACKNOWLEDGMENTS

A grant (GM22427) from the National Institutes of Health supported the authors' research on studies described in this review. We thank Drs. J. E. Escamilla and William R. Sherman for providing papers on their research in advance of publication. We also thank Subhash C. Gumber for valuable advice and comments.

Literature Cited

1. Andersen, J. R., Larsen, E., Harbo, H., Bertelsen, B., Christensen, J. E. J., Gregersen, G. 1982. Gas chromatographic mass spectrometric determination of *myo*-inositol in humans utilizing a deuterated internal standard. *Biomed. Mass Spectrom.* 9:135–40
1a. Anderson, L. 1972. The cyclitols. In *The Carbohydrates,* ed. W. Pigman, D. Horton, 1A:519–79. New York: Academic. 642 pp.
2. Anderson, L., Wolter, K. E. 1966. Cyclitols in plants: biochemistry and physiology. *Ann. Rev. Plant Physiol.* 17:209–22
3. ap Rees, T. 1980. Assessment of the contributions of metabolic pathways to plant respiration. In *Biochemistry of Plants,* Vol. 2, *Metabolism and Respiration,* ed. D. D. Davies, pp. 1–29. New York: Academic. 687 pp.
4. ap Rees, T. 1980. Integration of pathways of synthesis and degradation of hexose phosphates. In *The Biochemistry of Plants,* Vol. 3, *Carbohydrates: Structure and Function,* ed. J. Preiss, pp. 1–42. New York: Academic. 644 pp.
5. Asamizu, T., Nishi, A. 1979. Biosynthesis of cell-wall polysaccharides in cultured carrot cells. *Planta* 146:49–54
6. Bandurski, R. S. 1980. Homeostatic control of concentrations of indole-3-acetic acid. In *Plant Growth Substances 1979,* ed. F. Skoog, pp. 37–49. Berlin: Springer. 527 pp.
7. Barnett, J. E. G., Corina, D. L. 1968. The mechanism of glucose 6-phosphate-D-*myo*-inositol 1-phosphate cyclase of rat testis. The involvement of hydrogen atoms. *Biochem. J.* 108:125–29
8. Barnett, J. E. G., Rasheed, A., Corina, D. L. 1973. Partial reactions of D-glucose 6-phosphate-1L-*myo*-inositol 1-phosphate cyclase. *Biochem. J.* 131:21–30
9. Beveridge, R. J., Ford, C. W., Richards,

G. N. 1977. Polysaccharides of tropical pasture herbage. VII. Identification of a new pinitol galactoside from seeds of *Trifolium subterraneum* (subterranean clover) and analysis of several pasture legume seeds for cyclohexitols and their galactosides. *Aust. J. Chem.* 30:1583–90
10. Blaiche, F. M., Mukherjee, K. D. 1981. Radiolabeled phytic acid from maturing seeds of *Sinapus alba. Z. Naturforsch. Teil C* 36:383–84
11. Blaschek, W., Haass, D., Koehler, H., Franz, G. 1981. Cell wall regeneration by *Nicotiana tabacum* protoplasts: chemical and biochemical aspects. *Plant Sci. Lett.* 22:47–57
12. Burton, L. E., Wells, W. W. 1977. Studies on the effect of 5-thio-D-glucose and 2-deoxy-D-glucose on *myo*-inositol metabolism. *Arch. Biochem. Biophys.* 181:384–92
13. Byun, S. M., Jenness, R. 1981. Stereospecificity of L-*myo*-inositol-1-phosphate synthase for nicotinamide adenine dinucleotide. *Biochemistry* 20:5174–77
14. Byun, S. M., Jenness, R., Ridley, W. P., Kirkwood, S. 1973. The stereospecificity of D-glucose 6-phosphate: 1L-*myo*-inositol-1-phosphate cycloaldolase on the hydrogen atoms at C-6. *Biochem. Biophys. Res. Commun.* 54:961–67
15. Chakrabarti, S., Majumder, A. L. 1978. Phosphoinositol kinase from plant and avian sources. See Ref. 172, pp. 69–81. New York: Academic. 607 pp.
16. Charalampous, F. C., Chen, I-W. 1966. Inositol 1-phosphate synthase and inositol 1-phosphatase from yeast. *Methods Enzymol.* 9:698–704
17. Chen, C. H-J., Eisenberg, F. Jr. 1975. Myoinosose-2 1-phosphate: an intermediate in the myoinositol 1-phosphate synthase reaction. *J. Biol. Chem.* 250:2963–67
18. Chen, I-W., Charalampous, F. C. 1966.

Biochemical studies on inositol. IX. D-Inositol 1-phosphate as intermediate in the biosynthesis of inositol from glucose 6-phosphate, and characteristics of two reactions in this biosynthesis. *J. Biol. Chem.* 241:2194–99

19. Chen, I-W., Charalampous, F. C. 1967. Studies on the mechanism of cyclization of glucose 6-phosphate to D-inositol 1-phosphate. *Biochim. Biophys. Acta* 136: 568–70

20. Chen, M., Loewus, F. A. 1977. *myo*-Inositol metabolism in *Lilium longiflorum* pollen. Uptake and incorporation of *myo*-inositol-2-³H. *Plant Physiol.* 59:653–57

21. Chen, M., Loewus, M. W., Loewus, F. A. 1977. The effect of the *myo*-inositol antagonist, 2-*O,C*-methylene *myo*-inositol, on the metabolism of *myo*-inositol-2-³H and D-glucose-1-¹⁴C in *Lilium longiflorum* pollen. *Plant Physiol.* 59:658–63

21a. Cheryan, M. 1980. Phytic acid interactions in food systems. *CRC Crit. Rev. Food Sci. Nutr.* 13:297–335

21b. Clutterbuck, V. J., Briggs, D. E. 1974. Phosphate mobilization in grains of *Hordeum distichon. Phytochemistry* 13:45–54

22. Cohen, J. D., Bandurski, R. S. 1982. Chemistry and physiology of the bound auxins. *Ann. Rev. Plant Physiol.* 33: 403–30

23. Cosgrove, D. J. 1980. *Inositol Phosphates: their Chemistry, Biochemistry and Physiology.* Amsterdam: Elsevier. 191 pp.

24. Dalessandro, G., Northcote, D. H. 1977. Changes in enzymic activities of nucleoside diphosphate sugar interconversions during differentiation of cambium to xylem in sycamore and poplar. *Biochem. J.* 162:267–79

25. Dalessandro, G., Northcote, D. H. 1977. Changes in enzymic activities of nucleoside diphosphate sugar interconversions during differentiation of cambium to xylem in pine and fir. *Biochem. J.* 162:281–88

26. Dalessandro, G., Northcote, D. H. 1981. Xylan synthetase activity in differentiated xylem cells of sycamore trees (*Acer pseudoplatanus*). *Planta* 151:53–60

27. De, B. P., Biswas, B. B. 1979. Evidence for the existence of a novel enzyme system: *myo*-inositol-1-phosphate dehydrogenase in *Phaseolus aureus. J. Biol. Chem.* 254:8717–19

28. Deshusses, J., Gumber, S. C., Loewus, F. A. 1981. Sugar uptake in lily pollen: a proton symport. *Plant Physiol.* 67:

793–96

29. Deshusses, J., Reber, G. 1977. *myo*-Inositol transport in *Klebsiella aerogenes. Scyllo*-inositol, a nonmetabolizable substrate for the study of the *myo*-inositol transport system. *Eur. J. Biochem.* 72:87–91

30. Diamantoglou, S. 1974. On the physiological content of cyclitols in vegetative parts of higher plants. *Biochem. Physiol. Pflanz.* 166:511–23

31. Dickinson, D. B. 1982. Occurrence of glucuronokinase in various plant tissues and comparison of enzyme activity of seedlings and green plants. *Phytochemistry* 21:843–44

32. Dickinson, D. B., Hopper, J. E., Davies, M. D. 1973. A study of pollen enzymes involved in sugar nucleotide formation. See Ref. 87, pp. 29–48

33. Dickinson, D. B., Hyman, D., Gonzales, J. W. 1977. Isolation of uridine 5'-pyrophosphate glucuronic acid pyrophosphorylase and its assay using ³²P-pyrophosphate. *Plant Physiol.* 59: 1082–84

34. Donahue, T. F., Atkinson, K., Kolat, A., Henry, S. A. 1978. Inositol-1-phosphate synthase mutants of the yeast, *Saccharomyces cerevisiae.* See Ref. 15, pp. 311–16

35. Donahue, T. F., Henry, S. A. 1981. Inositol mutants of *Saccharomyces cerevisiae:* mapping the *ino 1* locus and characterizing alleles of the *ino 1, ino 2,* and *ino 4* loci. *Genetics* 98:491–503

36. Donahue, T. F., Henry, S. A. 1981. *myo*-Inositol-1-phosphate synthase. Characteristics of the enzyme and identification of its structural gene in yeast. *J. Biol. Chem.* 256:7077–85

37. Eisenberg, F. Jr. 1967. D-Myoinositol 1-phosphate as product of cyclization of glucose 6-phosphate and substrate for a specific phosphatase in rat testis. *J. Biol. Chem.* 242:1375–82

38. Eisenberg, F. Jr. 1978. Intermediates in the *myo*-inositol 1-phosphate synthase reaction. See Ref. 172, pp. 269–78

39. Eisenberg, F. Jr., Bolden, A. H. 1968. Biosynthesis of inositol from deuterated glucose 6-phosphates. *Fed. Proc. Fed. Am. Soc. Exp. Biol.* 27:595

40. Eisenberg, F. Jr., Bolden, A. H., Loewus, F. A. 1964. Inositol formation by cyclization of glucose chain in rat testis. *Biochem. Biophys. Res. Commun.* 14: 419–24

41. English, P. D., Deitz, M., Albersheim, P. 1966. Myoinositol kinase: partial purification and identification of product. *Science* 151:198–99

42. Ericsson, A. 1979. Effects of fertilization and irrigation on the seasonal changes of carbohydrate reserves in different age-classes of needles on 20-year-old Scots pine trees (*Pinus silvestris*). *Physiol. Plant* 45:270–80
43. Escamilla, J. E., Contreras, M., Martínez, A., Zentella-Piña, M. 1982. L-myo-Inositol-1-phosphate synthase from *Neurospora crassa:* purification to homogeneity and partial characterization. *Arch. Biochem. Biophys.* 218: 275–85
44. Fege, A. S., Brown, G. N. 1982. Oligosaccharide patterns in dormant *Populus* shoots. *Plant Physiol.* 69:6(Suppl)
45. Feingold, D. S. 1982. Aldo (and keto) hexoses and uronic acids. In *Encyclopedia of Plant Physiology, New Series*, Vol. 13A, *Carbohydrates I: Intracellular Carbohydrates*, ed. F. A. Loewus, W. Tanner, pp. 3–76. Heidelberg: Springer. 918 pp.
46. Ford, C. W. 1982. Accumulation of O-methyl-inositols in water-stressed *Vigna* species. *Phytochemistry* 21:1149–51
47. Ford, C. W., Wilson, J. R. 1981. Changes in levels of solutes during osmotic adjustment to water stress in leaves of four tropical pasture species. *Aust. J. Plant Physiol.* 8:77–91
48. Frydman, R. B., Neufeld, E. F. 1963. Synthesis of galactosylinositol by extracts from peas. *Biochem. Biophys. Res. Commun.* 12:121–25
49. Funkhouser, E. A., Loewus, F. A. 1975. Purification of myo-inositol 1-phosphate synthase from rice cell culture by affinity chromatography. *Plant Physiol.* 56:786–90
50. Gabriel, O., Darrow, R. A., Kalckar, H. M. 1975. UDP-galactose-4-epimerase. In *Subunit Enzymes*, ed. K. E. Ebner, pp. 85–135. New York: Marcel Dekker. 332 pp.
51. Gander, J. E. 1982. Polyhydroxy acids: relation to hexose phosphate metabolism. See Ref. 45, pp. 77–102
52. Gaudreault, P-R., Webb, J. A. 1981. Stachyose synthesis in leaves of *Cucurbita pepo. Phytochemistry* 20:2629–33
53. Gillard, D. F., Dickinson, D. B. 1978. Inhibition of glucuronokinase by substrate analogs. *Plant Physiol.* 62:706–9
54. Godfrey, D. A., Hallcher, L. M., Laird, M. H., Matschinsky, F. M., Sherman, W. R. 1982. Distribution of myo-inositol in the cat cochlear nucleus. *J. Neurochem.* 38:939–47
55. Gorham, J., Hughes, L., Jones, G. W. 1981. Low-molecular-weight carbohydrates in some salt-stressed plants. *Physiol. Plant.* 53:27–33
55a. Greenwood, J. S., Bewley, J. D. 1982. Involvement of the endoplasmic reticulum in phytin biosynthesis and deposition in protein bodies of developing castor bean seeds. *Abstr. Can. Bot. Soc./ Can. Soc. Plant Physiol.*, Ann. Meet. Regina, Sask., p. 25
56. Hallcher, L. M., Sherman, W. R. 1980. The effects of lithium ion and other agents on the activity of myo-inositol-1-phosphatase from bovine brain. *J. Biol. Chem.* 255:10896–901
57. Hankes, L. V., Politzer, W. M., Touster, O., Anderson, L. 1969. myo-Inositol catabolism in human pentosurics: the predominant role of the glucuronate-xylulose-pentose phosphate pathway. *Ann. NY Acad. Sci.* 165:564–76
58. Hanson, A. D., Hitz, W. D. 1982. Metabolic responses of mesophytes to plant water deficits. *Ann. Rev. Plant Physiol.* 33:163–203
59. Harran, W., Dickinson, D. B. 1978. Metabolism of myo-inositol and growth in various sugars of suspension-cultured tobacco cells. *Planta* 141:77–82
60. Hasegawa, R., Eisenberg, F. Jr. 1981. Selective hormonal control of myo-inositol biosynthesis in reproductive organs and liver of the male rat. *Proc. Natl. Acad. Sci. USA* 78:4863–66
61. Hauska, G., Hoffmann-Ostenhof, O. 1967. Studies on the biosynthesis of cyclitols. XVIII. On the mechanism of action of D-glucose-6-P to L-1-O-phospho-myo-inositol catalyzed by enzyme from *Candida utilis. Hoppe-Seyler's Z. Physiol. Chem.* 348:1558–59
62. Hawthorne, J. N. 1982. Is phosphatidylinositol now out of the calcium gate? *Nature* 295:281–82
63. Hayakawa, T., Kurasawa, F. 1976. Biochemical studies on inositol in rice seed. Part V. Some enzymic properties of myo-inositol 1-phosphate synthase in milky stage of rice seed. *Nippon Nogei Kagaku Kaishi* 50:339–44 [C.A. 85: 173240m]
64. Hoffmann-Ostenhof, O., Pittner, F., Koller, F. 1978. Some enzymes of inositol metabolism, their purification and their mechanism of action. See Ref. 172, pp. 233–47
64a. Hoffmann-Ostenhof, O., Pittner, F. 1982. The biosynthesis of myo-inositol and its isomers. *Can. J. Chem.* 60: 1863–71
65. Horecker, B. L., Tsolas, O., Lai, C. Y.

1972. Aldolases. In *The Enzymes,* ed. P. D. Boyer, 7:213–58. New York: Academic. 959 pp. 3rd ed.

66. Igaue, I., Miyauchi, S. 1982. Occurrence of organic acids from *myo*-inositol in cultured rice cells. *Agric. Biol Chem.* 46:1413–15

67. Igaue, I., Shimizu, M., Miyauchi, S. 1980. Formation of a series of *myo*-inositol phosphates during growth of rice plant cells in suspension culture. *Plant Cell Physiol.* 21:351–56

68. Imhoff, V., Bourdu, R. 1973. Formation d'inositol par les chloroplastes isolés de pois. *Phytochemistry* 12:331–36

69. Inhülsen, D., Niemeyer, R. 1978. Inositol phosphates from *Lemna minor* L. *Z. Pflanzenphysiol.* 88:103–16

70. IUPAC-IUB. 1973. Nomenclature of cyclitols. *Pure Appl. Chem.* 37:285–97

71. IUPAC-IUB. 1979. *Enzyme Nomenclature 1978,* pp. 246–47. New York: Academic. 606 pp.

72. Jackson, J. F., Jones, G., Linskens, H. F. 1982. Phytic acid in pollen. *Phytochemistry* 21:1255–58

73. Kandler, O., Hopf, H. 1980. Occurrence, metabolism, and function of oligosaccharides. See Ref. 4, pp. 221–70

74. Kandler, O., Hopf, H. 1982. Oligosaccharides based on sucrose (sucrosyl oligosaccharides). See Ref. 45, pp. 349–83

74a. Katayama, N., Suzuki, H. 1980. Possible effect of gibberellin on phytate degradation in germinating barley seeds. *Plant Cell Physiol.* 21:115–23

75. Kiely, D. E., Riordan, J. M., Sherman, W. R. 1978. Intramolecular aldol condensations of delta-dicarbonyl sugars: a novel approach to the syntheses of cycloses. See Ref. 172, pp. 13–22

76. Kiely, D. E., Sherman, W. R. 1975. A chemical model for the cyclization step in the biosynthesis of L-*myo*-inositol 1-phosphate. *J. Am. Chem. Soc.* 97:6810–14

77. Knee, M. 1978. Metabolism of polymethylgalacturonate in apple fruit cortical tissue during ripening. *Phytochemistry* 17:1261–64

78. Koller, E., Koller, F., Hoffmann-Ostenhof, O. 1976. *myo*-Inositol oxygenase from oat seedlings. *Mol. Cell. Biochem.* 10:33–39

79. Koller, F., Hoffmann-Ostenhof, O. 1979. *myo*-Inositol oxygenase from rat kidneys. I. Purification by affinity chromatography; physical and catalytic properties. *Hoppe-Seyler's Z. Physiol. Chem.* 360:507–13

80. Leavitt, A. L., Sherman, W. R. 1982. Direct gas-chromatographic resolution of DL-*myo*-inositol 1-phosphate and other sugar enantiomers as simple derivatives on a chiral capillary column. *Carbohydr. Res.* 103:203–12

81. Levitt, A. L., Sherman, W. R. 1982. Resolution of DL-*myo*-inositol 1-phosphate and other sugar enantiomers by gas chromatography. *Methods Enzymol.* 89:3–9

82. Leibowitz, M. D., Dickinson, D. B., Loewus, F. A., Loewus, M. W. 1977. Partial purification and study of pollen glucuronokinase. *Arch. Biochem. Biophys.* 179:559–64

83. Lim, P. E., Tate, M. E. 1971. The phytases. I. Lysolecithin-activated phytase from wheat bran. *Biochim. Biophys. Acta* 250:155–64

84. Lim, P. E., Tate, M. E. 1973. The phytases. II. Properties of phytase fractions F_1 and F_2 from wheat bran and the *myo*-inositol phosphates produced by fraction F_2. *Biochim. Biophys. Acta* 302:316–28

85. Loewus, F. A. 1964. Inositol metabolism in plants. II. The absolute configuration of D-xylose-5-t derived metabolically from *myo*-inositol-2-t in the ripening strawberry. *Arch. Biochem. Biophys.* 105:590–98

85a. Loewus, F. A. 1969. Metabolism of inositol in higher plants. *Ann. NY Acad. Sci.* 165:577–98

86. Loewus, F. A. 1971. Carbohydrate interconversions. *Ann. Rev. Plant Physiol.* 22:337–64

87. Loewus, F. A., ed. 1973. *Biogenesis of Plant Cell Wall Polysaccharides.* New York: Academic. 379 pp.

88. Loewus, F. A. 1974. The biochemistry of *myo*-inositol in plants. *Recent Adv. Phytochem.* 8:179–207

89. Loewus, F. A. 1983. Phytate metabolism with special reference to its *myo*-inositol component. In *Mobilization of Reserves During Germination,* ed. C. Nozzilillo, P. Lea, F. A. Loewus, pp. 173–92. New York: Plenum (*Recent Adv. Phyto-chem.,* Vol. 17). 306 pp.

90. Loewus, F. A., Dickinson, D. B. 1982. Cyclitols. See Ref. 45, pp. 193–216

91. Loewus, F. A., Kelly, S. 1962. Conversion of glucose to inositol in parsley leaves. *Biochem. Biophys. Res. Commun.* 7:204–8

92. Loewus, F. A., Kelly, S., Neufeld, E. F. 1962. The metabolism of *myo*-inositol in plants: conversion to pectin, hemicel-

lulose, xylose and sugar acids. *Proc. Natl. Acad. Sci. USA* 48:421–25

93. Loewus, F. A., Labarca, C. 1973. Pistil secretion product and pollen tube wall formation. See Ref. 87, pp. 175–93

94. Loewus, F. A., Loewus, M. W. 1980. *myo*-Inositol: biosynthesis and metabolism. See Ref. 4, pp. 43–76

95. Loewus, M. W. 1977. Hydrogen isotope effects in the cyclization of D-glucose 6-phosphate by *myo*-inositol 1-phosphate synthase. *J. Biol. Chem.* 252: 7221–23

96. Loewus, M. W., Loewus, F. A. 1971. The isolation and characterization of D-glucose 6-phosphate cycloaldolase (NAD dependent) from *Acer pseudoplatanus* L. cell culture. *Plant Physiol.* 48:255–60

97. Loewus, M. W., Loewus, F. A. 1973. Bound NAD$^+$ in glucose 6-phosphate cycloaldolase of *Acer pseudoplatanus*. *Plant Sci. Lett.* 1:65–69

98. Loewus, M. W., Loewus, F. A. 1973. D-Glucose 6-phosphate cycloaldolase: inhibition studies and aldolase function. *Plant Physiol.* 51:263–66

99. Loewus, M. W., Loewus, F. A. 1974. *myo*-Inositol 1-phosphate synthase inhibition and control of uridine diphosphate-D-glucuronic acid biosynthesis in plants. *Plant Physiol.* 54:368–71

100. Loewus, M. W., Loewus, F. A. 1980. The C-5 hydrogen isotope effect in *myo*-inositol 1-phosphate synthase as evidence for the *myo*-inositol oxidation pathway. *Carbohydr. Res.* 82:333–42

101. Loewus, M. W., Loewus, F. A. 1982. *myo*-Inositol-1-phosphatase from the pollen of *Lilium longiflorum* Thunb. *Plant Physiol.* 70:765–70

102. Loewus, M. W., Loewus, F. A., Brillinger, G-U., Otuska, H., Floss, H. G. 1980. Stereochemistry of the *myo*-inositol-1-phosphate synthase reaction. *J. Biol. Chem.* 255:11710–12

103. Loewus, M. W., Sasaki, K., Leavitt, A. L., Munsell, L., Sherman, W. R., Loewus, F. A. 1982. The enantiomeric for of *myo*-inositol 1-phosphate produced by *myo*-inositol 1-phosphate synthase and *myo*-inositol kinase in higher plants. *Plant Physiol.* 70:1661–63

104. Loewus, M. W., Wright, R. W. Jr., Bondioli, K. R., Bedgar, D. L., Karl, A. 1983. 1L-*myo*-Inositol 1-phosphate synthase in the epididymal spermatozoa of rams. Submitted for publication

105. Lott, J. N. A. 1980. Protein bodies. In *Biochemistry of Plants*, Vol. 1, *The Plant Cell*, ed. N. E. Tolbert, pp. 589–623. New York: Academic. 705 pp.

106. Lott, J. N. A., Greenwood, J. S., Vollmer, C. M. 1982. Mineral reserves of castor beans: the dry seed. *Plant Physiol.* 69:829–33

107. McComb, R. B., Bowers, G. N., Posen, S. 1979. *Alkaline Phosphatase.* New York: Plenum. 240 pp.

108. Madore, M., Webb, J. A. 1982. Stachyose synthesis in isolated mesophyll cells of *Cucurbita pepo*. *Can. J. Bot.* 60:126–30

109. Madore, M., Webb, J. A. 1982. Vein loading in leaves of *Cucurbita pepo* L. *Plant Physiol.* 69:97(Suppl)

110. Maeda, T., Eisenberg, F. Jr. 1980. Purification, structure and catalytic properties of L-*myo*-inositol-1-phosphate synthase from rat testis. *J. Biol. Chem.* 255:8458–64

111. Maga, J. A. 1982. Phytate: its chemistry, occurrence, food interactions, nutritional significance and methods of analysis. *J. Agric. Food Chem.* 30:1–9

112. Maiti, I. B., Loewus, F. A. 1978. Evidence for a functional *myo*-inositol oxidation pathway in *Lilium longiflorum* pollen. *Plant Physiol.* 62:280–83

113. Maiti, I. B., Loewus, F. A. 1978. *myo*-Inositol metabolism in germinating wheat. *Planta* 142:55–60

114. Maiti, I. B., Rosenfield, C-L., Loewus, F. A. 1978. *myo*-Inositol content of lily pollen. *Phytochemistry* 17:1185–86

115. Majumder, A. L., Biswas, B. B. 1973. Metabolism of inositol phosphates. Part V. Biosynthesis of inositol phosphates during ripening of mung bean (*Phaseolus aureus*) seeds. *Indian J. Exp. Biol.* 11:120–23

116. Majumder, A. L., Biswas, B. B. 1973. Further characterization of phosphoinositol kinase isolated from germinating mung bean seeds. *Phytochemistry* 12:315–19

117. Majumder, A. L., Duttagupta, S., Goldwasser, P., Donahue, T. F., Henry, S. A. 1981. The mechanism of interallelic complementation at the INO1 locus in yeast: immunological analysis of mutants. *Mol. Gen. Genet.* 184:347–54

118. Majumder, A. L., Mandel, N. C., Biswas, B. B. 1972. Phosphoinositol kinase from germinating mung bean seeds. *Phytochemistry* 11:503–8

119. Manthey, A. E., Dickinson, D. B. 1978. Metabolism of *myo*-inositol by germinating *Lilium longiflorum* pollen. *Plant Physiol.* 61:904–8

120. Matile, P. 1978. Biochemistry and function of vacuoles. *Ann. Rev. Plant Physiol.* 29:193–213

121. Mattoo, A. K., Lieberman, M. 1977.

Localization of the ethylene-synthesizing system in apple tissue. *Plant Physiol.* 60:794–99

122. Mauck, L. A., Wong, Y.-H., Sherman, W. R. 1980. L-*myo*-Inositol 1-phosphate synthase from bovine testis: purification to homogeneity and partial characterization. *Biochemistry* 19:3622–29

123. Miki-Hirosige, H., Nakamura, S. 1981. The metabolic incorporation of label from myoinositol-2-^{3}H by the growing young anther of *Lilium longiflorum. Acta Soc. Bot. Pol.* 50:77–82

123a. Michell, R. H. 1979. Inositol phospholipids in membrane function. *Trends Biochem. Sci.* 4:128–31

124. Mogyoros, M., Brunner, A., Piña, E. 1972. Behavior of cycloaldolase from *Neurospora crassa* toward substrate analogs and aldolase inhibitors. *Biochim. Biophys. Acta* 289:420–27

125. Molinari, E., Hoffmann-Ostenhof, O. 1968. Studies on the biosynthesis of cyclitols. XXI. On the enzyme system that can phosphorylate *myo*-inositol to phytic acid. *Hoppe-Seyler's Z. Physiol. Chem.* 349:1797–99

126. Moore, T. S. Jr. 1982. Phospholipid biosynthesis. *Ann. Rev. Plant Physiol.* 33:235–59

127. Naccarato, W. F., Ray, R. E., Wells, W. W. 1974. Biosynthesis of *myo*-inositol in rat mammary gland. Isolation and properties of the enzymes. *Arch. Biochem. Biophys.* 164:194–201

128. Nahapetian, A., Young, V. R. 1980. Metabolism of ^{14}C-phytate in rats: effect of low and high dietary calcium intakes. *J. Nutr.* 110:1458–72

128a. National Technical Information Service, U.S. Department of Commerce. 1975. Evaluation of the health aspects of inositol as a food ingredient. PB-660262. Springfield, VA. 14 pp.

129. Oberleas, D. 1973. Phytates. In *Toxicants Occurring Naturally in Foods*, pp. 363–71. Washington DC: Nat. Acad. Sci. 2nd ed.

130. Ogawa, M., Tanaka, K., Kasai, Z. 1979. Phytic acid formation in dissected ripening rice grains. *Agric. Biol. Chem.* 43:2211–13

131. Ogawa, M., Tanaka, K., Kasai, Z. 1979. Accumulation of phosphorus, magnesium and potassium in developing rice grains: followed by electron probe X-ray analysis focusing on the aleurone layer. *Plant Cell Physiol.* 20:19–27

132. Ogunyemi, E. O., Pittner, F., Hoffmann-Ostenhof, O. 1978. Studies on the biosynthesis of cyclitols. XXXVI. Purification of *myo*-inositol-1-phosphate synthase of duckweed, *Lemna gibba,* to homogeneity by affinity chromatography or NAD-Sepharose. *Hoppe-Seyler's Z. Physiol. Chem.* 359:613–16

133. Pernollet, J-C. 1978. Protein bodies of seeds: ultrastructure, biochemistry, biosynthesis and degradation. *Phytochemistry* 17:1473–80

134. Pharr, D. M., Sox, H. N., Locy, R. D., Huber, S. C. 1981. Partial characterization of the galactinol forming enzyme from leaves of *Cucumis sativius* L. *Plant Sci. Lett.* 23:25–33

135. Pittner, F., Fried, W., Hoffmann-Ostenhof, O. 1974. Biosynthesis of cyclitols. Purification of *myo*-inositol 1-phosphate synthase of rat testis to homogeneity by affinity chromatography on NAD-Sepharose. *Hoppe-Seyler's Z. Physiol. Chem.* 355:222–24

136. Pittner, F., Hoffmann-Ostenhof, O. 1976. Studies on the biosynthesis of cyclitols. XXXIII. Preparation of highly purified *myo*-inositol-1-phosphate synthase from bull testicles and comparison with the corresponding enzyme from rat testicles. *Monatsh. Chem.* 107:793–97

137. Pittner, F., Hoffmann-Ostenhof, O. 1976. Studies on the biosynthesis of cyclitols. XXXV. On the mechanism of action of *myo*-inositol 1-phosphate synthase from rat testicles. *Hoppe-Seyler's Z. Physiol. Chem.* 357:1667–71

138. Pittner, F., Hoffmann-Ostenhof, O. 1978. Studies on the biosynthesis of cyclitols. XXXVII. On mechanism and function of Schiff's base formation as an intermediary reaction step of *myo*-inositol-1-phosphate synthase from rat testicles. *Hoppe-Seyler's Z. Physiol. Chem.* 359:1395–1400

139. Pittner, F., Tovarova, I. I., Kornitskaya, E. Y., Khoklov, A. S., Hoffmann-Ostenhof, O. 1979. *myo*-Inositol 1-phosphate synthase from *Streptomyces griseus*. Studies on the biosynthesis of cyclitols. XXXVIII. *Mol. Cell. Biochem.* 25:43–46

140. Posternak, Th. 1965. *The Cyclitols.* San Francisco: Holden-Day. 431 pp.

141. Reddy, C. C., Pierzchala, P. A., Hamilton, G. A. 1981. *myo*-Inositol oxygenase from hog kidney. II. Catalytic properties of the homogeneous enzyme. *J. Biol. Chem.* 256:8519–24

142. Reddy, C. C., Swan, J. S., Hamilton, G. A. 1981. *myo*-Inositol oxygenase from hog kidney. I. Purification and characterization of the oxygenase and of an enzyme complex containing the oxyge-

nase and D-glucuronate reductase. *J. Biol. Chem.* 256:8510–18

143. Roberts, R. M. 1971. The formation of uridine diphosphate-glucuronic acid in plants. Uridine diphosphate-glucuronic acid pyrophosphorylase from barley seedlings. *J. Biol. Chem.* 246:4995–5002

144. Roberts, R. M., Loewus, F. A. 1968. Inositol metabolism in plants. VI. Conversion of *myo*-inositol to phytic acid in *Wolffiella floridana. Plant Physiol.* 43:1710–16

145. Roberts, R. M., Loewus, F. A. 1973. The conversion of D-glucose-6-^{14}C to cell wall polysaccharide material in *Zea mays* in presence of high endogenous levels of myoinositol. *Plant Physiol.* 52:646–50

146. Rosenfield, C-L., Fann, C., Loewus, F. A. 1978. Metabolic studies on intermediates in the *myo*-inositol oxidation pathway in *Lilium longiflorum* pollen. I. Conversion to hexoses. *Plant Physiol.* 61:89–95

147. Rosenfield, C-L., Loewus, F. A. 1978. Metabolic studies on intermediates in the *myo*-inositol oxidation pathway in *Lilium longiflorum* pollen. II. Evidence for the participation of uridine diphosphoxylose and free xylose as intermediates. *Plant Physiol.* 61:96–100

148. Rosenfield, C-L., Loewus, F. A. 1978. Metabolic studies on intermediates in the *myo*-inositol oxidation pathway in *Lilium longiflorum* pollen. III. Polysaccharidic origin of labeled glucose. *Plant Physiol.* 61:101–3

149. Roth, S. C., Harkness, D. R., Isaacks, R. E. 1981. Studies on avian erythrocyte metabolism: purification and properties of *myo*-inositol 1-phosphatase from chick erythrocytes. *Arch. Biochem. Biophys.* 210:465–73

150. Roughan, P. G., Slack, C. R. 1982. Cellular organization of glycerolipid metabolism. *Ann. Rev. Plant Physiol.* 33:97–132

151. Sakri, F. A. K., Shannon, J. C. 1975. Movement of ^{14}C-labeled sugars into kernels of wheat (*Triticum aestivum* L.). *Plant Physiol.* 55:881–89

152. Sasaki, K., Loewus, F. A. 1980. Metabolism of *myo*-[2-^{3}H]inositol and *scyllo*-[R-^{3}H]inositol in ripening wheat kernels. *Plant Physiol.* 66:740–45

153. Sasaki, K., Loewus, F. A. 1982. Redistribution of tritium during germination of grain harvested from *myo*-[2-^{3}H]inositol- and scyllo-[R-^{3}H]inositol-labeled wheat. *Plant Physiol.* 69:220–25

154. Scheiner, O., Pittner, F., Bollmann, O., Kandeler, R. 1978. Effect of nitrogen deficiency and other factors on phytic acid accumulation in *Lemna gibba* G1. *Z. Pflanzenphysiol.* 88:295–303

155. Schweizer, T. F., Horman, I., Würsch, P. 1978. Low molecular weight carbohydrates from leguminous seeds; a new disaccharide: galactopinitol. *J. Sci. Food Agr.* 29:148–54

156. Sebrell, W. H. Jr., Harris, R. S. 1971. Inositols. In *The Vitamins. Chemistry, Physiology, Pathology, Methods,* 3:340–415. New York: Academic. 601 pp. 2nd ed.

157. Sherman, W. R., Hipps, P. P., Mauck, L. A., Rasheed, A. 1978. Studies on enzymes of inositol metabolism. See Ref. 172, pp. 279–95

158. Sherman, W. R., Loewus, M. W., Piña, M. Z., Wong, Y-H. H. 1981. Studies on *myo*-inositol-1-phosphate synthase from *Lilium longiflorum, Neurospora crassa* and bovine testis. Further evidence that a classical aldolase step is not utilized. *Biochim. Biophys. Acta* 660:299–305

159. Sherman, W. R., Rasheed, A., Mauck, L. A., Wiecko, J. 1977. Incubations of testis *myo*-inositol-1-phosphate synthase with D-[5-^{18}O]glucose 6-phosphate and with $H_2{}^{18}$O show no evidence of Schiff base formation. *J. Biol. Chem.* 252:5672–76

160. Sherman, W. R., Stewart, M. A., Zinbo, M. 1969. Mass spectrometric study on the mechanism of D-glucose 6-phosphate-L-*myo*-inositol 1-phosphate cyclase. *J. Biol. Chem.* 244:5703–8

161. Smith, A. E., Phillips, D. V. 1982. Influence of sequential prolonged periods of dark and light on pinitol concentration in clover and soybean tissue. *Physiol. Plant.* 54:31–33

162. Stephen, A. M. 1980. Plant carbohydrates. In *Encyclopedia of Plant Physiology, New Series,* Vol. 8, *Secondary Plant Products,* ed. E. A. Bell, B. V. Charlwood, pp. 555–84. Heidelberg: Springer. 674 pp.

163. Streeter, J. G. 1980. Carbohydrates in soybean nodules. II. Distribution of compounds in seedlings during the onset of nitrogen fixation. *Plant Physiol.* 66:471–76

164. Streeter, J. G. 1981. Seasonal distribution of carbohydrates in nodules and stem exudate from field-grown soya bean plants. *Ann. Bot* 48:441–40

165. Strother, S. 1980. Homeostasis in germinating seeds. *Ann. Bot.* 45:217–18

166. Tanaka, K., Watanabe, K., Asada, K.,

Kasai, Z. 1971. Occurrence of *myo*-inositol monophosphate and its role in ripening rice grains. *Agric. Biol. Chem.* 35:314–20

167. Tanaka, K., Yoshida, T., Kasai, Z. 1976. Phosphorylation of myo-inositol by isolated aleurone particles of rice. *Agric. Biol. Chem.* 40:1319–25

168. Tolochka, V. V., Gamburg, K. Z. 1978. Possible functions of *myo*-inositol in plants. *Usp. Sovrem. Biol.* 85:50–62 (In Russian; translation available upon request to M. W. Loewus)

169. Ueno, Y., Hasegawa, A., Tsachiya, T. 1973. Isolation of *O*-metfiyl-*scyllo*-inositol from mung bean seeds. *Carbohydr. Res.* 29:520–21

170. Verma, D. C., Dougall, D. K. 1979. Biosynthesis of *myo*-inositol and its role as a precursor of cell-wall polysaccharides in suspension cultures of wild-carrot cells. *Planta* 146:55–62

171. Webb, J. A. 1973. Distribution, activity and synthesis of galactinol. *Plant Physiol.* 51:12(Suppl.)

171a. Webb, J. A. 1982. Partial purification of galactinol synthase from leaves of *Cucurbita pepo. Can. J. Bot.* 60: 1054–59

172. Wells, W. W., Eisenberg, F. Jr., eds. 1978. *Cyclitols and Phosphoinositides.* New York: Academic. 607 pp.

173. Yamagata, H., Tanaka, K., Kasai, Z. 1979. Isoenzymes of acid phosphatase in aleurone particles of rice grains and their interconversion. *Agric. Biol. Chem.* 43:2059–66

174. Yamagata, H., Tanaka, K., Kasai, Z. 1980. Purification and characterization of acid phosphatase in aleurone particles of rice grains. *Plant Cell Physiol.* 21:1449–60

175. Yoshida, T., Tanaka, K., Kasai, Z. 1975. Phytase activity associated with isolated aleurone particles of rice grains. *Agric. Biol. Chem.* 39:289–90

176. You, K., Arnold, L. J. Jr., Allison, W. S., Kaplan, N. O. 1978. Enzyme stereospecificities for nicotinamide nucleotides. *Trends Biochem. Sci.* 3:265–69

177. Zsindely, A., Szabolcs, M., Aradi, J., Schablik, M., Kiss, A., Szabó, G. 1977. Investigations on myo-inositol-1-phosphate synthase from wild-type and inositol-dependent mutant of neurospora-crassa. *Acta Biol. Acad. Sci. Hung.* 28:281–90

Ann. Rev. Plant Physiol. 1983. 34:163–97
Copyright © 1983 by Annual Reviews Inc. All rights reserved

THE BIOSYNTHESIS AND METABOLISM OF CYTOKININS[1]

D. S. Letham and L. M. S. Palni

Department of Developmental Biology, Research School of Biological Sciences, The Australian National University, Canberra City, A.C.T. 2601, Australia

CONTENTS

INTRODUCTION

Cytokinins, compounds first recognized by their ability to induce cell division in certain plant tissue cultures, are now known to evoke a diversity of responses in plants. The view that root-produced cytokinins move in the xylem to the shoot where they participate in the control of both development and senescence is now widely accepted. In addition to occurring in higher plants as free compounds, cytokinins also occur as component nu-

[1]This review is dedicated to Professor P. F. Wareing on the occasion of his nominal retirement.

0066-4294/83/0601-0163$02.00

cleosides in tRNA of plants, animals, and microorganisms. Recently cytokinins have been reported to occur in plant viral RNA (172). Production of free cytokinins by microorganisms was demonstrated many years ago, and these cytokinins are of great significance in certain relationships between plants and microorganisms. Cytokinins cause marked elongation of cultured fibroblasts and may mediate the regulatory role of mevalonate in DNA replication in mammalian cells (75). Hence, today, interest in cytokinins and their metabolism extends far beyond the boundaries of developmental botany.

This review has been prepared at a most appropriate time. This marks the close of a 20-year period in which the identities of naturally occurring cytokinins and their metabolites have been largely established and chemical procedures for their quantification have been devised. In the new era which is now beginning, we hope to see studies of cytokinin metabolism being closely related to normal plant development and providing new insight into many aspects of developmental botany. Relevant concepts concerning hormonal control of integrated plant development have been discussed recently in a classical review by Wareing (195). Space limitations do not permit discussion of certain topics. These are listed below with relevant review references: procedures for purifying, identifying, and quantifying cytokinins (11, 73); the incorporation of supplied cytokinins into tRNA and rRNA (77, 145); sites of cytokinin biosynthesis in plants (182).

METABOLISM OF EXOGENOUS CYTOKININS

The Identity of Metabolites of the Basic Purine Cytokinins

During the years 1973–1979, great advances were made in the identification and chemical synthesis of metabolites derived from the highly active synthetic cytokinin 6-benzylaminopurine (BAP) and from exogenous (externally applied) zeatin (Z) and zeatin riboside (9-β-D-ribofuranosylzeatin, [9R]Z), the first natural cytokinins to be isolated and now known to occur widely in plants. Metabolism studies which resulted in the conclusive identification of metabolites of Z and [9R]Z with an N^6-substituent are listed in Table 1. This table also lists critical studies of the metabolism of 6-(3-methylbut-2-enylamino)purine [N^6-Δ^2-isopentenyl)adenine, iP] and its 9-riboside, [9R]iP, naturally occurring cytokinins which are biosynthetic precursors of zeatin. It should be noted that an addendum to Table 1 lists all abbreviations for cytokinin metabolites used in this table and elsewhere in the chapter. The relative proportions of the metabolites formed from exogenous Z in some plant tissues have been tabulated elsewhere (44). Metabolites of BAP which have been identified unambiguously are: [9R]BAP (92), [9R-5'P]BAP (92), [3G]BAP (100, 143, 174), [7G]BAP (31, 49, 93, 103, 144, 201), [9G]BAP (31, 93, 103, 144, 174, 201), Ade (50), and

[9Ala]BAP (97). However, Ados, AMP, GMP, IMP, hypoxanthine, and ureides have also been tentatively identified as metabolites (105). Ureides are also prominent metabolites of kinetin in *Acer* cell cultures (36), and of Z in bean leaves (132).

Plant tissues convert exogenous cytokinin bases into a great diversity of metabolites which include products of ring substitution (ribosides, nucleotides, N-glucosides), and products of isoprenoid sidechain cleavage (Ade, Ados, AMP), reduction [(diH)Z, (diH)[9R]Z, (diH)[9R-5'P]Z] and substitution (O-glucosides). Some metabolites possess unique chemical structures. These include a group of N-glucosides in which the sugar moiety is linked to a purine ring nitrogen atom. Metabolites of this type which have been conclusively identified are the 3-, 7-, and 9-glucosides of BAP, the 7- and 9-glucosides of Z, and the 7-glucoside of iP. These were the first purine glucosides to be isolated from natural sources. The 3- and 7-glucosides are particularly unusual. The only other known naturally occurring purine derivatives with a sugar at position 7 are a few compounds related to vitamin B_{12}, and in these the sugar moiety is an α-D-ribofuranosyl residue. The other known natural purines with a sugar at N-3 are 3-ribosyluric acid and its 5'-monophosphate, both of which were purified from bovine blood. All the N-glucosyl metabolites of zeatin and BAP have been shown to be glucopyranosides; in the cases of the 7- and 9-glucosides this identification is based on comparisons with unambiguously synthesized glucofuranosides and pyranosides (31, 32, 93, 103). The glycosidic linkage of the metabolites [7G]BAP, [9G]BAP, and [3G]BAP was conclusively identified as β (31, 100).

Other unusual cytokinin metabolites are the three alanine conjugates, L-2-[6-(4-hydroxy-3-methylbut-*trans*-2-enylamino)purin-9-yl]alanine ([9Ala]Z), a metabolite of Z termed lupinic acid since it was first detected in *Lupinus* spp, the corresponding metabolite of BAP,[9Ala]BAP, and dihydro-lupinic acid, (diH)[9Ala]Z. Lupinic acid was the first compound isolated from plants with an amino acid moiety conjugated to a purine ring nitrogen atom. In addition, a number of metabolites have been identified in which a glucosyl moiety is conjugated to the oxygen on the isoprenoid sidechain. These O-glucosides, (OG)Z, (diH OG)Z, and their ribosides, are β-glucopyranosides and have been synthesized by unambiguous methods (38, 39, 95). These O-glucosides now appear to be widely distributed endogenous cytokinins.

Formation of Metabolites in Diverse Plant Tissues

In a few plant tissues (e.g. radish roots, derooted radish seedlings, ash embryos), the metabolism of Z is surprisingly simple, and in the principal metabolites formed the unsaturated isoprenoid sidechain is conserved (44; see also Table 1). In other tissues, e.g. derooted *Zea mays* seedlings (44,

Table 1.   Studies of the metabolism of zeatin, zeatin riboside, isopentenyladenine and isopentenyladenosine.

Cytokinin, plant tissue and references	Metabolites identified[d]
Zeatin	
1. Axes of Phaseolus vulgaris seed (162)	(diH)Z[b,c], (diH)[9R]Z[c]/[9R]Z, [9R-5'P]Z, (diH)[9R-5'P]Z
2. Ash embryos (180)	*[9R]Z*[b,c], [9R-5'P]Z[c]/[9R-5'PP]Z, [9R-5'PPP]Z
3. Radish seedlings, derooted (31,140)	*[7G]Z*[a,c], [9R-5'P]Z[c]/[9R]Z, Ade, Ados, Ade nucleotides
4. Radish roots (55,56)	*[7G]Z*[c], [9R-5'P]Z[c]/[9R]Z, Ade nucleotides
5. Zea mays roots (31,141,144)	[9G]Z[a,c]/[9R-5'P]Z, [9R]Z, Ade, Ados, Ade nucleotides
6. Lupin seedlings, derooted (39,142,143)	(OG)Z[a,c] [9Ala]Z[a,c]/[9R-5'P]Z, (diH)[9R-5'P]Z, (diH)Z, [9R]Z, (diH)[9R]Z, [7G]Z, [9G]Z, (diH OG)Z, (OG)[9R]Z, (diH OG)[9R]Z, (diH)[9Ala]Z, Ade, Ados, Ade nucleotides
7. Populus alba leaves (38,95)	(OG)Z[a,c] (diH OG)Z[a,c] (diH OG)[9R]Z[a,c], Ados[a,c] /[7G]Z, [9G]Z, (diH)Z, [9R]Z, AMP
8. Apple seeds, immature (44)	(diH)Z[c], (diH)[9R]Z[c], [9R]Z[c] [9G]Z[c], (OG)Z[c], (diH OG)Z[c], [9Ala]Z[b,c]/Ade, Ados, Ade nucleotides
9. Yellow lupin seeds, immature (44)	[9R]Z[c], (diH)[9R]Z[c], (OG)Z[c], (diH OG)Z[c], (OG)[9R]Z[c], (diH OG)[9R]Z[c], [9Ala]Z[c], (diH)[9Ala]Z[c]/AMP
10. Root nodules Alnus glutinosa (68)	[9R]Z[c]/(diH)Z, (OG)Z, (diH OG)Z, Ade, Ados
Zeatin riboside	
11. Populus nigra, green leaves (38,95)	(OG)[9R]Z[b,c] *(diH OG)[9R]Z*[c]/(diH)[9R]Z, Z, (diH)Z
12. Zea mays caryopses (166)	(diH OG)[9R]Z[c], (diH OG)Z[c], *Ados*[c]/ (diH)Z, Z, *Ade*, *Ade nucleotides*
Isopentenyladenine	
13. Tobacco cells (83,88)	[7G]iP[a], [9R-5'P]iP[c], [9R-5'PP]iP[c], [9R-5'PPP]iP[c]/ Ade, Ados, *Ade nucleotides*
Isopentenyladenosine	
14. Tobacco cells (87,88)	[9R-5'P]iP[c], [9R-5'PP]iP[c], [9R-5'PPP]iP[c]/Ade *Ados*, *Ade nucleotides*

a Identified principally by mass spectrometry.
b Identified principally by cocrystallization to constant specific activity.
c Identified by chromatography of the metabolite and usually of derivatives in diverse systems with authentic synthetic compounds as markers.
d Inconclusively identified metabolites occur after the bar in each sequence; compounds in italics were clearly the principal metabolites formed.

ADDENDUM TO TABLE 1

Abbreviations used in Table and Chapter

These are based on a system proposed previously in which iP and Z are regarded as basic compounds (92). Substitutions on the purine ring and on the isoprenoid sidechain are denoted in square brackets, i.e. [ ], and rounded brackets, i.e. ( ), respectively. 6-Benzylaminopurine (BAP) derivatives are denoted in an analogous manner.

Ade and Ados: adenine and adenosine

ip and its derivatives

iP: N^6-(Δ^2-isopentenyl)adenine
[7G]iP: 7-glucopyranosyl-iP
[2MeS]iP: 2-methylthio-iP
[2MeS 9R]iP: 2-methylthio-9-β-D-ribofuranosyl-iP
[9R]iP: 9-β-D-ribofuranosyl-iP
[9R-5'P]iP: 5'-monophosphate of [9R]iP
[9R-5'PP]iP and [9R-5'PPP]:   5'-di- and 5'-tri-phosphates of [9R]iP

Z and its derivatives

Z: zeatin, 6-(4-hydroxy-3-methylbut-trans-2-enylamino)purine
[9Ala]Z: L-β-[6-(4-hydroxy-3-methylbut-trans-2-enylamino)-purin-9-yl]alanine, lupinic acid
[7G]Z: 7-glucopyranosylzeatin
[9G]Z: 9-glucopyranosylzeatin
[2MeS]Z: 2-methylthiozeatin
[2MeS 9R]Z: 2-methylthio-9-β-D-ribofuranosylzeatin
(OG)Z: O-β-D-glucopyranosylzeatin
(OG)[9R]Z: 9-β-D-ribofuranosyl-(OG)Z
[9R]Z: 9-β-D-ribofuranosylzeatin
[9R-5'P]Z: 5'-monophosphate of [9R]Z
[9R-5'PP]Z and [9R-5'PPP]Z: 5'-di and 5'-tri-phosphates of [9R]Z

(diH)Z and its derivatives

(diH)Z: dihydrozeatin, 6-(4-hydroxy-3-methylbutylamino)-purine
(diH)[9Ala]Z: dihydrolupinic acid
(diH)[9G]Z: 9-glucopyranosyldihydrozeatin
(diH OG)Z: O-β-D-glucopyranosyldihydrozeatin
(diH OG)[9R]Z: 9-β-D-ribofuranosyl-(diH OG)Z
(diH)[9R]Z: 9-β-D-ribofuranosyldihydrozeatin
(diH)[9R-5'P]Z: 5'-monophosphate of (diH)[9R]Z

BAP and its derivatives

BAP: 6-benzylaminopurine
[9Ala]BAP: L-β-(6-benzylaminopurin-9-yl)alanine
[3G]BAP: 3-β-D-glucopyranosyl-BAP
[7G]BAP: 7-β-D-glucopyranosyl-BAP
[9G]BAP: 9-β-D-glucopyranosyl-BAP
(oOH)[9R]BAP: 6-(o-hydroxybenzylamino)purine riboside
[9R]BAP: 9-β-D-ribofuranosyl-BAP
[9R-5'P]BAP: 5'-monophosphate of [9R]BAP

141) and *Zea mays* caryopses (166), Ade, Ados, and Ade nucleotides are the dominant metabolites of Z, and hence sidechain cleavage is the main form of metabolism. However, in many tissues examined critically (e.g. studies 6-9 of Table 1), the metabolism of Z is very complex, and in lupin shoots, 16 metabolites were identified and three others were partially characterized (142). Some tissues which have very high levels of endogenous cytokinins, e.g. tumor tissues, metabolize supplied cytokinins very rapidly mainly by sidechain cleavage (74). Hence it is likely that exogenous cytokinins are subject to much more intense metabolism than are the endogenous compounds.

Ade and its derivatives are frequently reported to be major metabolites of Z, [9R]Z, iP and [9R]iP (Table 1), and this metabolism can be attributed to the enzyme cytokinin oxidase. In tobacco cells, the rate of degradation of [9R]iP due to sidechain cleavage is increased by preincubating the cells with 0.3 μM BAP. The rate of cleavage appears to be controlled to some extent by the cytokinins themselves (176). The benzyl group of BAP appears to be more resistant to cleavage in tissue than the isoprenoid moiety of Z and iP. However BAP and kinetin are degraded by cleavage of the N^6-substituent in a number of plant tissues (36, 47, 49, 50, 105, 110). Thus about 80% of BAP taken up by soybean tissue appeared to be degraded by benzyl cleavage to yield adenine and an unidentified acid which resembled benzoic acid (50). A substantial proportion of kinetin (50% after 8 hr) supplied to germinating lettuce seeds was degraded by cleavage of the furfuryl group giving AMP as the major purine metabolite (110). Such metabolism of BAP and kinetin cannot be accounted for by known enzymes (see below).

When cytokinin bases are supplied to plant tissues, the principal metabolites formed initially are usually the cytokinin riboside 5'-phosphates (36, 49, 50, 51a, 55, 74, 88, 110, 140). However, after the initial period of cytokinin metabolism, the proportion of the total radioactivity due to cytokinin nucleotides usually declines rapidly. As discussed later, nucleotide formation may be associated with cytokinin uptake. Nucleotides probably play a key role in cytokinin metabolism in many tissues; considerable evidence for such a role has been obtained with tobacco cell cultures (85). Conversion of cytokinin bases and nucleosides to riboside 5'-di- and 5'-triphosphates has been observed (Table 1). The significance of these triphosphates is not known, but their formation may result in the observed low incorporation of cytokinins into RNA, perhaps as a result of transcriptional errors. When extraction conditions were adopted which minimized enzyme activity, [9R-5'PPP]iP was found to be the major metabolite of [9R]iP in tobacco cells after uptake for 2 hr (85). Cytokinin nucleotides accumulated in these cells which appeared to be impermeable to these metabolites (86).

Low uptake probably accounts for the very low activity of cytokinin nucleo-tides in certain bioassays (92).

Some exogenous chemicals can markedly modify cytokinin metabolism. ABA inhibited the metabolism of kinetin to AMP in lettuce seed, and markedly lowered the level of kinetin nucleotide present, but did not inhibit conversion of kinetin to kinetin nucleotide by a crude enzyme preparation from lettuce seed (109). ABA also suppressed conversion of Z to dihydro derivatives in bean axes (162). These results are consistent with the proposal (92) that ABA has a cytokinin-sparing action. A synthetic purine derivative inhibited conversion of cytokinins to 7- and 9-glucosides both enzymically in vitro and in intact radish cotyledons (98).

Recently a number of critical studies of cytokinin metabolism in relation to plant development have been reported. These are relevant to leaf senes-cence and lateral shoot development in bean seedlings (130–132), oat leaf senescence (174), cell division in tobacco cell cultures (85, 86, 88), expan-sion of radish cotyledons (98, 100, 201), and lettuce seed germination (109, 110). While these studies are mentioned elsewhere in relation to specific metabolites formed, space limitations prevent discussion of their relevance to plant development. Less critical, but nevertheless useful, studies also concern bean (76) and maize seed (161, 181) germination and the cell cycle in tobacco cell cultures (126, 127). However, greater caution should be exercised in cell cycle studies because induction of synchrony is now known to generate metabolic artifacts (111).

Metabolism of Selected Synthetic Cytokinins

N,N'-Diphenylurea (DPU) exhibits weak cytokinin activity (92). Since DPU differs considerably from the purine cytokinins in structure, it has been suggested that the urea is not active per se, but is metabolized to an active purine derivative (18). In a study of DPU metabolism in tobacco callus, the major metabolite was identified as a β-D-glucopyranosyloxy derivative of DPU (18). This glucoside was much less active than DPU in the tobacco callus cytokinin bioassay. Hence a more critical search for highly active metabolites of DPU is required.

9-Alkyl derivatives of BAP have cytokinin activity although this is usu-ally much less than that of BAP (92). The possibility that this 9-substitution may prevent riboside and glucoside formation and yet permit cytokinin action stimulated interest in the metabolism of these 9-alkyl BAPs. Studies of the metabolism of 9-methyl-BAP yielded inconclusive results as no metabolites were identified conclusively (51, 147). A study of the metabo-lism of 9-(4-chlorobutyl)-BAP in radish cotyledons is more significant (96). Although free BAP was not identified definitely as a metabolite of this cytokinin which has activity similar to that of BAP (98), [7G]BAP and

[9G]BAP were isolated as metabolites and their identity established un-equivocally by mass spectrometry, TLC and GLC (96). Formation of these glucosides is dependent on prior cleavage of the chlorobutyl group at N-9 to give BAP. Without doubt certain other 9-alkyl derivatives of cytokinins are also dealkylated by plant tissue to yield free cytokinin bases which must exist at least transiently in the tissue. Hence the proposal (158) that the 9-alkyl group is metabolically stable and that the activity of 9-alkyl cytoki-nins does not depend on formation of nucleosides or nucleotide derivatives is invalid.

In view of the unique structure of cytokinin 7-glucosides, the structural requirements for acceptance of a glucosyl moiety by cytokinin-like purines are of interest. In a study with radish cotyledons, it was established that acceptance of a 7-glucosyl moiety by an adenine residue is greatly enhanced by, or is dependent on, the presence of an N^6-substituent of 5–7 carbon atoms, but this need not confer cytokinin activity (96). These in vivo studies are in accord with in vitro findings with the purified enzyme which catalyzes 7-glucosylation (46).

THE ENZYMES RESPONSIBLE
FOR METABOLITE FORMATION

Three enzymes which show specificity for cytokinins have been purified and characterized, namely, cytokinin oxidase, cytokinin-7-glucosyl transferase, and β-(9-cytokinin)-alanine synthase. These three enzymes are discussed first and then consideration is given to the enzyme systems involved in the interconversion of cytokinin bases, ribosides and nucleotides.

CYTOKININ OXIDASE AND RELATED ENZYMES   An enzyme (MW 88,000) which cleaves the isopentenyl sidechain from iP and [9R]iP to yield Ade and Ados, respectively, has been purified from *Zea mays* kernels (200). The aldehyde, 3-methylbut-2-enal, is the major product derived from the sidechain (12). Although Z and [9R]Z are substrates, N^6- isopentyladeno-sine with a saturated sidechain, BAP and kinetin are all not degraded appreciably by the enzyme. The enzyme reaction requires oxygen and hence is conveniently termed cytokinin oxidase. The mechanism of the enzyme reaction is not known, but an unstable intermediate appears to be the primary product. A possible mechanism is oxidation of the N^6 to C-1 bond to give an imino purine which would then hydrolyze to yield the C_5 alde-hyde and a 6-aminopurine moiety.

Cytokinin oxidase has also been purified from sources other than maize. [9R]Z is more effective than (*cis*)[9R]Z in inhibiting degradation of [9R]iP by an oxidase from tobacco tissue (128). An oxidase purified from *Vinca rosea* tumor tissue has recently been compared with the maize enzyme referred to above (157). Both enzymes show similar specificity for cytokinin substrates, and both degrade [7G]Z, [9G]Z, and [9Ala]Z. Hence, in plant tissues, these "stable" metabolites are probably compartmentalized so they are not exposed to cytokinin oxidase.

In plant tissues, cleavage of the sidechains of Z and iP and their ribosides is attributable to cytokinin oxidase. However, the analogous degradation of BAP and kinetin cannot be thus explained. Hence some plant tissues must contain an enzyme system, distinct from cytokinin oxidase, which cleaves benzyl and furfuryl groups from the N^6 position to yield Ade and its derivatives. In the case of BAP, a benzoic acid-like derivative appears to be a product of the cleavage (50). A more detailed characterization of this reaction is merited as this form of metabolism may limit the effectiveness of BAP applied to plants.

CYTOKININ 7-GLUCOSYLTRANSFERASE   In extracts of radish cotyledons, two enzymes were detected which converted BAP into 7- and 9-glucosides when UDPG was supplied as a glucose donor (43). The enzymes produced these two glucosides of BAP in markedly different proportions; the ratio of [7G]BAP/[9G]BAP was about 1.5 for the major enzyme activity and 10.5 for the other enzyme. The major enzyme was subsequently purified to a stage where kinetic behavior and substrate specificity could be studied critically (46). It was estimated that 1 kg of cotyledons yielded about 76 μg of the enzyme (MW 46,500). Kinetic studies with Z and UDPG established that both substrates combine with the enzyme to give a ternary complex which then undergoes reaction and sequentially releases the products. 3-Methyl-7-(n-pentylamino)pyrazolo[4.3-d]pyrimidine, a compound reported to inhibit cytokinin-induced growth, is an effective competitive inhibitor of the enzyme ($K_i = 2.2 \times 10^{-5}$ M) (44). This enzyme inhibition is the first direct evidence that this anticytokinin can compete per se with cytokinin at a site exhibiting specificity for cytokinin-active molecules.

β-(9-CYTOKININ)ALANINE SYNTHASE   In an initial study with a crude enzyme preparation derived from immature *Lupinus luteus* seeds, zeatin was converted into a ninhydrin-reacting product which cochromatographed with lupinic acid in numerous solvents (122). O-Acetylserine was supplied as donor of the alanine moiety. The enzyme preparations from lupin seeds were markedly more effective than similar preparations from numerous other plant sources in catalyzing formation of lupinic acid (122).

In a recent study, the enzyme from lupin seeds was further purified and the product derived from zeatin was identified unambiguously by mass spectrometry as lupinic acid (45). The ability of paired-ion reverse phase

HPLC to rapidly quantify purine substrates and products in the enzyme reaction made possible a detailed study of the enzyme. Kinetic studies established that the mechanisn is ping pong bi bi (Cleland nomenclature). The enzyme (MW 64,500) requires the unusual and unstable substrate O-acetylserine as donor of the alanine moiety; neither serine, N-acetylserine or O-phosphoserine, would substitute. Only one other enzyme (cysteine synthase) which uses O-acetylserine has been studied in detail. The names β-(9-cytokinin)alanine synthase and β-(6-alkylaminopurin-9-yl)alanine synthase have been proposed for this new enzyme.

Cytokinin 7-glucosides and alanine conjugates exhibit very low activity in bioassays (94). Hence cytokinin 7-glucosyltransferase, β-(9-cytokinin)alanine synthase, and cytokinin oxidase may provide alternative mechanisms for lowering cytokinin activity levels.

INTERCONVERSION OF BASES, NUCLEOSIDES, AND NUCLEOTIDES
The free base, nucleoside and nucleotide forms of cytokinins all appear to be interconvertible in plant tissues. Exogenous cytokinin bases are known to be converted into 9-ribosides and riboside 5'-phosphates (nucleotides); exogenous cytokinin ribosides yield cytokinin bases and nucleotides (see Table 1 and references elsewhere in this chapter). No studies of metabolism of exogenous cytokinin nucleotides have been reported, but conversion to nucleosides by phosphatases and 5'-nucleotidases would be expected to occur. It is uncertain whether plants contain cytokinin-specific enzymes which catalyze cytokinin base-nucleoside-nucleotide interconversions. These changes may well be caused by enzymes which catalyze analogous reactions for Ade, Ados, and AMP. Thus, Chen and co-workers (22, 25, 26, 28) purified the following five enzymes from wheat germ (K_m values in parentheses): adenosine phosphorylase, which strongly favors nucleoside formation (iP 57.1, Ade 32.2 μM); adenosine ribohydrolase ([9R]iP 2.4, Ados 1.4 μM), adenosine kinase ([9R]iP 31, Ados 8.7 μM), adenine phosphoribosyltransferase (iP 130, Ade 74 μM), and 5'-ribonucleotide phosphohydrolase ([9R-5'P]iP 3.5, AMP 3.2 μM). In each case, the K_m for Ade, Ados, or AMP was less than that for iP or its corresponding derivative. Crude preparations of some cytokinin-metabolizing enzymes have been obtained from other sources. For example, germinated rape seed yielded a crude enzyme which effectively converted [9R]iP into iP (129); from *Acer* cells a phosphoribosyl transferase was partially purified (146a).

The rapid conversion of exogenous cytokinin bases into cytokinin nucleotides has been mentioned previously. This could occur by two enzymic mechanisms: 1. a two-step process with a cytokinin riboside (formed by nucleoside phosphorylase action) as intermediate; 2. a one-step process involving a direct transfer of a ribose 5'-monophosphate group from 5-

phosphoribosyl-1-pyrophosphate to the base catalyzed by an adenine phosphoribosyl transferase. In *Acer pseudoplatanus* cell cultures, BAP nucleotide formation appears to occur by the latter route (36, 146a). The same situation very probably applies to kinetin nucleotide formation from kinetin in germinating lettuce seeds (109, 110).

NATURALLY OCCURRING CYTOKININS

Free Cytokinins

As a result of the continuing development of techniques in mass spectrometry and HPLC, cytokinin identifications are being reported with ever-increasing frequency, especially of free compounds. Zeatin and [9R]Z have now been identified unequivocally by mass spectrometry in 22 and 28 natural sources, respectively. The principal, free, natural cytokinins so far identified unambiguously in extracts of higher plants are[2]: Z (92)[3], [9R]Z (92)[3], (*cis*)[9R]Z (33, 80, 196, 198, 203), [9R-5'P]Z (91, 157, 170), (diH)Z (92, 139), (diH)[9R]Z (131, 134, 139, 166–168, 192), [9G]Z (134, 139, 155, 156, 165, 166), (diH)[9G]Z (139, 166), [7G]Z (139, 169), (OG)Z (116, 134*, 139, 156, 166–168, 202*, 203), (OG)[9R]Z (116, 134*, 139, 156, 166–168, 203), (diH OG)Z (139, 166–168, 194*), (diH OG)[9R]Z (134*, 139, 166–168), iP (92, 202), and [9R]iP (33, 34, 92, 179, 197, 202). In the culture media of microorganisms which form relationships with plants, some of these compounds are also present, namely, Z (78, 92), [9R]Z (78, 92, 106), (*cis*)[9R]Z (106), iP (78, 121), and [9R]iP (78), while Z and iP are produced by mosses (9, 193) and iP by slime mold (173). The following cytokinins have been identified unequivocally only in culture media of pathogenic bacteria: (*cis*)Z, (*cis*)[2MeS]Z, [2MeS 9R]Z, and [2MeS 9R]iP (78, 121). The slime mold *Dictyostelium discoideum* produces the unusual cytokinin 3-(3-amino-3-carboxylpropyl)-iP, discadenine (1). Eight more cytokinins have been identified in higher plants but are probably not widely distributed in them (92). These include hydroxylated forms of Z (92), (oOH)[9R]BAP (92), [9Ala]Z and (diH)[9Ala]Z (168), and 6-(o-hydroxybenzylamino)-2-methylthio-9-β-D-glucosylpurine (21). The glucosyl moiety of the last mentioned compound was assigned a furanosyl ring structure based on the relative intensities of fragment-ion peaks in the mass spectrum (21). However, this reasoning is invalid (103), and a 9-glucopyranoside structure is more probable by analogy with the structures of other 9-glucoside metabo-

---

[2]In the references bearing an asterisk, identifications of the cytokinin O-glucoside are not conclusive. A mass spectrum of the relevant intact glucoside was not obtained; only the purine aglycone, derived by treatment with β-glucosidase, was characterized by mass spectrometry. *However all other cited identifications appear to be unequivocal.*

[3]Identifications up to 1976 only are cited by reference to a previous review.

lites of cytokinins. In extracts of flowering plants, Z and its derivatives are the dominant cytokinins, and in extracts of many higher plants, iP and [9R]iP are not detectable (e.g. 131, 154). However, in certain bacteria (121), mosses, and slime mold, iP appears to be the principal free cytokinin produced.

It is noteworthy that all the metabolites of exogenous Z listed in Table 1 are now known to occur normally in plants. Identification of glucose and alanine conjugates as metabolites of supplied Z facilitated their recognition as endogenous compounds. Recent unambiguous identifications of cytokinins which especially merit note include the following: 1. the first identification of cytokinins Z and [9R]Z in the cambial region of a woody plant (101); 2. the identification of [9R]Z as a compound present in carrot roots which causes formation of tracheary elements in cultured carrot phloem tissue (114); 3. the identification of iP as a cytokinin produced by mosses (9, 193) and slime mold (173); 4. the detailed characteriztion of the cytokinin complex produced by cultured plant tumor tissues (74, 116, 134, 139, 156); 5. the identification of the cytokinins in lupin seeds and pod walls, 9 cytokinins being quantified by mass spectrometry using deuterium-labeled internal standards (167, 168); 6. the identification of O-glucosides of Z and (diH)Z and of their ribosides in diverse plant tissues; 7. the occurrence of (*cis*)[9R]Z as a minor cytokinin in extracts of several plant tissues. Although (*cis*)[9R]Z has been identified unambiguously in some plant extracts, the possibility that it was derived from tRNA by enzyme action during extraction has not been rigorously excluded. The relevant extractions were not performed after inactivation of plant enzymes, for example by the technique of Bieleski (11).

The O-glucoside cytokinins (OG)Z, (OG)[9R]Z, (diH OG)Z and (diH OG)[9R]Z have been identified conclusively in a number of plant tissues (see above), but these compounds now appear to be ubiquitous cytokinins. Thus, extracts from numerous and diverse plant organs contain cytokinins which are hydrolyzed by β-glucosidase to yield active products which cochromatograph with Z or [9R]Z (38, 166). However, some caution must be exercised in interpreting such results; also, the β-glucosidase used was frequently a crude enzyme preparation. O-Glucosides appear to account for most of the extractable cytokinin activity in mature or senescing leaves (38, 130, 183, 194) and in lupin pod-walls and seeds (35, 167, 168). The O-glucosides in leaves are probably formed in situ from [9R]Z and other cytokinins present in xylem sap (38, 67, 69, 130, 132). These glucosides appear to accumulate as leaves mature and can reach surprisingly high levels (130). O-Glucosides of cytokinins exhibit high activity in tissue culture and leaf senescence bioassays (94). Hence it is somewhat paradoxical that mature leaves should be subject to senescence. One obvious explanation is that the O-glucosides are strictly compartmentalized within the cells of leaves.

[9R-5'P]Z has been identified unequivocally only in *Zea mays* kernels and tumor tissues. However, chromatography and GC-MS of nucleosides produced by phosphatase hydrolysis indicate that nucleotides of zeatin may be widely distributed in higher plants (34, 114, 131, 152, 187). Dihydrozeatin nucleotide also appears to occur in some plant tissues (131, 139).

Recently, very significant progress has been made regarding the genetic control of cytokinin production by the pathogenic bacteria *Corynebacterium fascians* and *Agrobacterium tumefaciens.* High virulence of *C. fascians* strains and a high cytokinin activity level in the culture media are associated with the presence of a 63 Mdal plasmid (119, 121). Production of iP by *A. tumefaciens* appears to be an expression of the bacterial chromosome, but the nopaline T_i plasmid determines Z production (150). The principal cytokinins produced by *A. tumefaciens* are iP, Z, and [2MeS 9R]Z (*trans/cis* not known) (78). A fragment of the T_i plasmid, termed T-DNA (size 10 Mdal), is covalently joined to plant nuclear DNA and is replicated and transcribed in the plant cells. This confers upon the tissue cytokinin and auxin autotrophy the ability to synthesize cytokinins and to maintain high endogenous levels of these hormones (74, 117, 156, 164). It is important to determine how the genetic information of T-DNA results in an increased level of cytokinin. It is possible to mutagenize the T-DNA (transposon insertions) to produce tumors with altered morphology, e.g. overproduction of shoots and roots, and increased tumor size. The auxin/cytokinin ratio in the tumors was determined and appears to regulate morphology in accord with the classical concepts of Skoog and Miller (117).

Cytokinins in tRNA

In all tRNAs of known sequence, the base adjacent to the 3' end of the anticodon is always a purine which may be adenine, a simple methylated purine, or one of the following hypermodified purines: iP, [2MeS]iP, N-(purine-6-ylcarbamoyl)threonine termed t^6Ade (or occasionally its methyl derivative), the modified guanine termed "base Y," and some compounds related to base Y. All of these hypermodified purines appear to occur exclusively adjacent to the 3' end of the anticodon [for structures and other details see (4, 99)]. In sequenced tRNA species, hypermodified adenosines always occur in the sequence: Ados (third anticodon nucleoside)-modified Ados-Ados → 3' end. iP, [2MeS]iP, and t^6Ade are found in tRNAs of diverse organisms; iP and [2MeS]iP are of course cytokinins, but t^6Ade lacks cytokinin activity. Base Y and related compounds appear to occur only in eukaryotic tRNA[Phe] and are also inactive as cytokinins (4). Other cytokinins which have been isolated from tRNA hydrolysates and probably also occur adjacent to the 3' end of the anticodon are: (cis)Z and sometimes the $trans$ isomer Z, (cis)[2MeS]Z, and the $trans$ isomer [2MeS]Z.

N^6-Isoprenoid adenine bases have wide distribution as minor hyper-modified components of certain tRNA species isolated from microorganisms (bacteria, yeast, mycoplasma, viral RNA), animals, and plants. They do not occur in purified ribosomal RNA (for reviews see 57, 60, 61, 99). Unlike some modified nucleosides, e.g. pseudouridine and ribothymidine, which are common to most or all tRNAs, cytokinin moieties are restricted exclusively to tRNA species that respond to codons beginning with uridine (the "U group" tRNAs; isoacceptors for Cys, Leu, Phe, Ser, Trp, and Tyr), where they occur once per molecule. Most of the tRNA species within the U group in bacteria contain cytokinin active nucleosides (3, 7, 118). However, in other organisms relatively few tRNA species from this group contain cytokinin moieties.

Not all isoaccepting species contain cytokinin-active nucleosides. For example in $E.$ $coli,$ [2MeS 9R]iP occurs in $tRNA_1^{Ser}$ (UCU, UCA, UCG) but is absent in $tRNA_3^{Ser}$ (AGU, AGC) where it is replaced by t^6Ade (123); in $Glycine$ max cotyledons, only two of the six $tRNA^{Leu}$ isoaccepting forms (species 5 and 6) recognize U codons and also contain cytokinins (90). In $Drosophila$ $melanogaster,$ the bulk of cytokinin activity is associated with $tRNA^{Ser}_7$ (UCG) and is due to [9R]iP, while lesser activity is associated with $tRNA^{Ser}$ species 6 and 4 (199). Wheat germ $tRNA^{Ser}$ and minor $tRNA^{Leu}$ contain cytokinin activity, whereas all other tRNA species of the U group including the major peak of $tRNA^{Leu}$ lack cytokinin activity (163). A similar distribution has been observed for tRNA species from etiolated $Phaseolus$ $vulgaris$ seedlings (40). It is interesting that the major peak of $tRNA^{Phe}$ from etiolated seedlings lacks a cytokinin moiety, whereas chloroplast $tRNA^{Phe}$ from the same plant species contains [2MeS 9R]iP (58), like the chloroplast $tRNA^{Phe}$ from $Euglena$ $gracilis$ (64). It would appear that, in contrast to microbial systems, restriction of cytokinin occurrence to a relatively few tRNA species responding to codons with initial letter U may be a characteristic feature of higher eukaryotes.

Among the four chemically distinct cytokinin-active nucleosides which have been shown to occur in tRNA, there appears to be a certain species specificity. The simplest of the four, [9R]iP, has been reported to be the only cytokinin present in animal or yeast tRNA, and probably also occurs in the tRNA of most organisms. Bacterial tRNA usually contains predominantly the corresponding methylthio derivative [2MeS 9R]iP, together with smaller amounts of [9R]iP. The hydroxylated derivatives of these seem to be restricted to plant tRNAs, and the isoprenoid sidechain shows the cis-configuration in contrast to the $trans$-configuration exhibited by free Z and [9R]Z isolated from plants. (cis)[9R]Z is commonly the major cytokinin present in plant tRNA hydrolysates, with lesser amounts of [9R]iP, (*cis*)[2MeS 9R]Z, and [2MeS 9R]iP (16, 19, 40). However, exceptions to the above generalizations do exist, e.g. zeatin-type compounds are present in the tRNA of *Pseudomonas aeroginosa* (177) and some plant-associated bacteria (30, 41, 118, 121), and *trans*-isomers of the hydroxylated compounds have been isolated from tRNA of *A. tumefaciens* (118) and a few plant tRNAs (135, 189, 190).

The distribution of various cytokinin-active nucleosides within the tRNA species of the U group has recently been studied in *A. tumefaciens* (118). (*cis*)[9R]Z is almost exclusively present in tRNA^{Ser}, whereas [2MeS 9R]Z is the predominant cytokinin in tRNA$^{Phe, Ser, and Tyr}$, and tRNA^{Trp} contains only [9R]iP. Hence hydroxylation and methylthiolation would seem to occur with a high degree of specificity and to depend on tRNA structure or sequence.

Cytokinins have also been isolated from chloroplast tRNA (58, 171, 190), and a parallelism in the distribution of tRNA-bound cytokinins in chloroplast and prokaryotes has been drawn (190). (*cis*)[9R]Z and [9R]Z (*trans*) were associated exclusively with the cytoplasmic tRNA which also contained [9R]iP and [2MeS 9R]Z (190). The last two cytokinins were also present in the chloroplast tRNA; both (*cis*)- and (*trans*)-[2MeS 9R]Z were isolated from cytoplasmic tRNA whereas only the *cis*-isomer was present in chloroplast tRNA. It has been suggested (171) that the enzymes required for methylthiolation are localized primarily in the chloroplasts, while the hydroxylation reaction occurs in the cytoplasm.

The cytokinin molecules in tRNA appear to affect the ability of tRNAs to function in protein synthesis by influencing the binding of the aminoacyl tRNA molecules to the ribosome-mRNA complex (53, 92, 99). There is considerable evidence that methylthiolation of iP to yield [2MeS]iP in certain bacterial tRNAs is necessary for effective translation of the corresponding mRNA (20, 70, 108). However, discussion of such concepts is beyond the scope of this review.

STABILITY AND SIGNIFICANCE OF METABOLITES

Stability

Cytokinin bases and ribosides undergo very rapid and extensive metabolism in some plant tissues (Table 1), including those used as bioassays (49, 50, 56, 88, 174, 201), and this must greatly limit the sensitivity of such assays. There are no reported studies of the metabolism of exogenous cytokinin nucleotides. However, nucleotides formed from supplied cytokinin bases and ribosides appear to be metabolized rapidly (49, 51a, 55, 88, 110, 140), and in tobacco cells rapid isoprenoid sidechain cleavage appears to occur at the nucleotide level (85). In contrast, the 7-glucosides of cytokinins exhibit remarkable metabolic stability in tobacco cell cultures (51a), soybean callus tissue (49), and radish cotyledons (140). [9G]BAP and [9G]Z are also metabolites of high stability in radish cotyledons (98) and soybean callus tissue respectively (138); however, [3G]BAP is converted slowly by radish cotyledons to BAP which is subsequently metabolized to [7G]BAP and [9G]BAP (98). [3G]BAP is hydrolyzed by emulsin (β-glucosidase from almonds), but [7G]BAP and [9G]BAP are resistant to this enzyme (100). Although 7-glucosides are extremely resistant to enzymic degradation, the following minor metabolic transformations have been observed with labeled glucoside: conversion of [7G]BAP to BAP nucleotide in tobacco cells (51a) and to adenine-7-glucoside in mature radish cotyledons (98).

O-Glucosylation of Z markedly reduces degradative metabolism in lupin leaves (142) and bean leaves (132). In bean leaves, urea and ureides appear to be prominent metabolites of Z, and the following stability sequence has been determined: (diH OG)Z > (OG)Z > (diH)Z > Z (132). Hence O-glucosylation and sidechain reduction are structural modifications which in conjunction markedly enhance cytokinin stability. In this connection, it is relevant to note that when (diH)Z is supplied to *Alnus glutinosa* leaves, the cytokinin activity taken up is conserved in a polar metabolite which appears to be (diH OG)Z. In contrast, when Z is supplied to these leaves, cytokinin activity in the leaf declines rapidly, apparently due to sidechain cleavage, and significant amounts of O-glucosides do not accumulate (66).

While (OG)Z exhibits enhanced metabolic stability, it also appears to serve as a source of free Z. Thus hydrolysis of exogenous (OG)Z to Z appears to occur in several tissues (132, 142, 185). Similarly, (OG)[9R]Z is partially converted to Z and [9R]Z in soybean callus tissue in which it is much more stable than [9R]Z (138). It is relevant to note that while the 7- and 9-glucosides of cytokinins are resistant to hydrolysis by β-glucosidase (emulsin), the O-glucosides of Z, (diH)Z, and their ribosides are rapidly cleaved by this enzyme with release of the free cytokinin (38, 39). Thus hydrolysis of O-glucosides of cytokinins in tissue may result from the action of nonspecific β-glucosidases.

The metabolic stability of lupinic acid greatly exceeds that of Z in lupin leaves (142) and [9R]Z in soybean callus tissue (138). In these tissues, Z and [9R]Z are degraded rapidly by sidechain cleavage. While it is clear that 7-, 9-, and O-glucosylation and 9-alanine conjugation confer metabolic stability, it is uncertain to what extent this effect is due to resistance to degradative enzymes and to compartmentation of the "stable" metabolites so that they are not exposed to such enzymes. However, as noted before, [7G]Z, [9G]Z, and [9Ala]Z are probably compartmentalized so they are not exposed to cytokinin oxidase.

Some differences in activity exhibited by cytokinin metabolites may be rationalized from a consideration of the relative stabilities of the metabolites concerned. Thus, (diH)Z may be more active than Z in some bioassays, e.g. soybean callus (66), because inactivation by sidechain cleavage limits the level of Z in the tissue but not that of (diH)Z. Susceptibility to such cleavage appears to be the basis of the differing activities of (diH)[9R]Z and [9R]Z in *Phaseolus vulgaris* and *P. lunatus* callus cultures (115). 7- and 9-Glucoside metabolites, which are weakly active (94), may exhibit low activity because they are metabolically stable and not converted into the highly active aglycone; other glucosides (e.g. [3G]BAP, (OG)Z) are readily hydrolyzed enzymically to the aglycones, and this may account for their high activity in cytokinin bioassays (94).

While the resistance of cytokinins to enzymic attack can be enhanced by isoprenoid sidechain reduction, by glucosylation and alanine conjugation, cytokinins may also be stabilized by binding to proteins. Thus several workers (79, 148) have purified from wheat germ a cytokinin-binding protein (CBP) which has a high affinity for BAP, iP, and [9R]iP, but a low affinity for Z. The high concentration of CBP in wheat germ, 2.7 mg/g fresh weight (149), and the structural diversity of CBP ligands (148) suggest that CBP is not a receptor involved in cytokinin mechanism of action. CBP could be a sequestering protein for iP and [9R]iP. A large bound, and thus stabilized, pool of cytokinin could be in equilibrium with a pool of free cytokinin which is subject to metabolism but maintained by synthesis. When the level of the free pool drops, perhaps due to cessation of synthesis, free cytokinin would be released from CBP.

Significance

The functional significance of cytokinin metabolites is obscure, but these compounds could be: 1. active forms of cytokinin, i.e. the molecular species which bind to a perceptor to evoke a growth or physiological response; 2. translocation forms; 3. storage forms (these would release free cytokinin when required); 4. detoxification products formed when exogenous cytokinin levels are so high as to be toxic; 5. deactivation products formed to lower endogenous cytokinin levels; 6. inactivation products, formation of which is coupled with cytokinin action.

7- and 9-Glucosides and alanine conjugates are metabolites with enhanced metabolic stability, and yet they are much less active than the parent cytokinin. These metabolites per se are unlikely to be active forms of cytokinin. This conclusion is supported by experiments in which the intracellular concentrations of 9- and/or 7-glucosides were measured in relation to cytokinin-induced growth (82, 98). Thus when radish cotyledons were cultured on BAP solution and then transfered to a cytokinin-free medium, the growth rate declined rapidly as did the level of free BAP, [9R]BAP, and BAP nucleotide. However, the levels of [3G]BAP, [7G]BAP, and [9G]BAP remained essentially constant. This indicates that the base BAP, its riboside or nucleotide could be the active forms, but not the N-glucosides (98). In an attempt to determine whether 7- or 9-glycosylation is a prerequisite for the expression of cytokinin activity, Hecht et al (63) studied the cytokinin activity of substituted pyrazolo[4,3-d]pyrimidines and pyrazolo[3,4-d] pyrimidines. Some of these 7- or 9-deaza 8-aza purines were found to exhibit very weak cytokinin activity. From this work it is reasonable to conclude that 7- and 9-glucosylation and 9-ribosylation of purine cytokinin bases are not required for expression of some cytokinin activity, but these studies do not eliminate the 9-riboside or nucleotide as the principal active forms of cytokinin. Determination of the active form(s) of cytokinin is probably the most significant unsolved problem in cytokinin metabolism. Recently, growth studies with tobacco cell cultures of low cell density have indicated that conversion of cytokinin ribosides to bases is necessary for activity and that the latter may be the active form (85).

It has been suggested that 7-glucosides are stabilized, storage forms of cytokinin (51a, 140), and, as noted above, minor cleavage of the glucose moiety may occur in some tissues. Uncharacterized endogenous cytokinin metabolites which are rendered active by acid hydrolysis (92) could possibly be 7- and 9-glucosides. These unidentified metabolites have been termed "bound forms" of cytokinin and may be storage forms. The concept of a storage role can probably be applied with greater certainty to O-glucosides. Three types of evidence indicate that these glucosides are storage forms of cytokinin bases and ribosides. First, when plant tissues accumulate cytoki- nin activity, much of the increase in activity appears to be due to O- glucoside metabolites, the level of which may rise much more markedly than that of cytokinin bases and ribosides [(130, 183); numerous references in (166)]. Second, when O-glucosides are supplied to plant tissues, conver- sion to cytokinin bases and ribosides has been observed (see above). Third, following induction of some phases of plant development, the level of O-glucosides appears to fall rapidly, e.g. during germination of maize seed (161), lateral bud development in bean seedlings (130), and breaking of dormancy and apical bud growth in potato tubers (184). In the last-men- tioned case, decline in the level of a glucoside, possibly (OG) [9R]Z, appears to be accompanied by a rise in the level of a cytokinin which cochromato- graphs with [9R]Z. These observations are consistent with the view that cytokinin O-glucosides are storage forms of cytokinin which release free, active cytokinin bases and/or ribosides when these are required.

As mentioned before, when cytokinin bases are supplied exogenously to plant tissues, cytokinin nucleotides are often the principal metabolites formed initially. Nucleotide formation may be associated with cytokinin uptake and transport across membranes. It is noteworthy that the rate of purine base uptake into cultured mammalian cells is strictly a function of the rate of their phosphoribosylation (104). Phosphoribosylation plays a role in the uptake of adenine by isolated membrane vesicles of *E. coli* (71) and possibly by plant cell cultures (37). Similarly, in plant tissues, uptake of cytokinin bases and their phosphoribosylation may be coupled phenomena.

None of the metabolites listed in Table 1 are simply products of detoxification of excesses of exogenous cytokinin. 7- and 9-Glucosides could be possible metabolites of this type, but this view is clearly incorrect (51a, 98). Thus, when BAP was supplied to excised radish cotyledons at concentrations far below the optimum for a growth response, the percentages of metabolite radioactivity due to [3G]BAP, [7G]BAP, and [9G]BAP were almost the same as when the cytokinin was supplied at optimal levels (98). However, 7- and 9-glucosides may be deactivation products formed to lower physiological cytokinin levels, perhaps in some particular cell compartment. The results just mentioned are consistent with this view. Formation of [7G]BAP in radish cotyledons is not coupled to BAP action (98).

In summary, while ribosides are important translocation forms of cytokinin in the xylem (54, 92), no definite function can be assigned to other cytokinin metabolites. However, O-glucosides may well be storage forms; nucleotides may be associated with cytokinin uptake by tissues; formation of 7- and 9-glucosides and also of alanine conjugates, and sidechain cleavage to adenine and its derivatives, may constitute alternative mechanisms for lowering physiological levels of cytokinin activity in tissue; free bases per se may be active forms.

BIOSYNTHESIS OF CYTOKININS

Cytokinins in tRNA

In classical experiments conducted before 1970, it was established that formation of tRNA cytokinins, like other modified bases in tRNA, takes place at the polymer level during post-transcriptional processing. Δ^2-Isopentenyl pyrophosphate (IPP or dimethylallyl pyrophosphate) is the immediate precursor (donor) of the Δ^2-isopentenyl (i.e. 3-methylbut-2-enyl) sidechain of N^6-(Δ^2-isopentenyl)adenosine in tRNA (for reviews see 60, 61, 92, 99). Δ^2-IPP is formed from Δ^3-IPP and originally from mevalonic acid (MVA). These important conclusions were based on the following types of observations: (*a*) in vivo incorporation of radioactivity from MVA into isopentenyl groups of [9R]iP in tRNA of certain bacteria and of tobacco callus; (*b*) transfer of isopentenyl groups from Δ^2-IPP to tRNA species which lack this moiety, catalyzed by enzyme systems (Δ^2-IPP: tRNA-Δ^2-isopentenyl transferase) from diverse sources including tobacco cells.

Subsequent studies show incorporation of label from MVA into (cis)[9R]Z of tobacco callus tRNA (120). More recently, incorporation of [^{14}C]MVA into both (cis)[9R]Z and [9R]iP in total RNA isolated from cytokinin-autonomous and bacteria-transformed tobacco callus tissue has been demonstrated (17). Cultured human fibroblasts have also been shown to incorporate [^{3}H]MVA lactone into Δ^2-isopentenyl tRNA when the endogenous MVA concentration was reduced by compactin. The incorporation of label into Δ^2-isopentenyl tRNA was stimulated tenfold when its incorporation into cholesterol was decreased by addition of low-density lipoprotein to the culture medium (48).

The purified transferase enzyme from *E. coli* is specific for tRNA; adenosine, AMP, or oligoadenylic acids are neither substrates nor inhibitors of the reaction (6). However, studies with *E. coli* tRNA fragments indicate that a yeast enzyme may not require intact tRNA (186), and more recently a tRNA:isopentenyl transferase from *Zea mays* has been found to accept tRNA, oligo(A), other nucleic acids, and even adenosine as substrate (72). The low degree of isopentenylation of adenosine in *E. coli* tRNA^{Tyr} precursor in vitro, the low amounts of this modified nucleoside isolated from natural preparations of this precursor, and the high level of this modification in mature tRNA, together indicate that cleaved precursor is the most favored substrate for the transferase enzyme (153).

Although it would appear that the same enzyme catalyzes the isopentenylation of various tRNA species in vivo, the possible existence of multiple enzymes needs examination. Also the observations regarding nonspecific transferases (72) should be tested critically. However, in tRNA of known primary structure, the modified nucleoside (cytokinin) always occurs in a definite sequence, i.e. Ados-modified Ados-Ados with modified Ados adjacent to the anticodon. Whether a short oligonucleotide may be the recognition site and possibly a substrate for the enzyme has not been assessed.

It is of considerable interest that enzyme activities have also been detected in *Lactobacillus acidophilus* and liver homogenates that remove the isopentenyl moiety from [9R]iP residues in tRNA (107). The enzyme does not act when the free nucleoside is used as a substrate. Hence cells may have the ability to demodify species of tRNA which contain a [9R]iP moiety and thus regulate the level of such tRNA species at the degradative as well as the biosynthetic level.

Biosynthesis of the 2-methylthio moiety of [2MeS 9R]iP in *E. coli* tRNA occurs in a sequential manner, i.e. thiolation of [9R]iP followed by methylation (2). Isopentenylation of adenosine residues appears to take place prior to methylthiolation (53). In vitro work using methyl-deficient tRNA^{Tyr} from *E. coli* has shown that S-adenosyl-L-methionine is the methyl donor of [2MeS 9R]iP (52). The stage at which the hydroxylation of the isopentenyl sidechain occurs in the biosynthesis of [9R]Z and [2MeS 9R]Z, and whether the enzymes involved in the formation of *cis*- and *trans*-isomers are stereo-specific are not known.

Free cytokinins

In contrast to our understanding of the biosynthesis of tRNA cytokinins from a number of prokaryotic and eukaryotic systems, very limited progress has been made in elucidating the mechanism and rate of biosynthesis of free cytokinins. This lack of knowledge stems from a variety of causes. The study of cytokinin biosynthesis is particularly difficult and the two major problems associated with it are: the extremely low levels of the endogenous cytokinins in plant tissues, and the central role of the most likely precursors (adenine, its nucleoside or nucleotides) in cellular metabolism. Thus a major portion of supplied radioactivity is incorporated into the common purine metabolites and, maximally, only a very small fraction into cytokinins. This poses considerable technical problems regarding the isolation and purification of minute quantities of putatively labeled cytokinins from the complex of other highly labeled purines and their numerous degradation products.

The precursor of the sidechain of (oOH)[9R]BAP is unknown. However, MVA would appear to be the precursor of the isoprenoid sidechain present in naturally occurring cytokinins of the iP type. Although MVA may be considered a better precursor than Ade and its derivatives for biosynthetic studies of cytokinins because of less conversion into undesirable basic compounds, there are only a few reports of its incorporation into free cytokinins (5, 17, 65). The incorporation reported, if real, is very low and may largely reflect poor uptake of MVA by cells.

The problem of cytokinin biosynthesis is further exacerbated by the presence of cytokinin molecules in certain tRNA species. Although tRNA is a potential source of cytokinins, its hydrolysis to mononucleotides would presumably be required to release activity. It has been suggested that such hydrolysis may result in release of "kinins" in dying, autolyzing cells of differentiating vascular tissues (see 92). The free cytokinins of bacteria may arise from the turnover of tRNA, and some evidence was provided to suggest that the amounts of free iP extracted from *Agrobacterium* cultures could be attributed entirely to tRNA breakdown (59). Similarly, a hypothesis has been advanced that free cytokinin production could be accounted for by the turnover of cytokinin containing tRNAs in *L. acidophilus* and primary roots of *Zea mays* (81, 89) and by degradation of oligonucleotides in intact bean roots (102). These findings should, however, be viewed with a degree of caution until we have more substantial evidence.

While tRNA turnover may contribute to the free cytokinins, the following evidence strongly suggests biosynthesis of free cytokinins by some alternative mechanism:

1. The tRNA of cytokinin-dependent tissues contains cytokinins (13, 24) and yet supplied cytokinin is required for the growth of such tissues in culture.
2. The level of free cytokinins present in pea root tips exceeds by a factor of 27 the amount present in the tRNA (160). It should, however, be noted that the tRNA was hydrolyzed with acid, a treatment that partially destroys [9R]iP (62, 151). Thus the resultant biological activity might give an erroneously low cytokinin content.
3. The low rates of tRNA turnover in many plant tissues do not support the view that turnover contributes significantly to the pool of free cytokinins (61, 178).
4. When labeled adenine was fed to cells derived from moss, a labeled cytokinin with chromatographic behavior of [9R]iP was produced, while no labeled cytokinin could be isolated from the tRNA (8).
5. In some organisms, particularly higher plants, certain cytokinins in tRNA do not occur as free compounds and vice versa. Thus (diH)[9R]Z and (oOH)[9R]BAP, the free cytokinins in *Lupinus luteus* and *Populus robusta* respectively are not present in the tRNA of the respective species (14). Also, *cis*[9R]Z rarely seems to occur as a free compound in higher plants, yet it is the dominant cytokinin in plant tRNA in which [9R]Z (*trans*) does not occur or is a minor component (92).
6. If free cytokinins were derived simply by the hydrolysis of tRNA, compounds like (*cis*)[9R]Z and [2MeS 9R] would be expected to occur frequently as free cytokinins in plants.
7. The conversion of an adenosine derivative modified in the ribose ring to a cytokinin analog even though it could not be incorporated into tRNA (23).
8. Conversions of AMP to [9R-5'P]iP by cell-free extracts (27, 175).

Also there is evidence that the fungus *Rhizopogon roseolus* synthesizes N-(purin-6-ylcarbamoyl)threonine, a compound closely related to cytokinins, by a means independent of its synthesis in tRNA (84). Similarly 1-methyladenine, a minor base in nucleic acids, especially some tRNAs, induces oocyte maturation and spawning in starfish, and is synthesized independently of the tRNA (159).

There are some reports of in vivo conversions of labeled adenine and closely related compounds into free cytokinins. Systems[4] as diverse as the following have been used for such studies: a fungus, [9R]Z (113); cells derived from moss sporogonium, [9R]iP (8); cytokinin-autonomous tobacco callus, Z, *[9R]iP* (42), and [9R]Z, Z, *[9R]iP* and iP (15); synchronously dividing tobacco cells, Z, [9R]Z, iP, [9R]iP (124); tobacco crown-gall tissue, Z, [9R]Z, (*cis*)[9R]Z, iP, *[9R]iP* (17); *Vinca rosea* crown-gall tissue, Z, [9R]Z (146), and Z, [9R]Z, *[9R-5'P]Z*, (OG)Z, (OG)[9R]Z (137, 164); cultured rootless tobacco plants, Z, [9R]Z, iP, [9R]iP (29); roots of intact bean plants, Z (roots and leaves), [9R]iP (tRNA of roots), [9R]Z (oligonucleotides) (102); and mutants of the moss *Physcomitrella patens*, [9R]iP, [9R-5'P]iP in tissue and *[9R]iP* in culture medium (191). The duration of incubation ranged from less than one hour to a few weeks.

Some of these reports must be questioned because of lack of sufficient purification and positive identification of the cytokinins produced. In addition, studies based on low resolution chromatography are likely to give oversestimates of incorporation (74, 137). The level of incorporation obtained in most cases is very low, and even the most reliable identifications are usually based only on chromatography of the cytokinins produced or of their derivatives formed by enzymic or chemical modification. In a few cases cocrystallization to constant specific activity has been attempted. Investigations based on labeling with stable isotopes in conjunction with the available sophisticated mass-spectrometric techniques should now be exploited to overcome this problem. Furthermore, with the exception of a few studies (17, 137, 164, 191), no attention was given to cytokinin nucleotide production, which would seem to be of key importance in the biosynthetic pathway (see below).

The enzyme reactions which yield free cytokinins independently of tRNA turnover were first demonstrated using a cell-free preparation from the slime mold *Dictyostelium discoideum,* which contains high levels of free iP. AMP was shown to be the acceptor molecule of the isopentenyl group (175). This work has been extended to higher plants, and an enzyme (Δ^2-isopentenyl pyrophosphate: AMP-Δ^2-isopentenyl transferase) was partially purified from cytokinin-autonomous tobacco callus. This enzyme catalyzed the formation of [9R-5'P]iP from Δ^2-IPP and AMP; small amounts of [9R]iP and iP were also detected when less purified enzyme preparations were employed (27). Similarly, cytokinin synthase activity has also been demonstrated by other workers (117, 119, 125).

Although in vivo conversion of precursors (frequently adenine) to cytoki-

[4]Radioactive cytokinins formed are shown immediately after the experimental system used; cytokinins in italics had the majority of incorporated radioactivity.

nins has been reported, and although the in vitro studies do show the existence of enzymes that catalyze such reactions, the actual mechanism and rate of cytokinin biosynthesis and its control are far from being understood. A major reason for this lack of knowledge is the absence of a suitable system in which a study of this type is technically feasible. The bacteria transformed crown-gall tumor tissues produce very high levels of cytokinins and go part way to provide an experimental system for such a study (74).

Initial study on *Vinca rosea* crown-gall tumor tissue demonstrated the incorporation of labeled adenine into Z and [9R]Z, and the level of incorporation was shown to be affected by the nitrogen balance of the nutrient supply (146). This work has been considerably extended by a quantitative and qualitative time-course study (164). The maximum incorporation into cytokinins occurs after incubation for 8 hr; at all times radioactivity in zeatin nucleotide is considerably higher than in Z and [9R]Z (164). The peak of radioactivity in adenine nucleotides precedes that in zeatin nucleotides, and after 8 hr of incubation, [9R-5'P]Z is the only phosphate derivative of [9R]Z present in the tissue (137). Similar observations have been made with *Datura* crown-gall tissue (L. M. S. Palni, unpublished). This would suggest that the cytokinin biosynthesis occurs primarily at the nucleotide level, and the simplest interpretation would be a conversion of AMP to [9R-5'P]Z probably via [9R-5'P]iP, Z and [9R]Z being formed subsequently. All the available data from in vitro systems give credence to this suggestion. However, the possibility of the sidechain attachment at the free base or riboside level cannot be excluded at this stage.

In the above studies (137, 146, 164), incorporation of radioactivity into [9R-5'P]iP, its riboside or free base could not be detected, although they would appear to be the logical intermediates in the synthesis of Z-type cytokinins from adenine. This may, however, be due to extremely rapid conversion of iP-type cytokinins into Z-type compounds, a reaction involving stereospecific *trans*-hydroxylation of the terminal methyl group of the isopentenyl sidechain. This has now been demonstrated in *V. rosea* crown-gall tissue (136). Similar conversions occur in *Rhizopogon roseolus* (112, 113) and *Zea mays* endosperm (112). The stereospecificity of the *trans*-hydroxylation step resulting in the formation of [9R]Z was also established in the latter study.

The efficiency of incorporation of adenine (164) into the free cytokinins would suggest that modification and turnover of tRNA were not obligatory in the process. A pathway independent of tRNA turnover is again indicated. However, the possibility that the metabolism of tRNA in crown-gall tumor tissues and other cytokinin overproducers (e.g. mutants of the moss *Physcomitrella patens*) is abnormal cannot be ignored. Formation of appreciable quantities of cytokinins as a result of total or limited tRNA metabolism might occur if: (*a*) such tissues have very high levels of cytokinin containing tRNAs, (*b*) such tRNA species turn over with greater rapidity than the bulk of the tRNA, or (*c*) there was a means of selectively excising the cytokinin moiety (or a small oligonucleotide sequence including the cytokinin molecule) from tRNA, leaving the rest of the molecule intact.

The levels of cytokinins present in the tRNA of *V. rosea* crown-gall tissue are slightly greater than those in normal tissue, but there is a very pronounced difference between the two tRNAs in the ratio of (*cis*)[9R]Z to [9R]Z (135). Thus, (*cis*)[9R]Z isolated from crown-gall tRNA is about 1.5 times the amount present in normal tRNA. However, [9R]Z (*trans* configuration) is exclusively associated with the crown-gall tRNA and is also the major free cytokinin in this tissue. The *cis/trans* ratio is 2.3 which is very low compared to ratios of 40 in *Pisum sativum* (188) and 50 in *Spinacea oleracea* tRNAs (190). The high levels of the *trans* riboside (1 mole/800 moles tRNA) in the tRNA of crown-gall tissue of *V. rosea* may be of functional significance, and are most unlikely to have resulted from incorporation of free [9R]Z released into the medium by cultured tissue (133, 135). Thus biogenesis of free [9R]Z could perhaps result from an extremely rapid turnover of specific tRNA species containing the *trans* riboside [9R]Z. Abnormally high turnover rates of a subpopulation of tRNA have been observed in animal tumors (10). However, an unambiguous answer will be obtained only by direct measurement of the rate of turnover of cytokinin bases in individual species of tRNA under conditions where the rates of production of free cytokinins can be measured. Further work is necessary to assess the nature and extent of the role played by tRNA in the biogenesis of free cytokinins.

CONCLUDING REMARKS

During the last 8 years, a diversity of cytokinin metabolites, principally glucosides, have been identified unequivocally and synthesized chemically. Formation of Z metabolites is summarized in Figure 1, which also indicates the probable biosynthetic pathways leading to cytokinins. Nucleotides of cytokinins are of special significance in cytokinin metabolism—they appear to be the initial products of cytokinin biosynthesis and, in many tissues, the first metabolites formed in appreciable amounts from exogenous cytokinin bases. It is difficult to understand why many plant physiologists who are addicted to bioassaying plant extracts discard the fractions which could contain cytokinin nucleotides (e.g. the fraction not retained on a cation-exchange column). Identified natural cytokinins which occur in the free form, either in plants or in microorganisms, now total 28, but their activity varies from almost inactive to highly active.

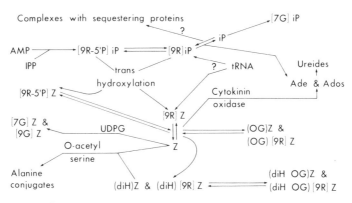

Figure 1 A simplified schematic representation of the biosynthesis and metabolism of natural cytokinins. For abbreviations see addendum to Table 1. According to the figure, both the riboside and nucleotide of iP are *trans* hydroxylated. However, the level(s) (base, riboside, nucleotide) at which this hydroxylation occurs is unclear.

While metabolic modifications which cytokinins undergo now appear to be largely defined, there are very few definitive studies involving rigorous chemical identifications and/or quantifications in which cytokinin metabolism has been closely related to plant development. Several critical studies of cytokinin metabolism in tissue cultures and excised organs which require supplied cytokinin for growth or retardation of senescence have been reported. The physiological relevance of these studies is obvious and need not be related to the endogenous cytokinin status of the tissue. However, studies of cytokinin metabolism in relation to "whole plant systems" are, with one possible exception (130–132, 194), of much less physiological relevance, since the endogenous cytokinins have not been identified rigorously and their physiological role is far from clear. Indeed, there are very few critical studies in which the metabolism of a cytokinin has been investigated in a tissue in which it is definitely known to occur. Environmental factors and chemical stimuli appear to alter cytokinin levels markedly in plant tissue (92) and thus influence development. It is important to relate these changes to specific aspects of cytokinin biosynthesis, metabolism, or translocation.

In a meticulous series of studies reported between 1974 and 1976, C. O. Miller established that *Vinca rosea* tumor tissue actively produces free cytokinins. Today, studies of cytokinin biosynthesis and metabolism in plant tumor cell cultures are attracting great interest and have provided very significant basic information. Attempts are in progress to gain insight into the basic mechanisms controlling production and turnover of cytokinins in such systems. These investigations should pave the way for meaningful biosynthetic studies in "normal" plant systems. These studies will undoubtedly be facilitated by the greater insight we now have regarding cytokinin metabolites. In some tissues it may be possible to detect precursor incorporation into stable metabolites but not into the immediate products of biosynthesis.

Several areas of endeavor which merit attention have been mentioned in the text. Further important problems are the following: cytokinin metabolism at the subcellular level; the distribution and metabolic fate of cytokinins moving in the xylem and phloem; the design of specific inhibitors of metabolizing enzymes which inactivate cytokinins as such inhibitors may influence plant development.

Studies of cytokinin metabolism, biosynthesis, and translocation in relation to plant development are really only in their infancy. In the future such studies should be integrated with precise studies of endogenous cytokinin levels. However, because of the new techniques now available for purifying, identifying, and quantifying cytokinins, our future progress should be both more rapid and more definitive and provide new insight into the integrated development of the flowering plant and into its relationships with microorganisms.

ACKNOWLEDGMENTS

We express thanks to numerous workers who provided preprints, to Mrs. P. A. Vest for her great patience in typing the manuscript, to Mr. D. J. Pianca, who assisted in compiling the reference list, and to Dr. M. V. Palmer for his comments both scientific and stylistic.

Literature Cited

1. Abe, H., Uchiyama, M., Tanaka, Y., Saito, H. 1976. Structure of discadenine, a spore germination inhibitor from the cellular slime mold, *Dictyostelium discoideum. Tetrahedron Lett.,* pp. 3807–10
2. Agris, P. F., Armstrong, D. J., Schäfer, K. P., Söll, D. 1975. Maturation of a hypermodified nucleoside in transfer RNA. *Nucleic Acids Res.* 2:691–98
3. Armstrong, D. J., Burrows, W. J., Skoog, F., Roy, K. L., Söll, D. 1969. Cytokinins: Distribution in transfer RNA species of *Escherichia coli. Proc. Natl. Acad. Sci. USA* 63:834–41
4. Barciszewska, M., Kaminek, M., Barciszewski, J., Wiewiórowski, M. 1981. Lack of cytokinin activity of Y-type bases isolated from phenylalanine specific tRNAs. *Plant Sci. Lett.* 20:387–92
5. Barnes, M. F., Tien, C. L., Gray, J. S. 1980. Biosynthesis of cytokinins by potato cell cultures. *Phytochemistry* 19:409–12

6. Bartz, J. K., Söll, D. 1972. N^6-Δ^2-isopentenyl)adenosine: Biosynthesis *in vitro* in transfer RNA by an enzyme purified from *Escherichia coli. Biochimie* 54:31–39
7. Bartz, J., Söll, D., Burrows, W. J., Skoog, F. 1970. Identification of the cytokinin-active ribonucleosides in *Escherichia coli* tRNA species. *Proc. Natl. Acad. Sci. USA* 67:1448–53
8. Beutelmann, P. 1973. Untersuchungen zur biosynthese eines cytokinins in calluszellen von Laubmoossporophyten. *Planta* 112:181–90
9. Beutelmann, P. 1977. Purification and identification of a cytokinin from moss callus cells. *Planta* 133:215–17
10. Borek, E., Baliga, B. S., Gehrke, C. W., Kuo, C. W., Belman, S., et al. 1977. High turnover rate of transfer RNA in tumor tissue. *Cancer Res.* 37:3362–66
11. Brenner, M. L. 1981. Modern methods for plant growth substance analysis. *Ann. Rev. Plant Physiol.* 32:511–38

12. Brownlee, B. G., Hall, R. H., Whitty, C. D. 1975. 3-Methyl-2-butenal: An enzymatic degradation product of the cytokinin, N^6-(Δ^2-isopentenyl)adenine. *Can. J. Biochem.* 53:37–41

13. Burrows, W. J. 1976. Mode of action of N,N'-diphenylurea: The isolation and identification of the cytokinins in the transfer RNA from tobacco callus grown in the presence of N,N'-diphenylurea. *Planta* 130:313–16

14. Burrows, W. J. 1978. Evidence in support of biosynthesis *de novo* of free cytokinins. *Planta* 138:53–57

15. Burrows, W. J. 1978. Incorporation of ^{3}H-adenine into free cytokinins by cytokinin-autonomous tobacco callus tissue. *Biochem. Biophys. Res. Commun.* 84:743–48

16. Burrows, W. J., Armstrong, D. J., Kaminek, M., Skoog, F., Bock, R. M., et al. 1970. Isolation and identification of four cytokinins from wheat germ transfer ribonucleic acid. *Biochemistry* 9:1867–72

17. Burrows, W. J., Fuell, K. J. 1981. Cytokinin biosynthesis in cytokinin-autonomous and bacteria-transformed tobacco callus tissues. In *Metabolism and Molecular Activities of Cytokinins*, ed. J. Guern, C. Péaud-Lenoël, pp. 44–55. Berlin: Springer. 352 pp.

18. Burrows, W. J., Leworthy, D. P. 1976. Metabolism of N,N'-diphenylurea by cytokinin-dependent tobacco callus: identification of the glucoside. *Biochem. Biophys. Res. Commun.* 70:1109–14

19. Burrows, W. J., Skoog, F., Leonard, N. J. 1971. Isolation and identification of cytokinins located in the transfer ribonucleic acid of tobacco callus grown in the presence of 6-benzylaminopurine. *Biochemistry* 10:2189–94

20. Buu, A., Menichi, B., Heyman, T. 1981. Thiomethylation of tyrosine transfer ribonucleic acid is associated with initiation of sporulation in *Bacillus subtilis:* Effect of phosphate concentration. *J. Bacteriol.* 146:819–22

21. Chaves das Neves, H. J., Pais, M. S. 1980. A new cytokinin from the fruits of *Zantedeschia aethiopica. Tetrahedron Lett.* 21:4387–90

22. Chen, C.-M. 1981. Biosynthesis and enzymic regulation of the interconversion of cytokinin. See Ref. 17, pp. 34–43

23. Chen, C.-M., Eckert, R. L. 1976. Evidence for the biosynthesis of transfer RNA-free cytokinin. *FEBS Lett.* 64:429–34

24. Chen, C.-M., Hall, R. H. 1969. Biosynthesis of N^6-(Δ^2-isopentenyl)adenosine

25. Chen, C.-M., Kristopeit, S. M. 1981. Dephosphorylation of cytokinin ribonucleotide by 5'-nucleotidases from wheat germ cytosol. *Plant Physiol.* 67:494–98

26. Chen, C.-M., Kristopeit, S. M. 1981. Deribosylation of cytokinin ribonucleoside by adenosine nucleosidase from wheat germ cells. *Plant Physiol.* 68:1020–23

27. Chen, C.-M., Melitz, D. K. 1979. Cytokinin biosynthesis in a cell-free system from cytokinin-autotrophic tobacco tissue cultures. *FEBS Lett.* 107:15–20

28. Chen, C.-M., Melitz, D. K., Clough, F. W. 1982. Metabolism of cytokinins: Phosphoribosylation of cytokinin bases by adenine phosphoribosyltransferase from wheat germ. *Arch. Biochem. Biophys.* 214:634–41

29. Chen, C.-M., Petschow, B. 1978. Cytokinin biosynthesis in cultured rootless tobacco plants. *Plant Physiol.* 62: 861–65

30. Cherayil, J. D., Lipsett, M. N. 1977. Zeatin ribonucleosides in the transfer ribonucleic acid of *Rhizobium leguminosarum, Agrobacterium tumefaciens, Corynebacterium fascians* and *Erwinia amylovora. J. Bacteriol.* 131:741–44

31. Cowley, D. E., Duke, C. C., Liepa, A. J., MacLeod, J. K., Letham, D. S. 1978. The structure and synthesis of cytokinin metabolites I. The 7- and 9-β-D-glucofuranosides and pyranosides of zeatin and 6-benzylaminopurine. *Aust. J. Chem.* 31:1095–1111

32. Cowley, D. E., Jenkins, I. D., MacLeod, J. K., Summons, R. E., Letham, D. S., et al. 1975. Structure and synthesis of unusual cytokinin metabolites. *Tetrahedron Lett.*, pp. 1015–18

33. Dauphin, B., Teller, G., Durand, B. 1979. Identification and quantitative analysis of cytokinins from shoot apices of *Mercurialis ambigua* by GC-MS computer system. *Planta* 144:113–19

34. Dauphin-Guerin, B., Teller, G., Durand, B. 1980. Different endogenous cytokinins between male and female *Mercurialis annua. Planta* 148:124–29

35. Davey, J. E., Van Staden, J. 1978. Cytokinin activity in *Lupinus albus* III. Distribution in fruits. *Physiol. Plant.* 43:87–93

36. Doree, M., Guern, J. 1973. Short-time metabolism of some exogenous cytoki-

nins in *Acer pseudoplatanus* cells. *Biochim. Biophys. Acta* 304:611–22

37. Dorée, M., Leguay, J. J., Terrine, C. 1970. Existence d'un double système de transport responsable de l'absorption de l'adénine par des suspensions cellulaires d'*Acer pseudoplatanus. C. R. Acad. Sci. Paris Ser. D* 271:1876–79

38. Duke, C. C., Letham, D. S., Parker, C. W., MacLeod, J. K., Summons, R. E. 1979. The complex of *O*-glucosylzeatin derivatives formed in *Populus* species. *Phytochemistry* 18:819–24

39. Duke, C. C., MacLeod, J. K., Summons, R. E., Letham, D. S., Parker, C. W. 1978. Lupinic acid and O-β-D-glucopyranosylzeatin from *Lupinus angustifolium. Aust. J. Chem.* 31:1291–1301

40. Edwards, C. A., Armstrong, D. J. 1981. Cytokinin-active ribonucleosides in *Phaseolus* RNA. II. Distribution in tRNA species from etiolated *P. vulgaris* L. seedlings. *Plant Physiol.* 67:1185–89

41. Einset, J. W., Skoog, F. K. 1977. Isolation and identification of ribosyl-*cis*-zeatin from transfer RNA of *Corynebacterium fascians. Biochem. Biophys. Res. Commun.* 79:1117–21

42. Einset, J. W., Skoog, F. 1973. Biosynthesis of cytokinins in cytokinin-autotrophic tobacco callus. *Proc. Natl. Acad. Sci. USA* 70:658–60

43. Entsch, B., Letham, D. S. 1979. Enzymic glucosylation of the cytokinin, 6-benzylaminopurine. *Plant Sci. Lett.* 14:205–12

44. Entsch, B., Letham, D. S., Parker, C. W., Summons, R. E., Gollnow, B. I. 1980. Metabolites of cytokinins. In *Plant Growth Substances 1979,* ed F. Skoog, pp. 109–18. Berlin: Springer. 527 pp.

45. Entsch, B., Parker, C. W., Letham, D. S. 1983. An enzyme from lupin seeds forming alanine derivatives of cytokinins. *Phytochemistry.* In press

46. Entsch, B., Parker, C. W., Letham, D. S., Summons, R. E. 1979. Preparation and characterization using HPLC of an enzyme forming glucosides of cytokinins. *Biochim. Biophys. Acta* 570:124–39

47. Erichsen, U., Knoop, B., Bopp, M. 1978. Uptake, transport and metabolism of cytokinin in moss protonema. *Plant Cell Physiol.* 19:839–50

48. Faust, J. R., Brown, M. S., Goldstein, J. L. 1980. Synthesis of Δ^2isopentenyl tRNA from mevalonate in cultured human fibroblasts. *J. Biol. Chem.* 255:6546–48

49. Fox, J. E., Cornette, J., Deleuze, G., Dyson, W., Giersak, C., et al. 1973. The formation, isolation and biological activity of a cytokinin 7-glucoside. *Plant Physiol.* 52:627–32

50. Fox, J. E., Dyson, W. D., Sood, C., McChesney, J. 1972. Active forms of the cytokinins. In *Plant Growth Substances 1970,* ed. D. J. Carr, pp. 449–58. Berlin: Springer. 837 pp.

51. Fox, J. E., Sood, C. K., Buckwalter, B., McChesney, J. D. 1971. Metabolism and biological activity of 9-substituted cytokinins. *Plant Physiol.* 47:275–81

51a. Gawer, M., Laloue, M., Terrine, C., Guern, J. 1977. Metabolism and biological significance of benzyladenine-7-glucoside. *Plant Sci. Lett.* 8:267–74

52. Gefter, M. L. 1969. The *in vitro* synthesis of 2'-O methylguanosine and 2-methylthio N^6-(γ, γ - dimethylallyl)adenosine in transfer RNA of *Escherichia coli. Biochem. Biophys. Res. Commun.* 36:435–41

53. Gefter, M. L., Russell, R. L. 1969. Role of modifications in tyrosine transfer RNA: A modified base affecting ribosome binding. *J. Mol. Biol.* 39:145–57

54. Goodwin, P. B., Gollnow, B. I., Letham, D. S. 1978. Phytohormones and growth correlations. In *Phytohormones and Related Compounds—A Comprehensive Treatise,* ed. D. S. Letham, P. B. Goodwin, T. J. V. Higgins, 2:215–49. Amsterdam: Elsevier-North Holland. 648 pp.

55. Gordon, M. E., Letham, D. S., Parker, C. W. 1974. The metabolism and translocation of zeatin in intact radish seedlings. *Ann. Bot.* 38:809–25

56. Gordon, M. E., Wilson, M. M., Parker, C. W., Letham, D. S. 1974. The metabolism of cytokinins by radish seedlings. In *Mechanisms of Regulation of Plant Growth,* ed. R. L. Bieleski, A. R. Ferguson, M. M. Cresswell, pp. 773–80. Wellington: Royal Soc. New Zealand. 934 pp.

57. Greene, E. M. 1980. Cytokinin production by microorganisms. *Bot. Rev.* 46:25–74

58. Guillemaut, P., Martin, R., Weil, J. H. 1976. Purification and base composition of a chloroplastic tRNA^{Phe} from *Phaseolus vulgaris. FEBS Lett.* 63:273–77

59. Hahn, H., Heitmann, I., Blumbach, M. 1976. Cytokinins: Production and biogenesis of N^6-Δ^2-isopentenyl)adenine in cultures of *Agrobacterium tumefaciens* strain B$_6$. *Z. Pflanzenphysiol.* 79:143–53

60. Hall, R. H. 1970. N^6-(Δ^2-isopentenyl)adenosine: chemical reactions, biosynthesis, metabolism, and significance to the structure and function of tRNA. *Prog. Nucleic Acid Res. Mol. Biol.* 10:57–86

61. Hall, R. H. 1973. Cytokinins as a probe of developmental processes. *Ann. Rev. Plant Physiol.* 24:415–44

62. Hall, R. H., Srivastava, B. I. S. 1968. Cytokinin activity of compounds obtained from soluble RNA. *Life Sci.* 7:7–13

63. Hecht, S. M., Frye, R. B., Werner, D., Hawrelak, S. D., Skoog, F., Schmitz, R. Y. 1975. On the activation of cytokinins. *J. Biol. Chem.* 250:7343–51

64. Hecker, L. I., Uziel, M., Barnett, W. E. 1976. Comparative base compositions of chloroplast and cytoplasmic tRNA$^{Phe's}$ from *Euglena gracilis. Nucleic Acids Res.* 3:371–80

65. Helbach, M., Klämbt, D. 1981. On the biogenesis of cytokinins in *Lactobacillus acidophilus* ATCC 4963. *Physiol. Plant.* 52:136–40

66. Henson, I. E. 1978. Types, formation and metabolism of cytokinins in leaves of *Alnus glutinosa* L. Gaertn. *J. Exp. Bot.* 29:935–51

67. Henson, I. E. 1978. Cytokinins and their metabolism in leaves of *Alnus glutinosa* L. Gaertn. *Z. Pflanzenphysiol.* 86:363–69

68. Henson, I. E., Wheeler, C. T. 1977. Metabolism of [8-^{14}C]zeatin in root nodules of *Alnus glutinosa* L. Gaertn. *J. Exp. Bot.* 106:1087–98

69. Hoad, G. V., Loveys, B. R., Skene, K. G. M. 1977. The effect of fruit-removal on cytokinins and gibberellin-like substances in grape leaves. *Planta* 136: 25–30

70. Hoburg, A., Aschhoff, H. J., Kersten, H., Manderschied, U., Gassen, H. G. 1979. Function of modified nucleosides 7-methylguanosine, ribothymidine, and 2-thiomethyl-N^6-(isopentenyl)adenosine in prokaryotic transfer ribonucleic acid. *J. Bacteriol.* 140:408–14

71. Hochstadt-Ozer, J., Stadtman, E. R. 1971. Adenine phosphoribosyltransferase in isolated membrane preparations and its role in transport of adenine across the membrane. *J. Biol. Chem.* 246:5304–11

72. Holtz, J., Klämbt, D. 1978. tRNA Isopentenyltransferase from *Zea mays* L. Characterization of the isopentenylation reaction of tRNA, oligo (A) and other nucleic acids. *Hoppe-Seyler's Z. Physiol. Chem.* 359:89–101

73. Horgan, R. 1981. Modern methods for plant hormone analysis. *Prog. Phytochem.* 7:137–70

74. Horgan, R., Palni, L. M. S., Scott, I., McGaw, B. 1981. Cytokinin biosynthesis and metabolism in *Vinca rosea* crown gall tissue. See Ref. 17, pp. 56–65

75. Huneeus, V. Q., Wiley, M. H., Siperstein, M. D. 1980. Isopentenyladenine as a mediator of mevalonate-regulated DNA replication. *Proc. Natl. Acad. Sci. USA* 77:5842–46

76. Hutton, M. J., Van Staden, J. 1982. Cytokinins in germinating seeds of *Phaseolus vulgaris* L. III. Transport and metabolism of 8[^{14}C]t-zeatin applied to cotyledons. *Ann. Bot.* 49:701–6

77. Jouanneau, J. P., Teyssendier de la Serve, B. 1981. Covalent insertion of N^6-benzyladenine into rRNA species of tobacco cells. See Ref. 17, pp. 228–38

78. Kaiss-Chapman, R. W., Morris, R. O. 1977. *trans*-Zeatin in culture filtrates of *Agrobacterium tumefaciens. Biochem. Biophys. Res. Commun.* 76:453–59

79. Keim, P., Erion, J., Fox, J. E. 1981. The current status of cytokinin-binding moieties. See Ref. 17, pp. 179–90

80. Kimura, K., Sugiyama, T., Hashizume, T. 1978. Isolation and identification of *cis*-zeatin riboside from the top of tobacco plant *Nicotiana tabacum. Nucleic Acid Res., Spec. Publ.* 5:339–42

81. Klemen, F., Klämbt, D. 1974. Half life of sRNA from primary roots of *Zea mays.* A contribution to the cytokinin production. *Physiol. Plant.* 31:186–88

82. Laloue, M. 1977. Cytokinins: 7-glucosylation is not a prerequisite of the expression of their biological activity. *Planta* 134:273–75

83. Laloue, M., Gawer, M., Terrine, C. 1975. Modalites de l'utilisation des cytokinines exogènes par les cellules de tabac culturées en milieu liquide agité. *Physiol. Veg.* 13:781–96

84. Laloue, M., Hall, R. H. 1973. Cytokinins in *Rhizopogon roseolus. Plant Physiol.* 51:559–62

85. Laloue, M., Pethe, C. 1982. Dynamics of cytokinin metabolism in tobacco cells. In *Plant Growth Substances 1982. Proc. 11th Int. Conf. Plant Growth Subst.,* ed. P. F. Wareing, pp. 185–95. New York: Academic. 683 pp.

86. Laloue, M., Pethe, C., Guern, J. 1981. Uptake and metabolism of cytokinins in tobacco cells: Studies in relation to the expression of their biological activities. See Ref. 17, pp. 80–96

87. Laloue, M., Terrine, C., Gawer, M. 1974. Cytokinins: formation of the nu-

cleoside-5'-triphosphate in tobacco and *Acer* cells. *FEBS Lett.* 46:45–50

88. Laloue, M., Terrine, C., Guern, J. 1977. Cytokinins: metabolism and biological activity of N^6-(Δ^2-isopentenyl)adenosine and N^6-(Δ^2-isopentenyl)adenine in tobacco cells and callus. *Plant Physiol.* 59:478–83

89. Leineweber, M., Klämbt, D. 1974. Half life of sRNA and its consequence for the formation of cytokinins in *Lactobacillus acidophilus* ATCC 4963 grown in the logarithmic and stationary phases. *Physiol. Plant.* 30:327–30

90. Lester, B. R., Morris, R. O., Cherry, J. H. 1979. Purification of leucine tRNA isoaccepting species from soybean cotyledons. II. RPC-2 purification, ribosome binding, and cytokinin content. *Plant Physiol.* 63:87–92

91. Letham, D. S. 1973. Cytokinins from *Zea mays. Phytochemistry* 12:2445–55

92. Letham, D. S. 1978. Cytokinins. In *Phytohormones and Related Compounds—A Comprehensive Treatise,* ed. D. S. Letham, P. B. Goodwin, T. J. V. Higgins, 1:205–63. Amsterdam: Elsevier/North-Holland. 641 pp.

93. Letham, D. S., Gollnow, B. I., Parker, C. W. 1979. The reported occurrence of 7-glucofuranoside metabolites of cytokinins. *Plant Sci. Lett.* 15:217–23

94. Letham, D. S., Palni, L. M. S., Tao, G.-Q., Gollnow, B. I., Bates, C. 1983. The activities of cytokinin glucosides and alanine conjugates in cytokinin bioassays. *J. Plant Growth Regul.* In press

95. Letham, D. S., Parker, C. W., Duke, C. C., Summons, R. E., MacLeod, J. K. 1977. O-Glucosylzeatin and related compounds—a new group of cytokinin metabolites. *Ann. Bot.* 41:261–63

96. Letham, D. S., Summons, R. E., Entsch, B., Gollnow, B. I., Parker, C. W., MacLeod, J. K. 1978. Glucosylation of cytokinin analogues. *Phytochemistry* 17:2053–57

97. Letham, D. S., Summons, R. E., Parker, C. W., MacLeod, J. K. 1979. Identification of an amino-acid conjugate of 6-benzylaminopurine formed in *Phaseolus vulgaris* seedlings. *Planta* 146:71–74

98. Letham, D. S., Tao, G.-Q., Parker, C. W. 1982. An overview of cytokinin metabolism. See Ref. 85, pp. 143–53.

99. Letham, D. S., Wettenhall, R. E. H. 1977. Transfer RNA and cytokinins. In *The Ribonucleic Acids,* ed. P. R. Stewart, D. S. Letham, pp. 129–93. New York: Springer. 374 pp.

100. Letham, D. S., Wilson, M. M., Parker, C. W., Jenkins, I. D., MacLeod, J. K., Summons, R. E. 1975. The identity of an unusual metabolite of 6-benzylaminopurine. *Biochim. Biophys. Acta* 399:61–70

101. Little, C. H. A., Andrew, D. M., Silk, P. J., Strunz, G. M. 1979. Identification of cytokinins zeatin and zeatin riboside in *Abies balsamea. Phytochemistry* 18:1219–20

102. Maab, H., Klämbt, D. 1981. On the biogenesis of cytokinins in roots of *Phaseolus vulgaris. Planta* 151:353–58

103. MacLeod, J. K., Summons, R. E., Letham, D. S. 1976. Mass spectrometry of cytokinin metabolites. Per(trimethylsilyl) and permethyl derivatives of glucosides of zeatin and 6-benzylaminopurine. *J. Org. Chem.* 41:3959–67

104. Marz, R., Wohlhueter, R. M., Plagemann, P. G. W. 1979. Purine and pyrimidine transport and phosphoribosylation and their interaction in overall uptake by cultured mammalian cells—a re-evaluation. *J. Biol. Chem.* 254:2329–36

105. McCalla, D. R., Moore, D. J., Osborne, D. J. 1962. The metabolism of a kinin, benzyladenine. *Biochim. Biophys. Acta* 55:522–28

106. McCloskey, J. A., Hashizume, T., Basile, B., Ohno, Y., Sonoki, S. 1980. Occurrence and levels of *cis*- and *trans*-zeatin ribosides in the culture medium of a virulent strain of *Agrobacterium tumefaciens. FEBS Lett.* 111:181–83

107. McLennan, B. D. 1975. Enzymatic demodification of transfer RNA species containing N^6-(Δ^2-isopentenyl)adenosine. *Biochem. Biophys. Res. Commun.* 65:345–51

108. Menichi, B., Heyman, T. 1976. Study of tyrosine transfer ribonucleic acid modification in relation to sporulation in *Bacillus subtilis. J. Bacteriol.* 127:268–80

109. Miernyk, J. A. 1979. Abscisic acid inhibition of kinetin nucleotide formation in germinating lettuce seeds. *Physiol. Plant.* 45:63–66

110. Miernyk, J. A., Blaydes, D. F. 1977. Short-term metabolism of radioactive kinetin during lettuce seed germination. *Physiol. Plant.* 39:4–8

111. Mitchison, J. M. 1981. Changing perspectives in the cell cycle. In *The Cell Cycle,* ed. P. C. L. John, pp. 1–10. Cambridge: Univ. Press. 276 pp.

112. Miura, G., Hall, R. 1973. *trans*-Ribosylzeatin: its biosynthesis in *Zea*

mays endosperm and the mycorrhizal fungus, *Rhizopogon roseolus. Plant Physiol.* 51:563–69

113. Miura, G. A., Miller, C. O. 1969. 6-(γ,γ-Dimethylallylamino)purine as a precursor of zeatin. *Plant Physiol.* 44:372–76

114. Mizuno, K., Komamine, A. 1978. Isolation and identification of substances inducing formation of tracheary elements in cultured carrot-root slices. *Planta* 138:59–62

115. Mok, M. C., Mok, D. W. S., Dixon, S. C., Armstrong, D. J., Shaw, G. 1982. Cytokinin structure-activity relationships and the metabolism of N^6-(Δ^2-isopentenyl)adenosine-[8-^{14}c] in *Phaseolus* callus tissues. *Plant Physiol.* 70:173–78

116. Morris, R. O. 1977. Mass spectroscopic identification of cytokinins. *Plant Physiol.* 59:1029–33

117. Morris, R. O. 1982. Cytokinins in crown gall tumors and in *Agrobacterium tumefaciens*. See Ref. 85, pp. 173–83

118. Morris, R. O., Regier, D. A., Olson, R. M. Jr., Struxness, L. A., Armstrong, D. J. 1981. Distribution of cytokinin-active nucleosides in isoaccepting transfer ribonucleic acids from *Agrobacterium tumefaciens. Biochemistry* 21:6012–17

119. Murai, N. 1981. Cytokinin biosynthesis and its relationship to the presence of plasmids in strains of *Corynebacterium fascians*. See Ref. 17, pp. 17–26

120. Murai, N., Armstrong, D. J., Skoog, F. 1975. Incorporation of mevalonic acid into ribosylzeatin in tobacco callus ribonucleic acid preparations. *Plant Physiol.* 55:853–58

121. Murai, N., Skoog, F., Doyle, M. E., Hanson, R. S. 1980. Relationships between cytokinin production, presence of plasmids, and fasciation caused by strains of *C. fascians. Proc. Natl. Acad. Sci. USA* 77:619–23

122. Murakoshi, I., Ikegami, F., Ookawa, N., Haginiwa, J., Letham, D. S. 1977. Enzymic synthesis of lupinic acid, a novel metabolite of zeatin in higher plants. *Chem. Pharm. Bull. Tokyo* 25:520–22

123. Nishimura, S. 1972. Minor components in transfer RNA: their characterization, location, and function. *Prog. Nucleic Acid Res. Mol. Biol.* 12:49–85

124. Nishinari, N., Syono, K. 1980. Biosynthesis of cytokinins by tobacco cell cultures. *Plant Cell Physiol.* 21:1143–50

125. Nishinari, N., Syono, K. 1980. Cell-free biosynthesis of cytokinins in cultured tobacco cells. *Z. Pflanzenphysiol.* 99: 383–92

126. Nishinari, N., Syono, K. 1980. Identification of cytokinins associated with mitosis in synchronously cultured tobacco cells. *Plant Cell Physiol.* 21: 383–93

127. Nishinari, N., Syono, K. 1980. Changes in endogenous cytokinin levels in partially synchronized cultured tobacco cells. *Plant Physiol.* 65:437–41

128. Paces, V., Kaminek, M. 1976. Effect of ribosylzeatin isomers on the enzymatic degradation of N^6-(Δ^2-isopentenyl)adenosine. *Nucleic Acids Res.* 3:2309–14

129. Paces, V., Rosenberg, I., Kaminek, M., Holy, A. 1977. Metabolism of cytokinins in rape seedlings. *Collect. Czech. Chem. Commun.* 42:2452–58

130. Palmer, M. V., Horgan, R., Wareing, P. F. 1981. Cytokinin metabolism in *Phaseolus vulgaris* L. I. Variations in cytokinin levels in leaves of decapitated plants in relation to lateral bud outgrowth. *J. Exp. Bot.* 32:1231–41

131. Palmer, M. V., Horgan, R., Wareing, P. F. 1981. Cytokinin metabolism in *Phaseolus vulgaris* L. III: Identification of endogenous cytokinins and metabolism of [8-^{14}C]-dihydrozeatin in stems of decapitated plants. *Planta* 153:297–302

132. Palmer, M. V., Scott, I. M., Horgan, R. 1981. Cytokinin metabolism in *Phaseolus vulgaris* L. II: Comparative metabolism of exogenous cytokinins by detached leaves. *Plant Sci. Lett.* 22: 187–95

133. Palni, L. M. S. 1980. *Biosynthesis and metabolism of cytokinins.* PhD thesis. Univ. Wales, U.K. 219 pp.

134. Palni, L. M. S., Horgan, R. 1982. Cytokinins from the culture medium of *Vinca rosea* crown gall tumour tissue. *Plant Sci. Lett.* 24:327–34

135. Palni, L. M. S., Horgan, R. 1983. Cytokinins in tRNA of normal and crown gall tissue of *Vinca rosea. Planta.* In press

136. Palni, L. M. S., Horgan, R. 1983. Cytokinin biosynthesis in crown-gall tissue of *Vinca rosea*: metabolism of isopentenyladenine. *Phytochemistry.* In press

137. Palni, L. M. S., Horgan, R., Darrall, N. M., Stuchbury, T., Wareing, P. F. 1983. Cytokinin biosynthesis in crown-gall tissue of *Vinca rosea*: the significance of nucleotides. *Planta.* In press

138. Palni, L. M. S., Palmer, M. V., Letham, D. S. 1983. The stability and biological

activity of cytokinin metabolites in soybean callus tissue. *Planta.* In press
139. Palni, L. M. S., Summons, R. E., Letham, D. S. 1983. Mass spectrometric analysis of cytokinins in plant tissues. V. Identification of the cytokinin complex of *Datura innoxia* crown-gall tissue. *Plant Physiol.* In press
140. Parker, C. W., Letham, D. S. 1973. Metabolism of zeatin by radish cotyledons and hypocotyls. *Planta* 114:199–218
141. Parker, C. W., Letham, D. S. 1974. Metabolism of zeatin in *Zea mays* seedlings. *Planta* 115:337–44
142. Parker, C. W., Letham, D. S., Gollnow, B. I., Summons, R. E., Duke, C. C., MacLeod, J. K. 1978. Metabolism of zeatin by lupin seedlings. *Planta* 142:239–51
143. Parker, C. W., Letham, D. S., Wilson, M. M., Jenkins, I. D., MacLeod, J. K., Summons, R. E. 1975. The identity of two new cytokinin metabolites. *Ann. Bot.* 39:375–76
144. Parker, C. W., Wilson, M. M., Letham, D. S., Cowley, D. E., MacLeod, J. K. 1973. The glucosylation of cytokinins. *Biochem. Biophys. Res. Commun.* 55:1370–76
145. Péaud-Lenoël, C., Jouanneau, J.-P. 1980. Presence and possible functions of cytokinins in RNA. See Ref. 44, pp. 129–43
146. Peterson, J. B., Miller, C. O. 1976. Cytokinins in *Vinca rosea* L. crown gall tumor tissue as influenced by compounds containing reduced nitrogen. *Plant Physiol.* 57:393–99
146a. Pethe-Sadorge, P., Signor, Y., Guern, J. 1972. Sur la synthèse des nucleosides 5'-monophosphates de cytokinines par les cellules d'*Acer pseudoplatanus. C. R. Acad. Sci. Paris Ser. D* 275:2493–96
147. Pietraface, W. J., Blaydes, D. F. 1981. Activity and metabolism of 9-substituted cytokinins during lettuce seed germination. *Physiol. Plant.* 53:249–54
148. Polya, G. M., Bowman, J. A. 1979. Ligand specificity of a high affinity cytokinin-binding protein. *Plant Physiol.* 64:387–92
149. Polya, G. M., Davis, A. W. 1978. Properties of a high-affinity cytokinin-binding protein from wheat germ. *Planta* 139:139–47
150. Regier, D. A., Morris, R. O. 1982. Secretion of *trans*-zeatin by *Agrobacterium tumefaciens:* a function determined by the nopaline Ti plasmid. *Biochem. Biophys. Res. Commun.* 104:1560–66

151. Robins, M. J., Hall, R. H., Thedford, R. 1967. N⁶-(Δ²-Isopentenyl)adenosine: A component of the transfer ribonucleic acid of yeast and of mammalian tissue, methods of isolation, and characterization. *Biochemistry* 6:1837–48
152. Rodriguez-Barrueco, C., Miguel, C., Palni, L. M. S. 1979. Cytokinins in root-nodules of the nitrogen-fixing non-legume *Myrica gale. Z. Pflanzenphysiol.* 95:275–78
153. Schaeffer, K.P., Altman, S., Söll, D. 1973. Nucleotide modification *in vitro* of the precursor of transfer RNA^Tyr of *Escherichia coli. Proc. Natl. Acad. Sci. USA* 70:3626–30
154. Scott, I. M., Browning, G., Eagles, J. 1980. Ribosylzeatin and zeatin in tobacco crown gall tissue. *Planta* 147:269–73
155. Scott, I. M., Horgan, R., McGaw, B. A. 1980. Zeatin-9-glucoside, a major endogenous cytokinin of *Vinca rosea* crown gall tissue. *Planta* 149:472–75
156. Scott, I. M., Martin, G. C., Horgan, R., Heald, J. K. 1982. Mass spectrometric measurement of zeatin glycoside levels in *Vinca rosea* L. crown gall tissue. *Planta* 154:273–76
157. Scott, I. M., McGaw, B. A., Horgan, R., Williams, P. E. 1982. The biochemistry of cytokinins in *Vinca rosea* crown gall tissue. See Ref. 85, pp. 165–74
158. Shaw, G., Smallwood, B. M., Steward, F. C. 1968. Synthesis and cytokinin activity of the 3-, 7- and 9-methyl derivatives of zeatin. *Experientia* 24:1089–90
159. Shirai, H., Kanatani, H., Taguchi, S. 1972. 1-Methyl-adenine biosynthesis in starfish ovary: Action of gonad-stimulating hormone by methylation. *Science* 175:1366–68
160. Short, K. C., Torrey, J. G. 1972. Cytokinins in seedling roots of pea. *Plant Physiol.* 49:155–60
161. Smith, A. R., Van Staden, J. 1978. Changes in endogenous cytokinin levels in kernels of *Zea mays* L. during imbibition and germination. *J. Exp. Bot.* 29:1067–75
162. Sondheimer, E., Tzou, D. 1971. The metabolism of 8-¹⁴C-zeatin in bean axes. *Plant Physiol.* 47:516–20
163. Struxness, L. A., Armstrong, D. J., Gillam, I., Tener, G. M., Burrows, W. J., Skoog, F. 1979. Distribution of cytokinin-active ribonucleosides in wheat germ tRNA species. *Plant Physiol.* 63:35–41
164. Stuchbury, T., Palni, L. M., Horgan, R., Wareing, P. F. 1979. The biosynthe-

sis of cytokinins in crown-gall tissue of *Vinca rosea*. *Planta* 147:97–102

165. Summons, R. E., Duke, C. C., Eichholzer, J. V., Entsch, B., Letham, D. S., et al. 1979. Quantitation of cytokinins in *Zea mays* kernels using deuterium labelled standards. *Biomed. Mass Spectrom.* 6:407–13

166. Summons, R. E., Entsch, B., Letham, D. S., Gollnow, B. I., MacLeod, J. K. 1980. Metabolites of zeatin in sweetcorn kernels: purifications and identifications using HPLC and chemical-ionization mass spectrometry. *Planta* 147:422–34

167. Summons, R. E., Entsch, B., Parker, C. W., Letham, D. S. 1979. Quantitation of the cytokinin glycoside complex of lupin pods by stable isotope dilution. *FEBS Lett.* 107:21–25

168. Summons, R. E., Letham, D. S., Gollnow, B. I., Parker, C. W., Entsch, B., et al. 1981. See Ref. 17, pp. 69–79

169. Summons, R. E., MacLeod, J. K., Parker, C. W., Letham, D. S. 1977. The occurrence of raphanatin as an endogenous cytokinin in radish seed. *FEBS Lett.* 82:211–14

170. Summons, R. E., Palni, L. M. S., Letham, D. S. 1983. Mass spectrometric analysis of cytokinins in plant tissues. IV. Determination of intact zeatin nucleotide by direct chemical ionization mass-spectrometry. *FEBS Lett.* 151:122–26

171. Swaminathan, S., Bock, R. M., Skoog, F. 1977. Subcellular localization of cytokinins in transfer ribonucleic acid. *Plant Physiol.* 59:558–63

172. Sziráki, I., Balázs, E. 1979. Cytokinin activity in the RNA of tobacco mosaic virus. *Virology* 92:578–82

173. Tanaka, Y., Abe, H., Uchiyama, M., Taya, Y., Nishimura, S. 1978. Isopentenyladenine from *Dictyostelium discoideum*. *Phytochemistry* 17:543–44

174. Tao, G.-Q., Letham, D. S., Palni, L. M. S., Summons, R. E. 1983. Cytokinin biochemistry in relation to leaf senescence. I. The metabolism of 6-benzylaminopurine and zeatin in oat leaf segments. *J. Plant Growth Regul.* In press

175. Taya, Y., Tanaka, Y., Nishimura, S. 1978. 5'-AMP is a direct precursor of cytokinin in *Dictyostelium discoideum*. *Nature* 271:545–47

176. Terrine, C., Laloue, M. 1980. Kinetics of N^6-(Δ^2-isopentenyl)adenosine degradation in tobacco cells. *Plant Physiol.* 65:1090–95

177. Thimmappaya, B., Cherayil, J. D. 1974. Unique presence of 2-methylthio-

ribosylzeatin in the transfer ribonucleic acid of the bacterium *Pseudomonas aeruginosa*. *Biochem. Biophys. Res. Commun.* 60:665–72

178. Trewavas, A. 1970. The turnover of nucleic acids in *Lemna minor*. *Plant Physiol.* 45:742–51

179. Tsui, C., Shao, L. M., Wang, C. M., Tao, G.-Q., Letham, D. S., et al. 1983. Identification of a cytokinin in water chestnuts (corms of *Eleocharis tuberosa*). *Plant Sci. Lett.* In press

180. Van Staden, J., Sondheimer, E. 1973. The effects and metabolism of zeatin in dormant and nondormant ash embryos. *Plant Physiol.* 51:894–97

181. Van Staden, J. 1981. Cytokinins in germinating maize caryopses. II. Transport and metabolism of 8[^{14}C]t-zeatin applied to the embryonic axis. *Physiol. Plant.* 53:275–78

182. Van Staden, J., Davey, J. E. 1979. The synthesis, transport and metabolism of endogenous cytokinins. *Plant Cell Environ.* 2:93–106

183. Van Staden, J., Davey, J. E. 1981. Seasonal changes in the levels of endogenous cytokinins in the willow *Salix babylonica* L. *Z. Pflanzenphysiol.* 104:53–59

184. Van Staden, J., Dimalla, G. G. 1978. Endogenous cytokinins and the breaking of dormancy and apical dominance in potato tubers. *J. Exp. Bot.* 29:1077–84

185. Van Staden, J., Papaphilippou, A. P. 1977. Biological activity of O-β-D-glucopyranosylzeatin. *Plant Physiol.* 60:649–50

186. Vickers, J. D., Logan, D. M. 1970. Isopentenyladenosine: synthesis in *E. coli* tRNA. *Biophys. Soc. Abstr.* 10:166a

187. Vonk, C. R., Davelaar, E. 1981. 8-^{14}C-Zeatin metabolites and their transport from leaf to phloem exudate of Yucca. *Physiol. Plant.* 52:101–7

188. Vreman, H. J., Schmitz, R. Y., Skoog, F., Playtis, A. J., Frihart, C. R., Leonard, N. J. 1974. Synthesis and biological activity of 2-methylthio-*cis*- and *trans*-zeatin and their ribosyl derivatives. Isolation of the *cis*- and *trans*-ribonucleosides from *Pisum* tRNA. *Phytochemistry* 13:31–37

189. Vreman, H. J., Skoog, F., Frihart, C. R., Leonard, N. J. 1972. Cytokinins in *Pisum* transfer ribonucleic acid. *Plant Physiol.* 49:848–51

190. Vreman, H. J., Thomas, R., Corse, J., Swaminathan, S., Murai, N. 1978. Cytokinins in tRNA obtained from *Spinacia oleracea* L. leaves and isolated

chloroplasts. *Plant Physiol.* 61:296–306
191. Wang, T. L., Beutelmann, P., Cove, D. J. 1981. Cytokinin biosynthesis in mutants of the moss *Physcomitrella patens.* *Plant Physiol.* 68:739–44
192. Wang, T. L., Horgan, R. 1978. Dihydrozeatin riboside, a minor cytokinin from the leaves of *Phaseolus vulgaris.* *Planta* 140:151–53
193. Wang, T. L., Horgan, R., Cove, D. 1981. Cytokinins from the moss *Physcomitrella patens.* *Plant Physiol.* 68: 735–38
194. Wang, T. L., Thompson, A. G., Horgan, R. 1977. A cytokinin glucoside from the leaves of *Phaseolus vulgaris.* *Planta* 135:285–88
195. Wareing, P. F. 1977. Growth substances and integration in the whole plant. In *Integration of Activity in the Higher Plant* (Symp. Soc. Exp. Biol. No. 31), ed. D. H. Jennings, pp. 337–64. Cambridge: Univ. Press
196. Watanabe, N., Yokota, T., Takahashi, N. 1978. *cis*-Zeatin riboside: its occurrence as a free nucleoside in cones of the hop plant. *Agric. Biol. Chem.* 42: 2415–16
197. Watanabe, N., Yokota, T., Takahashi, N. 1978. Identification of N^6-(3-methyl-but-2-enyl)adenosine, zeatin, zeatin riboside, GA_{19} and ABA in shoots of the hop plant. *Plant Cell Physiol.* 19:1263–70
198. Watanabe, N., Yokota, T., Takahashi, N. 1981. Variations in the levels of *cis*- and *trans*-ribosylzeatins and other minor cytokinins during development. *Plant Cell Physiol.* 22:489–500
199. White, B. N., Dunn, R., Gillam, I., Tener, G. M., Armstrong, D. J., et al. 1975. An analysis of five serine transfer ribonucleic acids from *Drosophila.* *J. Biol. Chem.* 250:515–21
200. Whitty, C. D., Hall, R. H. 1974. A cytokinin oxidase in *Zea mays.* *Can. J. Biochem.* 52:789–99
201. Wilson, M. M., Gordon, M. E., Letham, D. S., Parker, C. W. 1974. The metabolism of 6-benzylaminopurine in radish cotyledons and seedlings. *J. Exp. Bot.* 25:725–32
202. Yokota, T., Takahashi, N. 1980. Cytokinins in shoots of the chestnut tree. *Phytochemistry* 19:2367–69
203. Yokota, T., Ueda, J., Takahashi, N. 1981. Cytokinins in immature seeds of *Dolichos lablab. Phytochemistry* 20: 683–86

Ann. Rev. Plant Physiol. 1983. 34:199–224

ADENINE NUCLEOTIDE RATIOS AND ADENYLATE ENERGY CHARGE IN ENERGY METABOLISM

Alain Pradet and Philippe Raymond

Institut National de la Recherche Agronomique, Station de Physiologie Végétale, Centre de Recherches de Bordeaux, France

CONTENTS

INTRODUCTION

One of the very impressive characteristics of the living cell is its ability to satisfy its energy requirements through a tiny and nearly constant amount

199

0066-4294/83/0601-0199$02.00

of ATP.[1] However, plant cells whether photosynthesizing or not produce enormous quantities of ATP from ADP. Thus it can be calculated (53) that 1 g of actively metabolizing maize root tips regenerate about 5 g ATP per day. It follows that the utilization of ATP is extremely well adjusted to its regeneration.

Sometime after allosteric control of enzyme activity was recognized, the role of concentration ratios in metabolic regulation was emphasized (see 8, 9), because many enzymes are regulated in opposite ways by ATP and ADP or ATP and AMP. Atkinson (8, 11) argued that the balance between ATP, ADP, and AMP was essential for cellular homeostatis. The adenylate energy charge (AEC), which represents the relative saturation of the adenylate pool in phosphoanhydride bonds, is expressed by the ratio [ATP] + 0.5 [ADP] / [ATP] + [ADP] + [AMP]. This ratio was proposed (12) as a convenient means of comparing the physiological role of regulation of enzyme activities by adenine nucleotide (AdN) ratios. The AEC expression assumes that the terminal phosphate groups of ADP and ATP are metabolically equivalent, which is true only if the adenylate kinase (AK) maintains a near equilibrium state between the adenylates. The AEC will be representative of the state of the AdN pool only if this near-equilibrium is maintained. In these conditions, the AEC, ATP/ADP, and ATP/AMP ratios are correlated, and the knowledge of one of these ratios allows the others to be calculated.

From in vitro studies of the effect of AdN on enzyme activity, it was asserted that AdN ratios are the key factors of metabolic regulation (8, 11). Nevertheless, this view has been questioned extensively (125, 126), and the properties of some plant enzymes are not entirely consistent with the theory (164, 165).

At about the same time, Bomsel & Pradet (23) developed an equation which was related to AEC. The application of these equations to describe the energy status of tissues or cells makes no assumption about the regulatory roles of the AdN values, but emphasizes that the cellular energy status cannot be deduced from the level of ATP alone, since variation in the concentration of the total AdN pool as well as in the phosphorylation of the pool will affect the ATP value.

The first part of this review briefly considers AdN ratios and AEC as effectors of plant enzymes; it is followed by a discussion of the in vivo significance of AdN ratios. Progress in this field has been restricted by the methodology and this point is discussed extensively. Because of space limitations, we concentrate on the significance of AdN ratios and AEC in relation to energy metabolism. During recent years, AdN ratios and AEC

---

[1]ADP, AMP, ATP, respectively, indicate adenosine di-, mono-, triphosphate.

have been studied in relation to a number of physiological events, but these studies have not all been reviewed. We hope that enough information is given here to help the readers of these papers.

CONTROL OF ENZYME ACTIVITIES

The regulation of enzymes by competition between ATP and ADP or AMP for an active or regulatory site was considered by Atkinson (8–10) to be the simplest way of stabilizing the AdN ratios and so controlling the metabolic activity. If the binding site is nearly saturated, the enzyme activity depends on the value of the AdN ratios and the relative affinities for ATP and ADP or AMP. A steep increase of enzyme activity occurs in the high range of AEC if the affinity for ADP is about ten times higher than that for ATP (8). Many regulatory enzymes of ATP-utilizing pathways (called "U" enzymes) have been shown to be regulated according to this model (8, 12). Regulatory enzymes of ATP regenerating pathways ("R" enzymes) (150) are regulated in the converse way, decreasing their activity for increasing values of AEC (Figure 1). The steepness of the "U" and "R" curves in the high range of AEC values, where they intersect, was taken as evidence for the regulatory role of AEC (8).

In the early formulation of the AEC concept, it was shown that a number of additional effectors such as end products did not fundamentally change the situation described by the concept. However, other factors such as pH, Mg^{2+}, and reaction products so markedly affect the response of enzymes to AEC that in the view of Purich & Fromm (125, 126) there remains no theoretical basis for stabilization of AEC at high values. While the ratio

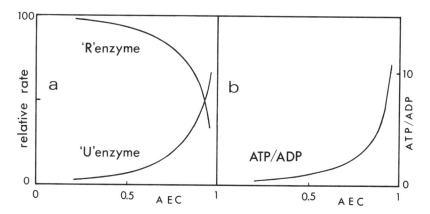

Figure 1 (a)   Response of "U" and "R" enzymes to AEC. Redrawn from (8). (b)   Value of the ATP/ADP ratio according to AEC, calculated for $K'_{app} = [ATP]\cdot[AMP]/[ADP]^2 = 1$.

ATP/ADP is a fundamental regulatory and regulated value, which is sensed by enzymes and related to the thermodynamic state of the AdN pool, the variation of enzyme activity is steep when activity is represented as a function of AEC. It can be seen that a typical "U" curve as defined by Atkinson roughly tracks the curve of the ATP/ADP ratio with respect to AEC (Figure 1). Although this ratio extrapolates to infinity at high AEC, within the physiological range (0.3–0.95) the plot of enzyme activity versus ATP/ADP approximates linearity. For an enzyme showing a typical "U" response to AEC the regulation by ATP/ADP can be effective over the physiological range of AdN ratios and AEC values (0.3–0.9) (130).

More general considerations of metabolic regulation suggest that undue importance has been attributed to the "proper levels of pools" as well as to regulatory enzymes for cellular homeostasis: control is a property of the system rather than that of individual enzymes or pacemakers (77). A model which considers the role of ATP in the activation of substrates for "R" pathways involves the stabilization and oscillation of AEC in the absence of enzyme regulation by effectors (145). Whatever may be the theoretical basis for the involvement of AEC in metabolic control, it is necessary to consider the relationship between AdN and the regulation of plant metabolism.

Photosynthetic CO_2 Fixation

The fixation of CO_2 is the first step of an ATP-consuming process which occurs in the cytosol of mesophyll cells of C_4 plants or in the stroma of chloroplasts in C_3 plants, where the values of the AdN ratios or the pyridine nucleotide ratios (65) change with illumination. Although these large changes do not conform to the principle of homeostasis, which according to Atkinson (8, 11) justifies the requirement for metabolic control by AdN, they can be expected to influence the activities of key enzymes of the CO_2 fixation pathway.

PEP carboxylase, the enzyme which catalyzes CO_2 fixation in the cytosol of C_4 mesophyll cells, is generally reported to be activated by sugar phosphates and inhibited by organic acids (39, 176), but its regulation by AdN is less clear. In a crude extract of etiolated seedlings of maize and in the presence of excess Mg^{2+}, the enzyme was activated by AMP and inhibited by ATP and ADP. An atypical response to AEC was observed in the range from 0.4 to 0.6, but the "R"-type response was unexpected for an enzyme of a "U" pathway, and was interpreted by considering that the enzyme might play an anaplerotic role (176). However, this response to AEC would favor CO_2 fixation in maize mesophyll cells if, as in spinach green cells, the ATP/ADP ratio decreases in the light (92, 93). PEP carboxylase extracted from green leaves of *Pennisetum purpureum* (39) was slightly affected by

AMP. The inhibition by ADP and ATP was overcome by using a concentration of Mg^{2+} high enough to saturate the adenylates; in vivo the enzyme would be regulated by the AEC only at unsaturating Mg^{2+} concentrations.

Pyruvate phosphate dikinase, which is present in the same compartment as PEP carboxylase, catalyzes the regeneration of PEP from pyruvate, requires ATP, and is inhibited by the reaction product AMP, which acts as a competitive inhibitor of ATP (5). Its response to AEC would be of the "U" type, as expected for an enzyme in this pathway.

3-Phosphoglycerate kinase and ribulose 5-phosphate kinase catalyze the two ATP-consuming reactions of the reductive pentose phosphate pathway. 3-Phosphoglycerate kinase is controlled by AEC (91, 112, 113), with ADP and AMP acting as competitive inhibitors of ATP in the phosphorylation of 3-phosphoglycerate to 1-3 bisphosphoglycerate (113). The properties of the enzyme from chloroplasts of *Pisum sativum* (113) and of *Beta vulgaris* (30) do not differ from those of the cytosolic enzymes which catalyze the reverse reaction both in plants (113) and in animals (30). The K_m value for ATP, first reported to be 2 mM (91), was found to be an artifact, and the real values vary from 0.15 to 0.2 mM. (30) and in vivo the enzyme is probably saturated by MgATP. The fivefold reduction of the activity of 3-phosphoglycerate kinase found in vitro when the ratio ATP/ADP was varied from its value in the stroma in the light to its value in the dark is insufficient to account for the dramatic change of the CO_2 fixation activity in light-dark transitions.

The regulation of ribulose 5-phosphate kinase appears to vary from plant to plant. The pea leaf enzyme has been reported (4) to have a high apparent affinity for ATP ($K_m = 0.069$ mM) and to be unaffected by AMP or ADP and, therefore, to be insensitive to the AEC. The spinach enzyme has a higher K_m for ATP (0.42 mM) and is controlled by the AEC (91). Hence, according to in vitro enzyme studies, the variation of AEC resulting from light-dark transitions affects the reductive pentose phosphate pathway at two steps in spinach leaves, but only at one step in pea leaves.

Possible mediators of the light stimulation of CO_2 fixation in chloroplasts include variations of pH (171), redox state (80), Mg^{2+} concentration (115, 116), phosphoglycerate concentration (133), and the ratio ATP/ADP (133). One consequence of this control is to avoid the consumption of ATP, produced by oxidative metabolism, for CO_2 fixation in the dark (171). This control could be similar to that of a "U" pathway. The importance of the AEC in metabolic control during the light-dark transition can be evaluated from its participation in the regulation of this pathway. The increase of pH by almost 1 unit, which occurs in the stroma after illumination, has been reported to be sufficient to switch CO_2 fixation from zero to maximal activity (171).

The role of the variation of the Mg^{2+} concentration which was considered essential (80) is now less clear. An effect of changes of Mg^{2+} levels on fructose- and sedoheptulosebiphosphatases was reported (115, 116), but it has been estimated that this effect could be minimal (29, 134). By using a reconstituted chloroplast system, it was shown that the rate of NADPH oxidation is proportional to the PGA·ATP/ADP ratio (133), a result which is in agreement with the reported regulation of phosphoglycerate kinase by AEC. However, the AdN do not act by controlling a rate-limiting step but rather by a mass action effect on this near equilibrium reaction (115, 133, 134). AdN ratios together with variation of pH appear to be the two main regulators of the reductive pentose phosphate pathway.

Starch Metabolism

Glucokinase, an enzyme involved in the synthesis of starch in developing pea seeds, is inhibited by ADP (163) and so could respond to the AEC as a "U" enzyme. ADP is also an inhibitor of ADP-glucose pyrophosphorylase (58). The formation of ADP-glucose is stimulated four- or fivefold by a change of the ATP concentration from 0.2 to 1.2 mM at levels of PGA and Pi that exist in whole chloroplasts (78). The variation of the ATP/ADP ratio in light-dark transitions would permit a quick regulation of starch synthesis (78) depending on the energy state of the cell, while the regulation of ADP-glucose pyrophosphorylase by the PGA/Pi ratio (40, 67, 154) is part of the control of the partition of carbon between starch synthesis in the chloroplast and its export for sucrose synthesis (64).

Glycolysis

The inhibition of muscle phosphofructokinase by ATP and its activation by AMP led to the concept of metabolic control by the AdN balance (8, 150). Phosphofructokinase from higher plants is inhibited by all three AdN (164). Although this seems to be inconsistent with regulation depending on the AEC, it is worth noting that pyruvate dehydrogenase, which is activated by ATP, ADP, or AMP when tested separately, is controlled as a "U" enzyme when the three AdN are present (150). However, PEP, Pi, and citrate are considered to be the main effectors of plant phosphofructokinase (164). Recently, two isoenzymes of phosphofructokinase have been separated from the developing endosperm of *Ricinus communis* (56, 57); in contrast to the cytosolic enzyme, the plastid isoenzyme is inhibited by ATP, as is the case for most other phosphofructokinases. Its inhibition by ATP and PEP was reversed by 1 mM P_i. The cytosolic isoenzyme is not inhibited by ATP; its main regulator is 3-phosphoglycerate, the inhibitory effect of which is only partially relieved by P_i (57).

Pyruvate kinase from higher plants, like that from most organisms, is inhibited by ATP (164). Because the enzyme from cotton seeds is activated by AMP, which permits an "R"-type regulation of the enzyme, it was proposed that plant pyruvate kinase is a regulatory enzyme (50). The variation of activity versus AEC was similar to that of an "R" enzyme, but this response was obtained at subsaturating concentrations of ADP, and it has been shown that it was mainly an effect of the ADP concentration (50). However, AMP has no effect on the pyruvate kinases of pea seeds or carrot tissues (164), soybean nodules (114), and germinating castor bean endosperm (72, 106). However, AMP does inhibit the cytosolic isoenzyme of the developing castor bean endosperm (72). Since phosphofructokinase and pyruvate kinase are considered to be the two main control points of glycolysis (164), it can be argued that there is no evidence for a control of this pathway by AEC in higher plants.

Tricarboxylic Acid Cycle

Because of increasing evidence indicating that the mitochondrial matrix lacks AK activity, no regulation of the enzymes of this compartment by AEC would be expected, although individual nucleotides or AdN ratios may act as effectors (164, 175). Pyridine nucleotides which are also related to the energy status of the mitochondria seem to play an important part in the control of the TCA cycle (175).

Other Enzymes

The arginosuccinate synthetase of soybean presents a typical "U"-type response to AEC (149). However, it has recently been proposed that this effect is probably not important relative to control by other effectors (162). The phosphoribosyl pyrophosphate synthetase of spinach increases in activity when the AEC is increased at low (0.2 mM), but not at high (4 mM), concentrations of nucleotides, so that control of this enzyme by AEC therefore seems unlikely (7). The methodology used in the study of the glutamine synthetase of sunflower roots (170) is inadequate to permit the conclusion that it is subjected to regulation by AEC.

Few studies of plant enzymes have shown "typical" variation of enzymatic activity with AEC. In many cases, the inhibition at high AEC was related to chelation of Mg^{2+} by ATP (7, 26, 43); in other cases pH or total AdN concentration have also been shown to affect the response to AEC (7). However, it should be noted that the response to AEC may depend on the presence of some other effectors in the medium (150). Because isoenzymes from different cellular compartments sometimes exhibit different regulatory properties (13, 56, 57), the effects of regulators of activity may not be apparent in studies using whole extracts. Available evidence suggests that AdN concentrations and ratios may play a role in the regulation of plant metabolism in concert with effectors such as pH and the redox state of the pyridine nucleotides to which they are connected through a number of near equilibrium reactions (86).

ADENYLATE KINASE

Adenylate kinase is a ubiquitous enzyme whose activity in most tissues is very high compared with that of many other enzymes (110). The soluble enzyme of lettuce seeds (118) and pea leaves (136) have been partially purified, and it exhibits many of the properties of the corresponding mammalian enzymes (97, 110).

The reaction catalyzed by AK is:

$$\text{ATP + AMP} \xrightleftharpoons{\text{Mg}^{2+}} \text{2 ADP}$$

The mass action ratio ($|\text{ATP}| \times |\text{AMP}|/|\text{ADP}|^2$) calculated from ATP, ADP, and AMP levels of whole extracts of rat liver tissues varied between 0.5 and 0.8. From these results and from in vitro values of the apparent equilibrium constant (K_{app}) of the AK reaction determined in vitro, which vary between 0.5 and 1.2 (see 110), it was postulated that AK catalyzes a state of near equilibrium (85). Similar values were found in wheat leaves and lettuce seeds (22, 23, 118, 120). This value varied only slightly, even when an imbalance was induced between the utilization and regeneration of ATP (23, 118, 120). The calculation of the mass action ratio of AK reaction from available AdN data of plants and animals suggested that the near equilibrium condition was general (23, 118). Two main problems which complicate the analysis have to be taken into account: factors which modify the K_{app} of AK reaction and the compartmentation of the enzyme and its substrates.

Modification of K_{app}

The Mg^{2+} concentration and the pH can greatly modify the K_{app} of the reaction (see 110). When the AEC of the cell is lowered, the concentration of free Mg^{2+} is increased because ADP and AMP form much weaker complexes than ATP (110). Blair (20) has calculated the effect of AEC on the K_{app} for AK in relation to Mg^{2+} levels. The concentration of the magnesium-bound nucleotides must be considered when evaluating the interactions of enzymes and nucleotides, but our lack of knowledge about free Mg^{2+} concentration limits our understanding of the phenomenon. The effect of the variation of the K_{app} on the proportion of AdN when the AEC increases from 0 to 1 was also described (23).

At Mg^{2+} concentration of 5 mM, it was calculated (20) that a lowering of the AEC from 0.96 to 0.24 induces a lowering of the K_{app} of AK from 1.22 to 0.72. In lettuce seeds, it was experimentally found (118) that the mass action ratio of AK decreased from 1.15 to 0.83 when the AEC was reduced from 0.81 to 0.35. In most materials (31), but not in lettuce seeds, a decrease of the AEC is often associated with a lowering of the total adenylates, which increases the concentration of free Mg^{2+} (20). These variations only slightly modify the relationship between AEC and the AdN ratios (23).

Compartmentation

In plant cells, the existence of different molecular species of AK has been suspected for a long time (44, 98). In green leaves, a large proportion (about 50%) of the AK activity is contained in the chloroplasts (19a, 105, 137, 141, 155), and most of this activity is found in the stroma (47, 79, 137). The AdN of the stroma are maintained near equilibrium by AK even when the different ATP regenerating mechanisms of green cells are rapidly changed (61, 141, 146, 147, 155). Some AK which is tightly bound to the envelope of the spinach chloroplast exhibits a lower activity than the enzyme of the stroma, but represents a significant part of the protein of the envelope (105). It was proposed that the bound enzyme is involved in the translocation of the AdN between the cytosol and the chloroplasts (105). Isoenzymes of AK have been described in mammalian cells. The most active of these, AK II, is located in the intermembrane space of the mitochondria (41, 42). The percentage of this isoenzyme in the nonparticulate fraction was proposed as a marker to evaluate the percentage of damaged mitochondria during the preparation of these organelles (41). It is generally agreed that the mitochondrial matrix does not contain AK and that the phosphorylation of AMP produced in the matrix is carried out by the GTP:AMP phosphotransferase (69, 88).

In plants, recent reports (19a, 61, 155) have shown that mitochondria of oat, wheat, spinach, and barley leaves contain a large percentage of the total AK of these tissues. According to Hampp et al (61) there is no AMP and therefore no equilibration by the AK reaction in the cytosol. However, Stitt et al (155) measured the amount of AMP in the cytosol and concluded that a near equilibrium of AdN was maintained in the cytosol by the AK located in the intermembrane space of the mitochondria to which the nucleotides have free access. It should also be noted that in white potato mitochondria an AK has been observed which is tightly bound to the inner membrane but which faces the intermembrane space (6).

METHODS FOR IN VIVO STUDIES
OF ADENINE NUCLEOTIDES

In vivo studies of AdN have been complicated by technical difficulties which are encountered at four steps of the analytical process: sampling, killing the tissue and extracting the nucleotides, assay, separation of cell compartments.

Sampling

The rate of turnover of the β and γ phosphate groups of nucleotides is very rapid, and the half life of ATP in whole plant cells is a matter of some seconds (18, 122). In chloroplasts (134) or in mitochondria (93) it is about 1 sec, as in microorganisms. This fact may have been obscured by the analogy between a battery and the cellular pools of AdN (8): a battery stores energy whereas the amount of energy stored in the AdN is very small when compared to the rate of its utilization (11). The measurement of AdN may be compared with taking a photograph of a fast-moving object. To avoid blurring the photograph, it is necessary to dramatically reduce the stop time. Similarly, to measure AdN, time is the essence. This fundamental point has not always been appreciated.

When studying the effect of environmental factors such as light, temperature, or gaseous environment, it is necessary to halt metabolic activity while the tissue is still being subjected to the particular environment. Thus, when etiolated wheat leaves were transferred from air to nitrogen, a delay of only 5 sec produced an increase in the ATP/ADP ratio from 0.28 to 0.74 (unpublished results). In maize root tips, if killing is effected within 2 sec of the transfer from anoxic conditions, the measured ratio is not affected (P. Saglio, unpublished results). The easiest way to instantaneously stop metabolic activity is to fast freeze the tissue. Liquid nitrogen is often used for this, but with diethyl ether at $-100°C$, kalefaction is avoided.

Enzyme Inactivation and Nucleotide Extraction

Two general methods have been used to inactivate enzymes and extract the nucleotides of plant tissues: boiling solvents or acid treatment.

A long time ago it was shown (14) that the extraction by boiling solvents should be avoided because it does not instantaneously halt enzyme activity. Nevertheless, the extraction of nucleotides by boiling solvents has been used in many laboratories during recent years. In most cases, the method yields AdN ratios lower than those obtained by acid extraction. Thus, the low AEC values found initially in maize root tips (94) probably originated from the use of boiling tris buffer, and higher values have been obtained with acid treatment (139). Bieleski (14) strongly advised destroying the enzymes by homogenizing the tissues in a cold solution of an acid in an organic solvent. Formic acid dissolved in ethanol gives good results (38). Trichloroacetic acid dissolved in diethyl ether, followed by extraction of the nucleotides with the same acid dissolved in water (172), is always used in our laboratory as a reference before using simpler methods. Sometimes the homogenization of the tissue in trichloroacetic or perchloric acid dissolved in water gives results which compare favorably with the results obtained by the preceding method. Nevertheless, the perchloric acid extraction has to be used cautiously, since some enzyme activity may be restored after neutralization (71). Cryodessication of the tissue before extracting the nucleotides with boiling buffer has been recommended (151), but in our hands this method always gives lower values of AdN ratios than acid extraction. One problem which remains unsolved is the fact that the treatment of tissues with acids sometimes extracts fewer nucleotides than boiling buffers.

Assays

Early methods used ion exchange chromatography for the separation of the nucleotides, but difficulties were caused by interference due to the UV absorbance of phenolic compounds. The labeling of tissues with ^{32}P avoided this difficulty, but these methods which gave the proportions of most phosphate esters in plant tissues (16) were extremely time consuming and therefore poorly adapted for metabolic studies (see 15, 19, 27). Two-dimensional paper or thin-layer chromatography allowed the separation of nucleotides from crude extracts (38). Today, with the additional use of electrophoresis, these methods still find use in studies involving radioactive labeling. Furthermore, recent developments in the purification of plant nucleotide extracts (109) coupled to improved ion exchange resins and HPLC should improve the versatility of these methods.

For assay in crude extracts, the most widely used methods involve enzymic coupling to oxidation or reduction of pyridine nucleotides. Such methods were often used to estimate AdN in animals and bacteria, but when the method was applied to plants, phenols presented difficulties because of their UV absorption and their action as enzyme inhibitors (95). These difficulties were overcome by the development of the bioluminescence assay of ATP (158, 159) for the estimation of AdN in plant extracts (117). It involves the firefly assay in which ATP is measured directly from light emission and ADP is measured after conversion to ATP with pyruvate kinase and PEP. After determining the total AdN in the presence of AK, AMP is obtained by difference. This method permits quantities of about 0.1 pmol per assay to be measured. To increase the sensitivity 1000-fold, purified luciferase (108) and synthetic luciferin must be used, but at this low level ADP and AMP cannot be assayed. The concentration of AdN in plant tissues which varies from 5 to 200 nmol per gram in carrot culture and in imbibed lettuce seeds respectively, is sufficiently high to allow the use of crude extracts of firefly even in diluted extracts of tiny quantities of tissues (1 mg). When using an apparatus which measures the peak height of light emission which reaches a maximum within one or two seconds after adding crude extracts of firefly to the extract, the determination is specific for ATP (128). Purified luciferase may be necessary to increase the specificity if the apparatus used integrates the light emitted for longer periods of time. In aerobic tissues, the content of AMP is low compared to that of ATP and ADP. Consequently, it is very difficult to obtain accurate values of AMP by this method. Since the concentration of AMP appears only in the denominator of the AEC ratio where it is part of the total AdN, it is not necessary to determine directly the concentration of AMP for the determination of AEC. The method does not permit the degree of accuracy necessary to study the variation of AdN ratios in tissues exhibiting high AEC values because the AMP and ADP concentrations are low.

^{31}P Nuclear magnetic resonance has been used for the determination of cellular energy parameters in animal cells and microorganisms [see references in (53, 97a, 131a)]. The adaptation of this method to plant cells and tissues presents special difficulties because of the high vacuolization of plant cells. Nevertheless, ^{31}P NMR spectrum made it possible to estimate the repartition of inorganic phosphate in the cytosol and the vacuole of some plant cells and tissues. These data are necessary to calculate the important thermodynamic parameter: [ATP]/[ADP] X [P$_i$].

Separation of Cell Compartments

The estimation of the values of AdN in the different compartments superimposes difficulties in obtaining pure fractions (174) and those originating from the fast turnover of these molecules. Heber and his co-workers were pioneers in this field; they were able to compare the in vivo values of ATP and ADP of chloroplasts and whole leaves (141, 142) at a time when data from whole cells (animal or plant) were extremely scarce (23) and when studies with microorganisms were hardly beginning (84).

FRACTIONATION OF INTACT TISSUE IN APOLAR MEDIUM   In this method (51, 160), the metabolic activity of the intact tissue is halted by fast freezing followed by cryodessication. The tissue is then homogenized in nonaqueous solvents and centrifuged on a density gradient of an apolar solution. Finally, enzymatic activity is destroyed by treatments with acids (141).

SILICONE OIL FAST CENTRIFUGATION OF PROTOPLASTS   The protoplasts are placed on a nylon net and are broken by centrifugation. The chloroplasts are sedimented through a layer of silicone oil and killed by perchloric acid at the bottom of the tube (132, 156, 174). A modification of this method has been used to study the metabolites in the mitochondria and the cytosol (60).

FAST FILTRATION OF PROTOPLASTS   The protoplasts contained in a syringe are disrupted by forcing them through a nylon net. The use of filters of various sizes produces filtrates corresponding to mainly cytosol, or cytosol and mitochondria. The whole content of the broken protoplasts is obtained by omitting the filters. In this method, the enzymes are inactivated within 0.1 sec after the disruption of the protoplasts instead of requiring several seconds as in the preceding method (92, 93). In all these methods, the contamination of one compartment by another is corrected by using marker enzymes.

ADENINE NUCLEOTIDES IN ACTIVELY METABOLIZING WHOLE PLANT TISSUES

The term "actively metabolizing" refers to most plant materials which are used in laboratories as controls; they are not starved of nutrients, and in particular not of oxygen. From experimental work with plants, and using data from animal cells (23), it has been asserted that the AEC of actively metabolizing tissues is higher than 0.8, and very often around 0.85 (17, 118). The view that the AEC of normal cells lies between 0.85 and 0.94 has been widely accepted for animal cells and microorganisms (10, 11, 33, 84). However, lower values, which may reflect technical errors, have been reported frequently for plant cells, and it has sometimes been asserted that the AEC in plants is not regulated as in other cells (33, 94). It should be noted that significant differences in AEC are likely to occur between green and nongreen cells.

High AEC or ATP/ADP values in the same range as those found in animals and microorganisms were found in germinating seeds of crested wheat grass (172), lettuce (122, 130), rice (123), wheat (36), squash (127), in wheat (111, 135) and soybean embryos (3, 135), in etioled coleoptiles of rice (103), and in developing wheat grains (74). In roots, low AEC values have been reported (94), but using acid extraction, values of around 0.9 were found in root tips of maize (129, 139, 140) and rice (129). High values have also been reported in potato slices when the concentration of AdN was corrected for a metabolically inactive pool of AMP (70a). Low values of AEC were found in suspension culture of sycamore cells (28), but in the same material, the trichloroacetic acid extraction method gave values around 0.9, and similar values have been found in tobacco cells (100). Usually AEC values between 0.7 and 0.8, or the corresponding ATP/ADP and ATP/AMP ratios, were found in green leaves of wheat (21, 22, 35, 146–148), barley (96), winter rape (151), spinach (25), and pinto beans (169). Transferring green leaves from the dark to the light produces very small changes in the AEC or in the corresponding ATP/ADP ratios of whole leaves (22, 92, 93, 99, 146, 147).

It thus appears that AdN ratios of nongreen plant cells are stabilized at high values as is the case in other living cells, but these ratios are stabilized at lower values in green cells.

COMPARTMENTATION OF ADENINE NUCLEOTIDES

The main pathways of ATP regeneration: glycolysis, oxidative phosphorylation, and photophosphorylation, are located in three compartments: cytosol, mithochondria, and chloroplasts. The nucleotides cannot move freely across the inner membranes of mitochondria (62, 175) and chloroplasts (47, 63, 64) which separate the matrix and the stroma from the cytosol.

Relationship between the Mitochondria and the Cytosol

In mitochondria, the ATP synthase is bound to the inner membrane, where, under the influence of the electrical gradient generated by the electron transport in the respiratory chain, it phosphorylates ADP into ATP in the matrix (see 62, 157). The properties of an AdN translocator inserted in the inner membrane, which specifically catalyzes the exchange (ADP^{3-}/ATP^{4-}), have been studied in mitochondria of rat liver and beef heart (see 82, 83, 166). Corresponding work with plant mitochondria has been reviewed by Hanson & Day (62), and some differences between the animal and plant translocators have been observed (167).

In 1969, Klingenberg predicted that the ATP/ADP ratio had to be higher in the extramitochondrial space than in the intramitochondrial space (see 82, 83). The first indirect evidence for this view is to be found in the results of Scholz & Bücher (143) and Krebs & Veech (87), who measured substrates and products (including ATP and ADP) of near equilibrium reactions in the mitochondria and in the cytosol of rat liver. The first direct evidence was obtained by using subcellular, nonaqueous fractionation. It was demonstrated that in vivo the mitochondrial ATP/ADP ratios were considerably lower than the cytosolic ratios (68, 152). The nonaqueous method of fractionation has not been used to separate mitochondria from the cytosol of plant cells. Recently, in a study of AdN of *Avena* protoplasts which was carried out by a modification (60) of the fast centrifugation method (132), Hampp et al (61) found that the ATP/ADP ratios in the dark were similar in the mitochondria and the cytosol. Nevertheless, the authors expressed some doubt about their results because the separation of the mitochondria needs about one minute. A very short time later, the method of "fast filtration of protoplasts" (93) gave the ratios of ATP/ADP in mitochondria in the dark to be 0.6 compared with 9 in the cytosol (155).

Consequently, as in the case of animal cells, it appears that ATP is transported into the cytosol of plant cells against a concentration gradient. In spite of the general acceptance of the existence of large gradients of ATP/ADP ratios across the mitochondrial membrane, it should be noted that recently Erecinska & Wilson (54), analyzing published data of animal extra- and intramitochondrial ATP/ADP ratios, conclude that the intramitochondrial ratios have been underestimated. They argue that a significant part of the ADP is enzyme-bound and that the methodology used can lead to low values of the ratio in the mitochondria.

Relationship Between the Chloroplasts and the Cytosol

In chloroplasts ATP synthase is located in the thylakoids. An AdN carrier transports ATP very slowly from the stroma to the cytosol (66). The bulk of the energy and the reducing power are transported from the stroma to the cytosol by a shuttle system, involving dihydroxyacetone phosphate and 3-phosphoglycerate, by means of the phosphate translocator (see 47, 63, 64, 75). The ATP/ADP ratio in the stroma is higher in the light (ca 1.5–3) than in the dark (ca 1). This result was obtained by using the three methods of fractionation and applying them to green leaves of various species of plants (23, 24, 55, 59, 61, 81, 141, 142, 146, 147). The consensus of these experiments is that that ATP/ADP ratio in the chloroplast (whether photosynthesizing or not) is lower than in the cytosol plus mitochondria. These experiments led to the belief that the ATP/ADP ratio of the cytosol decreases in the dark; but more recent experiments, in which mitochondria were separated from the cytosol, gave ATP/ADP ratios in the cytosol which increase in the dark (155). Although this point has yet to be confirmed, it is extremely important in terms of respiratory control, because it has often been asserted that the very high cytosolic ATP/ADP ratios in the light induce a nonphosphorylating state IV in the mitochondria (34, 64). The thermodynamic significance of the gradients between the AdN ratios in the different compartments has been discussed elsewhere (83, 155).

In animal cells, it is widely admitted that the AdN ratios of whole cells or tissues give a good approximation to the values in the cytosol because about 90% of the nucleotides are contained in this compartment (52, 86, 152, 173). There is only one report on the AdN distribution among cellular compartments of a plant (155). In protoplasts of green wheat leaves, the distribution of AdN was as follows: 47% in the chloroplast, 44% in the cytosol (including the nuclear compartment), and 9% in the mitochondria. These results explain why the values of AEC in whole green tissues, even in the light, are usually lower than in nonphotosynthetic tissues (see preceding paragraph). It also makes the values of AdN in whole green cells difficult to interpret. The level and the variations of AdN concentrations and ratios are very similar in nongreen plants and in animal cells or tissues. It is possible to make the assumption that AdN ratios in seeds, roots, and etiolated leaves reflect the values in the cytosol. The AdN content of mitochondria extracted from potatoes was found to be very low compared to the nucleotide content of animal mitochondria (167). If this fact is general, it would reinforce the preceding assertion. In nongreen materials, the plastid compartments can be large, but there is no data concerning their AdN content.

NONGREEN CELLS UNDER CONDITIONS LIMITING THE METABOLIC ACTIVITY

The regeneration of ATP and its utilization are so closely connected that a control of energy metabolism can be achieved by methods which limit one or both of these functions. An inhibition of the respiratory rate (e.g. by limiting the oxygen availability or using an inhibitor of the cytochrome oxidase) automatically produces a lowering of the ATP-requiring processes of the cell. The converse is equally true: an inhibition (e.g. of protein synthesis) is followed by a lowering of the respiratory rate as a consequence of the reduced needs for ATP (e.g. 37). The question which arises is: do variations of the metabolic activity affect the AEC value? Sometimes it has been suggested that high AEC values are related to high metabolic activities, but Atkinson argues (11, 32) that the tendency of tissues to reestablish a high AEC implies that if a variation of AEC according to the metabolic activity occurs, it would be in such a narrow range of high AEC values that it could not be demonstrated experimentally.

However, in most experiments with microbial and animal cells (124, 161), the relationship between the AEC and metabolic activity has been studied under conditions which mainly limited the utilization of ATP or its simultaneous utilization and regeneration. Under these conditions, the AEC remains at a high level. Similar results have been obtained with plant tissues; for instance, a limitation of the metabolic rate of lettuce seeds by lowering the temperature (118), or of maize root tips by reducing the availability of sugar (139), induced no observable change in the AEC which remained at values close to 0.9, when the metabolic activity was drastically reduced. The lowering of the temperature caused a balanced reduction of the activity of ATP-utilizing and ATP-regenerating pathways. Under carbon-limited metabolizing conditions, in which the carbon source provides both carbon and energy, it was shown that in microorganisms energy is not normally a limiting factor (107). Consequently, a limitation of the metabolic rate by carbon availability affects primarily the ATP-utilizing pathways. On the contrary, conditions such as hypoxia or anoxia, which clearly limit the energy production, simultaneously reduce the respiratory rate and the value of the AdN ratios.

Anoxia

The AEC of obligatory anaerobic microorganisms such as spiroplasma under actively growing conditions is close to 0.9 (138). When facultative anaerobic organisms such as *E. coli* are transferred from an aerated to an anoxic medium, there is a slight and transient decrease of the AEC, which recovers to its initial value within about 30 minutes (33). However, the same treatment caused no change in the AEC of *Saccharomyces cerevisiae* growing aerobically on glucose. The transfer of animal (48, 131, 144) and plant (22, 45, 118, 130, 140) tissues to anoxic conditions always induces a large decrease in AEC and, after a few minutes, the AEC stabilizes at a value which can be very different from one tissue to another (121). Values of around 0.6 are found in many germinating starchy seeds subjected to anoxia (2). These seeds exhibit a very active fermentative metabolism, as estimated by the rate of ethanol production.

The AEC of the embryo of rice seeds in air increases from the very low value of 0.2 in dry seeds to values of about 0.9 within one or two hours of imbibition. Under anoxia, the AEC increases more slowly and stabilizes around 0.8 after a few hours (123). When aerobically grown seedlings are transferred to anaerobic conditions, the AEC stabilizes at values between 0.5 and 0.6. It then increases slowly, to stabilize after about 10 hours at 0.8. (103, 123). Rice coleoptiles can grow under anaerobic conditions with a synthesis of DNA, RNA, and proteins which is accompanied by an increase in total AdN and an increase of the AEC from 0.6 to 0.8 (101–103). On the other hand, very low AdN ratios (AEC between 0.2 and 0.3) were found in wheat leaves (148) and in many fatty seeds, including radish seeds (2, 130). In previous work (104) it was found that in germinating radish seeds high AEC values (about 0.85) were obtained under aerobic and anaerobic conditions, but this may have been due to a slow inactivation of metabolic activity at the end of the anaerobic treatment. The value of the AEC was similar when fatty seeds were imbibed under anoxia or previously imbibed under aeration and then transferred to anoxia. Fermentative activity and the value of the AEC under anoxia are much lower in fatty seeds than in starchy seeds.

In excised maize roots, the fermentative capacity can be reduced by depleting the sugar content by aerobic aging of the tips. In this material a correlation between AEC and fermentative activity was observed (140) similar to that reported in blood platelets when the respiratory metabolism was inhibited by cyanide (1). From these results it seems that the measurement of the AEC in nongreen cells and tissues gives a very good indication of fermentative activity under anoxia (124). The factors which determine this relationship remain to be established.

Hypoxia

Under hypoxia, respiratory activity is limited by the availability of oxygen. As reported in the preceding paragraph, in many cells or tissues, anoxia produces low AEC values. A gradual decrease of oxygen partial pressure below the value which characterizes the normoxic state (the critical oxygen pressure) (49, 121) induces a gradual decrease of AdN ratios until the point is reached where the tissues are anoxic. This has been observed in lettuce seeds (118, 130, 130a), in excised maize root tips provided with sugars (unpublished data from our laboratory), and in cultures of rat kidney cells (173) as well as in rat hepatocytes (76). A correlation between oxygen availability, AEC, and protein synthesis was also observed in squash cotyledons isolated from germinating seeds (127). In *Candida utilis,* when the AEC was lowered to 0.69 by reducing the oxygen partial pressure to 400 Pa, the growth of this organism was reduced but not stopped (161); the germination of lettuce seeds was also slowed under 2.5 kPa oxygen and the AEC of this material was 0.65 (2, 130). These results disagree with Atkinson's interpretation which asserts that growth is only possible in the high range of AEC values (11). In *Rhodotorula gracilis,* on the other hand, it was observed that oxygen partial pressure slowed growth and protein synthesis without lowering the AEC (36a). This result needs confirmation because, from all other available data, it appears that the level of the AEC under hypoxia is clearly correlated with metabolic activity.

Correlations between AdN ratios and respiratory rate have also been obtained with animal mitochondria in vitro. In most cases, respiration was negatively correlated to ATP/ADP and depended on factors such as Pi (46), the activity of ATP-consuming (90) or regenerating pathways (89, 168), and the substrate used (89). In these in vitro systems, the ATP/ADP ratios were much higher than those generally found in vivo, and the rate of respiration was limited by the rate of ATP-consuming reactions. On the other hand, a positive correlation has also been reported (73). Controversial conclusions concerning the controlling factors of respiration have been drawn from these results (see 54). The variability of these results shows that the relationship between respiration limited by hypoxia and the ATP/ADP ratio may be affected by a number of factors. A relationship established under well-defined conditions may be used as an indicator of metabolic states and would permit an in vivo approach to many aspects of energy metabolism (124). In our laboratory, this approach has been used to demonstrate the occurrence of oxidative phosphorylation in seeds during the first minutes of imbibition when phosphorylating mitochondria still cannot be extracted (70, 119), and to study the efficiency of the ATP regeneration by cyanide-insensitive respiration in the same material (121a). It was also used to measure the respiratory rate of roots (129) and rhizomes (153) placed in anaerobic environment, which were supplied with oxygen "transported" from the aerial parts through the aerenchyma.

CONCLUDING REMARKS

The importance of AdN in metabolic regulation is presently a matter of debate. It is extremely difficult to determine if the stabilization of AdN ratios and consequently of AEC are the cause or the consequence of regulation. However, the accurate determination of AdN ratios has proved useful to plant physiologists by providing an in vivo approach to energy metabolism. The analysis of AdN nucleotides at the level of the cellular compartments should provide valuable information for understanding their regulatory metabolic interactions and thus extend the usefulness of AdN ratios as metabolic indicators.

ACKNOWLEDGMENTS

We wish to thank Professor D. D. Davies and Dr. P. Saglio for stimulating discussions and very helpful comments on the manuscript, and Professor R. Bieleski, Dr. M. Erecínska, and Professor H. Heldt for providing unpublished information. We are also grateful to E. Barade for her patience and cooperative help during the preparation of the manuscript.

Literature Cited

1. Akkerman, J. W. N., Gorter, G. 1980. Relation between energy production and nucleotide metabolism in human blood platelets. *Biochim. Biophys. Acta* 590:107–16
2. Al-Ani, A., Leblanc, J. M., Raymond, P., Pradet, A. 1982. Effet de la pression partielle d'oxygène sur la vitesse de germination des semences à réserves lipidiques et amylacées: rôle du métabolisme fermentaire. *C. R. Acad. Sci. Paris.* 295:271–74
3. Anderson, J. D. 1977. Adenylate metabolism of embryonic axes from deteriorated soybean seeds. *Plant Physiol.* 59:610–14
4. Anderson, L. E. 1974. Regulation of pea leaf ribulose-5-phosphate kinase activity. *Biochim. Biophys. Acta* 321: 484–88
5. Andrews, T. J., Hatch, M. D. 1969. Properties and mechanism of action of pyruvate, phosphate dikinase from leaves. *Biochem. J.* 114:117–25
6. Arron, G. P., Day, D. A., Laties, G. G. 1978. Intramitochondrial location of

adenylate kinase. *Plant Physiol.* 61S:84 (Abstr.)
7. Ashihara, H. 1977. Regulation of the activity of spinach phosphoribosylpyrophosphate synthetase by "energy charge" and end products. *Z. Pflanzenphysiol.* 85:383–92
8. Atkinson, D. E. 1968. The energy charge of the adenylate pool as a regulatory parameter. Interaction with feedback modifiers. *Biochemistry* 7:4030–34
9. Atkinson, D. E. 1971. Enzymes as control elements in metabolic regulation. *Enzymes* 1:461–89
10. Atkinson, D. E. 1976. Adaptation of enzymes for regulation of catalytic functions. *Biochem. Soc. Symp.* 41:205–23
11. Atkinson, D. E. 1977. *Cellular Energy Metabolism and its Regulation.* New York: Academic. 293 pp.
12. Atkinson, D. E., Walton, G. M. 1967. Adenosine triphosphate conservation in metabolic regulation. *J. Biol. Chem.* 242:3239–40
13. Axelrod, B., Beevers, M. 1972. Differential response of mitochondrial and

glyoxysomal citrate synthase to ATP. *Biochim. Biophys. Acta* 256:175–78

14. Bieleski, R. L. 1964. The problem of halting enzyme action when extracting plant tissue. *Anal. Biochem.* 9:431–42
15. Bieleski, R. L. 1965. Separation of phosphate esters by thin-layer chromatography and electrophoresis. *Anal. Biochem.* 12:230–35
16. Bieleski, R. L. 1968. Levels of phosphate esters in *Spirodela. Plant Physiol.* 43:1297–308
17. Bieleski, R. L. 1973. Phosphate pools, phosphate transport, and phosphate availability. *Ann. Rev. Plant Physiol.* 24:225–52
18. Bieleski, R. L., Laties, G. G. 1963. Turnover rates of phosphate esters in fresh and aged slices of potato tuber tissue. *Plant Physiol.* 38:586–94
19. Bieleski, R. L., Young, R. E. 1963. Extraction and separation of phosphate esters from plant tissues. *Anal. Biochem.* 6:54–68
19a. Birkenhead, K., Walker, D., Foyer, C. 1982. The intracellular distribution of adenylate kinase in the leaves of spinach, wheat and barley. *Planta* 156:171–75
20. Blair, J. McD. 1970. Magnesium, potassium and the adenylate kinase equilibrium: Magnesium as a feedback signal from the adenine nucleotide pool. *Eur. J. Biochem.* 13:384–90
21. Bomsel, J. L. 1973. Influence de la charge énergétique sur la vitesse de synthèse in vivo du saccharose. *C.R. Acad. Sci. Paris* 277:1753–56
22. Bomsel, J. L., Pradet, A. 1967. Etude des adénosine 5'-mono-, di- et triphosphates dans les tissus végétaux. II. Evolution in vivo de l'ATP, l'ADP et l'AMP dans les feuilles de Blé en fonction de différentes conditions de milieu. *Physiol. Vég.* 5:223–36
23. Bomsel, J. L., Pradet, A. 1968. Study of adenosine 5'-mono-, di- and triphosphates in plant tissues. IV. Regulation of the level of nucleotides in vivo by adenylate kinase: theoretical and experimental study. *Biochim. Biophys. Acta* 162:230–42
24. Bomsel, J. L., Sellami, A. 1975. In vivo measurement of the rate of transfer of ~ P from adenylate through the chloroplast envelope. *Proc. 3rd Int. Congr. Photosynth.*, ed. M. Avron, pp. 1363–67. Amsterdam: Elsevier
25. Bonzon, M., Hug, M., Wagner, E., Greppin, H. 1981. Adenine nucleotides and energy charge evolution during the induction of flowering in spinach leaves. *Planta* 152:189–94
26. Bowman, E. J., Ikuma, H. 1976. Regulation of malate oxidation in mung bean mitochondria. II. Role of adenylates. *Plant Physiol.* 58:438–46
27. Brown, E. G. 1962. The acid-soluble nucleotides of mature pea seeds. *Biochem. J.* 85:633–40
28. Brown, E. G., Short, K. C. 1969. The changing nucleotide pattern of sycamore cells during culture in suspension. *Phytochemistry* 8:1365–72
29. Buchanan, B. B. 1980. Role of light in the regulation of chloroplast enzymes. *Ann. Rev. Plant Physiol.* 31:341–74
30. Cavell, S., Scopes, R. K. 1976. Isolation and characterization of the "photosynthetic" phosphoglycerate kinase from *Beta vulgaris. Eur. J. Biochem.* 63:483–90
31. Chapman, A., Atkinson, D. E. 1973. Stabilization of adenylate energy charge by the adenylate deaminase reaction. *J. Biol. Chem.* 248:8309–12
32. Chapman, A. G., Atkinson, D. E. 1977. Adenine nucleotide concentration and turnover rates. Their correlation with biological activity in bacteria and yeast. *Adv. Microb. Physiol.* 15:254–307
33. Chapman, A. G., Fall, L., Atkinson, D. E. 1971. Adenylate energy charge in *Escherichia coli* during growth and starvation. *J. Bacteriol.* 108:1072–86
34. Chevalier, D., Douce, R. 1976. Interactions between mitochondria and chloroplasts in cells. *Plant Physiol.* 57:400–2
35. Ching, T. M., Hedtke, S., Garay, A. E. 1975. Energy state of wheat leaves in ammonium nitrate-treated plants. *Life Sci.* 16:603–10
36. Ching, T. M., Kronstadt, W. E. 1972. Varietal differences in growth potential adenylate energy level and energy charge of wheat. *Crop Sci.* 12:785–89
36a. Cocucci, M., Cocucci, M. C., Marré, E. 1973. Growth inhibition without ATP decrease in *Rhodotorula gracilis* cultured under reduced oxygen pressure. *Plant Sci. Lett.* 1:425–31
37. Cocucci, M. C., Marré, E. 1973. The effect of cycloheximide on respiration, protein synthesis and adenosine nucleotide levels in *Rhodotorula gracilis. Plant Sci. Lett.* 1:293–301
38. Cole, C. V., Ross, C. 1966. Extraction, separation, and quantitative estimation of soluble nucleotides and sugar phosphates in plant tissues. *Anal. Biochem.* 17:526–39
39. Coombs, J., Maw, S. L., Baldry, C. W. 1974. Metabolic regulation in C_4 photo-

synthesis: PEP-carboxylase and energy charge. *Planta* 117:279–92

40. Copeland, L., Preiss, J. 1981. Purification of spinach leaf ADP-glucose pyrophosphorylase. *Plant Physiol.* 68:996–1001

41. Criss, W. E. 1970. Rat liver adenosine triphosphate: adenosine monophosphate phosphotransferase activity. II. Subcellular localization of adenylate kinase isoenzymes. *J. Biol. Chem.* 245: 6352–56

42. Criss, W. E., Pradhan, T. K. 1978. Purification and characterization of adenylate kinase from rat liver. *Methods Enzymol.* 51:459–67

43. Crompton, M., Laties, G. G. 1971. The regulatory function of potato pyruvate dehydrogenase. *Arch. Biochem. Biophys.* 143:143–50

44. Davies, D. D. 1956. Soluble enzymes from pea mitochondria. *J. Exp. Bot.* 7:203–10

45. Davies, D. D. 1980. Anaerobic metabolism and the production of organic acids. In *Biochemistry of Plants, Vol. 2: A comprehensive treatise/Metabolism and respiration*, ed. D. D. Davies, pp. 581–611. New York: Academic. 687 pp.

46. Davis, E. J., Davis van Thienen, W. I. A. 1978. Control of mitochodrial metabolism by the ATP/ADP ratio. *Biochem. Biophys. Res. Commun.* 83: 1260–66

47. Douce, R., Joyard, J. 1979. Structure and function of the plastid envelope. *Adv. Bot. Res.* 7:2–116

48. Drewes, L. R., Gilboe, D. D. 1973. Cerebral metabolite and adenylate energy charge recovery following 10 min of anoxia. *Biochim. Biophys. Acta* 320:701–7

49. Ducet, G., Rosenberg, A. J. 1962. Leaf respiration. *Ann. Rev. Plant Physiol.* 13:171–200

50. Duggleby, R. G., Dennis, D. T. 1973. Pyruvate kinase, a possible regulatory enzyme in higher plants. *Plant Physiol.* 52:312–17

51. Elbers, R., Heldt, H. W., Schmucker, P., Soboll, S., Wiese, H. 1974. Measurement of the ATP/ADP ratio in mitochondria and in the extramitochondrial compartment by fractionation of freeze-stopped liver tissue in non aqueous media. *Hoppe-Seyler's Z. Physiol. Chem.* 355:378–93

52. Elbers, R., Heldt, H. W., Schmucker, P., Wiese, H. 1972. Messung der ATP/ADP quotienten in mitochondrien und im extramitochondrialen raum durch fraktionierung von geffriergestopp-ten lebergewebe mit nichtwässrigen medien. *Hoppe-Seylers Z. Physiol. Chem.* 353:702–3

53. Erecínska, M., Wilson, D. F. 1982. Regulation of cellular energy metabolism. *J. Membr. Biol.* 70:1–14

54. Erecínska, M., Wilson, D. F. 1978. Homeostatic regulation of cellular energy metabolism. *Trends Biochem. Sci.* 3:219–23

55. Flügge, U. I., Stitt, M., Freisl, M., Heldt, H. W. 1982. On the participation of phosphofructokinase in the light regulation of CO_2 fixation. *Plant Physiol.* 69:263–67

56. Garland, W. J., Dennis, D. T. 1980. Plastid and cytosolic phosphofructokinases from the developing endosperm of *Ricinus communis*. I. Separation, purification and initial characterization of the isoenzymes. *Arch. Biochem. Biophys.* 204:302–9

57. Garland, W. J., Dennis, D. T. 1980. Plastid and cytosolic phosphofructokinases from the developing endosperm of *Ricinus communis*. II. Comparison of the kinetic and regulatory properties of the isoenzymes. *Arch. Biochem. Biophys.* 204:310–17

58. Ghosh, H. P., Preiss, J. 1966. Adenosine diphosphate glucose pyrophosphorylase. A regulatory enzyme in the biosynthesis of starch in spinach leaf chloroplasts. *J. Biol. Chem.* 241:4491–4504

59. Giersch, C., Heber, U., Kobayashi, Y., Inoue, Y., Shibata, K., Heldt, H. W. 1980. Energy charge, phosphorylation potential and proton motive force in chloroplasts. *Biochim. Biophys. Acta* 590:59–73

60. Hampp, R. 1980. Rapid separation of the plastid, mitochondrial, and cytoplasmic fractions from intact leaf protoplasts of *Avena*. Determination of in vivo ATP pool sizes during greening. *Planta* 150:291–98

61. Hampp, R., Goller, M., Ziegler, H. 1982. Adenylate levels, energy charge and phosphorylation potential during dark-light and light-dark transition in chloroplasts mitochondria and cytosol of mesophyl protoplasts from *Avena sativa* L. *Plant Physiol.* 69:448–55

62. Hanson, J. B., Day, D. A. 1980. Plant mitochondria. In *Biochemistry of Plants, Vol. 1: A comprehensive treatise/The plant cell*, ed. N. E. Tolbert, pp. 315–58. New York: Academic. 705 pp.

63. Heber, U. 1974. Metabolite exchange between chloroplasts and cytoplasm. *Ann. Rev. Plant Physiol.* 25:393–421

64. Heber, U., Heldt, H. W. 1981. The chloroplast envelope: Structure, function, and role in leaf metabolism. *Ann. Rev. Plant Physiol.* 32:139–68

65. Heber, U., Santarius, K. A. 1965. Compartmentation and reduction of pyridine nucleotides in relation to photosynthesis. *Biochim. Biophys. Acta* 109:390–408

66. Heldt, H. W. 1969. Adenine nucleotide translocation in spinach chloroplasts. *FEBS Lett.* 5:11–14

67. Heldt, H. W., Chon, C. J., Maronde, D., Herold, A., Stakovic, Z. S., et al. 1977. Role of orthophosphate and other factors in the regulation of starch formation in leaves and isolated chloroplasts. *Plant Physiol.* 59:1146–55

68. Heldt, H. W., Klingenberg, M., Milovancev, M. 1972. Differences between the ATP/ADP ratios in the mitochondrial matrix and the extramitochondrial space. *Eur. J. Biochem.* 30:434–40

69. Heldt, H. W., Schwalbach, K. 1967. The participation of GTP-AMP-P transferase in substrate level phosphate transfer of rat liver mitochondria. *Eur. J. Biochem.* 1:199–206

70. Hourmant, A., Pradet, A. 1981. Oxidative phosphorylation in germinating lettuce seeds (*Lactuva sativa*) during the first hours of imbibition. *Plant Physiol.* 68:631–35

70a. Hourmant, A., Pradet, A., Penot, M. 1979. Action de la benzylaminopurine sur l'absorption du phosphate et le métabolisme des composés phosphorylés des disques de tubercule de pomme de terre en survie. *Physiol. Vég.* 17:483–99

71. Ikumah, H., Tetley, R. M. 1976. Possible interference by an acid-stable enzyme during the extraction of nucleoside di- and triphosphates from higher plant tissues. *Plant Physiol.* 58:320–23

72. Ireland, R. J., Deluca, V., Dennis, D. T. 1980. Characterization and kinetics of isoenzymes of pyruvate kinase from developing castor bean endosperm. *Plant Physiol.* 65:1188–93

73. Jacobus, W. E., Moreadith, R. W., Vandegaer, K. M. 1982. Mitochondrial respiratory control. Evidence against the regulation of respiration by extramitochondrial phosphorylation potentials or by ATP/ADP ratios. *J. Biol. Chem.* 257:2397–402

74. Jenner, C. F. 1968. The composition of soluble nucleotides in the developing wheat grain. *Plant Physiol.* 43:41–49

75. Jensen, R. G. 1980. Biochemistry of the chloroplast. See Ref. 62, pp. 273–313

76. Jones, D. P., Mason, H. S. 1978. Gradients of O_2 concentration in hepatocytes. *J. Biol. Chem.* 253:4874–80

77. Kacser, H., Burns, J. A. 1973. The control of flux. *Symp. Soc. Exp. Biol.* 27:65–104

78. Kaiser, W. M., Bassham, J. A. 1979. Light-dark regulation of starch metabolism in chloroplasts. II. Effect of chloroplastic metabolite levels on the formation of ADP-glucose by chloroplast extracts. *Plant Physiol.* 63:109–13

79. Karu, A. E., Moundrianakis, E. N. 1969. Fractionation and comparative studies of enzymes in aqueous extracts of spinach chloroplasts. *Arch. Biochem. Biophys.* 129:655–71

80. Kelly, G. J., Latzko, E., Gibbs, M. 1976. Regulatory aspects of photosynthetic carbon metabolism. *Ann. Rev. Plant Physiol.* 27:181–205

81. Keys, A. J., Whittingham, C. P. 1969. Nucleotide metabolism in chloroplast and non-chloroplast components of tobacco leaves. In *Progress in Photosynthesis Research,* ed. H. Metzner, 1:352–58. Tübingen: Laupp

82. Klingenberg, M. 1980. The ADP/ATP translocation in mitochondria a membrane potential controlled transport. *J. Membr. Biol.* 56:97–105

83. Klingenberg, M., Heldt, H. W. 1982. The ADP/ATP translocation in mitochondria and its role in extracellular compartmentation. In *Metabolic Compartmentation,* ed. H. Sies, pp. 101–22. London: Academic. 561 pp.

84. Knowles, C. J. 1977. Microbial metabolic regulation by adenine nucleotide pools. In *Microbial Bioenergetics,* ed. B. A. Haddock, W. A. Hamilton, pp. 240–83. Cambridge: Cambridge Univ. Press. 442 pp.

85. Krebs, H. A. 1965. Control of energy metabolism. In *Control of Energy Metabolism,* ed. B. Chance, R. W. Estabrook, J. R. Williamson, p. 198. New York: Academic. 441 pp.

86. Krebs, H. A. 1973. Pyridine nucleotides and rate control. *Symp. Soc. Exp. Biol.* 23:299–318

87. Krebs, H. A., Veech, R. C. 1970. Regulation of the redox state of the pyridine nucleotides in the rat liver. In *Pyridine Nucleotide Dependent Dehydrogenases,* ed. H. Sund, pp. 413–38. Berlin: Springer

88. Krebs, H. A., Wiggins, D. 1978. Phosphorylation of adenosine monophosphate in the mitochondrial matrix. *Biochem. J.* 174:297–301

89. Kunz, W., Bohnensack, R., Böhme, G., Kuster, U., Letko, G., Schonfeld, P. 1981. Relations between extramitochondrial adenine nucleotide systems. *Arch. Biochem. Biophys.* 209:219–29

90. Kunz, W., Bohnensack, R., Gellerich, F. N., Böhme, G., Letko, G., et al. 1982. Dependence of respiratory control at mitochondrial and cellular level on the conditions. *2nd Eur. Bioenerg. Conf.*, pp. 523–24. Lyon: LBTM, CNRS. 660 pp.

91. Lavergne, D., Bismuth, E., Champigny, M. L. 1974. Further studies of phosphoglycerate kinase and ribulose-5-phosphate kinase of the photosynthetic carbon reduction cycle: Regulation of the enzymes by the adenine nucleotides. *Plant Sci. Lett.* 3:391–97

92. Lilley, R. McC., Stitt, M., Heldt, H. W. 1982. Determination of adenine nucleotide levels in the mitochondria, cytosol and chloroplasts of plant protoplasts. See Ref. 90, pp. 567–68

93. Lilley, R. McC., Stitt, M., Mader, G., Heldt, H. W. 1982. Rapid fractionation of wheat leaf protoplasts using membrane filtration. The determination of metabolic levels in the chloroplasts, cytosol and mitochondria. *Plant Physiol.* 70:965–70

94. Lin, W., Hanson, J. B. 1974. Phosphate absorption rates and adenosine 5'-triphosphate concentrations in corn root tissues. *Plant Physiol.* 54:250–56

95. Loomis, W. D., Bataille, J. 1966. Plant phenolic compounds and the isolation of plant enzymes. *Phytochemistry* 5:423–38

96. Lüttge, U., Ball, E. 1976. ATP levels and energy requirements of ion transport in cells of slices of greening barley leaves. *Z. Pflanzenphysiol.* 80:50–59

97. Markland, F., Wadkins, C. L. 1977. Adenosine triphosphate-adenosine-5'-monophosphate phosphotransferase of bovine liver mitochondria. I. Isolation and chemical properties. *J. Biol. Chem.* 241:4124–35

97a. Martin, J. B., Bligny, R., Rebeille, F., Douce, R., Leguay, J. J., et al. 1982. A ^{31}P Nuclear magnetic study of intracellular pH of plant cells cultivated in liquid medium. *Plant Physiol.* 70:1156–61

98. Mazelis, M. 1956. Particulate adenylic kinase in higher plants. *Plant Physiol.* 31:37–43

99. Miginiac-Maslow, M., Hoarau, A. 1979. The adenine nucleotide levels and the adenylate energy charge of different *Triticum* and *Aegilops* species. *Z. Pflanzenphysiol.* 93:387–94

100. Miginiac-Maslow, M., Mathieu, Y., Nato, A., Hoarau, A. 1981. Contribution of photosynthesis to the growth and differentiation of cultured tobacco cells. The role of inorganic phosphate. In *Photosynthesis*, ed. G. Akoyunoglou, 5:977–84. Philadelphia: Balaban

101. Mocquot, B., Aspart, L., Delseny, M., Pradet, A. 1980. Synthese d'ARN dans les embryons de riz en condition d'anoxie. *Physiol. Vég.* 18:395

102. Mocquot, B., Pradet, A., Litvak, S. 1977. DNA synthesis and anoxia in rice coleoptiles. *Plant Sci. Lett.* 9:365–71

103. Mocquot, B., Prat, C., Mouches, C., Pradet, A. 1981. Effect of anoxia on energy charge and protein synthesis in rice embryo. *Plant Physiol.* 68:636–40

104. Moreland, D. E., Hussey, G. G., Shriner, C. R., Farmer, F. S. 1974. Adenosine phosphates in germinating radish (*Raphanus sativus* L.) seeds. *Plant Physiol.* 54:560–63

105. Murukami, S., Strotman, H. 1978. Adenylate kinase bound to the envelope membranes of spinach chloroplasts. *Arch. Biochem. Biophys.* 185:30–38

106. Nakayama, H., Fujii, M., Miura, K. 1976. Partial purification and some regulatory properties of pyruvate kinase from germinating castor bean endosperm. *Plant Cell Physiol.* 17:653–60

107. Neijssel, O. M., Tempest, D. W. 1976. The role of energy spilling reactions in the growth of *Klebsiella aerogenes* NCTC 418 in aerobic chemostat culture. *Arch. Microbiol.* 110:305–11

108. Nielsen, R., Rasmussen, N. 1968. Fractionation of firefly tails by gel filtration. *Acta Chem. Scand.* 22:1757–62

109. Nieman, R. H., Pap, D. L., Clark, R. A. 1978. Rapid purification of plant nucleotide extract with XAD-2, polyvinylpyrrolidone and charcoal. *J. Chromatogr.* 161:137–46

110. Noda, L. 1973. Adenylate kinase. *Enzymes* 8:279–305

111. Obendorf, R. L., Marcus, A. 1974. Rapid increase in adenosine 5'-triphosphate during early wheat embryo germination. *Plant Physiol.* 53:779–81

112. Pacold, I., Anderson, L. E. 1973. Energy charge control of the Calvin cycle enzyme 3-phosphoglyceric acid kinase. *Biochem. Biophys. Res. Commun.* 51:139–43

113. Pacold, I., Anderson, L. E. 1975. Chloroplast and cytoplasmic enzymes. VI. Pea leaf 3-phosphoglycerate kinases. *Plant Physiol.* 55:168–71

114. Peterson, J. B., Evans, H. J. 1978. Properties of pyruvate kinase from soybean

nodule cytosol. *Plant Physiol.* 61: 909–14

115. Portis, A. R., Chon, C. J., Mosbach, A., Heldt, H. W. 1977. Fructose- and sedo-heptulosebisphosphatase. The sites of a possible control of CO_2 fixation by light dependent changes of the stromal Mg^{2+} concentration. *Biochim. Biophys. Acta* 461:313–25

116. Portis, A. R., Heldt, H. W. 1976. Light-dependent changes of the Mg^{2+} concentration in the stroma in relation to the Mg^{2+} dependency of CO_2 fixation in intact chloroplasts. *Biochim. Biophys. Acta* 449:434–46

117. Pradet, A. 1967. Etude des adenosine 5'-mono-, di- et triphosphates. I. Dosage enzymatique. *Physiol. Vég.* 5: 209–21

118. Pradet, A. 1969. Etude des adenosine 5'-mono-, di- et triphosphates dans les tissus végétaux. V. Effet in vivo sur le niveau de la charge énergétique d'un déséquilibre induit entre fourniture et utilisation de l'énergie dans les semences de laitue. *Physiol. Vég.* 7: 261–75

119. Pradet, A. 1982. Oxidative phosphorylation in seeds during the initial phases of germination. In *The Physiology and Biochemistry of Seed Development, Dormancy and Germination,* ed. A. A. Khan, pp. 347–69. Elsevier Biomedical Press. 547 pp.

120. Pradet, A., Bomsel, J. L. 1968. Contrôle in vivo des nucléotides adényliques par l'adénylate kinase. *C. R. Acad. Sci. Paris* 266:2416–18

121. Pradet, A., Bomsel, J. L. 1978. Energy metabolism in plants under hypoxia and anoxia. In *Plant Life in Anaerobic Environments,* ed. D. D. Hook, R. M. M. Crawford, pp. 89–118. Ann Arbor, Mich: Ann Arbor Sci. 564 pp.

121a. Pradet, A., Hourmant, A. 1982. In vivo studies of oxidative phosphorylation. See ref. 90, pp. 575–76

122. Pradet, A., Narayanan, A., Vermeersch, J. 1968. Etude des adenosine 5'-mono-, di- et triphosphates dans les tissus végétaux. III. Métabolisme au cours des premiers stades de la germination des semences de laitue. *Bull. Soc. Fr. Physiol. Vég.* 14:107–14

123. Pradet, A., Prat, C. 1976. Métabolisme énergétique au cours de la germination du riz en anoxie. In *Etudes de Biologie Végétale,* éd. R. Jacques, pp. 561–74. Gif/Yvette France: Phytotron CNRS. 600 pp.

124. Pradet, A., Raymond, P. 1983. Adenylate energy charge, an indicator of en-

ergy metabolism. In *Physiology and Biochemistry of Plant Respiration,* ed. J. M. Palmer. Cambridge: Cambridge Univ. Press. In press

125. Purich, D. L., Fromm, H. J. 1972. Studies on factors influencing enzyme responses to adenylate energy charge. *J. Biol. Chem.* 247:249–55

126. Purich, D. L., Fromm, H. J. 1973. Additional factors influencing enzyme responses to the adenylate energy charge. *J. Biol. Chem.* 248:461–66

127. Rasi-Caldogno, F., De Michelis, M. I. 1978. Correlation between oxygen availability, energy charge and protein synthesis in squash cotyledons isolated from germinating seeds. *Plant Physiol.* 61:85–88

128. Rasmussen, H., Nielsen, R. 1968. An improved analysis of adenosine triphosphate by the luciferase method. *Acta Chem. Scand.* 22:1745–56

129. Raymond, R., Bruzau, F., Pradet, A. 1978. Etude du transport d'oxygene des parties aériennes aux racines à l'aide d'un paramètre du métabolisme: la charge énergétique. *C. R. Acad. Sci. Paris* 286:1061–63

130. Raymond, P., Pradet, A. 1980. Stabilization of adenine nucleotide ratios at various values by an oxygen limitation of respiration in germinating lettuce (*Lactuca sativa*) seeds. *Biochem. J.* 190:39–44

130a. Raymond, P., Pradet, A. 1982. Modifications of ATP/ADP ratios reflect limitations of the activity of ATP-regenerative pathways. See Ref. 90, pp. 577–78

131. Ridge, J. W. 1972. Hypoxia and the energy charge of the cerebral adenylate pool. *Biochem. J.* 127:351–53

131a. Roberts, J. K. M., Ray, P. M., Wade-Jardetzky, N., Jardetzky, O. 1980. Estimation of cytoplasmic and vacuolar pH in higher plant cells by ^{31}P NMR. *Nature* 283:870–72

132. Robinson, S. P., Walker, D. A. 1979. Rapid separation of the chloroplast and cytoplasmic fractions from intact leaf protoplasts. *Arch. Biochem. Biophys.* 196:319–23

133. Robinson, S. P., Walker, D. A. 1979. The controls of 3-phosphoglycerate reduction in isolated chloroplasts by the concentrations of ATP, ADP and 3-phosphoglycerate. *Biochim. Biophys. Acta* 545:528–36

134. Robinson, S. P., Walker, D. A. 1981. Photosynthetic carbon reduction cycle. In *Biochemistry of Plants, Vol. 8: A comprehensive treatise/ Photosynthesis,* ed. M. D. Hatch, N. K. Boardman, pp.

193–236. New York: Academic. 521 pp.

135. Rodaway, S., Huang, B. F., Marcus, A. 1979. Nucleotide metabolism and the germination of seed embryonic axes. In *The Plant Seed Development, Preservation, and Germination*, ed. I. Rubenstein, R. L. Phillips, C. E. Green, B. G. Gengenbach, pp. 203–18. New York: Academic. 266 pp.

136. Rodionova, M. A., Kholodenko, N. Ya., Makarov, A. D. 1976. Adenylate kinase of plants. Properties of adenylate kinase from pea leaves. *Biochemistry (USSR)* 41:1568–72

137. Rodionova, M. A., Kholodenko, N. Ya., Makarov, A. D. 1978. Localization of adenylate kinase in cellular organelles and chloroplast substructures in spinach leaves. *Sov. Plant Physiol.* 25:569–72

138. Saglio, P. H. M., Daniels, M. J., Pradet, A. 1979. ATP and energy charge as criteria of growth and metabolic activity of mollicutes: Application to *Spiroplasma citri*. *J. Gen. Microbiol.* 110:13–20

139. Saglio, P. H. M., Pradet, A. 1980. Soluble sugars, respiration and energy charge during aging of excised maize root tips. *Plant Physiol.* 66:516–19

140. Saglio, P. H. M., Raymond, P., Pradet, A. 1980. Metabolic activity and energy charge of excised maize root tips under anoxia. *Plant Physiol.* 66:1053–57

141. Santarius, K. A., Heber, U. 1965. Changes in intracellular levels of ATP, ADP, AMP and Pi and regulatory function of the adenylate system in leaf cells during photosynthesis. *Biochim. Biophys. Acta* 102:39–54

142. Santarius, K. A., Heber, U., Ullrich, W., Urbach, W. 1964. Intracellular translocation of ATP, ADP and inorganic phosphate in leaf cells of *Elodea densa* in relation to photosynthesis. *Biochem. Biophys. Res. Commun.* 15:139–46

143. Scholz, R., Bücher, T. 1965. Hemoglobin free perfusion of rat liver. See Ref. 85, pp. 393–414

144. Schöttler, U. 1978. The influence of anaerobiosis on the levels of adenosine nucleotides and some glycolytic metabolites in Tubifex sp. (*Annelida oligochaeta*). *Comp. Biochem. Physiol. B* 61:29–32

145. Sel'kov, E. 1975. Stabilization of energy charge, generation of oscillations and multiple steady states in energy metabolism as a result of purely stoichiometric regulation. *Eur. J. Biochem.* 59:151–57

146. Sellami, A. 1976. Evolution des adénosine phosphates et de la charge énergétique dans les compartiments chloroplastique et non chloroplastique des feuilles de blé. *Biochim. Biophys. Acta* 423:524–39

147. Sellami, A. 1977. *Evolution des nucléotides libres dans les compartiments chloroplastique et non chloroplastique des feuilles de blé*. Thèse Doctorat ès Sciences. Université Paris VI. 146 pp.

148. Sellami, A., Bomsel, J. L. 1975. Evolution de la charge énergétique du pool adénylique des feuilles de blé au cours de l'anoxie. Etude de la réversibilité des phénomènes observés. *Physiol. Vég.* 13:611–17

149. Shargool, P. D. 1973. The response of soybean arginosuccinate synthetase to different energy charge values. *FEBS Lett.* 33:348–50

150. Shen, L. C., Fall, L., Walton, G. M., Atkinson, D. E. 1968. Interaction between energy charge and metabolite modulation in the regulation of enzymes of amphibolic sequences. Phosphofructokinase and pyruvate dehydrogenase. *Biochemistry* 7:4041–45

151. Sobczyk, E. A., Kacperska-Palacz, A. 1978. Adenine nucleotide changes during cold acclimatation of winter rape plants. *Plant Physiol.* 62:875–78

152. Soboll, S., Scholz, R., Heldt, H. W. 1978. Subcellular metabolite concentrations. Dependence of mitochondrial and cytosolic ATP systems on the metabolic state of perfused rat liver. *Eur. J. Biochem.* 87:377–90

153. Steinmann, F., Brändle, R. 1981. Die überflutungstoleranz der Seebinse (*Schoenoplectus lacustris* (L.) Palla): III. Beziehungen zwischen der sauerstoffversorgung in der umbegung. *Flora* 171:307–14

154. Steup, M., Peavey, D. G., Gibbs, M. 1976. The regulation of starch metabolism by inorganic phosphate. *Biochem. Biophys. Res. Commun.* 72:1554–61

155. Stitt, M., Lilley, R. McC., Heldt, H. W. 1982. Adenine nucleotide levels in the cytosol, chloroplasts and mitochondria of wheat leaf protoplasts. *Plant Physiol.* 70:971–77

156. Stitt, M., Wirtz, W., Heldt, H. W. 1980. Metabolite levels during induction in the chloroplast and extrachloroplast compartments of spinach protoplasts. *Biochim. Biophys. Acta* 593:85–102

157. Storey, B. T. 1980. Electron transport and energy coupling in plant mitochondria. See Ref. 45, pp. 125–95

158. Strehler, B. L. 1953. Firefly luminescence in the study of energy transfer mechanisms. II. Adenosine triphos-

phate and photosynthesis. *Arch. Biochem. Biophys.* 43:67–79

159. Strehler, B. L., Totter, J. R. 1952. Firefly luminescence in the study of energy transfer mechanisms. I. Substrate and enzyme determination. *Arch. Biochem. Biophys.* 40:28–41

160. Thalacker, V. R., Behrens, M. 1959. Über den reinheitsgrad der in einem nichtwässrigen spezifischen gewitchtsgradienten gewonnenen chloroplasten. *Z. Naturforsch. Teil B* 14:443–46

161. Thomas, K. C., Dawson, P. S. S. 1977. Energy charge values of *Candida utilis* growing under energy-excess and energy-limiting conditions. *FEMS Microbiol. Lett.* 1:347–49

162. Thompson, J. F. 1980. Arginine synthesis, proline synthesis, and related processes. In *Biochemistry of Plants, Vol. 5: A comprehensive treatise/Amino acids and derivatives,* ed. B. J. Miflin, pp. 375–402. New York: Academic. 670 pp.

163. Turner, J. F., Chensee, Q. J., Harrison, D. D. 1977. Glucokinase of pea seeds. *Biochim. Biophys. Acta* 480:367–75

164. Turner, J. F., Turner, D. H. 1975. The regulation of carbohydrate metabolism. *Ann. Rev. Plant Physiol.* 26:159–86

165. Turner, J. F., Turner, D. H. 1980. The regulation of glycolysis and the pentose phosphate pathway. See Ref. 45, pp. 279–316

166. Vignais, P. V. 1976. Molecular and physiological aspects of adenine nucleotide transport in mitochondria. *Biochim. Biophys. Acta* 456:1–38

167. Vignais, P. V., Douce, R., Lauquin, G. J. M., Vignais, P. M. 1976. Binding of radioactively labeled carboxyatractyloside, atractyloside and bongkrekic acid to the ADP translocator of potato mitochondria. *Biochim. Biophys. Acta* 440:688–96

168. Wanders, R. J. A., Groen, A. K., Meijer, A. J., Tager, J. M. 1981. Determination of the free-energy difference of the adenine nucleotide translocator reaction in rat liver mitochondria using intra- and extra-mitochondrial ATP-utilizing reactions. *FEBS Lett.* 132:201–6

169. Weinstein, J. H., McCune, D. C., Mancini, J. F., van Leuken, P. 1969. Acid-soluble nucleotides of pinto bean leaves at different stages of development. *Plant Physiol.* 44:1499–1510

170. Weissman, G. S. 1976. Glutamine synthetase regulation by energy charge in sunflower roots. *Plant Physiol.* 57:339–43

171. Werdan, K., Heldt, H. W., Milovancev, M. 1975. The role of pH in the regulation of carbon fixation in the chloroplast stroma. Studies on CO_2 fixation in the light and dark. *Biochim. Biophys. Acta* 396:276–92

172. Wilson, A. M., Harris, G. A. 1966. Hexose-, inositol-, and nucleoside phosphate esters in germinating seeds of crested wheatgrass. *Plant Physiol.* 41:1416–19

173. Wilson, D. E., Erecinska, M., Drown, C., Silver, I. A. 1977. Effect of oxygen tension on cellular energetics. *Am. J. Physiol.* 233:C135–C40

174. Wirtz, W., Stitt, M., Heldt, H. W. 1980. Enzymic determination of metabolites in the subcellular compartments of spinach protoplasts. *Plant Physiol.* 66:187–93

175. Wiskich, J. T. 1980. Control of the Krebs cycle. See Ref. 45, pp. 243–78

176. Wong, K. F., Davies, D. D. 1973. Regulation of phosphoenolpyruvate carboxylase of *Zea mays* by metabolites. *Biochem. J.* 131:451–58

Ann. Rev. Plant Physiol. 1983. 34:225–40
Copyright © 1983 by Annual Reviews Inc. All rights reserved

REGULATION OF PEA INTERNODE EXPANSION BY ETHYLENE

William Eisinger

Department of Biology, University of Santa Clara, Santa Clara, California
95053

CONTENTS

INTRODUCTION

Within the past 10 years, several excellent overview articles (8, 38) and reviews (1, 2, 28, 30) have dealt with certain aspects of the effects of ethylene on cell expansion. In order to develop a more in-depth analysis of the topic, I have elected to limit this review to studies on etiolated pea (*Pisum sativum*) internode tissues and to those responses which can be detected within 24 hours.

Ethylene-induced lateral expansion can serve as an interesting model for the study of hormonal regulation of cell expansion in plants. Historically, lateral expansion or swelling was one of the earliest physiological effects of ethylene described and has been studied by several labs during the past 20 years. The response is easily recognizable, especially with the intact, etiolated seedlings. Within a few hours of treatment, seedlings are noticeably shorter than controls and the region immediately below the hook is markedly thicker. In addition, treated seedlings feel noticeably more turgid.

225

0066-4294/83/0601-0225$02.00

Changes in many kinds of processes have been reported associated with this ethylene-induced change in the mode of growth from primarily elongation growth to a more isodiametric expansion. These changes include basic cell and wall structure (3, 4, 9, 16, 18, 22, 26, 29, 33, 36, 38, 43, 44, 48–50, 52, 55, 57), hormone balance (8), water relations (17, 43), metabolism and pool sizes (16, 18, 26, 31, 33, 36, 44, 49, 51, 57), intercellular localization of enzyme activity (51), and wall extensibility (35, 38, 43, 48, 55). It is difficult to prove whether or not there is a cause-effect relationship between these changes and the altered mode of growth. However, precise kinetic analyses of the ethylene-induced growth changes have helped to clarify the problem (36, 55). Altered growth patterns similar to those seen with ethylene treatment can be induced by a variety of factors including colchicine, supraoptimal auxin, supraoptimal cytokinins, inhibitors of cellulose synthesis, and deuterium oxide (9, 16, 21, 23, 33, 52, 58). Because the action of many of these agents is known, they have been useful in attempting to elucidate the mode of action of ethylene (16, 33, 52).

Most researchers use the third internode tissue of 7-day-old etiolated pea (*Pisum sativum* L.) seedlings. The seedlings are easy to grow and the internode tissue is very sensitive to ethylene. However, there are certain problems associated with this system. Ethylene responsiveness of the tissue is very light-sensitive and necessitates working under dim safe lights which can impede careful, in-progress observation and measurement of responding tissues.

Many workers have chosen to study the effects of ethylene on the internodes of intact seedlings. This system is more "natural" than using excised tissue and responds more quickly, for a longer duration, and produces more total lateral expansion in response to the gas. However, with the intact system it is difficult to distinguish the effects of ethylene on cell expansion from its effects on other processes (4, 8, 26). For example, internode elongation in the untreated internode is normally a combination of expansion of existing cells and the addition of newly produced cells from divisions in the apical hook region. Ethylene interferes with the expansion mode of existing cells, but also inhibits DNA synthesis and cell division in the apical hook (4). In addition, ethylene inhibits auxin biosynthesis in the hook region and its transport into the internode tissue (8). Since auxin is required for both longitudinal expansion and ethylene-induced lateral expansion, the effects of the gas itself on cell expansion per se in the intact system are difficult to assess.

The very presence of the apical hook can interfere with accurate measurements of internode elongation in the presence of ethylene. The gas induces curvature of the hook, which can confuse mechanical or optical systems for measurement of elongation.

With the intact system it is difficult to administer chemical treatments or radioactive precursors to the internode tissue because of the cuticular barrier and other mechanical factors. Many workers have resorted to applying chemical treatments in a lanolin paste onto the surface of the tissue. Although this method is effective, it does not allow for quantitative application.

An alternative to experimenting with intact seedlings is to work with excised internode segments. The growth of such segments is generally easier to measure manually or with high resolution systems. Because the segments have cut end surfaces and are floated on aqueous media, application of chemicals and radioactive precursors is less of a problem than with intact seedlings. Exchange through the cuticle can be facilitated by abrading the surface. However, labeling studies can be confounded by cut surface phenomena, and cuticle abrasion may damage living cells. There are several important points to consider about working with excised segments. The system is less "natural" than the intact system and responds differently to ethylene. For example, the induction lag for the change in the mode of growth is about 6 minutes with intact seedlings but is about an hour with excised segments. Excised segments respond for a shorter period of time and show considerably less lateral growth.

As an intermediate system, the entire apex (embryonic leaves, apical hook, and upper internode tissue) can be excised and incubated in a flask with growth medium. The internode tissue attached to these apices responds rather like internodes of intact seedlings, but the system carries along many of the problems as well. Conversely, the apical hook alone can be removed to produce decapitated seedlings. This system eliminates the problems associated with ethylene responses in the apical hook. However, having removed the natural auxin source, little growth or ethylene response is seen in the epicotyl tissue unless exogenous auxin is applied.

INDUCTION KINETICS

The accurate measurement of hormone induction kinetics can be very useful in establishing cause-effect relationships of various parameters altered by the presence of the hormone. Although we normally think of the study of induction kinetics as a recent development, Van der Laan (59) was using rather sophisticated photographic techniques about 50 years ago to determine that the lag period for ethylene-induced inhibition of elongation in intact seedlings was less than 30 minutes. These results have been subsequently confirmed (8, 26). Warner & Leopold (62), using a position transducer to measure elongation of decapitated 4-day-old (probably second internode) pea seedlings, reported inhibition within 7 minutes for seedlings pretreated with 10 μl ethylene, but the lag period was more than doubled with 5 μl. Nee et al (36) proposed a distinct three-phase induction process based on their kinetics of elongation inhibition of excised third internode segments. During the first hour no inhibition was detected (phase I); the second hour showed 50% of maximal inhibition (phase II); and the third hour brought full inhibition (about 20% of the control expansion rate). Recent work with a laser optical lever system is essentially consistent with this proposal (17).

Precise kinetics of the induction of lateral expansion have been measured only recently. Eisinger et al (17) have shown a close correlation between the induction of lateral expansion and the inhibition of elongation using a dual laser optical system to measure growth of excised internode segments. Lateral expansion was detected after about a one-hour exposure to ethylene and continued at a linear rate thereafter. Barkley & DiFrancesco (6) and Barkley & Rayl (7), using a position transducer on decapitated seedlings, found essentially the same results. They also reported that auxin increases the lag period to about 2 hours but accelerates the rate of lateral expansion once it has begun.

Why intact seedlings show a much shorter induction period than excised segments when exposed to ethylene remains a question. The apical hook seems to play a very important role since internode segments excised with apical hooks show induction lag periods more similar to those seen with intact seedlings (36). Although ethylene has been shown to interfere with auxin transport and metabolism (5, 25), gibberellic acid responsivness of the tissue (53), and with DNA synthesis and cell division (4), these changes can be detected only hours after inhibition is at its maximum. The role of the apical hook in the regulation of internode elongation is probably much more complex than previously thought. For example, if the internode tissue immediately below the hook is excised as two 5 mm segments (apical and basal, respectively), the basal segment will elongate about 15% more than the apical segment (36). However, if the equivalent 5 mm regions are marked on an intact seedling, the apical region will elongate about 750% more than the basal region after 18 hours (W. Eisinger, unpublished data). Thus, the rapid inhibition kinetics seen with intact seedlings might well reflect an effect of ethylene on the apical hook rather than a direct effect on the internode tissue itself.

MICROTUBULE AND CELLULOSE MICROFIBRIL ORIENTATION

Cellulose microfibril orientation has been proposed as the regulator of the mode of cell expansion (19, 20, 31, 42, 47); transversely oriented cellulose microfibrils may enforce polarized elongation despite a two-to-one stress ratio favoring lateral expansion. This hypothesis requires that ethylene treatment which induces lateral expansion must also alter cellulose microfibril orientation away from a predominately transverse order. If the cellulose in the wall regulates the mode of expansion, then it follows that ethylene should alter cellulose orientation. If the cells of this tissue are considered to be thin-walled cylinders with an inherent two-to-one stress ratio favoring lateral expansion, then ethylene-induced lateral expansion could be merely a release from constraint. It does not require per se the imposition of a restraint on elongation in order to bring about lateral expansion.

Much of our understanding of the role of the cell wall in the regulation of growth comes from studies with *Nitella,* a proven model system (20, 41, 42). However, information obtained from studies with these large algal cells may not relate directly to multicellular, higher plant tissues. For example, *Nitella* does not appear to respond to higher plant hormones, although elongation appears to be regulated by wall acidification (32).

Burg & Burg (10) first proposed that ethylene-induced lateral expansion is mediated through a change in cellulose microfibril orientation. Viewing ethylene-treated cells with the polarizing microscope, they reported the appearance of broad, longitudinal bands assumed to be groupings of cellulose microfibrils similar to those reported by Probine (40) with benzimidazole-treated cells (which were stimulated to expand laterally). Eisinger & Burg (16) reported that these banded regions were not localized sites of deposition of label following incubation of tissues with labeled glucose. Ridge (43) also reported uniform deposition of new wall material following ethylene treatment. The appearance of the banded cells is probably not directly related to ethylene-induced lateral expansion. Careful polarizing microscope analysis of large numbers of both macerated and thin-sectioned, living cells reveal that banded cells can be found in initial (freshly excised) and incubated control tissue. The banding is seen only in small, perhaps immature cells and is probably an optical effect which results from the size and shape of the cell (W. Eisinger, unpublished data). Ethylene treatment which inhibits expansion may merely be maintaining larger numbers of this size cell in the tissue.

Ethylene treatment does indeed affect cellulose microfibril orientation (3, 22, 29, 43, 60, 61). Apelbaum & Burg (3) were first to provide ultrastructural evidence of ethylene-induced longitudinal cellulose microfibrils, although Veen (60, 61) had reached the same conclusion from his work with polarizing light microscopy. Veen induced lateral expansion in pea internode tissue with supraoptimal auxin (which induces ethylene production) rather than treating with ethylene directly. He found that the deposition of longitudinal microfibrils occured even when cell expansion was blocked by 8% sucrose. Thus, the longitudinal orientation is not merely the result of passive reorientation of existing microfibrils caused by the vectorial forces of elongation. Ridge (43), using microscopy techniques similar to those of Veen, confirmed that longitudinal microfibrils can occur independent of cell expansion. Ridge concluded that there was a close link between the effect of ethylene on cell expansion and its effect on microfibrils after she observed that ethylene treatment induced the deposition of longitudinally orientated microfibrils after only 6 hours of treatment. She proposed that long-term ethylene treatment (2 or 4 days) results in deposition of equal layerings of transverse and longitudinal microfibrils. Sargent et al (50) used the electron microscope to confirm much of the work of Ridge, but also reported that ethylene treatment significantly increased total wall thickness of epidermal cells (49, 50). However, wall thickening may not be directly related to the induction of the altered mode of growth, because it was not detected until after 12 hours of treatment (50).

Microtubules are believed responsible for the orientation of the deposition of new cellulose microfibrils (37, 39). Steen & Chadwick (52) and Lang et al (29) have shown that ethylene treatment alters the orientation of microtubules from predominately transverse to longitudinal. After 5 hours of treatment, ethylene-treated cortical cells show a complete reversal of microtubule (and microfibril) polarity (29). The gas does not affect the normal, parallel relationship between microtubules and microfibrils (29).

Ridge (personal communication), treating decapitated seedlings with supraoptimal IAA in lanolin paste, observed lateral expansion 8 to 10 hours before any deposition of longitudinal cellulose microfibrils. The lag period for supraoptimal auxin-induced deposition of longitudinal microfibrils is apparently much slower than that with ethylene (29, 43).

Thus, the data available is generally consistent with the hypothesis that ethylene inhibits elongation and induces lateral expansion by altering the cellulose microfibril orientation (via microtubules). However, if this hypothesis is to be accepted, we must assume that the mode of cell expansion is controlled by only the most recently synthesized (innermost) cellulose microfibrils of the wall (36). This appears to be the case with *Nitella* (41, 42). Deposition of longitudinally orientated microfibrils may be a normal part of cell aging and the subsequent cessation of growth (56). Ethylene treatment may be just hastening the onset of that normal process.

EFFECTS OF ETHYLENE ON CELL WALL BIOCHEMISTRY

Ethylene does not appear to induce any dramatic or sweeping changes in the overall biochemistry of pea internode tissue associated with the onset of lateral cell expansion (9, 16). However, a number of more subtle changes in wall components have been reported. For example, ethylene treatment affects the hydroxyproline-rich protein associated with the cell wall (16, 36, 45, 46, 48). Wall glycoproteins in general have been recently reviewed by Lamport & Catt (28). Eisinger & Burg (16) reported that ethylene treatment inhibited incorporation of label from ^{14}C-proline into that wall fraction during the period when lateral expansion began. Ethylene also lowered the ^{14}C-hydroxyproline/^{14}C-proline ratio of the tissue. The effects seen in these labeling studies may be the result of ethylene affecting free proline pool sizes since the former effect is not seen when the label was supplied at low specific activity. In addition, no differences were found in total proline or hydroxy-proline associated with the wall (36). However, Sadava & Chrispeels (48), using nearly identical tissue preparations, found a 20% increase in wall-associated hydroxyproline after 6 hours of Ethrel (ethylene) treatment. Elevated levels of hydroxyproline (per gram fresh weight) have been reported with ethylene treatment of intact seedlings (45, 46) and excised internodes with intact apices (36); however, no ethylene effect is seen if these data are presented on a per segment basis. Increases in wall hydroxyproline have been correlated with tissue age and the subsequent cessation of elongation (12). Ridge & Osborne (46) proposed that increased wall-associated hydroxyproline may be part of the mechanism of ethylene inhibition of elongation. Sadava & Chrispeels (48) have used the chelator α, α-dipyridyl, an inhibitor of proline hydroxylation, to test their proposal. In the presence of dipyridyl, they found no increase in wall-associated hydroxyproline and growth occurred as if Ethrel were not present. If segments were pretreated with Ethrel to inhibit elongation, they were able to show some reversal with dipyridyl. In other experiments they found no elongation inhibition with supraoptimal auxin when the chelator was present, indicating a complete inhibition of the usual ethylene effect.

Mondal & Nance (33) reported several interesting effects of ethylene on wall composition. A 24 hour ethylene treatment resulted in an increase in weak acid extractable materials and pectic uronic acids, although there was no detectable effect on total hemicellulose or hemicellulosic uronic acids. They proposed that ethylene-induced increases in wall pectic substances might constitute a mechanism by which ethylene alters the mode of cell expansion. Unfortunately, it may not be valid to relate their results directly to the work of others because they chose to work with first (not third) internode tissue from 4-(not 7-)day-old peas.

Terry et al (57) investigated the effects of ethylene on release of wall components in the free space (wall polymer turnover) of internode segments from 7-day-old peas. Ethylene treatment depressed the release of xylose and glucose, suggesting an inhibition of xyloglucan turnover. They reported that such changes could be detected after as little as a 1 hour exposure to the gas. Ethylene affected increases in many other wall components including rhamnose, arabinose, galactose, and hydroxyproline-rich proteins. They proposed that changes in wall xyloglucan turnover are causally related to the effect of ethylene on the mode of cell expansion. Time course studies of the neutral sugars reveal that all decline rapidly in air controls with time and ethylene affects these rates of decline. Auxin stimulates xyloglucan turnover in peas (27, 57), and ethylene is known to affect auxin synthesis and transport, but only over long periods of time (8). Therefore, some of the effects of ethylene on xyloglucan may be indirect.

Ethylene has been reported to affect several enzymes whose activities may be related to the regulation of cell expansion. Lamport & Catt (28) have recently reviewed the literature on wall-related enzymes. Ridge & Osborne (45, 46) reported increases in peroxidases associated with ethylene-induced hydroxyproline rich wall protein synthesis and lateral cell expansion. They found that the wall-bound peroxidase showed a similar DEAE cellulose elution profile to wall-bound hydroxyproline. However, the wall-bound peroxidases do not appear to be acting as proline hydroxylases; the hydroxylation step appears to occur while the proline-containing peptides are still in the cytoplasm. Thus, the site of action of ethylene for affecting the hydroxyproline content of the wall must be the cytoplasm. Supraoptimal auxin did not affect peroxidase activity (49). However, supraoptimal auxin greatly increases cellulase activities of pea internode tissue (18), but ethylene treatment alone does not affect cellulase activities (44, 49). Maclachlan (31) reported only minor effects of ethylene alone on cellulase activities (and only after 3 days of treatment), but ethylene treatment reduced the auxin-enhanced cellulase increase by about half.

Ethylene has been reported to affect several enzymes which could play a role in wall metabolism. Wong & Maclachlan (63) treated intact pea seedlings with ethylene for 3 days and found repressed levels of both 1,3 β-glucanases found in this tissue. Ethylene inhibited the normal auxin stimulation of these enzymes. Shore et al (51) found that a 12 hour ethylene treatment of intact seedlings resulted in about 50% lower activities (compared with untreated controls whose activities of these enzymes nearly doubled) in cell surface and intracellular, alkali-insoluble β-1,4-glucan synthetase (cellulose synthetase). Intracellular activity of this enzyme remained at initial levels and cell surface activity declined slightly. Although ethylene-treated tissues had less total cellulose than equivalent controls, the amount of cellulose per unit fresh weight remained unchanged. Ethylene treatment resulted in a dramatic change in the apparent intracellular localization of the alkali-insoluble β-1,4-glucan synthetase activity. Using sucrose density gradients, Shore et al (51) found a single large peak of activity at density of about 1.15 gm/cc in homogenates from initial (time zero) seedlings. The density profile for control seedlings remained largely unchanged after 12 hours. However, with ethylene treatment, an additional, large peak appeared at a density of about 1.11 gm/cc. A large peak at this density was seen with controls after 48 hours when the growth rate of the tissue is declining. The lower density peak corresponds to the density of endoplasmic reticulum-derived vesicles and the higher density peak corresponds to Golgi-derived vesicles. Shore et al (51) propose that the Golgi-derived vesicles are the source of precursors for the synthesis of transverse microfibrils and the endoplasmic reticulum-derived vesicles are the source of precursors for the synthesis of longitudinal microfibrils. This bold proposal certainly deserves further investigation because of its far-ranging implications to the regulation of cell expansion.

WALL EXTENSIBILITY AND PROTON SECRETION

Many mechanisms have been proposed for the action of ethylene in inhibiting expansion and inducing lateral cell expansion. For nearly all mechanisms, whether they involve cellulose microfibril reorientation (3, 22, 29, 43, 60, 61), xyloglucan turnover (57), or some other mechanism (17, 33, 46, 51), a change in longitudinal wall extensibility is implied. The effects of ethylene on wall extensibility have been measured by several different methods, and the results are generally consistent in showing an ethylene-induced reduction of extensibility. There are three reports in the literature where wall extensibility was measured using the classic Instron technique (13). Although Morré & Eisinger (35) reported no effect of ethylene, Osborne (38) reported a reduction in extensibility after 18 hours of ethylene treatment (but no effect after 3 hours), and Sadava & Chrispeels (48) reported a significant reduction in extensibility after a 5 hour exposure to the gas. Ridge (43), using a plasmometric technique, found about a 50% reduction in elastic extensibility following a 3 hour ethylene treatment. Taiz et al (55) pretreated tissue for 5 hours with ethylene and found no effect of the gas on the longitudinal extensibility of freeze-thaw killed tissue at neutral pH, but a 40% reduction in extensibility at acid pH. They also measured transverse wall extensibility at acid pH's and were able to detect an increase after a 3 hour pretreatment. With a 5 hour pretreatment, transverse extensibility was increased 300%. As with lateral extensibility, researchers were not able to detect an effect at neutral pH. Thus, ethylene appears to affect both longitudinal and transverse acid-induced wall extensibility by about the time when ethylene begins to affect the growth mode of isolated segments.

Wall acidification appears to play a critical role in the ethylene-induced changes in cell expansion. Taiz et al (55) found that treatment with strong neutral buffers or orthovanadate [an inhibitor of proton secretion (24)] inhibited ethylene-induced lateral cell expansion, and they suggest that lateral cell expansion requires an acidified wall just as longitudinal cell expansion does. In other experiments they showed that ethylene pretreatment strongly inhibited the normal burst of elongation when abraded tissue is transferred from neutral to acid buffers. Thus, ethylene pretreatment appears to reduce the in vivo acid-induced, longitudinal extensibility. However, like Barkley & DiFrancesco (6) and Barkley & Rayl (7), Taiz et al (55) were unable to demonstrate acid-induced lateral cell expansion when abraded tissues were transferred from neutral to acid buffers. Although Barkley and his co-workers view the latter experiments as evidence that proton secretion is not required for lateral expansion, these results could be viewed as evidence that lateral expansion requires other factors in addition to an acidified wall. These factors, which may not be present under the conditions of these experiments, could include elevated turgor pressure (17) or differential expansion of certain cell types (29).

Ethylene treatment does not appear to affect proton secretion by the tissue (14, 55). The rate of proton secretion in the presence or absence of auxin is not affected by pretreatment with gas (14, 55), and the kinetics of auxin-stimulated secretion are unaffected (55). De Michelis & Lado (14) also reported that ethylene does not affect fusicoccin-stimulated proton secretion. Ethylene was shown to inhibit fusicoccin-stimulated elongation to the same degree as auxin-stimulated elongation (14).

OTHER AGENTS WHICH INDUCE LATERAL EXPANSION

A comparatively wide variety of agents other than ethylene can inhibit elongation and induce lateral expansion (9, 16, 21, 23, 33, 58). In most cases, when the effects of treatment with these agents were compared to those with ethylene, there were marked similarities with respect to the nature, magnitude, induction kinetics, and biochemistry of these effects (16, 33). Because the mode(s) of action of many of these agents is thought to be known, information from these experiments has been used to try to understand the mechanism of action of ethylene. Colchicine and DCB (2,6-dichlorobenzonitrile) induce a change in the mode of pea internode expansion very similar to that seen with ethylene (16, 17, 36). Colchicine is believed to operate by disrupting wall-associated microtubules, which results in randomly ordered cellulose microfibrils (20). DCB is a specific inhibitor of cellulose synthesis (34). These results, combined with electron microscopic evidence (3, 22, 29), have been used to argue that ethylene is primarily acting by affecting cellulose microfibrils. It should be noted that although colchicine is not known to affect cell wall synthesis (17), DCB could be thought of as directly affecting wall composition by preferentially enriching the noncellulosic portion of the cell wall.

Steen & Chadwick (52) observed that deuterium oxide, which stabilizes microtubules, mimicked many physiological effects of ethylene. Low temperature (6°C), which destabilizes microtubules, reversed many effects of the gas. They also reported partial reversal of ethylene inhibition of elongation with colchicine (I have not been able to repeat these results in my laboratory). Based on these results, Steen & Chadwick (52) proposed that the primary action of ethylene is to overstabilize microtubules, which results in a cellulose microfibril orientation that favors the altered mode of cell expansion seen with ethylene.

Cytokinins at relatively high concentrations (often about 1.0 mM) induce a change in the mode of growth of pea internode segments which is very similar to that seen with ethylene (16, 33, 40). Although cytokinins stimulate ethylene production by the tissue [reviewed by Lieberman (30)], cytokinins apparently induce lateral expansion independently of ethylene (17, 33). Mondal & Nance (33) proposed that cytokinins, as well as calcium ions and ethylene, induce lateral expansion by increasing the pectic fraction of the cell wall.

Application of supraoptimal auxin to intact or decapitated pea seedlings induces lateral cell expansion which persists for several days, resulting in a tumor-like mass of tissue (16, 18, 44). Treatment of excised internode segments with supraoptimal auxin inhibits elongation and induces lateral cell expansion (9, 10, 17, 18, 44). The role of ethylene in this response is a point of controversy. Burg & Burg (9) have proposed that the supraoptimal auxin effect is an indirect one, the result of auxin-induced ethylene production and subsequent response by the tissue. Pea seedlings treated with supraoptimal auxin are little effected by the treatment if grown under hypobaric conditions (5, 8) which are believed to remove endogenous ethylene from the tissue. The differences between the long-term effects of auxin- and ethylene-induced lateral growth with intact seedlings could result from the effects of the gas on auxin metabolism (8). Ethylene-induced lateral expansion requires auxin (49).

In direct opposition, Ridge & Osborne (44) proposed that auxin and ethylene operate by two different mechanisms based on the effects of the two hormones on cellulase activities. Sargent et al (49, 50) found differences in cell wall thickness, microfibril orientation, and hydroxyproline, as well as peroxidase and cellulase activities with the two hormones. During the first 8 to 10 hours of auxin-induced lateral expansion, no longitudinal microfibrils were detected (I. Ridge, unpublished data). Longitudinal microfibrils do appear later in response to auxin treatment, but their deposition may not be mediated via ethylene because they are unaffected by hypobaric conditions (I. Ridge, unpublished data). Ethylene-induced inhibition of cell expansion in excised internode segments is not reversible as is the inhibition with supraoptimal auxin (W. Eisinger, unpublished data).

Whether auxin and ethylene operate by the same or different mechanisms remains an open question and one deserving of further study. Future studies should concentrate on comparing the early biochemical and physiological changes induced by the two hormones because these are probably the ones most directly related to the mechanism(s) of action. In studies with intact or decapitated seedlings, more "physiological" concentrations of auxin (preferably a stable form like 2,4-D) should be used to avoid artifacts unrelated to its hormonal action (11).

In his recent review, Lieberman (30) noted several instances of an antagonism (or "buffering") between auxin and ethylene in the regulation of plant growth and development. In reviewing the literature on lateral cell expansion, many reports appear which could be viewed as cases of antagonism with respect to cell wall morphology (49, 50), cell wall composition (16, 33, 48) and turnover (57), and enzyme activities (31, 49, 50, 63). However, in many other processes, such as auxin-stimulated RNA metabolism (16), induction of lateral expansion (15) and uptake of many labeled solutes (16) ethylene had no detectable affect.

Stewart et al (53) have reported an antagonism between gibberellic acid (GA) and ethylene in regulation of internode elongation of intact pea seedlings; ethylene exposure retards GA-stimulated elongation. If seedlings were pretreated with both ethylene and GA, seedlings treated with GA recovered more quickly when the gas was removed than did equivalent water controls.

CONCLUSIONS

There are two critical questions which remain to be answered about ethylene-induced inhibition of elongation and promotion of lateral expansion: By what mechanism(s) does ethylene bring about this altered mode of growth? Is this the same mechanism by which other agents (especially supraoptimal auxin) bring about lateral cell expansion? All proposed mechanisms of action center on ethylene having its ultimate effect on the mechanical properties of cell wall, via altered cellulose microfibril orientation (3, 22, 29, 43, 60, 61), cross-linking within the wall (16, 45, 46, 48, 57), or enrichment of the pectic fraction (33). Eisinger et al (17) have proposed that ethylene operates by a complex mechanism involving both altered microfibrils and wall composition. This model seems to be consistent with nearly all of the data reported in the literature. Their specific application was for excised internode segments, but the basic elements could be applied to the intact system. In the intact system, changes in wall composition appear to occur more rapidly than with excised segments, and changes in composition must be more important than changes in microfibril orientation (I. Ridge, personal communication). The presence of the apical hook may be responsible for these differences although it is not clear just how that might occur.

It is important to note that lateral expansion is strongly favored by the geometry of the cell and, therefore, can be thought of as the release from constraint. There must be a number of mechanisms by which this constraint can be released, because such a wide variety of agents and conditions can be used to induce lateral expansion. Those agents such as ethylene which actually restrain elongation [decreased wall extensibility (54)] may also accentuate lateral expansion through increases in osmotic concentration [turgor pressure (17)].

An area of investigation that has been largely ignored is the concentration dependence of the various agents on the lag time of the response. Over 10 years ago Warner & Leopold (62) reported a nearly threefold reduction in the inhibition lag time with a mere doubling of ethylene concentration. Since that time no one has followed up these studies. It would be interesting to see whether other agents show such a concentration dependency, especially comparing those agents believed to primarily affect wall composition with those believed to primarily affect microfibril orientation.

Another area which deserves future study is the possibility of interaction among the various agents which induce lateral expansion. Colchicine and ethylene have been reported to have a negative interaction (52), and supraoptimal auxin and ethylene have been reported to have no interaction (16). It would be very interesting to observe the kinetics of lateral expansion in the presence of pairs of agents, one of which primarily affects wall composition and the other which primarily affects microfibril orientation. Such experiments might give us a better appreciation of the relative contribution of the two agents when compared with the kinetics of each agent applied separately.

During the past 10 years a considerable amount of solid work has been done toward understanding the mechanism of action of ethylene in the regulation of cell expansion. Although ethylene has been shown to affect several significant cellular processes, many of these changes were measured at time periods long after the altered growth pattern had begun. If cause-effect relationships are to be established, such changes must be correlated with the induction kinetics. These studies could lay the foundation for establishing the mechanism of action of ethylene.

ACKNOWLEDGMENTS

I wish to thank Research Corporation and the University of Santa Clara for their continued support of my research. I also wish to thank Dr. Lincoln

Taiz for his critical reading of an earlier draft of this manuscript. I am grateful to Kitty Sherburne and Joan Goetze for their assistance in preparing the manuscript.

Literature Cited

1. Abeles, F. B. 1972. Biosynthesis and mechanism of action of ethylene. *Ann. Rev. Plant Physiol.* 23:259–92
2. Abeles, F. B. 1973. *Ethylene in Plant Biology.* New York: Academic. 320 pp.
3. Apelbaum, A., Burg, S. P. 1971. Altered cell microfibril orientation in ethylene-treated *Pisum sativum* stems. *Plant Physiol.* 48:648–52
4. Apelbaum, A., Burg, S. P. 1972. Effect of ethylene on cell division and deoxyribonucleic acid synthesis in *Pisum sativum. Plant Physiol.* 50:117–24
5. Apelbaum, A., Burg, S. P. 1972. Effects of ethylene and 2,4-dichlorophenoxyacetic acid on cellular expansion in *Pisum sativum. Plant Physiol.* 50:125–31
6. Barkley, G. M., DiFrancesco, A. 1981. Effects of ethylene on pea epicotyl tissue II. The response of pea epicotyl tissue to auxin and hydrogen ions after ethylene treatment. *Ohio J. Sci.* 81:116 (Abstr.)
7. Barkley, G. M., Rayl, D. R. 1981. Kinetics of ethylene-induced radial expansion in pea epicotyls and modification of the auxin, hydrogen-ion and carbon dioxide fast-growth responses. *Plant Physiol.* 67:99 (Abstr.)
8. Burg, S. P. 1973. Ethylene in plant growth. *Proc. Natl. Acad. Sci. USA* 70:591–97
9. Burg, S. P., Burg, E. A. 1966. The interaction between auxin and ethylene and its role in plant growth. *Proc. Natl. Acad. Sci. USA* 55:262–96
10. Burg, S. P., Burg, E. A. 1967. Auxin stimulated ethylene formation: its relationship to auxin inhibited growth, root geotropism and other plant processes. In *Biochemistry and Physiology of Plant Growth Substances,* ed. F. Wightman, G. Setterfield, pp. 1275–94. Ottawa: Runge
11. Chadwick, A. V., Burg, S. P. 1970. Regulation of root growth by auxin-ethylene interaction. *Plant Physiol.* 45:192–200
12. Cleland, R. 1968. Distribution and metabolism of protein-bound hydroxyproline in an elongating tissue, the *Avena* coleoptile. *Plant Physiol.* 43:865–70
13. Cleland, R. 1971. Cell wall extension. *Ann. Rev. Plant Physiol.* 22:197–222
14. De Michelis, M. I., Lado, P. 1973. Effect of ethylene on auxin- and fusicoccin-induced growth of pea internode segments and on the secretion of H+ in the incubation medium. *Rend. Accad. Naz. Lincei* 56:808–13
15. Eisinger, W. R. 1971. *Role of ethylene in the pea internode swelling response.* PhD thesis. Univ. Miami, Coral Gables, Fla.
16. Eisinger, W. R., Burg, S. P. 1972. Ethylene-induced pea swelling: its relation to ribonucleic acid metabolism, wall protein synthesis and cell wall structure. *Plant Physiol.* 50:510–17
17. Eisinger, W. R., Croner, L., Taiz, L. 1983. Ethylene-induced lateral expansion in etiolated pea epicotyls: analysis of kinetics. *Plant Physiol.* In press
18. Fan, D., Maclachlan, G. A. 1967. Massive synthesis of RNA and cellulose in the pea epicotyl in response to indoleacetic acid with and without concurrent cell division. *Plant Physiol.* 42:1114–23
19. Frey-Wyssling, A. 1976. *The Plant Cell Wall.* Berlin: Borntraeger
20. Green, P. B. 1963. On mechanisms of elongation. In *Cytodifferentiation and Macromolecular Synthesis,* ed. M. Locke, pp. 203–34. New York: Academic
21. Hara, M., Umetsu, M., Miyamoto, C., Tamari, K. 1973. Inhibition of the biosynthesis of plant cell wall materials, especially cellulose biosynthesis, by coumarin. *Plant Cell Physiol.* 14:11–28
22. Henry, E. W. 1978. An ultrastructural study of ethylene-treated stem segments of Alaska pea (*Pisum sativum*). *Cytologia* 43:423–32
23. Hogetsu, T., Shibaoka, H., Shimokoriyama, M. 1974. Involvement of cellulose synthesis in actions of gibberellin and kinetin on cell expansion. Gibberellin-coumarin and kinetin-coumarin interactions on stem elongation. *Plant Cell Physiol.* 15:262–72
24. Jacobs, M., Taiz, L. 1980. Vanadate inhibition of auxin-enhanced H+ secretion and elongation in pea epicotyls and oat coleoptiles. *Proc. Natl. Acad. Sci. USA* 77:7242–46
25. Kang, B. G., Burg, S. P. 1972. Relation of phytochrome-enhanced geotropic

sensitivity to ethylene production. *Plant. Physiol.* 50:132–35

26. Kang, B. G., Burg, S. P. 1973. Influence of ethylene on nucleic acid synthesis in etiolated *Pisum sativum. Plant Cell Physiol.* 14:981–88

27. Labavitch, J. M., Ray, P. M. 1974. Turnover of cell wall polysaccharides in elongating pea stem segments. *Plant Physiol.* 53:669–73

28. Lamport, D. T. A., Catt, J. W. 1981. Glycoproteins and enzymes of the cell wall. In *Plant Carbohydrates,* ed. W. Tanner, F. A. Loewus, 2:133–65. *Encyclopedia of Plant Physiology,* Vol. 13B

29. Lang, J. M., Eisinger, W. R., Green, P. B. 1962. Effects of ethylene on the orientation of microtubules and cellulose microfibrils of pea epicotyl cells with polylamellate cell walls. *Protoplasm* 110:5–14

30. Lieberman, M. 1979. Biosynthesis and action of ethylene. *Ann. Rev. Plant Physiol.* 30:533–91

31. Maclachlan, G. A. 1977. Cellulose metabolism and cell growth. In *Plant Growth Regulation,* ed. P. I. Pilet, pp. 13–20. Heidelberg: Springer

32. Metraux, J-P., Richmond, P. A., Taiz, L. 1980. Control of cell elongation in *Nitella* by endogenous cell wall pH gradients. Multiaxial extensibility and growth studies. *Plant Physiol.* 65: 204–10

33. Mondal, M. H., Nance, J. F. 1975. Effects of ethylene, kinetin, and calcium on growth and wall composition of pea epicotyls. *Plant Physiol.* 55:450–54

34. Montezinos, D., Delmer, D. P. 1980. Characterization of inhibitors of cellulose synthesis in cotton fibers. *Planta* 148:305–11

35. Morré, D. J., Eisinger, W. R. 1968. Cell wall extensibility: its control by auxin and relationship to cell elongation. See Ref. 10, pp. 625–45

36. Nee, M., Chiu, L., Eisinger, W. 1978. Induction of swelling in pea internode tissue by ethylene. *Plant Physiol.* 62: 902–6

37. Newcomb, E. H. 1969. Plant microtubules. *Ann. Rev. Plant. Physiol.* 20:253–88

38. Osborne, D. J. 1977. Auxin and ethylene and the control of cell growth. The identification of three classes of target cells. See Ref. 31, pp. 161–71

39. Pickett-Heaps, J. C. 1967. The effects of colchicine on the ultrastructure of dividing cells, xylem wall differentiation, and distribution of cytoplasmic microtubules. *Dev. Biol.* 15:206–36

40. Probine, M. C. 1964. Chemical control of plant cell wall structure and cell shape. *Proc. R. Soc. London Ser. B* 161: 526–37

41. Richmond, P. A. 1977. *Control of plant cell morphogenesis by the cell wall: analysis in Nitella.* PhD thesis. Univ. Penn., Philadelphia, Pa.

42. Richmond, P. A., Metraux, J-P., Taiz, L. 1980. Cell expansion patterns and directionality of wall mechanical properties in *Nitella. Plant Physiol.* 65:211–17

43. Ridge, I. 1973. The control of cell shape and rate of cell expansion by ethylene: effects on microfibril orientation and cell wall extensibility in etiolated peas. *Acta. Bot. Neerl.* 22:144–58

44. Ridge, I., Osborne, D. J. 1969. Cell growth and cellulose: regulation by ethylene and indole-3-acetic acid in shoots of *Pisum sativum. Nature* 223:318–19

45. Ridge, I., Osborne, D. J. 1970. Regulation of peroxidase activity by ethylene in *Pisum sativum:* regulation by ethylene. *J. Exp. Bot.* 21:843–56

46. Ridge, I., Osborne, D. J. 1971. Role of peroxidase when hydroxyproline-rich protein in plant cell walls is increased by ethylene. *Nature New Biol.* 229:205–8

47. Roelofsen, P. A. 1965. Ultrastructure of the wall of growing cells in relation to the direction of growth. *Adv. Bot. Res.* 2:69–149

48. Sadava, D., Chrispeels, M. J. 1973. Hydroxyproline-rich cell wall protein (extensin): Role in the cessation of elongation in excised pea epicotyls. *Dev. Biol.* 30:49–55

49. Sargent, J. A., Atack, A. V., Osborne, D. J. 1973. Orientation of cell growth in the etiolated pea stem, effects of ethylene and auxin on wall deposition. *Planta* 109:185–92

50. Sargent, J. A., Atack, A. V., Osborne, D. J. 1974. Auxin and ethylene control of growth in epidermal cells of *Pisum sativum:* A biphasic response to auxin. *Planta* 115:213–25

51. Shore, G., Raymond, Y., Maclachlan, G. A. 1975. The site of cellulose synthesis, cell surface and intracellular β-1,4-glucan (cellulose) synthetase activities in relation to the stage and direction of cell growth. *Plant Physiol.* 56:34–38

52. Steen, D. A., Chadwick, A. V. 1981. Ethylene effects in pea stem tissue, evidence of microtubule mediation. *Plant Physiol.* 67:460–66

53. Stewart, R. N., Lieberman, M., Kunishi, A. T. 1974. Effects of ethylene and gibberellic acid on cellular growth and development in apical and subapical re-

gions of etiolated pea seedling. *Plant Physiol.* 54:1–5

54. Taiz, L., Metraux, J-P. 1979. The kinetics of bidirectional growth of stem sections from etiolated pea seedlings in response to acid, auxin and fusiococcin. *Planta* 146:171–78

55. Taiz, L., Rayle, D. L., Eisinger, W. 1983. Ethylene-induced lateral expansion in etiolated pea epicotyls: the role of acid secretion. *Plant Physiol.* In press

56. Takeda, K., Shibaoka, H. 1981. Changes in microfibril arrangement on the inner surface of the epidermal cell walls in the epicotyl of *Vigoa angularis* Ohwi et Ohiashi during cell growth. *Planta* 151:385–92

57. Terry, M. E., Rubinstein, B., Jones, R. L. 1981. Soluble cell wall polysaccharides released from pea stems by centrifugation II. Effect of ethylene. *Plant Physiol.* 68:538–42

58. Umetsu, N., Satoh, S., Masuda, K. 1976. Effects of 2,6-dichlorobenzonitrile on suspension-culture soybean cells. *Plant Cell Physiol.* 17:1071–75

59. van der Laan, P. A. 1934. Der einfluss von Aethylen auf die Wachsstoffbilduns bei *Avena* und *Vicia. Rec. Trav. Bot. Neerl.* 31:691

60. Veen, B. W. 1970. Orientation of microfibrils in parenchyma cells of pea stem before and after longitudinal growth. *K. Ned. Akad. Wet.* 73:113–17

61. Veen, B. W. 1970. Control of plant cell shape by cell wall structure. *K. Ned. Akad. Wet.* 73:118–21

62. Warner, L. N., Leopold, A. C. 1971. Timing of growth regulator responses in peas. *Biochem. Biophys. Res. Commun.* 44:989–94

63. Wong, Y-S., Maclachlan, G. A. 1980. 1,3-β-D-Glucanases from *Pisum sativum* seedlings. *Plant Physiol.* 65:222–28

Ann. Rev. Plant Physiol. 1983. 34:241–78

CHLOROPHYLL BIOSYNTHESIS:
Recent Advances and Areas of Current Interest

Paul A. Castelfranco

Department of Botany, University of California, Davis, California 95616

Samuel I. Beale

Division of Biology and Medicine, Brown University, Providence, Rhode Island 02912

CONTENTS

241

INTRODUCTION[1]

The field of chlorophyll biosynthesis was last reviewed in this series 10 years ago (183). The progress that has occurred in the intervening decade has been recorded in several general reviews (19, 26, 47, 53, 54, 107, 111, 123, 124) and in specialized articles dealing with the origin of plant tetrapyrrole precursors (25, 112), the common porphyrin pathway (48, 120), regulation of the branched pathway (141), and the state of protochlorophyll in cells (215).

The present review is not intended to be comprehensive. Rather, several topics have been selected on the basis of current research activity or recent breakthroughs in understanding. Other topics are discussed with the goals of pointing out major information gaps and suggesting important targets for future research. Coverage is restricted to higher plant chlorophylls of the *a* and *b* types, and readers are directed elsewhere for information on the *c* type algal chlorophylls (26, 47) and the bacteriochlorophylls (123, 141). Attention is drawn to two recent comprehensive reviews (26, 53) which have served as our starting point in this endeavor.

OUTLINE OF THE PATHWAY

The earliest Chl precursor that has been identified unequivocally is ALA, a 5-carbon compound that arises from condensation of succinyl-CoA and glycine in animals and bacteria. Plants and algae form ALA by another route, utilizing the intact carbon skeleton of glutamic acid. All of the carbon and the nitrogen atoms in the porphyrin nucleus of Chls are derived from ALA.

The pathway from ALA to Chl *a* is illustrated in Figure 1. Assembly of the tetrapyrrole skeleton begins by condensation of groups of two ALA molecules to form PBG units, which are the first pyrrolic intermediates in the pathway. Next, four PBG units are linked together in a head-to-tail sequence to produce an unstable linear tetrapyrrole. The linear molecule is then enzymatically closed to form the first cyclic tetrapyrrole, uroporphyrinogen III. This isomer has pyrrole ring D inverted, so that the sequence of substituents at the ß positions is reversed with respect to those on the other rings. Conversion to coproporphyrinogen III is accomplished by decarboxylation of the four acetic acid substituents, leaving methyl groups, and then two of the four propionic acid moieties (on rings A and B) are oxidatively decarboxylated to vinyl groups, forming protoporphyrinogen IX, which has vinyl substituents at these positions. Removal of six electrons from the porphyrinogen macrocycle con-

---

[1]*Abbreviations used:* ALA, δ-aminolevulinic acid; Chl(ide), chlorophyll(ide); DOVA, γ,δ-dioxovaleric acid; GSA, Glutamic acid-1-semialdehyde; MgDVP, Mg-2,4-divinylpheoporphyrin a_5; Mg-Proto(-Me), Mg-protoporphyrin IX (6-methyl ester); PBG, porphobilinogen; Pchl(ide), protochlorophyll(ide); Proto, protoporphyrin IX.

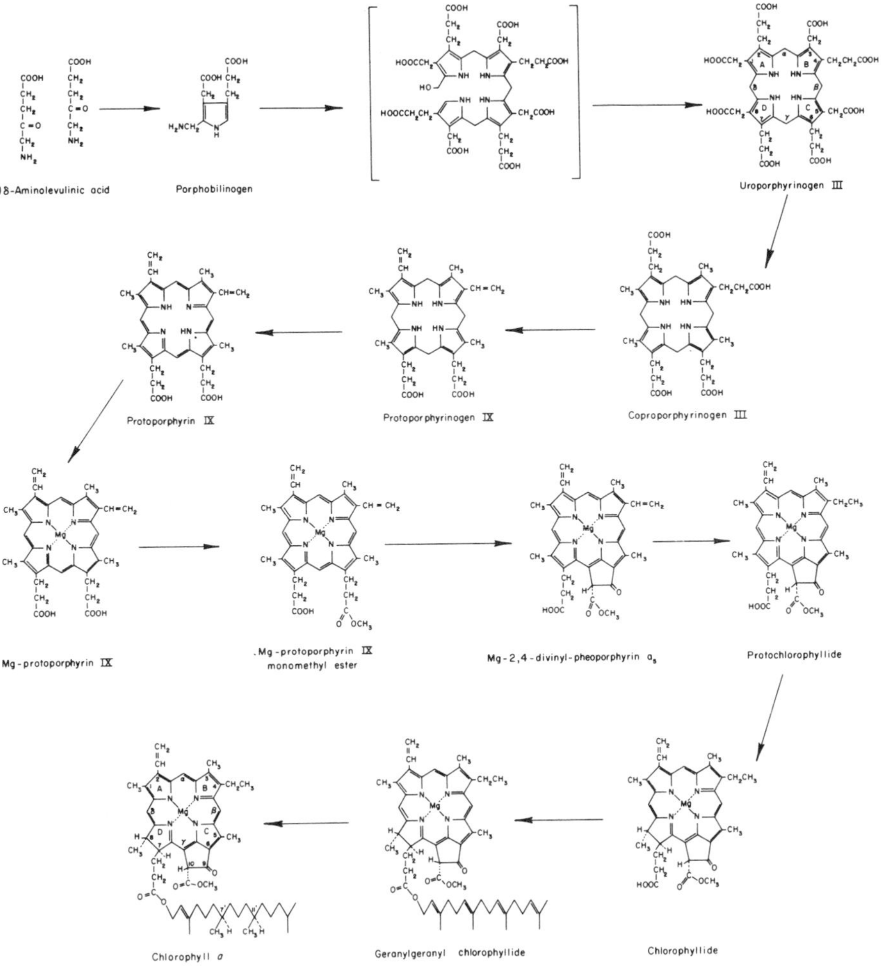

Figure 1. The pathway from ALA to Chl *a* with the principal intermediates. Uroporphyrinogen III and Chl *a* are drawn with the conventional capital letter designations for the pyrrole rings, the number designations for the substituent positions, and Greek lower-case letter designations for the meso bridge positions. Reproduced from (53) with permission of the publisher.

fers aromatic properties to Proto. Proto is the last common precursor to hemes, bilins, and chlorophylls.

The Chl branch of the pathway begins with insertion of magnesium into the Proto nucleus. This is followed by methylation of the propionic acid group on ring C. Next, in an unknown order, the vinyl group on ring B is reduced to an ethyl group, and the newly methylated propionic acid group is oxidized at its ß position to a carbonyl and joined at its α-carbon to the γ-meso carbon of the porphyrin, forming Pchlide, which contains a fifth, isocyclic, ring that is present in all Chls. Conversion of Pchlide to Chlide usually requires light in angiosperms, although some algae and higher plant tissues are able to carry out Pchlide reduction in the absence of light.

The final step of Chl *a* formation involves addition of the long chain polyisoprene phytyl moiety. This process is initiated by esterification of the propionic acid on ring D with geranylgeraniol (activated as the pyrophosphate ester) and subsequent reduction of the geranylgeranyl group to phytyl. The available evidence suggests that Chl *b* is derived directly from Chl *a* by oxidation of the methyl group on ring B to a formyl group.

ALA FORMATION

The diversion of the general metabolic intermediates toward the tetrapyrrole pathway is accomplished by enzymes that catalyze the formation of ALA, the first recognized compound that is committed to tetrapyrrole formation. The mechanism of ALA synthesis is currently one of the least understood and most controversial aspects of the pathway; no fewer than three proposed ALA biosynthetic routes (Figure 2) are being investigated in a number of laboratories.

From Glycine and Succinyl-CoA

ALA formation was first studied in avian erythrocyte preparations which have heme as the major tetrapyrrole end product (100) and in anaerobically grown photosynthetic bacteria, which form large quantities of bacteriochlorophyll as well as lesser amounts of heme and corrinoids (134, 196). In both of the experimental systems, ALA is formed by the condensation of succinyl-CoA and glycine, mediated by the pyridoxal phosphate-requiring enzyme, ALA synthase. In the reaction, the carboxyl carbon of glycine is lost as CO_2.

Although ALA synthase activity has been detected in extracts from a wide range of bacterial, animal, and fungal cells, its presence in plants and algae has remained doubtful. Certain atypical plant cells, such as nongreening callus cultures (14, 15, 225, 226) and green peels of cold-stored potatoes (176–178), have been reported to contain ALA synthase. Indirect evidence supporting the existence of ALA synthase has been reported in dark-grown barley (158), some algae (139, 140, 159, 173), and one moss (112). These studies are based on the relative label incorporation rates into ALA from specifically labeled exogenous presumptive substrates in vivo. It is not known whether the ALA that was labeled from ^{14}C-glycine or ^{14}C-succinate represented the pool that serves as precursor to Chl or whether, instead, this pool of ALA might be destined exclusively for nonplastid cytochrome heme. Until recently, ALA synthase had not been detected by direct in vitro methods in any greening photosynthetic plant or algal (including bluegeen algal) cell extract.

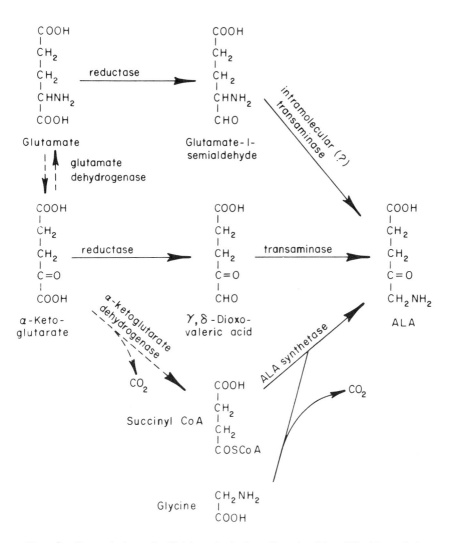

Figure 2.   Proposed schemes for ALA formation in plants. Reproduced from (53) with permission of the publisher.

Recently, in vitro ALA synthase activity was detected in extracts of the green photosynthetic phytoflagellate *Euglena gracilis* (30). The physical and kinetic properties of the enzyme and the regulation of its activity were examined in detail (29, 75, 87). The investigators concluded that in *Euglena*, ALA synthase functions to provide tetrapyrrole precursors exclusively for nonplastid porphyrins. This conclusion was based on the following results: (*a*) the enzyme was present at high activity in extracts of dark-grown wild-type cells, which have no

Chl, and in aplastidic mutant cells, which are devoid of all plastid components, including DNA; (b) when ALA synthase was induced to high levels by specific inhibitors of heme formation, Chl formation remained unaffected; (c) the activity of ALA synthase in cell extracts was inversely related to the rate of Chl formation in the extracted cells, suggesting that ALA synthase functions to provide nonplastid ALA during occasions when the cells are not capable of forming ALA via another mechanism that is present in greening plastids. ALA synthase activity has also been reported recently to be present in extracts of *Scenedesmus obliquus* pigment mutants, but the reports were too brief to permit evaluation of the nature of the reaction and product (118, 128).

From Five-Carbon Precursors

ALA can be made to accumulate in vivo in greening plant and algal cells by treatment with levulinic acid (24, 27) or 4,6-dioxoheptanoic acid (87, 157), which are structural analogs of ALA and competitive inhibitors of ALA dehydrase. ALA that accumulates in the presence of these inhibitors is found to incorporate label from ^{14}C-labeled exogenous compounds. Among tested compounds, ^{14}C-labeled glycine and succinate were able to contribute label to ALA only very poorly, while the five-carbon compounds glutamate, α-ketoglutarate, and glutamine were much better contributors of label to ALA (28, 127, 156, 178). Based on the uniform degree of incorporation of all of the carbons of glutamate, and the carbon-to-carbon correspondence of label position in precursor and product molecules, a number of hypothetical routes were proposed for the transformation of the intact carbon skeleton of glutamate or α-ketoglutarate into ALA (31, 156). Preferential and specific transfer of label from glutamate or α-ketoglutarate to ALA formed in vivo, in the presence of levulinic acid, has now been found in a variety of plant cell types, including bluegreen (13, 135, 159), red (127), and green (159) algae, and many higher plant tissues (28, 31, 156, 178).

It appears that ALA derived from five-carbon sources is the precursor of most or all tetrapyrroles in plants and most algae. In *Cyanidium caldarium*, glutamate specifically labels the small amounts of ALA that accumulate in the presence of levulinic acid, even in mutant cells that have lost the ability to form Chl and phycobilins (212). Castelfranco & Jones (55) reported that in greening barley leaves, glutamate was superior to glycine in supplying label to both Chl and newly formed heme. Similarly, ^{14}C-labeled α-ketoglutarate was superior to glycine or succinate in contributing label to phycocyanobilin in growing cultures of *Anacystis nidulans* (144). Based on the relative abilities of C_1- or C_2-labeled acetate to contribute label to Chl and glutamate in *Synechococcus* 6301, McKie et al (155) concluded that the 5-carbon pathway operates exclusively in tetrapyrrole precursor formation. Finally, Oh-hama et al (168) performed ^{13}C-NMR analysis on Chl formed from ^{13}C-labeled glycine or glutamate in *Scenedesmus obliquus*. Glycine was shown to contribute label only to the methoxyl group adjacent to the isocyclic ring, while glutamate contributed label in a manner that was consistent with the exclusive operation of the C_5 pathway in ALA formation. This conclusion was confirmed by Klein & Porra (138).

Of the many mechanisms postulated for the incorporation of the intact carbon skeleton of glutamate into ALA, two have been investigated in considerable detail.

VIA DOVA   Conversion of α-ketoglutarate is postulated to proceed via the intermediate compound DOVA, which could be formed by reduction of C_1 (Figure 2). Subsequent transamination of the intermediate ketoaldehyde at C_1 yields ALA. The second step, transamination of DOVA, has been demonstrated in extracts from a number of higher plants and algae (86, 98, 99, 192. 193), and it occurs even in bacterial and animal cell extracts that have ALA synthase activity as well (101, 136, 163, 214). Although ^{14}C-labeled α-ketoglutarate was incorporated into ALA by extracts of greening spinach (103), wheat (149), maize (113), and *Euglena* chloroplasts (191, 193), the initial step in this scheme, the formation of DOVA from α-ketoglutarate, has not been directly observed in cell extracts. Traces of ^{14}C-DOVA were detected in greening *Scenedesmus obliquus* cells that were incubated with 1-^{14}C-glutamate in the presence of levulinic acid (66). This result, however, could have been caused by catabolic deamination of ALA that was formed in some other way, and that accumulated to high levels in the presence of levulinic acid.

DOVA transaminase from *Euglena gracilis* was examined with respect to physical, kinetic, and regulatory properties (86). It was concluded that in this organism, DOVA transaminase is a subsidiary reaction catalyzed by glyoxylate transaminase and has no relation to ALA formation in vivo. The conclusion was based on the following observations: *(a)* wild-type and aplastidic mutant cells had identical levels of DOVA transaminase whether grown in the light or dark; *(b)* DOVA transaminase was inhibited competitively by glyoxylate; *(c)* DOVA transaminase and glyoxylate transaminase co-purified on gel-filtration and DEAE-cellulose column chromatography; *(d)* glyoxylate transaminase activity was at least 100-fold higher than DOVA transaminase, even in those fractions exhibiting the highest DOVA transaminase activity.

VIA GSA   ALA formation from glutamate has been reported in extracts from *Euglena* (191, 193), greening cucumber cotyledons (220), and leaves from barley (104), wheat (89), spinach (129), and maize (160). In this route, a proposed intermediate is GSA (130; Figure 2). This compound could arise from reduction of C_1 of glutamate; subsequent removal of the amino group from C_2 and addition of (the same?) amino group at C_1 would yield ALA. Chemically synthesized GSA is efficiently converted to ALA by plastid extracts from greening barley (130, 131). The conversion required no other nitrogen donor. A radioactive spot that co-chromatographed with GSA was detected in barley extracts that were incubated with ^{14}C-glutamate plus ATP, NADPH, Mg^{2+}, and glyoxylate (132). Wang et al (218) have reported the resolution of barley plastid preparations into a component that can convert GSA to ALA, and another component that converts glutamate to some intermediate which is transformed to ALA by the first enzymic component. Neither the chemically synthesized GSA nor the plastid extract incubation product has been subjected to rigorous structural analysis.

ALA formation from glutamate by intact plastids appears to require ATP, Mg^{2+}, and NADPH, but not pyridoxal phosphate (104, 220), whereas formation of ALA from α-ketoglutarate was reported to require Mg^{2+}, NADH rather than NADPH, pyridoxal phosphate, and amino donor (alanine), but not ATP (113). Finally, the α-ketoglutarate system is two to three times more active in dark-grown tissue than in tissue that has been greened for several hours (113), while the glutamate system has very low activity in barley and cucumber etioplasts, is very active in greening chloroplasts, and is much less active in chloroplasts isolated from fully greened leaves (131, 219).

In summary, the key initial steps of the Chl biosynthetic pathway are only poorly understood. Available evidence suggests that the ALA synthase step which is employed by animals, bacteria (including phtosynthetic bacteria), and *Euglena* is not involved in the biosynthesis of ALA that is destined for Chl in plants and algae. Although rapid progress is being made toward elucidating the pathway from five-carbon precursors to ALA, the details of this transformation are still uncertain.

ALA TO PROTO

The steps of the Chl biosynthetic pathway that are shared with the pathway to hemes operating in nonphotosynthetic organisms, i.e. from ALA to Proto, are now understood in considerable detail. Recent findings have been discussed in detail elsewhere (26, 53), and are only listed briefly here. These include: *(a)* determination of the order of addition of the two ALA molecules to form PBG by ALA dehydrase (126); *(b)* identification of a linear hydroxymethylbilane porphyrinogen precursor (18, 201) which is the product of condensation of four PBG units in ring order A, B, C, D (18, 125) by the enzyme PBG deaminase; *(c)* characterization of the reaction catalyzed by uroporphyrinogen III cosynthase as being inversion of the D ring of the linear hydroxymethylbilane, followed by macrocycle closure to yield the correct type III isomer of uroporphyrinogen (17); *(d)* determination that the order of the pyrrole ring acetic acid decarboxylations in the conversion of uroporphyrinogen III to coproporphyrinogen III by uroporphyrinogen decarboxylase is A,B,C,D (120); and that *(e)* the ring order of the two propionic acid oxidative decarboxylations to vinyl groups in the conversion of coproporphyrinogen III to protoporphyrinogen IX by coproporphyrinogen oxidase is A then B (96); and finally *(f)* detection of protoporphyrinogen oxidase, a specific enzymatic activity that catalyzes aromatization of protoporphyrinogen IX by removal of six electrons to form Proto (121, 174).

All of the steps from ALA to Proto have been reported to occur in the soluble stroma phase of developing plastids (208). This point should be accepted with caution until confirmed, because in bacteria, yeast, and animal cells, both coproporphyrinogen oxidase (229) and protoporphyrinogen oxidase (174) are membrane-associated enzymes.

Mg CHELATION

Incorporation of ^{14}C-ALA into Mg-Proto-Me by intact etiolated cucumber cotyledons and homogenates of the cotyledons was reported by Rebeiz et al (179), and of ^{28}Mg into Mg-Proto in extracts of etiolated wheat seedlings by Ellsworth & Lawrence (82). Smith & Rebeiz (207) reported Proto-dependent formation of Mg-Proto plus other Mg-porphyrins by chloroplasts derived from etiolated or greening cucumber cotyledons. The identity of the product was inferred from spectrofluorometric data only. Mg chelatase activity in cucumber cotyledon plastid preparations was recently established with much greater certainty by Pardo et al (171), Richter & Rienits (185, 186), and Fuesler et al (95). Whereas earlier studies were hampered by high endogenous levels of Mg^{2+} and Mg-Proto, and by considerable Zn chelatase activity, pretreatment of the plastids with EDTA eliminated these difficulties. The product was identified as Mg-Proto by low-temperature spectrofluorometry and high-pressure liquid chromatography. Activity depended on added ATP, Mg^{2+}, and Proto (57). Maximal activity occurred at 10 mM ATP, 10 mM Mg^{2+}, and 10 µM Proto (95). The ATP requirement was not satisfied by other nucleoside triphosphates (171). Because the reaction in vitro required intact plastids, activity could not be assigned to soluble or membrane phases of the plastids, nor could meaningful K_m values be measured, although half-maximal activities occurred at 2 mM Mg^{2+}, 3.5 mM ATP, and 3.5 µM Proto.

Membrane preparations obtained from osmotically shocked cucumber cotyledon plastids have been reported to synthesize Mg-Proto at rates which are very low compared to those of intact plastids (186). There was an absolute requirement for ATP, and maximal rates of Mg-Proto formation were obtained at 0.1 µM Proto, a considerably lower concentration than required for intact plastids. In view of the low rates of Mg-Proto formation, however, it is difficult to exclude the possibility that the observed activity was caused by a small fraction of intact plastids or plastid vesicles that had escaped total disruption.

ISOCYCLIC RING FORMATION

The first step in the formation of the isocyclic ring from the 6-propionate moiety of Mg-Proto is esterification of the carboxylic acid, which is necessary to prevent later spontaneous decarboxylation when a keto group is introduced at the β position of the propionate. A specific S-adenosyl-L-methionine: Mg-Proto methyl transferase has been detected in several plant and algal extracts (76, 84, 175).

Ellsworth & Aronoff (79, 80) described mutant strains of *Chlorella* that accumulated Mg-porphyrins containing acrylic, β-hydroxypropionic, and β-ketopropionic groups. These correspond to the likely intermediates in the conversion of the 6-propionic ester into the isocyclic ring.

The conversion of Mg-Proto and Mg-Proto-Me to MgDVP by a preparation of developing chloroplasts was studied by Castelfranco and co-workers (59, 60, 228). The product was identified by chromatographic and spectrofluorometric comparison with authentic MgDVP obtained from etiolated cucumber cotyledons (38) and *Rhodopseudomonas spheroides* mutant V-3 cells (109). More recently, the identity of the product was confirmed by NMR and secondary ion mass spectroscopy (B. M. Chereskin and P. A. Castelfranco, in preparation).

MgDVP formation showed an absolute requirement for molecular oxygen. The reaction was also totally dependent on added S-adenosyl-L-methionine when Mg-Proto was used as a substrate. Even with Mg-Proto-Me, the conversion was stimulated by S-adenosyl-L-methionine because of the presence of a methyl esterase that converted Mg-Proto-Me to Mg-Proto during the incubation. Mg-Proto then had to be remethylated before MgDVP could be formed. NADP(H) was stimulatory. The reaction was inhibited by phenazine methosulfate, methylene blue, and methyl viologen. The nature of these inhibitions is still under study. Other common metabolic inhibitors had no effect on MgDVP formation.

Overall, formation of the isocyclic ring requires the removal of six electrons from the molecule. Spiller et al (210) found that plants which could germinate anaerobically from seeds, tubers, or rhizomes were chlorotic and accumulated Mg-Proto(-Me) during anaerobic germination. Iron deficiency that was induced by iron chelators (69, 216) or germination in low Fe medium (52) also caused the accumulation of Mg-porphyrin Pchlide precursors. Chereskin & Castelfranco (58) proposed that both Fe and O_2 may be specifically involved in β-hydroxylation of the 6-propionate group.

PROTOCHLOROPHYLLIDE REDUCTION

Light Catalyzed

In the photoreduction of Pchlide the substrate molecule apparently serves as its own light receptor: the action spectrum for photoreduction is nearly identical to the absorption spectrum of phototransformable Pchlide in intact etiolated tissues (77, 167). Within 0.5 microsecond after illumination with an intense flash of light, a nonfluorescent intermediate is generated from phototransformable Pchlide (90, 119). This intermediate absorbs maximally at 690–695 nm and is spontaneously transformed to Chlide within a few microseconds in the dark. The 690–695 nm absorbing form is reported to be phototransformable back to Pchlide by an intense laser flash (148).

Photochemically active Pchlide-protein complexes, named Pchlide-holochromes, have been isolated from etiolated tissues. The term holochrome has been superseded now that the enzymatic nature of the reaction is becoming better understood: The smallest isolated photoactive holochrome particles probably correspond to ternary enzyme-substrate complexes of NADPH-Pchlide oxidoreductase. Like Pchlide itself, Pchlide reductase is localized in the prothylakoids (150). Electrophoresis of detergent-denatured purified Pchlide reductase preparations reveals the presence of polypeptides of 35,000 to 38,000 molecular weight (10, 32). The sulfhydryl reagent ^{3}H-N-phenylmaleimide labels Pchlide reductase irreversibly and specifically (169, 170). Both NADPH and Pchlide are required to protect the enzyme from inhibition by N-phenylmaleimide. It has been calculated that two or three Pchlide molecules are bound to each polypeptide molecule (10).

Continuous illumination of etiolated plants, or a short light treatment followed by darkness, resulted in a drastic decrease in extractable Pchlide reductase activity (110, 151, 152). This finding was initially attributed to an alteration in the redox state of the plastid $NADP^+$ pool (110). However, since the enzyme is assayed in the presence of saturating NADPH concentration, it was soon realized that the observed decrease in Pchlide reductase activity must be an indication of a drop in the concentration of the enzyme itself (151, 152). A subsequent dark incubation of the intact plant restored the reductase activity in the isolated plastids (152).

Further studies on the decline of Pchlide reductase upon illumination in vivo (10, 194) have revealed that besides the light-induced breakdown of the enzyme already present in the dark-grown tissue, there is also a light-induced decrease in a specific mRNA which is translatable in vitro to a polypeptide that is precipitated by a rabbit antiserum against purified Pchlide reductase (10). This in vitro synthesized peptide has a molecular weight of 44,000, that is, approximately 9000 larger than Pchlide reductase, and was, therefore, deemed to be a precursor of the reductase found in etiolated tissue. The existence of the higher molecular weight precursor form is consistent with the reports from two laboratories that Pchlide reductase is made on cytoplasmic 80S ribosomes (16, 108).

A pulse of red light, followed by dark, is sufficient to bring about the decrease in the specific mRNA. However, if the red light is followed by a far-red pulse, no decrease was observed. Therefore, the synthesis of Pchlide reductase was considered to be under regulation by phytochrome (10).

In continous light, by the time the maximum rate of Chl accumulation is reached, the level of Pchlide reductase is only a few percent of the level in etiolated tissue. Therefore, it has been suggested that the function of this enzyme may be restricted to the initial phase of greening and that, later on, the conversion of Pchlide to Chlide is brought about by another, as yet unknown, biosynthetic mechanism (194). The available spectrophotometric methods do not permit us to follow the formation of Chlide in a tissue that already contains high amounts of Chl. Therefore, the important question of whether or not Pchlide reductase is responsible for Chlide formation throughout the greening process will have to await the development of new experimental approaches.

Dark Reaction

Unlike the etiolated tissues of angiosperms, other plant tissues and algae can carry out Pchlide reduction without the aid of light. Among the gymnosperms, the primitive *Metasequoia glyptostroboides* is able to form Chl in the dark in new leaves at about half the rate observed in the light (143). Within the genus *Pinus*, dark Chl formation is apparently restricted to the cotyledons of germinating seedlings (44, 46). Dark reduction, but not light reduction, is inhibitied in pine cotyledons by lowering the temperature to 10°C (227).

Bogorad (46) observed that pine cotyledons can form Chl in the dark only when they are in contact with megagametophyte tissue. When the latter has been excised, little (200) or no (46) dark Chl synthesis takes place even when sucrose and nitrogen sources are provided. Excised cotyledons are capable of light Chl synthesis. The existence of a specific diffusible factor that is necessary for dark Chl formation is indicated by the results of Bogdanovíc (45), who found that wheat embryos are capable of dark Chl formation when grafted onto black pine megagametophyte tissue.

Among the angiosperms, traces of Chl were reported in dark-grown *Arabidopsis thaliana* (187), but the possibility of Chl carryover from the seeds was not excluded. More recently, it has been reported that expanding leaves of *Tradescantia albiflora* are able to form Chl in the dark at one-third the rate observed in the light (1, 5). Subsequently, Adamson & Hiller (4) have extended these observations to another member of the same genus *(T. blossfeldiana)* and another species of the same family *(Pollia crispata)*. Two sea grasses *(Zostera capricorni* and *Posidonia australis)* were also shown to synthesize Chl in the dark. This ability was limited to young leaves that had been initiated in the light. Chl accumulation was inhibited if leaves were exposed to red light prior to transferring to dark; the red light inhibition was abolished if the leaves were exposed to far-red light after the red light. This behavior suggests a typical phytochrome response.

When light-grown barley leaves were placed in the dark, Chl accumulation (2) and chloroplast development (3) continued for a few hours before coming to a stop. This phenomenon was more pronounced in the young cells containing partly developed chloroplasts that are found in the basal portion of the leaf nearer to the intercalary meristem. The apical cells accumulated little or no Chl in the dark.

These observations are in full agreement with the earlier work of Popov & Dilova (172) and suggested to the author that barley, a typical species adapted to full sunlight, exhibits to a limited extent the behavior previously demonstrated in the shade-adapted plants *Tradescantia, Pollia, Zostera,* and *Posidonia.*

Members of Hase's group have observed recently that tobacco callus cells grown on low agar plates can synthesize Chl(ide) in total darkness (E. Hase, personal communication). Perhaps the preponderance of studies with etiolated seedlings has led to an unjustified generalization that angiosperms cannot ever form Chl in the dark.

PHYTYLATION

Until recenlty, the only candidate for a Chlide esterifying enzyme was chlorophyllase. This enzyme is apparently ubiquitous in green plants and algae. It can catalyze three types of reactions: (*a*) hydrolysis of Chl to Chlide and phytol (137); (*b*) transesterification, e.g. methyl Chlide plus phytol yields methanol plus Chl (222); (*c*) under some conditions, free phytol and Chlide are joined in ester linkage by chlorophyllase (78). Chlorophyllase has the unusual property of being quite active in solutions containing high proportions of organic solvents, e.g. 40% acetone (114). A methoxycarbonyl group at position 10 on the substrate is required, but the central Mg atom is not: pheophytins as well as Chls can serve as substrates (115). In aqueous solution, the direction of reaction favors Chl hydrolysis rather than ester bond formation. Thus, chlorophyllase is thought to be important primarily in Chl catabolism.

Ogawa et al (166) reported the transient occurence of a pigment that they suggested was an intermediate between Chlide and Chl. Liljenberg (146) identified the alcohol group of Pchlide ester in etiolated barley as geranylgeraniol instead of phytol. Wellburn (223) also showed that a small amount of geranylgeraniol was present on Chl of young horse chestnut leaves. This alcohol is also known to be present instead of phytol in the bacteriochlorophyll of some species of photosynthetic bacteria (49, 133). Rüdiger et al (189) found that the herbicide 3-amino-1,2,4-triazole, which interferes with carotenoid metabolism, causes the accumulation in wheat leaves of Chl containing geranylgeranyl and dihydrogeranylgeranyl groups instead of phytol. Subsequently, Schoch et al (198) were able to detect in etiolated oat leaves given one minute of light and then two hours of dark, Chl species containing C_{20} alcohols with all degrees of hydrogenation between geranylgeraniol and phytol. The kinetics of accumulation and disappearance of the various Chl types indicates that Chlide is first esterified with geranylgeraniol, and then the alcohol is hydrogenated in three steps (corresponding to dihydro- and tetrahydro-geranylgeraniol, and finally phytol). The order of hydrogenation steps appears to proceed in a direction toward the distal end of the molecule (198, 199).

Intact spinach chloroplasts (209), maize homogenates (190), and oat etio-plast membranes (188) were capable of esterifying Chlide in vitro with gerany-lgeranyl pyrophosphate. In the absence of any additional energy source, the reaction went 80–90% to completion in the direction of esterification. There was a specific requirement for pyrophosphorylated forms of the alcohol subs-trate, and Pchlide could not be substituted for Chlide (188). Chemically synthesized pyrophosphates of farnesol and phytol could serve as substrates, but with only one-half to one-sixth the activity of geranylgeranyl pyrophos-phate. When ATP was added to the esterifying preparation derived from oat etioplast membranes, non-pyrophosphorylated phytol and farnesol, but not geranylgeraniol, could serve as substrates (188).

In vitro partial hydrogenation of pigment-bound geranylgeraniol was de-tected in illuminated oat etioplast fragments (43). Only one of the three hydrogenation steps (geranylgeraniol to dihydrogeranylgeraniol) necessary for conversion to phytol was observed. In the absence of ATP, the hydrogenation specifically required NADPH. When ATP was included in the incubation mixture, NADH could be substituted for NADPH. Wellburn (221) had reported earlier that intact bean and oat seedlings could incorporate label from $4R$-[^{3}H]-NADPH into phytol. Finally, Schoch et al (197) reported that etiolated oat seedlings, illuminated under strictly anaerobic conditions, accumulate geranyl-geranyl Chlide rather than the corresponding phytyl ester, suggesting that oxygen is essential for in vivo hydrogenation of the newly esterified alcohol. This last observation may be related to the presence of geranylgeranyl bacter-iochlorophyllide in some species of anaerobic photosynthetic bacteria (49, 133).

CHLOROPHYLL *b* AND CHLOROPHYLLIDE *b*

The earlier work on this subject was reviewed by Shlyk (202), and more recent findings were discussed by Beale (26). A short summary of some current thinking is presented here.

The accepted view that Chl b is formed from a special pool of "young" Chl a molecules rests on pulse-labeling experiments (6, 7, 74, 83, 203). The action spectrum for Chl b formation in etiolated wheat leaves was found to be identical with the action spectrum for Chl a formation, and both were similar to the absorption spectrum of Pchlide (167). Light does not appear to be required directly for the conversion of Chl a to Chl b (203). Exogenous Chl a was converted to Chl b even in tissue that had not been previously exposed to light (204).

Aronoff & Kwok (12) have obtained results which cast doubt on many previous claims of in vitro Chl a to b conversion. They showed that Chls a and b form mixed aggregates which migrate together during paper adsorption chromatography. The aggregates survive repeated rechromatography and give the appearance of purification to constant purity. Moreover, upon degradation of mixtures of pure ^{14}C-Chl a plus unlabeled Chl b, considerable radioactivity from ^{14}C-Chl a degradation products copurifies with the rhodin g_7 fraction which is derived from the tetrapyrrole moiety of Chl b. Aronoff & Kwok (12) were able to attain improved yields of Chl b in vitro by employing ruptured chloroplasts derived from etiolated cucumber seedlings and young spinach leaves. Highest yields were 2 percent in terms of added ^{14}C-Chl a, or 19 percent in terms of recovered substrate.

A major gap in our knowledge of Chl b synthesis concerns the identity of the oxidizing agent and the mechanism of the methyl to formyl conversion. Although NADP$^+$ was reported to stimulate Chl b formation (83, 204), its essentiality could not be demonstrated in the relatively unrefined preparations. The fact that all Chl b-containing organisms are aerobic is consistent with the hypothesis that the Chl a to Chl b conversion is a mixed-function oxidase type reaction. If this is the case, then there will be requirements for both molecular oxygen and a reductant such as NADPH rather than NADP$^+$. In one report (12), Chl a to Chl b conversion was observed to take place at a diminshed rate in an atmosphere of nitrogen, but strict anaerobiosis was not achieved. The conversion would be expected to consist of two steps, with a hydroxymethyl intermediate. The detection of an intermediate between Chls a and b has not been reported.

The prevailing hypothesis that Chl b is made from Chl a was recently questioned by Oelze-Karow and co-workers (164, 165). These investigators observed that Chlide a and Chl b accumulation are both more sensitive to inhibition by far-red light than is Chl a accumulation, and they postulated that the precursor of Chl b might be Chlide a rather than Chl a, with conversion of the 3-methyl moiety to a formyl group ocurring before the phytylation step. This scheme is consistent with the observation that Chlide b undergoes enzymatic esterification with geranylgeranyl pyrophosphate as readily as Chlide a (42). The source of enzyme was a broken etioplast preparation from dark-grown oat seedlings.

The incorporation of ^{14}C-ALA into Chlides a and b in the presence of a chloroplast suspension from young pea leaves exceeded the incorporation into Chls a and b (11). In 10 to 60 minute incubations, the ^{14}C-Chlide a/^{14}C-Chl a ratio varied between 4.6 and 12.7; likewise, the ^{14}C-Chlide b/^{14}C-Chl b ratio varied between 2.7 and 7.2. Specific radioactivities were not reported. Chlorophyllase activity was tested in this chloroplast preparation under identical incubation conditions. The chlorophyllase activity present could have accounted, at most, for 1% of the ^{14}C-Chlide b being formed from ^{14}C-Chl b during the incubation.

Duggan & Rebeiz (73) detected Chlide b by fluorescence spectroscopy in extracts from greening and photoperiodically grown cucumber cotyledons, and confirmed their identification by synthesizing the oxime and the methyl ester, which had the expected spectrofluorometric and chromatographic properties. ^{14}C-Chl b synthesized in vivo from ^{14}C-ALA and purified by TLC was used to check how much of the observed Chlide b could have arisen by chlorophyllase activity during pigment extraction. This approach indicated that no more than 13% of the Chlide b detected spectrofluorometrically could have been an artifact of in vitro hydrolysis. Fuesler & Castelfranco (94) also noticed Chlide b in greening cucumber cotyledons. Their observations agree with the results of Duggan & Rebeiz. These findings suggest that Chlide b may be a biosynthetic intermediate on the route to Chl b.

An alternative explanation for the occurrence of Chlide b in greening tissues is that Chls and Chl-protein complexes are subject to turnover before they become established at stable sites within the thylakoid membranes. It is known that Chls do turn over, Chl b more than Chl a, in seedlings that are greening under intermittent illumination (211), or when transferred to darkness after a period of continuous light (41). Both the protein and pigment components of newly formed light-harvesting Chl a/b complex are subject to degradation in the dark (41) or when Chl b synthesis is blocked by mutation (40). During the early stages of greening, Chl a may not be made at a rate sufficient to saturate the requirements of the simultaneously forming light-harvesting complexes, and incomplete complexes might then be degraded, along with any attached molecules of Chl b. The first step of Chl b degradation is likely to be dephytylation to Chlide b.

Thus, presently it is not certain whether phytylation preceeds the oxidation of the ring B methyl group, or vice versa. If phytylation comes first, then Chl a would be a biosynthetic intermediate. If the opposite is true, the reaction sequence would be Chlide $a \rightarrow$ Chlide $b \rightarrow$ Chl b.

CHEMICAL HETEROGENEITY OF CHLOROPHYLLS AND PRECURSORS

The heterogeneity of Chl by virtue of its position in the membrane, which manifests itself by multiple in vivo spectral forms of Chl a and Chl b (50, 51), has long been known. This type of heterogeneity will not be discussd here. Instead, this section addresses the heterogeneity resulting from structural modifications of the tetrapyrrole molecule. This topic was reviewed recently from the viewpoint of one of the most active experimental groups in this field (181, 184).

4-Vinyl vs 4-Ethyl

The two compounds, Pchlide and Pchlide ester, have both been shown to undergo phototransformation resulting ultimately in the synthesis of Chl a (63, 146, 195). The conversion of Pchlide to Chl a via Chlide a has been studied much more extensively than the alternative route to Chl a through Pchlide ester. Both Pchlide and Pchlide ester can exit in two forms depending on whether the side chain in the 4 position is an ethyl or a vinyl (35, 36, 38). Strictly speaking, Pchlide is *monovinyl* Pchlide, that is, Mg-2-vinyl, 4-ethyl pheoporphyrin a_5, and its divinyl analog is MgDVP. MgDVP has long been regarded as the immediate precursor of Pchlide in the Chl biosynthetic scheme. Griffiths & Jones (109) showed that MgDVP could be photoreduced by a Pchlide reductase preparation; however, they did not determine whether the reaction product was Chlide a or its 4-vinyl analog.

Belanger & Rebeiz (36) showed that the Pchlide pool found in etiolated plant tissues contained a greater or lesser percentage of the 4-vinyl analog, depending on the species and on the history of the tissue. Moreover, the Pchlide pool that reformed after phototransformation was enriched in MgDVP and the Pchlide pool made from exogenous ALA consisted mostly of MgDVP. These studies utilized separation of the two components of the Pchlide pool on polyethylene thin layers, spectrofluorometry, catalytic hydrogenation of the side chains, and comparison of the Pchlide components from plants with authentic MgDVP produced by a previously characterized *Rhodopseudomonas spheroides* mutant (36).

The accumulation of MgDVP in greening cucumber cotyledons has been confirmed by Chereskin et al (59, 60). These workers found that the Pchlide pool in dark-grown cucumber seedlings, exposed to white light for 20 hours, consisted exclusively of MgDVP, which was identified by chromatography on polyethylene thin layers and spectrofluorometry in comparison with authentic MgDVP accumulated by $R.$ *spheroides* mutant V-3 [a gift of Dr. O.T.G. Jones; see (109)].

Phototransformation of the 4-ethyl and 4-vinyl Pchlide species gave rise to the corresponding 4-ethyl or 4-vinyl Chlide a (34, 35, 37, 72). In these cases, the optical properties of the 4–vinyl analog differed from those of the better-known 4-ethyl analog by being shifted toward longer wavelengths, particularly in the Soret region. In vivo, the red-shifted forms of Chl(ide) a were transient. A few minutes after the phototransformation, both Chlide a and Chl a pools appeared to consist mostly of the 4-ethyl components, together with blue-shifted forms which could have arisen from dark transformation of the 4-vinyl components (37). The chemical identity of these blue-shifted Chl(ide) *a* forms is at present a matter of speculation. Similarly, the phototransformation of a mixed pool of Pchlide ester yielded a heterogeneous pool of Chl *a* consisting of a normal 4-ethyl component and a red-shifted 4-vinyl component. Within minutes, the latter disappeared with a concomitant appearance of a blue-shifted Chl *a* species (37). A subsequent study on the phototransformation of 4-vinyl Pchlide has shown that the resulting 4-vinyl Chlide is rapidly converted to 4-ethyl Chlide (73a).

Examination of the Chls *a* and *b* found in light-grown cucumber cotyledons, spinach, and *Euglena* (180) confirmed the heterogeneity of the Chl *a* and *b* pools observed upon light exposure of dark-grown tissue (37). The heterogeneity of Chls *a* and *b* arising from the difference of the C_4 side chain gained support from a study done with a maize mutant that is partly Chl deficient (20, 21). Both Chls *a* and *b* were red-shifted with respect to the wild types. The optical properties of the mutant Chl *a* were similar to those of the initial Chl(ide) *a* derived from 4-vinyl Pchlide and 4-vinyl Pchlide ester (37). By NMR spectroscopy, the Chl *a* of the mutant was shown to have an additional vinyl group at position 4 that was absent in the wild type (23). Moreover, mass spectroscopy confirmed that the molecular weight mutant Chl *a* was two units smaller than that of the wild type (22).

Heterogeneity based on the oxidation level of the C_4 side chain was also reported in the Mg-Proto-Me (39), where a minor component, estimated to about 8% of the total, was deemed to consist of the 4-ethyl analog. The same heterogeneity was found also in the Mg-Proto and in the Mg-Proto diester pools (see below). An esterase that converts Mg-Proto-Me to Mg-Proto has been found in cucumber cotyledons (228). It appears, therefore, that the 4-ethyl-Mg-Proto could be an enzymatic breakdown product of 4-ethyl-Mg-Proto-Me. These observations are, therefore, in agreement with the report of Ellsworth & Hsing (81), who observed the reduction of the 4-vinyl side chain of Mg-Proto-Me by NADH in a cell-free homogenate from etiolated wheat seedlings.

In summary, the heterogeneity of Chls and Chl precursors based on differences in the oxidation state of the C_4 side chain appears to be firmly established. It has been demonstrated in Mg-tetrapyrroles ranging from Mg-Proto to Chl *b*. This heterogeneity has been investigated using a number of experimental approaches that have yielded, on the whole, convergent results. The physiological reason for this heterogeneity, and its biosynthetic origin, are still unknown.

Other Optically Detectable Heterogeneity

Another approach has been used by Rebeiz and co-workers to study the heterogeneity of chromophores. Fluorescence emission spectra were determined in ether at 77°K, while the excitation was varied by steps of a few nanometers over a certain interval. It was observed that as the excitation wavelength varied, the emission maximum also varied in the same direction. A matrix was constructed and a mathematical analysis of this matrix, according to published methods, yielded the excitation and emission maxima of the fluorescent chromophores in the mixture (36, 62, 180, 181).

The above approach is based on the theory that chromophores in dilute solution should behave as independent oscillators and that the shape of the emission spectrum should not vary with the excitation wavelength. However, as shown by Singhal et al (205), these assumptions may not hold in organic solvent glasses at 77°K, where a single chemical entity may give rise to several stable solvates. This seems to be particularly true for Mg-tetrapyrroles.

Using this approach, Rebeiz and co-workers reported a heterogeneity of extracted Pchl and Chl pools far greater than would be expected simply on the basis of the oxidation level of the C_4 side chain. For example, they reported four fluorometrically distinct Pchlide components (36), four Chl a species, and at least four Chl b species (180). As to the validity of this analytical method, certain reservations must be pointed out. The following examples are taken from the publications of Rebeiz and co-workers.

The matrix analysis failed to resolve mixtures of standards if the excitation maxima of the components were too close together (36). Four compounds, E399, E402, E408, and E414, differing respectively in their excitation maxima by 3, 6, and 6 nm, could not be resolved. However, if the last compound (E414) was replaced by another that was excited maximally at 425 nm (E425), then the spectra of the four components could be resolved by matrix analysis. It is difficult to understand how the resolution between E399 and E402, only 3 nm apart, could be improved by replacing E414 in the mixture by another compound which is shifted 11 nm still farther toward the red.

Cohen & Rebeiz (62) analyzed the heterogeneity of Pchl in etiolated cucumber cotyledons and etiolated bean leaves. In both tissue extracts, as the excitation was varied from 440 to 455 nm, the emission maximum due to phototransformable Pchl changed from 630 to 640 nm. Matrix analysis revealed the presence of three short-wavelength, phototransformable components. Yet, as the excitation was varied, the emission maximum caused by the nonphototransformable Pchl likewise underwent a shift from 630 to 635 nm in cucumber [Figure 2 in (62)] and from 630 to 636 nm in bean [Figure 3 in (62)]. The authors chose to ignore this 5 to 6 nm shift and treated the nonphototransformable Pchl as a single optical species, E430F630 [Table III in (62)], in line with the consensus of prior published evidence. It appears somewhat contradictory that a comprehensive hypothesis is built on a 10nm shift, while a 5 to 6 nm shift observed in the very same experiments is passed over as insignificant.

The physical separation of the four putative Chl a chromophores was attempted using high performance liquid chromatography and met with only partial success. Four chl *a* peaks were obtained that appeared to be mixtues of the four chromophores in different proportions. Physical separation of the Chl *b* components was not reported.

An attempt was made to correlate the four forms of Chl *a* and Chl *b* with specialized areas within the photosynthetic membranes. When etiolated cucumber seedlings were subjected to an alternating light regime to decrease the formation of the light-harvesting Chl protein complex relative to photosystem I and II particles, the resulting Chl *a* was enriched in the short wavelength forms (181). P_{700}, photosystem II particles, and light-harvesting Chl protein complex particles were prepared from spinach leaves by standard procedures. The Chl *a* extracted from P_{700} and photosystem II particles was found to be relatively enriched in the short wavelength forms, whereas the light-harvesting Chl protein complex particles were relatively enriched in long-wavelength Chl *a* (91).

In summary, Rebeiz and co-workers have presented preliminary evidence for the chemical heterogeneity of Chl and Pchl chromophores beyond the heterogeneity arising from the oxidation level of the C_4 side chain. However, because of the ambiguities inherent to the basic methodology, this second level of chemical heterogeneity cannot be considered as proved, and must await confirmation by independent experimental means.

Dörnemann & Senger (67) studied the mutant C-6E of *Scenedesmus obliquus*. The dark-grown cells of this organism have very high photosystem I activity, but no photosystem II activity. Moreover, they lack Chl *b* and carotenoids, and produce a very high P_{700}/Chl *a* ratio (67, 224). It was observed that the Chl *a* extracted from this organism contained a long-wavelength component that could be separated from the major Chl *a* component by thin layer chromatography. The amount of the new Chl (Chl RCI) varied with the enrichment of P_{700} relative to bulk Chl *a*. When Chl RCI was acidified, the resulting pheophytin (Pheo RCI) was also red-shifted with respect to the conventional Pheo *a*.

More recently (68), Chl RCI was extracted from a photosystem I-enriched fraction prepared from digitonin-treated spinach chloroplasts. The molecular weights of Chl RCI and Pheo RCI, determined by gel filtration and mass spectrometry, were 35 units larger than conventional Chl *a* and Pheo *a*. The chemical identity of Chl RCI is still unknown. Because the extraction involved boiling methanol at alkaline pH [$Mg(OH)_2$], the photosystem I chromophore could have been derivatized during isolation. It is too early to speculate on the relationship between Chl RCI and the short wavelength Chl *a* forms reported by Rebeiz's group to be associated with photosystem I and II particles (91).

Degree of Esterification of Intermediates

Esterification of the two carboxyls derived from the propionic acid substituents of Proto does not affect the optical properties of tetrapyrrole molecules. Pchlide and Pchlide ester differ chromatographically but not optically. Both Pchlide and Pchlide ester are Chl precursors and are made from ALA (182). Where do the two pathways diverge, leading to Pchlide and Pchlide ester? This question was probed in two recent papers by McCarthy et al (153, 154).

First, they prepared Mg-Proto-Me by incubating etiolated cucumber cotyledons with ALA plus α,α'-dipyridyl, and showed that in addition to Mg-Proto-Me, Mg-Proto diesters were also formed. Segregation of the mixed diester pool by high performance liquid chromatography showed that it consisted of one relatively polar diester with the chromatographic properties of Mg-Proto dimethyl ester, plus three less polar peaks, which, upon fluorescence analysis, were each shown to be mixtures of Mg-Proto and Pchlide esterified with three major terpenoid alcohols. Each peak appeared to contain both Mg-Proto-Me and Pchlide esterified with the same alcohol or alcohols. Therefore, the authors concluded, the Mg-Proto diesters are probably intermediates in the synthesis of the corresponding Pchlide esters. Small amounts of Mg-Proto diesters were also found in untreated dark-grown *Euglena* cells and in etiolated cucumber cotyledons treated with ALA but without α,α'-dipyridyl.

In a subsequent paper, McCarthy et al (154) examined the pattern of label incorporation into Pchlide and Pchlide ester from ^{14}C-ALA, ^{14}C-Proto, ^{14}C-Mg-Proto-Me, and ^{14}C-Pchlide. The incubations were carried out with isolated plastids. The labeling ratios suggested that the two pathways diverge somewhere between Proto and Mg-Proto-Me. The data presented do not allow a more precise localization of the branching point.

In conclusion, the presence of chemical heterogeneity in the naturally occurring Mg-tetrapyrroles has been established. However, the extent of this heterogeneity, its role within the physiology of the photosynthetic apparatus, and the way it comes about biosynthetically, are not understood at the present time.

REGULATION

At the Enzyme Level

The regulation of Chl biosynthesis is unquestionably a very complex process. Setting aside those elements of Chl regulation that have to do with transcription, translation, movement of proteins across the chloroplast envelope, phytochrome-mediated and hormonal responses, one can draw a simplified scheme for the regulation of Chl biosynthesis in terms of the specific activities of certain key enzymes. This scheme, which is still largely hypothetical, is given in Figure 3. The key regulatory steps indicated by numbers 1 through 9 below refer to the figure.

1. ALA SYNTHESIS   The first evidence that heme may inhibit ALA formation in plants was obtained by Hoober & Stegeman (116), who reported that

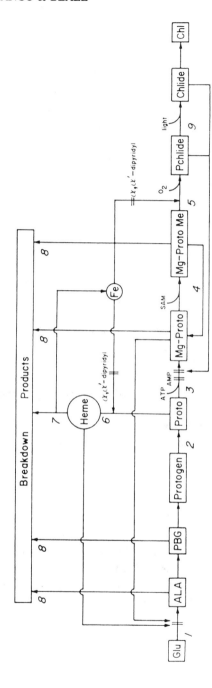

Figure 3. Postulated scheme for the regulation of Chl *a* biosynthesis in higher plants.

exogenous heme inhibited Chl synthesis in greening cultures of *Chlamydomonas reinhardtii*. Exogenous heme and Mg-Proto were found to inhibit ALA formation in isolated cucumber cotyledon plastids (54, 58). With these two inhibitors, half-maximal inhibition occurred at 2 and 3.5 μM, respectively; however, complete inhibition was not attained.

Added heme and Mg-Proto also inhibited ALA formation by a soluble enzyme system extracted from greening barley plastids. The approximate I_{50} values in this case were 17 μM for heme and 30 μm for Mg-Proto (102). No inhibition of ALA synthesis by the added metalloporphyrins was observed with intact barley plastids (S. P. Gough, personal communication). Inhibition of ALA synthesis by Proto, which was observed in some of the early work (54, 105), turned out to be mediated through heme formation. Exogenous Pchlide had no direct inhibitory effect on the cell-free ALA-synthesizing system from either cucumber or barley. It is interesting that heme and Mg-Proto also inhibit the synthesis of ALA from glycine and succinyl-CoA in *Rhodopseudomonas spheroides* (142, 230).

2. PROTOPORPHYRINOGEN OXIDATION   There do not seem to be any regulatory steps between ALA and protoporphyrinogen. The enzymes involved in this portion of the pathway seem to be present in large excess, their activities more than sufficient to sustain maximal rates of Chl biosynthesis. Protoporphyrinogen can be converted to Proto nonenzymatically, particularly in the presence of O_2 and light. However, it has been shown that this conversion is catalyzed by a specific enzyme, protoporphyrinogen oxidase, which is found in both etioplasts and chloroplasts (121). The presence of an enzyme to carry out a reaction that can also proceed spontaneously suggests a regulatory role. However, there are other possible explanations. For example, the O_2 concentration in developing plastids might not be high enough to sustain the spontaneous reaction, and the enzyme might be more efficient in coupling the removal of hydrogen from protoporphyrinogen to the reduction of O_2.

3. MG CHELATION   Mg chelation would be expected to be a site of metabolic regulation simply because it is the first step in the Mg-porphyrin pathway. Mg-chelatase has stubbornly resisted isolation, as all attempts at plastid fractionation by mechanical means (freeze-thawing, sonication, French-press, etc), osmotic shock, detergents, and chaotropic salts have resulted in loss of activity (T. P. Fuesler and P. A. Castelfranco, unpublished). Therefore, this reaction appears to have some requirement for structural integrity. In fact, electron microscopy of an active preparation revealed the presence of many intact plastids (171). On the other hand, this reaction requires ATP and is inhibited by AMP. These nucleotides, as a rule, do not penetrate readily through plastid envelopes. Mg chelation is also inhibited by

p-chloromercuribenzene sulfonic acid (T. P. Fuesler and P. A. Castelfranco, unpublished), a sulfhydryl reagent that is supposedly unable to cross biological membranes (213). All of these properties imply either that Mg chelatase is located on the outer surface of the developing plastids, or that the envelope of partially greened cucumber plastids is atypically permeable to such polar molecules as ATP, AMP, and *p*-chloromercuribenzene sulfonic acid. Possibly, in order for in vitro activity to occur, the plastid envelope membranes must be made sufficiently permeable to allow the entry of the exogenous substrates, but yet not so permeable as to allow the loss of other compounds or an internal environmental state that is necessary for activity. The significance of the inhibition by AMP is not known. It has been interpreted as an "energy charge" phenomenon (53). For example, AMP could, in the presence of adenylate kinase, act as an ATP scavenger.

There is both indirect and direct evidence that Pchlide, under some conditions, can inhibit Mg chelation. Wang et al (217) have presented evidence for feedback inhibition of Mg chelation by Pchlide. If Pchlide was made to accumulate in mutant cells of *Chlamydomonas reinhardtii*, either by feeding ALA or by introducing a second mutation that deregulates ALA synthesis, Pchlide reached a plateau level, even though Proto continued to rise and eventually attained a concentration many times greater than Pchlide.

Cohen et al (61) found that in photoperiodically grown cucumber cotyledons Mg-Proto(-Me) was detectable only during the light cycles. Pchlide levels, as one might expect, were higher in the dark than in the light, and Proto did not accumulate during either light or dark cycles. These observations are consistent with the postulated feedback inhibition of Mg chelatase by Pchlide. Proto did not accumulate in this tissue, presumably, because it was converted to heme by ferrochelatase, which has a high affinity for Proto [$K_m = 0.2 \, \mu M$ (122)], and, presumably, the resulting protoheme shut off ALA formation. Inhibiton of Mg chelatase in vitro by added Pchlide was observed by T. P. Fuesler & P. A. Castelfranco (unpublished). Fifty percent inhibition occurred at 3 μM Pchlide. Chlide also inhibited to the same extent.

The observation that etiolated higher plant tissues treated with ALA accumulate Pchlide argues against the ability of Pchlide to act as a feedback inhibitor of Mg chelatase. However, the "ALA-Pchlide" which is formed in tissues that are treated with exogenous ALA might be sequestered in some lipophilic domain of the etioplast, where it cannot reach the locale of Mg chelation and therefore cannot inhibit the Mg chelatase.

4. MG-PROTO METHYL ESTERASE ACTIVITY   The recently demonstrated hydrolysis of Mg-Proto-Me (60, 228) may function to recycle excess Mg-Proto-Me back to Mg-Proto, because, as was mentioned, Mg-Proto acts as

a feedback inhibitor of ALA synthesis. However, Mg-Proto-Me does accumulate to very high levels in etiolated tissues when they are treated with transition metal ion chelators, with (106) or without (69) added ALA. This observation suggests the presence of an esterase with a heavy metal requirement, a possibility that has not yet been investigated.

5. ISOCYCLIC RING FORMATION   The conversion of Mg-Proto-Me to Pchlide exhibits an in vivo Fe requirement. Mg-Proto(-Me) accumulates in Fe-deficient plants (52, 210), and abnormally high Mg-Proto-Me/Pchlide ratios have been found in etiolated tissues treated with compounds such as α,α'-dipyridyl that decrease the activity of Fe by forming a bidentate complex (69, 106, 216).

Although cell-free systems are now available that convert Mg-Proto-Me to MgDVP (60, 228), the Fe-requiring enzyme has not been identified. Molecular oxygen is also required in the oxidative cyclization of the C_6 methyl propionate side chain, as shown by in vitro and in vivo studies. Mg-Proto(-Me) accumulates in tissues of submerged aquatic plants grown under anaerobic conditions (52, 210).

6. HEME SYNTHESIS   Ferrochelatase has been demonstrated in spinach chloroplasts (122) and more recently in barley etioplasts (147), where it was shown to be different from the mitochondrial enzyme. Its high affinity for free Proto makes heme a likely candidate for feedback regulation of ALA synthesis. In the photosynthetic bacteria, heme appears to be a key regulatory feedback modulator of ALA synthesis for both the heme and bacteriochlorophyll branches of the tetrapyrrole pathway (141, 142). The concentration of noncovalently bound heme drops when plant tissues are incubated with levulinic acid (162), α,α'-dipyridyl (69), or N-methyl mesoporphyrin IX, which is a specific competitive inhibitor of ferrochelatase (29). However, administration of N-methyl mesoporphyrin IX to light-grown *Euglena* cells did not cause any increase (or decrease) in the rate of Chl synthesis over a 6-hour period, even though the cellular heme content dropped to 75% of the control value and the extractable ALA synthase activity was increased 2.5-fold (29).

Administration of exogenous ALA in the dark does not normally result in an overaccumulation of heme, suggesting that heme synthesis is already saturated with respect to Proto, even though Proto and Pchlide undergo large increases (55, 162). However, when 5 mM $FeSO_4$ was included in the incubation mixture, a 50% increase in noncovalently bound heme was observed (P. A. Castelfranco and J. Shinoda, unpublished). In the same incubation, Pchlide increased 50-fold.

7. HEME TURNOVER   Evidence for rapid heme turnover during greening comes from a variety of experimental approaches. Castelfranco & Jones (55) observed that upon administration of different ^{14}C-precursors to greening barley, Chl and extractable protoheme acquired essentially the same specific radioactivity. This observation has two implications: first, both products share the same pools of precursors and thus share a single pathway; second, because the total extractable heme did not increase during the first 12 hours of greening, while Chl levels increased 35-fold, a large fraction of the heme must turn over. Additional evidence for rapid heme turnover was obtained by Duggan & Gassman (69) and Gassman et al (97), who found that intracellular heme levels in etiolated and greening plant tissues dropped during incubation with α,α'-dipyridyl, which prevents heme formation by chelating iron.

Nasrulhaq-Boyce & Jones (162) extended the results of Castelfranco & Jones (55) and showed by labeling studies that heme was being synthesized in greening barley leaves even though there was no net accumulation. They concluded that the heme level in greening barley is regulated neither by product inhibition of the ferrochelatase nor by feedback regulation of ALA synthesis, but by turnover of the heme. The enzyme heme oxygenase, which is responsible for heme turnover in animal cells, has not been reported in plants.

8. TURNOVER OF CHL PRECURSORS   The Chl biosynthetic pathway of algae and higher plants appears to be remarkably tidy. Except in metabolically deranged mutants, biosynthetic intermediates do not accumulate spontaneously. Some intermediates do accumulate in response to experimental manipulations. For example, treatment with inhibitors of ALA dehydrase causes the accumulation of ALA, and treatment with certain chelating agents causes the accumulation of Mg-Proto and its monomethyl ester. Higher plant tissues growing in the dark accumulate Pchlide, and the accumulation is enormously enhanced by feeding ALA.

Plant cells possess several systems for the catabolism of Chl biosynthetic precursors. Probably these catabolic sequences help to regulate the level of very small pools of biosynthetic intermediates, and thus play an important role in the regulation of tetrapyrrole metabolism in plants. It may be essential for plant cells to eliminate any excess of Chl precursors, because many of these compounds have photodynamic properties that could cause lethal damage under strong light.

Added ^{14}C-labeled ALA is readily metabolized to nonpyrrolic products by barley tissue (70, 97). A PBG oxygenase was isolated by Frydman et al (92, 93) from wheat germ, and Duggan et al (71) described the breakdown of ^{14}C-PBG by segments of barley seedlings. The catabolism of PBG was extensive: less than 0.5% of the administered label ended up in the porphyrin-phorbin fraction.

The Mg-Proto-Me that accumulated in etiolated bean leaves under the influence of α,α'-dpyridyl disappeared when the external chelating agent was washed out and 1 mM Co^{2+}, Zn^{2+}, or Fe^{2+} salts were added to tie up intracellular α,α'-dipyridyl (216). The disappearance of Mg-Proto-Me required O_2 and was not inhibited by the usual poisons of mitochondrial respiration. The involvement of a specific oxygenase was postulated.

Hougen et al (117) found that cell-free homogenates prepared from etiolated bean leaves contain an enzyme that degrades Mg-Proto-Me. The activity was totally dependent on O_2. The reaction product was not identified. More recently it was observed that the same homogenate also catalyzes the breakdown of Mg-Proto (M. L. Gassman, personal communication).

9. PCHLIDE REDUCTASE AND BEYOND   There is evidence that the activity of Pchlide reductase responds to the state of oxidation of the plastidic $NADP^+$ (110). The greater the reduction of this pool, the higher the activity of the enzyme. During continuous illumination both the $NADPH/NADP^+$ ratio and enzyme activity drop. It has already been mentioned that this drop in enzyme activity is accompanied by a degradation of the actual enzyme protein (9, 19, 151, 152) and by a decrease in the mRNA coding for the enzyme (9, 194). The significance of these changes has not been elucidated.

Other regulatory mechanisms must be operating in the later portion of the pathway, between Pchl(ide) and Chl. These mechanisms must be complex because the newly formed Chl molecules need to be inserted at specific sites in the membrane, in association with proteins and lipids. These processes are just beginning to be unraveled, and very little is known about their regulation.

Light and Hormone Effects

PHYTOCHROME   Light has other effects in addition to that of promoting Pchlide reduction. Several of these effects exhibit the hallmark of phytochrome-mediated responses: triggering by red light and reversibility of the effect of red light by a subsequent far-red light exposure.

When completely dark-grown seedlings are first exposed to light, the initial photoconversion of Pchl(ide) to Chl(ide) is followed by a lag phase which may last for several hours, depending on the species and age of the seedlings. Only after completion of the lag phase does Chl synthesis begin in earnest. It is generally thought that during the lag phase, the enzyme system responsible for synthesizng ALA is formed de novo, because: (a) exogenous ALA eliminates the lag phase (56, 206); (b) inhibition of protein synthesis prevents entry into the rapid phase (161); and exogenous ALA permits Chl synthesis even in the presence of inhibitors of protein synthesis (161).

If completely dark-grown tissues are briefly preilluminated with white or red light and then placed back in the dark for a few hours, Chl synthesis begins immediately upon reillumination. It is believed that the elimination of the lag phase is caused by the preillumination-triggered formation in the dark of the ALA synthesizing system, which can then become active immediately upon reillumination. The ability of red light preillumination to abolish the lag phase of greening is reversed by far-red light applied shortly after the end of the red light (206).

Other red light inducible, far-red light reversible effects that have already been mentioned are the turnover of the mRNA for Pchlide reductase (9), the inhibition of dark Pchlide reduction in light-initiated leaves (4), the synthesis of Chl b (164). Yet another phytochrome-mediated effect that has a bearing on chlorophyll synthesis is induction of the mRNA for the light-harvesting Chl a/b complex apoprotein (8).

In the cases where the levels of specific mRNA species are affected, we are at the threshold of a mechanistic understanding of phytochrome action. It remains to be seen whether all of the phytochrome effects on Chl synthesis are mediated at the level of mRNA.

CYTOKININ PRETREATMENT   Pretreatment of etiolated wheat leaves (33) or cucumber cotyledons (85) with cytokinins abolishes the lag phase of a subsequent light-induced greening period. In sunflower cotyledons, 6-benzylaminopurine had no effect on the steady state rate of Chl synthesis achieved after the end of the lag phase, but the cytokinin greatly enhanced the rate during the initial lag phase (88). Another effect of cytokinins was to increase the rate of Pchlide regeneration after removal of the tissue to the dark (145). Overall, it appears that exogenous cytokinins act by stimulating the formation of the ALA-synthesizing system, so that after exposure to light, Chl formation can proceed immediately rather than having to wait until the ALA-synthesizing system is made de novo. In this respect, cytokinins mimic the effect of a brief preillumination, which also abolishes the lag phase if given within a few hours before the continuous greening illumination. Since the preillumination effect is photoreversible by far-red light, it is tempting to propose that some or all of the phytochrome effects on greening are mediated via cytokinins. It would be interesting indeed if this were the case.

The ability of etioplasts to accumulate Pchlide from ALA was greatly increased if the tissue was pretreated with kinetin for 20 hours prior to plastid isolation (64). Direct addition of kinetin to the etioplasts during the incubation had no effect, while incubation of the intact tissue with kinetin gave only a slight increase in the accumulation of Pchl in vivo. The authors concluded that the kinetin treatment promotes the formation of plastid membranes, which act as acceptors for the accumulation of Chl precursors synthesized in vitro. Plastids isolated from cucumber cotyledons pretreated with kinetin and gibberellic acid accumulated Chlide a and Chl a at particularly high rates when incubated in the presence of ALA under intermittent light (65).

SUMMARY

Considerable progress has been achieved in the last decade. All of the enzymatic steps of the pathway have now been demonstrated in cell-free systems, and many of them have also been detected in solubilized plastid-free preparations. Several heretofore enigmatic steps, including type III porphyrinogen formation, porphyrinogen aromatization, Pchlide photoreduction, and phytylation, are at the point of being understood at the mechanistic level. Work on other steps, such as ALA synthesis, Mg chelation, and isocyclic ring formation, has progressed to the point where rapid advances can be expected in the near future, as the recently available cell-free reactions are being studied intensively in several laboratories. A tantalizing array of information has now been amassed concerning effects of exogenous chemicals, hormones, and light on whole cell preparations, and, in some cases, on isolated enzymes as well. However, at this time, it cannot be said that we know how Chl synthesis is regulated in a coordinated fashion during greening. Nor do we yet know how plant cells adjust their final Chl level to be in tune with the particular physiological role of the cells within the plant and with the requirements of the prevailing photic environment. We look forward to answers to these important questions in the coming decade.

ACKNOWLEDGMENTS

We thank the many authors who generously supplied copies of papers in press and allowed us to quote unpublished results. Research support from the NSF and the USDA Competitive Grants Office is gratefully acknowledged. We are indebted to Academic Press for providing permission to reproduce Figures 1 and 2, to T. P. Fuesler for drawing Figure 3 and to J. D. Weinstein for carefully reading and criticizing the manuscript.

Literature Cited

1. Adamson, H. 1978. Evidence for the accumulation of both chlorophyll *a* and *b* in darkness in an angiosperm. In *Chloroplast Development*, ed. G. Akoyunoglou, J. H. Argyroudi-Akoyunoglou, pp. 135–40. North Holland: Elsevier. 888 pp.
2. Adamson, H. 1983. Evidence for a light-independent protochlorophyllide reductase in green barley leaves. In *Cell Function and Differentiation*, 2:33–41. Proc. Spec. FEBS Meet., Athens, 1982. New York: Liss
3. Adamson, H. 1983. Chloroplast development in green barley leaves transferred to darkness. See Ref. 2, pp. 189–99
4. Adamson, H., Hiller, R. H. 1981. Chlo-

rophyll synthesis in the dark in angiosperms. In *Photosynthesis*, ed. G. Akoyunoglou, 5:213–21. Philadelphia: Balaban. 1018 pp.
5. Adamson, H. Y., Hiller, R. G., Vesk, M. 1980. Chloroplast development and the synthesis of chlorophyll *a* and *b* and chlorophyll protein complexes I and II in the dark in *Tradescantia albiflora* (Kunth). *Planta* 150:269–74
6. Akoyunoglou, G., Argyroudi-Akoyunoglou, J. H., Michel-Wolwertz, M. R., Sironval, C. 1966. Effect of intermittent and continuous light on chlorophyll formation in etiolated plants. *Physiol. Plant.* 19:1101–4
7. Akoyunoglou, G., Argyroudi-Akoyuno-

glou, J. H., Michel-Wolwertz, M. R., Sironval, C. 1967. Chlorophyll *a* as a precursor for chlorophyll *b*. Synthesis in barley leaves. *Chim. Chron.* 32:5–8

8. Apel, K. 1979. Phytochrome-induced appearance of mRNA activity for the apoprotein of the light-harvesting chlorophyll *a*/*b* protein of barley *(Hordeum vulgare)*. *Eur. J. Biochem.* 97:183–88

9. Apel, K. 1981. The protochlorophyllide holochrome of barley *(Hordeum vulgare* L.). Phytochrome-induced decrease of translatable mRNA coding for the NADPH: protochlorophyllide oxidoreductase. *Eur. J. Biochem.* 120:89–93

10. Apel, K., Santel, H.-J., Redlinger, T. E., Falk, H. 1980. The protochlorophyllide holochrome of barley *(Hordeum vulgare* L.). Isolation and characterization of the NADPH:protochlorophyllide oxidoreductase. *Eur. J. Biochem.* 111:251–58

11. Aronoff, S. 1981. Chlorophyllide *b*. *Biochem. Biophys. Res. Commun.* 102:108–12

12. Aronoff, S., Kwok, E. 1977. Biosynthesis of chlorophyll *b*. *Can. J. Biochem.* 55:1091–95

13. Avissar, Y. J. 1980. Biosynthesis of 5-aminolevulinate from glutamate in *Anabaena variabilis*. *Biochim. Biophys. Acta* 613:220–28

14. Barreiro, O. L. C. de. 1975. Effect of pyridoxal phosphate and cysteine on δ-aminolaevulinate synthetase and dehydratase in soya. *Phytochemistry* 14:2165–68

15. Batlle, A. M. del C., Llambias, E. B. C., Wider de Xifra, E., Tigier, H. A. 1975. Porphyrin biosynthesis in the soybean callus tissue system—XV. The effect of growth conditions. *Int. J. Biochem.* 6:591–606

16. Batschauer, A., Santel, H.-J., Apel, K. 1982. The presence and synthesis of the NADPH-protochlorophyllide oxidoreductase in barley leaves with a high temperature-induced deficiency of plastid ribosomes. *Planta* 154:459–64

17. Battersby, A. R., Fookes, C. J. R., Matcham, G. W. J., Pandey, P. S. 1981. Biosynthesis of natural porphyrins: studies with isomeric hydroxymethylbilanes on the specificity and action of cosynthetase. *Angew. Chem. Int. Ed. Engl.* 20:293–95

18. Battersby, A. R., Fookes, C. J. R., McDonald, E., Matcham, G.W.J. 1979. Chemical and enzymatic studies on biosynthesis of the natural porphyrin macrocycle: formation and role of unrearranged hydroxymethylbilane and

order of assembly of the pyrrole rings. *Bioorg. Chem.* 8:451–63

19. Battersby, A. R., McDonald, E. 1975. Biosynthesis of porphyrins, chlorins and corrins. In *Porphyrins and Metalloporhyrins*, ed. K. M. Smith, pp. 61–122. Amsterdam: Elsevier. 910 pp.

20. Bazzaz, M.B. 1981. New chlorophyll *a* and *b* chromophores isolated from a mutant of *Zea mays* L. *Naturwissenschaften* 68:94

21. Bazzaz, M. B. 1981. New chlorophyll chromophores isolated from a chlorophyll-deficient mutant of maize. *Photobiochem. Photobiophys.* 2:199–207

22. Bazzaz, M. B., Bradley, C. V., Brereton, R. G. 1982. 4-Vinyl-4-desethyl chlorophyll *a*: Characterization of a new naturally occurring chlorophyll using fast atom bombardment, field desorption and "in beam" electron impact mass spectroscopy. *Tetrahedron Lett.* 23:1211–14

23. Bazzaz, M. B., Brereton, R. G. 1982. 4-Vinyl-4-desethyl chlorophyll *a*: a new naturally occurring chlorophyll. *FEBS Lett.* 138:104–8

24. Beale, S. I. 1971. Studies on the biosynthesis and metabolism of δ-aminolevulinic acid in *Chlorella*. *Plant Physiol.* 48:316–19

25. Beale, S. I. 1978. δ-Aminolevulinic acid in plants: its biosynthesis, regulation, and role in plastid development. *Ann. Rev. Plant Physiol.* 29:95–120

26. Beale, S. I. 1983. Biosynthesis of photosynthetic pigments. In *Topics in Photosynthesis*, ed. J. Barber, N. R. Baker, Vol. 5. Amsterdam: Elsevier. In press

27. Beale, S. I., Castelfranco, P. A. 1974. The biosynthesis of δ-aminolevulinic acid in plants. I. Accumulation of δ-aminolevulinic acid in greening plant tissues. *Plant Physiol.* 53:291–96

28. Beale, S. I., Castelfranco, P. A. 1974. The biosynthesis of δ-aminolevulinic acid in plants. II. Formation of ^{14}C-δ-aminolevulinic acid from labeled precursors in greening plant tissues. *Plant Physiol.* 53:297–303

29. Beale, S. I., Foley, T. 1982. Induction of δ-aminolevulinic acid synthase activity and inhibition of heme synthesis in *Euglena gracilis* by N-methyl mesoporphyrin IX. *Plant Physiol.* 69:1331–33

30. Beale, S. I., Foley, T., Dzelzkalns, V. 1981. δ-Aminolevulinic acid synthase from *Euglena gracilis*. *Proc. Natl. Acad. Sci. USA* 78:1666–69

31. Beale, S. I., Gough, S. P., Granick, S. 1975. The biosynthesis of δ-aminolevulinic acid from the intact car-

bon skeleton of glutamic acid in greening barley. *Proc. Natl. Acad. Sci. USA* 72:2719–23

32. Beer, N. S., Griffiths, W. T. 1981. Purification of the enzyme NADPH: protochlorophyllide oxidoreductase. *Biochem J.* 195:83–92

33. Beevers, L., Loveys, B., Pearson, J. A., Wareing, P. F. 1970. Phytochrome and hormonal control of expansion and greening of etiolated wheat leaves. *Planta* 90:286–94

34. Belanger, F. C., Duggan, J. X., Rebeiz, C. A. 1982. Chloroplast biogenesis. Identification of chlorophyllide *a* (E458F674) as a divinyl chlorophyllide *a* *J. Biol. Chem.* 257:4849–58

35. Belanger, F. C., Rebeiz, C. A. 1979. Chloroplast biogenesis XXVII. Detection of novel chlorophyll and chlorophyll precursors in higher plants. *Biochem Biophys. Res. Commun.* 88:365–72

36. Belanger, F. C., Rebeiz, C. A. 1980. Chloroplast biogenesis. Detection of divinyl protochlorophyllide in higher plants. *J. Biol. Chem.* 255:1266–72

37. Belanger, F. C., Rebeiz, C. A. 1980. Chloroplast biogenesis 30. Chlorophyll(ide) (E459F675) and chlorophyll(ide) (E449F675) the first detectable products of divinyl and monovinyl protochlorophyll photoreduction. *Plant Sci. Lett.* 18:343–50

38. Belanger, F. C., Rebeiz, C. A. 1980. Chloroplast biogenesis: Detection of divinylprotochlorophyllide ester in higher plants. *Biochemistry* 19:4875–83

39. Belanger, F. C., Rebeiz, C. A. 1982. Chloroplast biogenesis. Detection of monovinyl magnesium-protoporphyrin monoester and other monovinyl magnesium-porphyrins in higher plants. *J. Biol. Chem.* 257:1360–71

40. Bellemare, G., Bartlett, S. G., Chua, N.-H. 1982. Biosynthesis of chlorophyll *a/b*-binding polypeptides in wild type and the chlorina f2 mutant of barley. *J. Biol. Chem.* 257:7762–67

41. Bennett, J. 1981. Biosynthesis of the light-harvesting chloropyll *a/b* protein. Peptide turnover in darkness. *Eur. J. Biochem.* 118:61–70

42. Benz, J., Rüdiger, W. 1981. Chlorophyll biosynthesis: Various chlorophyllides as exogenous substrates for chlorophyll synthetase. *Z. Naturforsch. Teil C* 36:51–57

43. Benz, J., Wolf, C., Rüdiger, W. 1980. Chlorophyll biosynthesis: Hydrogenation of geranylgeraniol. *Plant Sci. Lett.* 19:225–30

44. Bogdanovíc, M. 1973. Chlorophyll formation in the dark I. Chlorophyll in pine seedlings. *Physiol. Plant.* 29:17–18

45. Bogdanovíc, M. 1973. Chlorophyll formation in the dark II. Chlorophyll in wheat leaves transplanted to pine megagametophytes. *Physiol. Plant* 29:19–21

46. Bogorad, L. 1950. Factors associated with the synthesis of chlorophyll in the dark in seedlings of *Pinus jeffreyi. Bot. Gaz.* 111:221–41

47. Bogorad, L. 1976. Chlorophyll biosynthesis. In *Chemistry and Biochemistry of Plant Pigments,* ed. T. W. Goodwin, 1:64–148. New York: Academic. 2nd ed., 870 pp.

48. Bogorad, L. 1979. In *The Porphyrins,* ed D. Dolphin, 6A:179–232. New York: Academic. 932 pp.

49. Brockmann, H. Jr., Knoblock, G., Schweer, I., Trowitzsch, W. 1973. The esterifying alcohol of the bacteriochlorophyll *a* from *Rhodospirillum rubrum. Arch. Mikrobiol.* 90:161–64

50. Brown, J. 1972. Forms of chlorophyll in vivo. *Ann. Rev. Plant Physiol.* 23:73–86

51. Brown, J. S. 1977. Spectroscopy of chlorophyll in biological and synthetic systems. *Photochem. Photobiol.* 26:319–26

52. Castelfranco, P. A. 1981. Chlorophyll biosynthesis: Mg insertion; O_2 and Fe requirements. See Ref. 4, pp. 171–76

53. Castelfranco, P. A., Beale, S. I. 1981. Chlorophyll biosynthesis. In *The Biochemistry of Plants,* ed. M. D. Hatch, N. K. Boardman, 8:375–421. New York: Academic

54. Castelfranco, P. A., Chereskin, B. M. 1982. Biosynthesis of chlorophyll *a.* In *On the Origins of Chloroplasts,* ed. J. A. Schiff, pp. 199–218. Amsterdam: Elsevier

55. Castelfranco, P. A., Jones, O. T. G. 1975. Protoheme turnover and chlorophyll synthesis in greening barley tissue. *Plant Physiol.* 55:485–90

56. Castelfranco, P. A., Rich, P. M., Beale, S. I. 1974. The abolition of the lag phase in greening cucumber cotyledons by exogenous δ-aminolevulinic acid. *Plant Physiol.* 53:615–18

57. Castelfranco, P. A., Weinstein, J. D., Schwarcz, S., Pardo, A. D., Wezelman, B. E. 1979. The Mg insertion step in chlorophyll biosynthesis. *Arch. Biochem. Biophys.* 192:592–98

58. Chereskin, B. M., Castelfranco, P. A. 1982. Effects of iron and oxygen on chlorophyll biosynthesis. II. Observations on the biosynthetic pathway in iso-

lated etiochloroplasts. *Plant Physiol.* 69:112–16

59. Chereskin, B. M., Castelfranco, P. A., Wong, Y.-S. 1982. Formation of the chlorophyll isocyclic ring. II. Stimulation and inhibition of in vitro synthesis of Mg-divinyl pheoporphyrin a_5. *Plant Physiol.* 69S:68

60. Chereskin, B. M., Wong, Y.-S., Castelfranco, P. A. 1982. In vitro synthesis of the chlorophyll isocyclic ring. Transformation of magnesium-protoporphyrin IX and magnesium-protoporphyrin mono methyl ester into magnesium-2,4-divinyl pheoporphyrin a_5. *Plant Physiol.* 70:987–93

61. Cohen, C. E., Bazzaz, M. B., Fullett, S. H., Rebeiz, C. A. 1977. Chloroplast biogenesis XX. Accumulation of porphyrin and phorbin pigments in cucumber cotyledons during photoperiodic greening. *Plant Physiol.* 60:743–46

62. Cohen, C. E., Rebeiz, C. A. 1981. Chloroplast biogenesis 34. Spectrofluorometric characterization in situ of the protochlorophyll species in etiolated tissues of higher plants. *Plant Physiol.* 67:98–103

63. Cohen, C. E., Schiff, J. A. 1976. Events surrounding the early development of *Euglena* chloroplasts-XI protochlorophyll(ide) and its photoconversion. *Photochem. Photobiol.* 24:555–66

64. Daniell, H., Rebeiz, C. A. 1982. Chloroplast culture VII. A new effect of kinetin in enhancing the synthesis and accumulation of protochlorophyllide in vitro. *Biochem. Biophys. Res. Commun.* 104:837-43

65. Daniell, H., Rebeiz, C. A. 1982. Chloroplast culture IX. Chlorophyll(ide) *a* biosynthesis in vitro at rates higher than in vivo. *Biochem Biophys. Res. Commun.* 106:466–70

66. Dörnemann, D., Senger, H. 1980. The synthesis and properties of 4,5-dioxovaleric acid, a possible intermediate in the biosynthesis of 5-aminolaevulinic acid, and its in vivo formation in *Scenedesmus obliquus*. *Biochim. Biophys. Acta* 628:35–45

67. Dörnemann, D., Senger, H. 1981. Isolation and partial characterization of a new chlorophyll associated with the reaction centre of photosystem I of *Scenedesmus*. *FEBS Lett.* 126:323–27

68. Dörnemann, D., Senger, H. 1982 Physical and chemical properties of chlorophyll RCI extracted from photosystem I of spinach leaves and from green algae. *Photochem. Photobiol.* 35:821–26

69. Duggan, J., Gassman, M. 1974. Induc-tion of porphyrin synthesis in etiolated bean leaves by chelators of iron. *Plant Physiol.* 53:206–15

70. Duggan, J. X., Meller, E., Gassman, M. L. 1982. The catabolism of 5-aminolevulinic acid to CO_2 by etiolated barley leaves. *Plant Physiol.* 69:19–22

71. Duggan, J. X., Meller, E., Gassman, M. L. 1982. Catabolism of porphobilinogen by etiolated barley leaves. *Plant Physiol.* 69:602–8

72. Duggan, J.X., Rebeiz, C.A. 1982. Chloroplast biogenesis 37. Induction of chlorophyllide *a* (E459F675) accumulation in higher plants. *Plant Sci. Lett.* 24:27–37

73. Duggan, J.X., Rebeiz, C.A. 1982. Chloroplast biogenesis 38: quantitative detection of a chlorophyllide *b* pool in higher plants. *Biochim. Biophys. Acta* 679:248–60

73a. Duggan, J. X., Rebeiz, C. A. 1982. Chloroplast biogenesis 42. Conversion of divinyl chlorophyllide *a* to monovinyl chlorophyllide *a* in vivo and in vitro. *Plant Sci. Lett.* 27:137–45

74. Duranton, J., Galmiche, J.-M., Roux, E. 1958. Métabolisme des pigments chlorophylliens chez le Tabac. *C.R. Acad. Sci.* 246:992–95

75. Dzelzkalns, V., Foley, T., Beale, S. I. 1982. δ-Aminolevulinic acid synthase of *Euglena gracilis:* Physical and kinetic properties. *Arch. Biochem. Biophys.* 216:196–203

76. Ebbon, J. G., Tait, G. H. 1969. Studies on S-adenosylmethionine-magnesium protoporphyrin methyltransferase in *Euglena gracilis* strain Z. *Biochem. J.* 111:573–82

77. Egan, J. M. Jr., Schiff, J. A. 1974. A reexamination of the action spectrum for chlorophyll synthesis in *Euglena gracilis*. *Plant Sci. Lett.* 3:101–5

78. Ellsworth, R. K. 1971. Studies on chlorophyllase I. Hydrolytic and esterification activities of chlorophyllase from wheat seedlings. *Photosynthetica* 5:226–32

79. Ellsworth, R. K., Aronoff, S. 1968. Investigations on the biogenesis of chlorophyll *a* III. Biosynthesis of Mg-vinylpheoporphine a_5 methylester from Mg-protoporphine IX monomethylester as observed in *Chlorella* mutants. *Arch. Biochem. Biophys.* 125:269–77

80. Ellsworth, R.K., Aronoff, S. 1969. Investigations on the biogenesis of chlorophyll *a* IV. Isolation and partial characterization of some biosynthetic intermediates between Mg-protoporphine IX monomethylester and Mg-vinylpheoporphine a_5, obtained from

Chlorella mutants. *Arch. Biochem. Biophys.* 130:374–83

81. Ellsworth, R. K., Hsing, A. S. 1973. The reduction of vinyl side-chains of Mg-protoporphyrin IX monomethyl ester in vitro. *Biochim. Biophys. Acta* 313:119–29

82. Ellsworth, R. K., Lawrence, G. D. 1973. Synthesis of magnesium-protoporphyrin IX in vitro. *Photosynthetica* 7:73–86

83. Ellsworth, R. K., Perkins, H. J., Detwiler, J. P., Liu, K. 1970. On the enzymatic conversion of ^{14}C-labeled chlorophyll *a*. to ^{14}C-labeled chlorophyll *b*. *Biochim. Biophys. Acta* 223:275–80

84. Ellsworth, R. K., St. Pierre, L. A. 1976. Biosynthesis and inhibition of (-)-*S*-adenosyl-l-methionine: magnesium protoporphyrin methyltransferase of wheat. *Photosynthetica* 10:291–301

85. Fletcher, R. A., McCullagh, D. 1971. Benzyladenine as a regulator of chlorophyll synthesis in cucumber cotyledons. *Can. J. Bot.* 49:2197–2201

86. Foley, T., Beale, S. I. 1982. δ-Aminolevulinic acid formation from γ, δ-dioxovaleric acid in extracts of *Euglena gracilis*. *Plant Physiol.* 70:1495–1502

87. Foley, T., Dzelzkalns. V., Beale, S. I. 1982. δ-Aminolevulinic acid synthase of *Euglena gracilis*: Regulation of activity. *Plant Physiol.* 70:219–26

88. Ford, M. J., Alhadeff, M., Chapman, J. M., Black, M. 1979. A rapid and selective action of 6-benzylaminopurine on 5-aminolevulinate production in excised sunflower cotyledons. *Plant Sci. Lett.* 16:397–402

89. Ford, S. H., Friedmann, H. C. 1979. Formation of δ-aminolevulinic acid from glutamic acid by a partially purified enzyme system from wheat leaves. *Biochim. Biophys. Acta* 569:153–58

90. Franck, F., Mathis, P. 1980. A short-lived intermediate in the photoenzymatic reduction of protochlorophyll(ide) into chlorophyll(ide) at a physiological temperature. *Photochem. Photobiol.* 32:799–803

91. Freyssinet, G., Rebeiz, C. A., Fenton, J. M., Khanna, R., Govindjee. 1980. Unequal distribution of novel chlorophyll *a* and *b* chromophores in subchloroplast particles of higher plants. *Photobiochem. Photobiophys.* 1:203–12

92. Frydman, R. B., Tomaro, M. L., Wanschelbaum, A., Anderson, E. M., Awruch, J., Frydman, B. 1973. Porphyrinogen oxygenase from wheat germ: isolation, properties and products formed. *Biochemistry* 12:5253–62

93. Frydman, R. B., Tomaro, M. L., Wanschelbaum, A., Frydman, B. 1972. The enzymatic oxidation of porphobilinogen. *FEBS Lett.* 26:203–6

94. Fuesler, T. P., Castelfranco, P. A. 1982. Chlide *b* in greening cucumber cotyledons. *Plant Physiol.* 69S: 100

95. Fuesler, T. P. Wright, L. A. Jr., Castelfranco, P. A. 1981. Properties of magnesium chelatase in greening etioplasts. Metal ion specificity and effect of substrate concentrations. *Plant Physiol.* 67:246–49

96. Games, D. E., Jackson, A. H., Jackson, J. R., Belcher, R. V., Smith, S. G. 1976. Biosynthesis of protoporphyrin-IX from coproporphyrinogen-III. *J. Chem. Soc. Chem. Commun.* pp. 187–89

97. Gassman, M. L., Duggan, J. X., Stillman, L. C., Vlcek, L. M., Castelfranco, P. A., Wezelman, B. 1978. Oxidation of chlorophyll precursors and its relation to the control of greening. See Ref. 1, pp. 167–81

98. Gassman, M. L., Pluscec, J., Bogorad, L. 1966. δ-Aminolevulinic acid transaminase from *Chlorella* and *Phaseolus*. *Plant Physiol.* 41:xiv

99. Gassman, M., Pluscec, J., Bogorad, L. 1968. δ-Aminolevulinic acid transaminase in *Chlorella vulgaris*. *Plant Physiol.* 43:1411–14

100. Gibson, K. D., Laver, W. G., Neuberger, A. 1958. Initial stages in the biosynthesis of porphyrins 2. The formation of δ-aminolevulinic acid from glycine and succinyl-coenzyme A by particles from chicken erythrocytes. *Biochem. J.* 70:71–81

101. Gibson, K. D., Neuberger, A., Tait, G. H. 1962. Studies on the biosynthesis of porphyrin and bacteriochlorophyll by *Rhodopseudomonas spheroides* 1. The effect of growth conditions. *Biochem. J.* 83:539–49

102. Gough, S. P., Girnth, C., Kannangara, C. G. 1981. δ-Aminolevulinate synthesis in greening barley. 1. Regulation. See Ref. 4, pp. 107–16

103. Gough, S. P., Kannangara, C. G. 1976. Synthesis of δ-aminolevulinic acid by isolated plastids. *Carlsberg Res. Commun.* 41:183–90

104. Gough, S. P., Kannangara, C. G. 1977. Synthesis of δ-aminolevulinate by a chloroplast stroma preparation from greening barley leaves. *Carlsberg Res. Commun.* 42:459–64

105. Gough, S. P., Kannangara, C. G. 1979. Biosynthesis of δ-aminolevulinate in greening barley leaves III: The formation of δ-aminolevulinate in *tigrina* mutants

of barley. *Carlsberg Res. Commun.* 44:403–16

106. Granick, S. 1961. Magnesium protoporphyrin monoester and protoporphyrin monomethyl ester in chlorophyll biosynthesis. *J. Biol. Chem.* 236:1168–72

107. Granick, S., Beale, S. I. 1978. Hemes, chlorophylls, and related compounds: biosynthesis and metabolic regulation. *Adv. Enzymol.* 46:33–203

108. Griffiths, W. T., Beer, N. S. 1982. Site of synthesis of NADPH protochlorophyllide oxidoreductase in rye *(Secale cereale)*. *Plant Physiol.* 70:1014–18.

109. Griffiths, W. T., Jones, O. T. G. 1975. Magnesium 2,4-divinylphaeoporphyrin a_5 as a substrate for chlorophyll biosynthesis in vitro. *FEBS Lett.* 50:355–58

110. Griffiths, W. T., Mapleston, R. E. 1978 NADPH - protochlorophyllide oxidoreductase. See Ref. 1, pp. 99–104

111. Harel, E. 1978. Chlorophyll biosynthesis and its control. *Prog. Phytochem.* 5:127–80

112. Harel, E. 1978. Initial steps in chlorophyll synthesis—problems and open questions. See Ref. 1, pp. 33–44

113. Harel, E., Meller, E., Rosenberg, M. 1978. Synthesis of 5-aminolevulinic acid–[^{14}C] by cell-free preparations from greening maize leaves. *Phytochemistry* 17:1277–80

114. Holden, M. 1961. The breakdown of chlorophyll by chlorophyllase. *Biochem. J.* 78:359–64

115. Holden, M. 1963. The purification and properties of chlorophyllase. *Photochem. Photobiol.* 2:175–80

116. Hoober, J. K., Stegeman, W. J. 1973. Control of the synthesis of a major polypeptide of chloroplast membranes in *Chlamydomonas reinhardi. J. Cell Biol.* 56:1–12

117. Hougen, C. L., Meller, E., Gassman, M. L. 1982. Magnesium protoporphyrin monoester destruction by extracts of etiolated red kidney bean leaves. *Plant Sci. Lett.* 24:289–94

118. Humbeck, K., Senger, H. 1981. Chlorophyll formation and ALA biosynthesis via two different pathways during the cell cycle of wild type cells of *Scenedesmus.* See Ref. 4, pp. 161–70

119. Inoue, Y., Kobayashi, T., Ogawa, T., Shibata, K. 1981. A short lived intermediate in the photoconversion of protochlorophyllide to chlorophyllide *a. Plant Cell Physiol.* 22:197–204

120. Jackson, A. H., Sancovich, H. A., Ferramola, A. M., Evans, N., Games, D. E., et al. 1976. Macrocyclic intermediates in the biosynthesis of porphyrins. *Philos. Trans. R. Soc. London Ser.* B 273:191–206

121. Jacobs, J. M., Jacobs, N. J., De Maggio, A. E. 1982. Protoporphyrinogen oxidation in chloroplasts and plant mitochondria, a step in heme and chlorophyll synthesis. *Arch. Biochem. Biophys.* 218:233–39

122. Jones, O. T. G. 1968. Ferrochelatase of spinach chloroplasts. *Biochem. J.* 107:113–19

123. Jones, O. T. G. 1978. Biosynthesis of porphyrins, hemes, and chlorophylls. In *The Photosynthetic Bacteria,* ed. R. K. Clayton, W. R. Sistrom, pp.751–77. New York: Plenum. 946 pp.

124. Jones, O. T. G. 1979. Chlorophyll biosynthesis. See Ref. 48, pp. 179–232

125. Jordan, P. M., Seehra, J. S. 1979. The biosynthesis of uroporphyrinogen III: order of assembly of the four porphobilinogen molecules in the formation of the tetrapyrrole ring. *FEBS Lett.* 104:364–66

126. Jordan, P. M., Seehra, J. S. 1980. Mechanism of action of 5-aminolevulinic acid dehydratase: stepwise order of addition of the two molecules of 5-aminolevulinic acid in the enzymatic synthesis of porphobilinogen. *J. Chem. Soc. Chem. Commun.* pp. 240-42

127. Jurgenson, J. E., Beale, S. I., Troxler, R. F. 1976. Biosynthesis of δ-aminolevulinic acid in a unicellular Rhodophyte, *Cyanidium caldarium. Biochem. Biophys. Res. Commun.* 69:149–57

128. Kah, A., Dörnemann, D., Rühl, D., Senger, H. 1981. The influence of light and levulinic acid on the regulation of enzymes for ALA-biosynthesis in two pigment mutants of *Scenedesmus obliquus.* See Ref. 4, pp. 137-44

129. Kannangara, C. G., Gough, S. P. 1977. Synthesis of δ-aminolevulinic acid and chlorophyll by isolated chloroplasts. *Carlsberg Res. Commun.* 42:441–57

130. Kannangara, C. G., Gough, S. P. 1978. Biosynthesis of δ-aminolevulinate in greening barley leaves: Glutamate 1-semialdehyde aminotransferase. *Carlsberg Res. Commun.* 43:185–94

131. Kannangara, C. G., Gough, S. P. 1979. Biosynthesis of δ-aminolevulinate in greening barley leaves II: Induction of enzyme synthesis by light. *Carlsberg Res. Commun.* 44:11–20.

132. Kannangara, C. G., Gough, S. P., von Wettstein, D. 1978. The biosynthesis of δ-aminolevulinate and chlorophyll and its genetic regulation. See Ref. 1, pp. 147–60

133. Katz, J. J., Strain, H. H., Harkness, A. L., Studier, M. H., Svec, W. A., et al. 1972. Esterifying alcohols in the chlorophylls of purple photosynthetic bacteria. A new chlorophyll, bacteriochlorophyll(gg), all-trans-geranylgeranyl bacteriochlorophyllide a. J. Am. Chem. Soc. 94:7938–39

134. Kikuchi, G., Kumar, A., Talmage, P., Shemin, D. 1958. The enzymatic synthesis of δ-aminolevulinic acid. J. Biol. Chem. 233:1214–19

135. Kipe-Nolt, J. A., Stevens, S. E. Jr. 1980. Biosynthesis of δ-aminolevulinic acid from glutamate in Agmenellum quadruplicatum. Plant Physiol. 65:126–28

136. Kissel, H. J., Heilmeyer, L. Jr. 1969. Nachweis und Bestimmung von γ,δ-Dioxovaleriansäure. Reversible Umwandlung von γ,δ-Dioxovaleriansäure und δ-Aminolävulinsäure in Ratten. Biochim. Biophys. Acta 177:78–87

137. Klein, A. O., Vishniac, W. 1961. Activity and partial purification of chlorophyllase in aqueous systems. J. Biol. Chem. 236:2544–47

138. Klein, O., Porra, R. J. 1982. The participation of the Shemin and C_5 pathways in 5-aminolaevulinate and chlorophyll formation in higher plants and facultative photosynthetic bacteria. Z. Physiol. Chem. 363:551–62

139. Klein, O., Senger, H. 1978. Biosynthetic pathways to δ-aminolevulinic acid induced by blue light in the pigment mutant C-2A' of Scenedesmus obliquus. Photochem. Photobiol. 27:203–8

140. Klein, O., Senger, H. 1978. Two biosynthetic pathways to δ-aminolevulinic acid in a pigment mutant of the green alga, Scenedesmus obliquus. Plant Physiol. 62:10–13

141. Lascelles, J. 1978. Regulation of pyrrole synthesis. See Ref. 123, pp. 795–808

142. Lascelles, J. A., Hatch, T. P. 1969. Bacteriochlorophyll and heme synthesis in Rhodopseudomonas spheroides: possible role of heme in regulation of the branched biosynthetic pathway. J. Bacteriol. 98:712–20

143. Laudi, G., Manzini, M. L. 1975. Chlorophyll content and plastid ultrastructure in leaflets of Metasequoia glyptostroboides. Protoplasma 84:185–90

144. Laycock, M. V., Wright, J. L. C. 1981. The biosynthesis of phycocyanobilin in Anacystis nidulans. Phytochemistry 20:1265–68

145. Lew, R., Tsuji, H. 1982. Effect of benzyladenine treatment duration on δ-aminolevulinic acid accumulation in the

dark, chlorphyll lag phase abolition, and long-term chlorophyll production in excised cotyledons of dark-grown cucumber seedlings. Plant Physiol. 69:663–67

146. Liljenberg, C. 1974. Chracterization and properties of a protochlorophyllide ester in leaves of dark grown barley with geranylgeraniol as esterifying alcohol. Physiol. Plant. 32:208–13

147. Little, H. N., Jones, O. T. G. 1976. The subcellular localization and properties of the ferrochelatase of etiolated barley. Biochem. J. 156:309–14

148. Litvin, F. F., Ignatov, N. V., Belyaeva, O. B. 1981. Photoreversibility of transformations of photochlorophyllide into chlorophyllide. Photobiochem. Photobiophys. 2:233–37

149. Lohr, J. B., Friedmann, H. C. 1976. New pathway for δ-aminolevulinic acid biosynthesis: formation from α-ketoglutaric acid by two partially purified plant enzymes. Biochem. Biophys. Res. Commun. 69:908–13

150. Lütz, C., Roper, U., Beer, N. S., Griffiths, T. 1981. Sub-etioplast localization of the enzyme NADPH: protochlorophyllide oxidoreductase. Eur. J. Biochem. 118:347–53

151. Mapleston, R. E., Griffiths, W. T. 1977. Effects of illumination of whole barley plants on the protochlorophyllide-activating system in the isolated plastids. Biochem. Soc. Trans. 5:319–21

152. Mapleston, R. E., Griffiths, W. T. 1980. Light modulation of the activity of protochlorophyllide reductase. Biochem. J. 189:125–33

153. McCarthy, S. A., Belanger, F. C., Rebeiz, C. A. 1981. Chloroplast biogenesis: Detection of a magnesium protoporphyrin diester pool in plants. Biochemistry 20:5080–87

154. McCarthy, S. A., Mattheis, J. R., Rebeiz, C. A. 1982. Chloroplast biogenesis: Biosynthesis of protochlorophyll(ide) via acidic and fully esterified biosynthetic branches in higher plants. Biochemistry 21:242–47

155. McKie, J., Lucas, C., Smith, A. 1981. δ-Aminolaevulinate biosynthesis in the cyanobacterium Synechococcus 6301. Phytochemistry 20:1547–49

156. Meller, E., Belkin, S., Harel, E. 1975. The biosynthesis of δ-aminolevulinic acid in greening maize leaves. Phytochemistry 14:2399–402

157. Meller, E., Gassman, M. L. 1981. The effects of levulinic acid and 4,6-dioxoheptanoic acid on the metabolism of etiolated and greening barley leaves. Plant Physiol. 67:728–32

158. Meller, E., Gassman, M. L. 1982. Biosynthesis of 5-aminolevulinic acid: two pathways in higher plants. *Plant Sci. Lett.* 26:23–29
159. Meller, E., Harel, E. 1978. The pathway of 5-aminolevulinic acid synthesis in *Chlorella vulgaris* and in *Fremyella diplosiphon.* See Ref. 1, pp. 51–57
160. Meller, E., Harel, E., Kannangara, C.G. 1979. Conversion of glutamic-l-semialdehyde and 4,5-dioxovaleric acid to 5-aminolevulinic acid by cell-free preparations from greening maize leaves. *Plant Physiol.* 63S:98
161. Nadler, K., Granick, S. 1970. Controls on chlorophyll synthesis in barley. *Plant Physiol.* 46:240–46
162. Nasrulhaq-Boyce, A., Jones, O. T. G. 1981. Tetrapyrrole biosynthesis in greening etiolated barley seedlings. *Phytochemistry* 20:1005–9
163. Neuberger, A., Turner, J. M. 1963. γ,δ-Dioxovalerate aminotransferase activity in *Rhodopseudomonas spheroides.* *Biochim Biophys. Acta* 67:342–45
164. Oelze-Karow, H., Kasemir, H., Mohr, H. 1978. Control of chlorophyll *b* formation by phytochrome and a threshold level of chlorophyllide *a.* See Ref. 1, pp. 787–92
165. Oelze-Karow, H., Mohr, H. 1978. Control of chlorophyll *b* biosynthesis by phytochrome. *Photochem. Photobiol.* 27:189–93
166. Ogawa, T., Bovey, F., Shibata, K. 1975. An intermediate in the phytylation of chlorophyllide *a* in vivo. *Plant Cell Physiol.* 16:199–202
167. Ogawa, T., Inoue, Y., Kitajima, M., Shibata, K. 1973. Action spectra for the biosynthesis of chlorophylls *a* and *b* and β-carotene. *Photochem. Photobiol.* 18:229–35
168. Oh-hama, T., Seto, H., Otake, N., Miyachi, S. 1982. ^{13}C-NMR evidence for the pathway of chlorophyll biosynthesis in green algae. *Biochem. Biophys. Res. Commun.* 105:647–52
169. Oliver, R. P., Griffiths, W. T. 1980. Identification of the polypeptides of NADPH-protochlorophyllide oxidoreductase. *Biochem. J.* 191:277–80
170. Oliver, R. P., Griffiths, W. T. 1981. Covalent labelling of the NADPH:protochlorophyllide oxidoreductase from etioplast membranes with [^{3}H]*N*-phenylmaleimide. *Biochem. J.* 195:93-101
171. Pardo, A. D., Chereskin, B. M., Castelfranco, P. A., Franceschi, V. R., Wezelman, B. E. 1980. ATP requirement for

172. Popov, K., Dilova, S. 1969. On the dark synthesis and stabilization of chlorophyll. In *Progress in Photosynthesis Research,* ed. H. Metzner, 2:606–10. Tubingen: Int. Union Biol. Sci. 1127 pp.
173. Porra, R. J., Grimme, L. H. 1974. Chlorophyll synthesis and intracellular fluctuations of 5-aminolaevulinate formation during the regreening of nitrogen-deficient *Chlorella fusca.* *Arch. Biochem. Biophys.* 164:312–21
174. Poulson, R., Polglase, W. J. 1975. The enzymatic conversion of protoporphyrinogen IX to protoporphyrin IX. Protoporphyrinogen oxidase activity in mitochondrial extracts of *Saccharomyces cerevisiae.* *J. Biol. Chem.* 250:1269–74
175. Radmer, R. J., Bogorad, L. 1967. (-)S-Adenosyl- L- methionine- magnesium protoporphyrin methyltransferase, an enzyme in the biosynthetic pathway of chlorophyll in *Zea mays. Plant Physiol.* 42:463–65
176. Ramaswamy, N. K., Nair, P. M. 1973. δ-Aminolevulinic acid synthetase from cold-stored potatoes. *Biochim. Biophys. Acta* 293:269–77
177. Ramaswamy, N. K., Nair, P. M. 1974. Temperature and light dependency of chlorophyll synthesis in potatoes. *Plant Sci. Lett.* 2:249–56
178. Ramaswamy, N. K., Nair, P. M. 1976. Pathway for the biosynthesis of delta-aminolevulinic acid in greening potatoes. *Indian J. Biochem. Biophys.* 13:394–97
179. Rebeiz, C. A., Abou Haidar, M., Yaghi, M., Castelfranco, P.A. 1970. Porphyrin biosynthesis in cell-free homogenates from higher plants. *Plant Physiol.* 46:543–49
180. Rebeiz, C. A., Belanger, F. C., Freyssinet, G., Saab, D. G. 1980. Chloroplast biogenesis XXIX. The occurrence of several novel chlorophyll *a* and *b* chromophores in higher plants. *Biochim. Biophys. Acta* 590:234–47
181. Rebeiz, C. A., Belanger, F. C., McCarthy, S. A., Freyssinet, G., Duggan, J. X., et al. 1981. Biosynthesis and accumulation of novel chlorophyll *a* and *b* chromophoric species in green plants. See Ref. 4, pp. 197–212
182. Rebeiz, C. A., Castelfranco, P. A. 1971. Protochlorophyll biosynthesis in a cell-free system from higher plants. *Plant Physiol.* 47:24–32
183. Rebeiz, C. A., Castelfranco, P. A. 1973. Protochlorophyll and chlorophyll biosynthesis in cell-free systems from

Mg chelatase in developing chloroplasts. *Plant Physiol.* 56:956–60 higher plants. *Ann. Rev. Plant Physiol.* 24:129–72

184. Rebeiz, C. A., Smith, B. B., Mattheis, J. R., Cohen, C. E., McCarthy, S. A. 1978. Chlorophyll biosynthesis: the reactions between protoporphyrin IX and phototransformable protochlorophyll in higher plants. See Ref. 1, pp. 59–76.

185. Richter, M. L., Rienits, K. G. 1980. The synthesis of magnesium and zinc protoporphyrin IX and their monomethyl esters in etioplast preparations studied by high pressure liquid chromatography. *FEBS Lett.* 116:211–16

186. Richter, M. L., Rienits, K. G. 1982. The synthesis of magnesium-protoporphyrin IX by etiochloroplast membrane preparations. *Biochim. Biophys. Acta* 717:255–64

187. Röbbelen, G. 1956. Über die Protochlorophyllreduktion in einer Mutante von *Arabidopsis thaliana* (L) Heynh. *Planta* 26:532–46

188. Rüdiger, W., Benz, J., Guthoff, C. 1980. Detection and partial characterization of activity of chlorophyll synthetase in etioplast membranes. *Eur. J. Biochem.* 109:193–200

189. Rüdiger, W., Benz, J., Lempert, U., Schoch, S., Steffens, D. 1976. Inhibition of phytol accumulation with herbicides. Geranylgeraniol and dihydrogeranylgeraniol-containing chlorophyll from wheat seedlings. *Z. Pflanzenphysiol.* 80:131–43

190. Rüdiger, W., Hedden, P., Köst, H. -P., Chapman, D. J. 1977. Esterification of chlorophyllide by geranylgeranyl pyrophosphate in a cell-free system from maize shoots. *Biochem. Biophys. Res. Commun.* 74:1268–72

191. Salvador, G. F. 1978. La synthèse d'acide δ-aminolévulinique par des chloroplastes isolés d' *Euglena gracilis*. *C. R. Acad. Sci.* 286:49–52

192. Salvador, G. F. 1978. δ-Aminolevulinic acid synthesis from γ,δ-dioxovaleric acid by acellular preparations of *Euglena gracilis*. *Plant Sci. Lett.* 13:351–55

193. Salvador, G. F. 1978. δ-Aminolevulinic acid synthesis during greening of *Euglena gracilis*. See Ref. 1, pp. 161–65

194. Santel, H.-J., Apel, K. 1981. The protochlorophyllide holochrome of barley (*Hordeum vulgare* L.). The effect of light on the NADPH: protochlorophyllide oxidoreductase. *Eur. J. Biochem.* 120:95–103

195. Sasa, T., Sugahara, K. 1976. Photoconversion of protochlorophyll to chlorophyll *a* in a mutant of *Chlorella regularis*. *Plant Cell Physiol.* 17:273–79

196. Sawyer, E., Smith, R. A. 1958. δ-Aminolevulinate synthesis in *Rhodopseudomonas spheroides*. *Bacteriol. Proc.*, p.111

197. Schoch, S., Hehlein, C., Rüdiger, W. 1980. Influence of anaerobiosis on chlorophyll biosynthesis in greening oat seedlings (*Avena sativa* L.). *Plant Physiol.* 66:576–79

198. Schoch, S., Lempert, U., Rüdiger, W. 1977. On the last steps of chlorophyll biosynthesis. Intermediates between chlorophyllide and phytol-containing chlorophyll. *Z. Pflanzenphysiol.* 83:427–36

199. Schoch, S., Schäfer, W. 1978. Tetrahydrogeranylgeraniol, a precursor of phytol in the biosynthesis of chlorophyll *a*—localization of the double bonds. *Z. Naturforsch. Teil C* 33:408–12

200. Schou, L. 1951. On chlorophyll formation in the dark in excised embryos of *Pinus jeffreyi*. *Physiol. Plant.* 4:617–20

201. Scott, A. I., Burton, G. Jordan, P. M., Matsumoto, H., Fagerness, P. E., Pryde, L. M. 1980. N. M. R. spectroscopy as a probe for the study of enzyme-catalysed reactions. Further observations of preuroporphyrinogen, a substrate for uroporphyrinogen III cosynthetase. *J. Chem. Soc. Chem. Commun.* pp. 384–87

202. Shlyk, A. A. 1971. Biosynthesis of chlorophyll *b*. *Ann. Rev. Plant Physiol.* 22:169–84

203. Shlyk, A. A., Prudnikova, I.V. 1967. Dark biosynthesis of chlorophyll *b* by fractions of barley chloroplasts. *Photosynthetica* 1:157–70

204. Shlyk, A. A., Prudnikova, I.V., Malashevich, A.V. 1971. Dark conversion of externally introduced chlorophyll *a* to chlorophyll *b* in a homogenate of etiolated corn seedlings. *Dokl. Akad. Nauk SSSR* 201:1481–84

205. Singhal, G. S., Williams, W.P., Rabinowitch, E. 1968. Fluorescence and absorption studies on chlorophyll *a* in vitro at 77°K. *J. Phys. Chem.* 72:3941–51

206. Sisler, E. C., Klein, W. H. 1963. The effect of age and various chemicals on the lag phase of chlorophyll synthesis in dark grown bean seedlings. *Physiol. Plant.* 16:315–22

207. Smith, B. B., Rebeiz, C. A. 1977. Chloroplast biogenesis: detection of Mg-protoporphyrin chelatase in vitro. *Arch. Biochem. Biophys.* 180:178–85

208. Smith, B. B., Rebeiz, C. A. 1979. Chloroplast biogenesis XXIV. Intrachloroplastic localization of the biosynthesis and accumulation of protoporphyrin IX, magnesium-protoporphyrin monoester,

and longer wavelength metalloporphyrins during greening. *Plant Physiol.* 63:227–31

209. Soll, J., Schultz, G. 1981. Phytol synthesis from geranylgeraniol in spinach chloroplasts. *Biochem. Biophys. Res. Commun.* 99:907–12

210. Spiller, S. C., Castelfranco, A. M., Castelfranco, P. A. 1982. Effects of iron and oxygen on chlorophyll biosynthesis. I. In vivo observations on iron and oxygen-deficient plants. *Plant Physiol.* 69:107–11

211. Thorne, S. W., Boardman, N. K. 1971. Formation of chlorophyll *b*, and the fluorescence properties and photochemical activites of isolated plastids from greening pea seedlings. *Plant Physiol.* 47:252–61

212. Troxler, R. F., Offner, G. D. 1979. δ-Aminolevulinic acid synthesis in a *Cyanidium caldarium* mutant unable to make chlorophyll *a* and phycobiliproteins. *Arch. Biochem. Biophys.* 195:53–65

213. Van Steveninck, J., Weed, R. I., Rothstein, A. 1965. Localization of erythrocyte membrane sulfhydryl groups essential for glucose transport. *J. Gen. Physiol.* 48:617–32

214. Varticovski, L., Kushner, J. P., Burnham, B. F. 1980. Biosynthesis of porphyrin precursors. Purification and characterization of mammalian L-alanine:γ,δ-dioxovaleric acid aminotransferase. *J. Biol. Chem.* 255:3742–47

215. Virgin, H. I. 1981. The physical state of protochlorophyll(ide) in plants. *Ann. Rev. Plant Physiol.* 32:451–63

216. Vlcek, L. M., Gassman, M. L. 1979. Reversal of α, α'-dipyridyl-induced porphyrin synthesis in etiolated and greening red kidney bean leaves. *Plant Physiol.* 64:393–97

217. Wang, W.-Y, Boynton, J. E., Gillham, N. W. 1977. Genetic control of chlorophyll biosynthesis: Effect of increased δ-aminolevulinic acid synthesis on the phenotype of the *y-1* mutant of *Chlamydomonas*. *Mol. Gen. Genet.* 152:7–12

218. Wang, W.-Y., Gough, S. P., Kannangara, C. G. 1981. Biosynthesis of δ-aminolevulinate in greening barley leaves IV. Isolation of three soluble enzymes required for the conversion of glutamate to δ-aminolevulinate. *Carlsberg Res. Commun.* 46:243-57

219. Weinstein, J. D. 1979. *The biosynthesis of δ-aminolevulinic acid and Mg-protoporphyrin-IX in greening chloroplasts.* PhD thesis. Univ. Calif., Davis. 121 pp.

220. Weinstein, J. D., Castelfranco, P. A. 1978. Mg-Protoporphyrin-IX and δ-aminolevulinic acid synthesis from glutamate in isolated greening chloroplasts. δ-Aminolevulinic acid synthesis. *Arch. Biochem. Biophys.* 186:376–82

221. Wellburn, A. R. 1968. The stereochemistry of hydrogen transfer during the reduction of C-20 isoprenoids in higher plants. *Phytochemistry* 7:1523–28

222. Wellburn, A. R. 1970. Studies on the esterification of chlorophyllides. *Phytochemistry* 9:2311–13

223. Wellburn, A. R. 1976. Evidence for chlorophyllide esterified with geranylgeraniol in newly greened leaves. *Biochem. Physiol. Pflanz.* 169:265–71

224. Wellburn, F. A. M., Wellburn, A. R., Senger, H. 1980. Changes in ultrastructure and photosynthetic capacity within *Scenedesmus obliquus* mutants C-2A', C-6D and C-6E on transfer from dark grown to illuminated conditions. *Protoplasma* 103:35–54

225. Wider de Xifra, E. A., Batlle, A. M. del C., Tigier, H. A. 1971. δ-Aminolaevulinate synthetase in extracts of cultured soybean cells. *Biochim. Biophys. Acta* 235:511–17

226. Wider de Xifra, E. A., Stella, A. M., Batlle, A. M. del C. 1978. Porphyrin biosynthesis—immobilized enzymes and ligands. IX. Studies on δ-aminolaevulinate synthetase from cultured soybean cells. *Plant Sci. Lett.* 11:93–98

227. Wolwertz, M.-R. 1978. Two alternative pathways of chlorophyll biosynthesis in *Pinus jeffreyi*. See Ref. 1, pp. 111–18

228. Wong, Y.-S., Castelfranco, P. A., Chereskin, B. M. 1982. Formation of the Chl isocyclic ring. I. Synthesis of Mg-2,4 divinyl pheoporphyrin a-5 by isolated plastids. *Plant Physiol.* 69S:68

229. Yoshinaga, T., Sano, S. 1980. Coproporphyrinogen oxidase. I. Purification, properties, and activation by phospholipids. *J. Biol. Chem.* 255:4722–26

230. Yubisui, T., Yoneyama, Y. 1972. δ-Aminolevulinic acid synthetase of *Rhodopseudomonas spheroides:* Purification and properties of the enzyme. *Arch. Biochem. Biophys.* 150:77–85

Ann. Rev. Plant Physiol. 1983. 34:279–310

ORGANIZATION AND STRUCTURE OF CHLOROPLAST GENES

Paul R. Whitfeld and Warwick Bottomley

Division of Plant Industry, CSIRO, Canberra City, A.C.T. 2601, Australia

CONTENTS

0066–4294/83/0601–0279$02.00

INTRODUCTION

Chloroplast DNA (cpDNA) ranges in size from 85 kb to more than 190 kb, depending on the organism from which it is derived, but for most of the higher plants so far examined it is around 150 kb (9). The DNA molecules of such a size lend themselves readily to analysis by the application of recombinant DNA techniques, and the preparation of libraries of cloned restriction fragments which in toto represent the whole chloroplast genome has become standard practice in many laboratories. The availability of recombinant plasmids bearing segments of cpDNA has simplified the process of identifying genes and of mapping them onto a physical restriction map of cpDNA and has led to an explosion of information on the structure and organization of the chloroplast genome. In view of the current high level of activity in this area, we considered it appropriate in this review to concentrate our attention largely on papers that through the analysis of cpDNA sequences have yielded interesting information on the molecular architecture of the organelle genome. In so doing, we have been unable to cover many other important facets of chloroplast molecular biology research. Fortunately, however, over the past few years there have been many excellent reviews wherein these topics are discussed at length (4, 9, 12, 18, 29, 40, 46, 54, 59, 114, 142, 151).

GENERAL ORGANIZATION OF CHLOROPLAST DNA SEQUENCES

Digestion of cpDNA with restriction enzymes yields a spectrum of different sized fragments identifiable in terms of their relative electrophoretic mobility on agarose or acrylamide gels. A physical map of cpDNA can be constructed by determining the consecutive order of the restriction fragments in the intact molecule. Restriction maps of cpDNA from numerous plant species (Table 1) have now been published, and they provide a framework for the mapping of individual genes and for comparing the sequence organization of cpDNAs from different species (Figure 1).

cpDNA from the majority of species studied contains a large (20–28 kb) inverted repeat sequence, the segments of which are separated on the circular cpDNA molecule by a large and a small single-copy DNA region (see 4, 9, and references therein). There is no inverted repeat sequence in the cpDNAs of *Vicia faba* (72), *Pisum sativum* (75), or *Euglena gracilis,* although in *Euglena* there are three copies of a 5.6 kb DNA segment arranged tandemly in a cluster (48, 62, 109).

The restriction patterns of cpDNAs from plants within a species, and in some cases even between species, are identical or similar (e.g. 35, 43, 111). On the other hand, restriction patterns of cpDNA from more distantly related plants

Table 1 Species for which restriction maps are available and the genes which have been located on them[a]

Species	Genes mapped	References
Atriplex triangularis	23S, 16SrDNA; *rbcL*; *psbA*	101
Atropa belladonna	23S, 16S, 5S, 4.5S rDNA	35
Chlamydomonas reinhardtii	23S, 16S, 7S, 5S, 3S *rDNA*; *rbcL*; *psbA*; *tufA*	89, 90, 113–116 150
Cucumis sativa	23S, 16SrDNA; *rbcL*; *psbA*	101, 104
Euglena gracilis	23S, 16S, 5S rDNA; tRNA; *rbcL*; *psbA*	44, 47, 48, 54, 97, 98, 131
Lycopersicon spp.	23S, 16S, rDNA	105
Nicotiana spp.	23S, 16S, 5S, 4.5S rDNA; +RNA; *rbcL*	36, 64, 68, 78, 126, 127, 135, 139, 144
Oenothera spp.	23S, 16S, 5S rDNA	42, 43
Osmunda cinnamonea	23S, 16S, rDNA; *atpA,B*; *rbcL*; psbA	102
Pennisetum americanum	23S, 16S, 5S rDNA	108
Petunia spp.	23S, 16S, 5S, 4.5S rDNA; *rbcL*	15, 35, 77, 104
Pisum sativum	23S, 16S rDNA; *cytF*; *rbcL*; *psbA*	20, 103, 104, 156a
Sinapis alba	23S, 16S rDNA; *rbcL*; *psbA*	82, 84
Solanum spp.	23S, 16S, rDNA	105
Spinacia oleracea	23S, 16S, 5S, 4.5S rDNA; tRNA; *atpA,B,E*; *rbcL*: *psbA*	10, 26, 27, 120, 152–154
Spirodela oligorrhiza	tRNA	51, 149
Triticum vulgaris	23S, 16S rDNA; *atpB,E,H*; *rbcL*	17, 60, 61
Vicia faba	23S, 16S, rDNA	72, 104
Vigna radiata	*23S, 16S, rDNA*; *rbcL*; *psbA*	103, 104
Zea mays	23S, 16S, 5S, 4.5S rDNA; tRNA; *atpB,E*; *rbcL*; *psbA*	2, 5a, 76, 83, 121, 124

[a]Abbreviations used here and throughout this chapter:

cpDNA = chloroplast DNA;
DCCD = dicyclohexylcarbodiimide;
LSU = large subunit;
RuBP = ribulose 1,5-bisphosphate;
atpA, atpB, atpE = genes for the subunits α, β, and ε respectively of ATP synthase;
atpH = gene for subunit III of the membrane embedded part of ATP synthase (DCCD binding polypeptide);
rbcL = gene for the LSU of RuBP carboxylase;
rrn = operon for ribosomal RNAs;
psbA = gene for the "32 kd" thylakoid membrane protein of Photosystem II;
tufA = gene for the protein synthesis elongation factor EF_t.

may show no similarity (111). Nevertheless, DNA-DNA hybridization studies reveal that even among unrelated plant species at least 30% of the cpDNA sequences are shared (7, 79). These conserved sequences are interspersed with divergent ones (80) and probably correlate with coding and noncoding regions. There is strong evidence for a marked conservation of positional arrangement of sequences along the cpDNA molecule. The cpDNAs of tobacco, spinach,

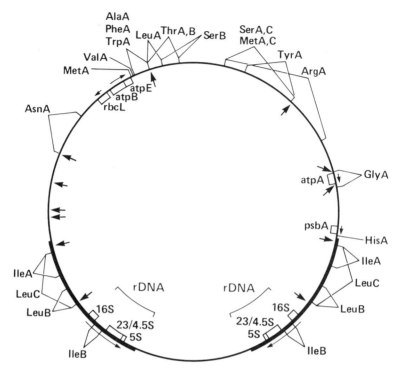

Figure 1. Map of spinach cpDNA showing the location of genes for ribosomal RNAs, tRNAs, and five proteins. Direction of transcription of the ribosomal RNA operon and the five protein genes are indicated with arrows. Heavy lines represent the inverted repeat segments. *Sal*I restriction sites are marked with arrows. Map compiled from data given in references for *Spinacia oleracea* in Table I. Location of tRNA genes in most cases is to a restriction fragment.

Atriplex, petunia, and cucumber, for instance, are essentially colinear (35, 101, 104). Palmer & Thompson (104) showed that the genomes of spinach, petunia, and cucumber differ from those of the legumes, mung bean, pea, and broad bean, by virtue of a 50 kb inversion (104). A smaller, second inversion within the 50 kb inversion distinguishes the maize chloroplast genome from the spinach group (104). In contrast to these simple switches, comparison of the pea, broad bean, and mung bean chloroplast genomes reveals the existence of extensive rearrangements (103, 104). Palmer & Thompson (104) correlate the relative stability of chloroplast genome organization with the presence of the inverted repeat region, and the more dynamic arrangement seen in pea and broad bean with the lack of this feature.

In the final analysis, the extent of sequence divergence and rearrangement between the cpDNAs of different species will be fully appreciated only when the DNAs are completely sequenced. In the following sections of this review it will be seen that quite large regions of the maize, spinach, tobacco, and *Euglena* chloroplast genomes have already been sequenced and, as more data become available, comparisons of a significant proportion of these four genomes will be possible.

GENES FOR RIBOSOMAL RNAS

Organization

The ribosomal RNA genes of chloroplasts are arranged as they are in *E. coli* and the blue-green alga *Anacystis nidulans* (146), in the order 16S rDNA-spacer-23S rDNA-spacer-5S rDNA (e.g. 5, 24, 49, 116, 154). In flowering plants (16, 137, 156) and in at least one species of fern (134), a small (4.5S) RNA species (ranging in size from 65–103 nucleotides), in addition to the usual 5S rRNA, is found associated with the 50S ribosomal subunit (155). The gene for this RNA lies close to the 3' end of the 23S rDNA (3, 32, 135). Two low molecular weight (3S and 7S) RNAs occur also in the 50S ribosomal subunit of *Chlamydmonas reinhardii,* and they are encoded by sequences immediately before the 5' end of the 23S rDNA (116). In *Euglena* cpDNA the size of the 16S–23S rDNA spacer [259 bp (44)] is less than that of the spacer in *rrnD* of *E. Coli* (437 bp), but in both organisms this spacer contains the genes for $tRNA^{Ile}$ and $tRNA^{Ala}$ (44). The 16S–23S rDNA spacer region in many higher plants, however, is much greater (>2 kb) (5, 20, 72, 78), the larger size, at least in the case of maize and tobacco, being attributable mainly to the presence of introns in the tRNA genes (see tRNA section below).

The *rrn* operon in *Chlamydomonas* and most higher plant chloroplasts is located in the inverted repeat region of the DNA (Figure 1), and there are consequently two copies per genome (e.g. 5, 17, 36, 64, 116, 154). Two exceptions to this generalization have been reported. Electron microscopic analysis of rRNA hybridized to *Vicia faba* cpDNA and to restriction fragments thereof showed there to be only one *rrn* operon present and no large inverted repeat sequence (72). Likewise, *Pisum sativum* cpDNA does not contain an inverted repeat sequence (75), and contrary to an earlier report (143), it has now been unequivocally established that there is only one copy of the rDNA genes present in the genome (20, 103). The organization of the *rrn* operon in *Euglena* is quite different from that of higher plants. In the two most frequently studied strains, *Z* and *bacillaris,* there are three copies of the *rrn* operon arranged tandemly in a cluster of repeating 5.6 kb units (48, 58, 62, 109). Minor differences in the length and sequence of the region separating the repeating units in the two strains have been noted (58, 157), and strain Z carries an extra copy of the 16S rDNA 3 kb upstream from the main cluster (63). This extra 16S rDNA has the same orientation, length, and probably sequence as the 16S rDNA in the complete operon (73). The occurrence of three complete sets of rDNA genes in *Euglena* is not invariant as it has recently been shown that *Euglena gracilis* strain Z-S carries only one set (157).

Sequencing of cloned cpDNA fragments bearing the rDNA genes has now produced a detailed picture of the complete *rrn* operon of maize (32, 122, 123) and tobacco (133, 140, 145) and of a substantial portion of the *Euglena* operon (45). In addition to providing the primary structure of the genes, these studies have contributed greatly to validating the secondary structure models proposed for both 16S and 23S rRNAs (41, 123).

16S rDNA

The first large rRNA gene from chloroplasts to be sequenced was that for the 16S rRNA of maize (122, 123). Since the 5' terminal sequence of the 16S chloroplast rRNA was not known, the start point of the gene was defined by aligning the sequence with that of 16S rDNA from *E. coli*. The size of the chloroplast gene was thus estimated to be 1491 bp (123) (cf 1541 bp for *E. coli*); 1144 of the 1491 positions are the same as in *E. coli* (74% homology), the maximum number of consecutive identical nucleotides being 53. The 16S rRNA genes in *N. tabacum* and *Euglena gracilis* have also been sequenced now. They are 1486 bp (145) and 1491 bp (45) respectively, the 5' terminus being precisely determined in the case of tobacco by S1 nuclease mapping. The tobacco and *Euglena* sequences have 96% and 80% homology with the maize sequence and 74% and 72% homology with the *E. coli* sequence. The extensive homology between the maize and *E. coli* sequences includes the colicin E3 cleavage site and 10 of the 11 methylated positions in *E. coli* 16S rRNA but stops abruptly at the junction of the structural gene with the flanking regions (123).

In common with all organisms examined so far, including the cyanobacterium *Synechococcus,* maize, tobacco, and *Euglena* chloroplasts contain sequences close to the 3' end of the 16S rDNA which code for a conserved hairpin structure (11, 148). However, it is noteworthy that it is only the chloroplast 16S rDNAs which have the *E. coli*-like 5'CCTCCTTT$_{OH}$ (5'CTCC$_{OH}$ in *Euglena*) 3' terminal sequence (148). In bacterial 16S rRNA, part of this sequence is known to base-pair with the "Shine-Dalgarno" (128) sequence upstream from the first methionine codon in the mRNA, forming a structure important in the initiation of translation.

Apart from base changes and single or double base-pair deletions or insertions, the major difference between the chloroplast and *E. coli* 16S rDNAs lies in the presence of several deletions ranging from 5 bp to 26 bp in the organelle gene. The same deletions occur in both the maize and tobacco genes, and two of them are common to the genes of all three chloroplast species and result in possible hairpin structures in the *E. coli* sequence being precisely cut out (145). Comparison of the sequences of the *E. coli* and chloroplast genes with reference to the secondary structure model of 16S RNA (145) shows that where a base change occurs in one side of a stem structure there is usually a corresponding change in the complementary nucleotide on the other side of the stem. This sort of analysis demonstrates that compensating base substitutions between functionally related RNA species can provide a valuable criterion for the evaluation of the hypothetically or experimentally deduced secondary structure (45, 123, 145).

One other similarity between the secondary structures of the *E. coli* and tobacco chloroplast 16S rDNAs was noted by Tohdoh & Suguira (145). The 5' flanking nucleotides between positions −8 and −146 of the tobacco gene can base-pair extensively with 3' flanking nucleotides between positions 1485 and 1610 and generate a structure comparable to that formed between the complementary 5' and 3' flanking sequences of the 16S rDNA in *E. coli rrn* operons. In *E. coli* this base-paired stem structure is a substrate for RNase III in the processing of precursor rRNA into mature 16S rRNA (158). The implication of this observation is that RNA-processing enzymes with properties similar to those in bacteria also occur in chloroplasts.

23S rDNA

Cloned cpDNA fragments carrying the genes for the 23S and 4.5S rRNAs of maize and tobacco chloroplasts have been sequenced (32, 140). The 5' and 3' ends of tobacco 23S rDNA were determined by S1 nuclease mapping and the size of the gene shown to be 2804 bp (cf 2904 bp for *E. coli* 23S rDNA). The extremities of the maize gene have not been defined experimentally, but by aligning the sequence with the *E. coli* and tobacco sequences it is possible to identify the ends tentatively, allowing an estimate of 2887 bp for the overall size. Following the 3' end of the 23S rDNA there is a 101 bp spacer in tobacco (84 bp in maize) and then the 103 bp 4.5S rDNA (95 bp in maize). The tobacco chloroplast 23S rDNA shows 92% homology with the maize species, as does also the 4.5S rDNA if a 7 bp deletion from the maize 4.5S rDNA sequence is ignored. However, the 23S–4.5S rDNA spacer sequence is not so well conserved (32, 140).

Homology between the chloroplast and *E. coli* 23S rDNA sequences (67% for tobacco, 71% for maize) is slightly less than it is for the 16S rDNAs. There are three significant insertions (25 bp, 65 bp, and 78 bp) in the maize sequence compared to *E. coli* (32). Each features an inverted and a tandem repeat sequence and thus shows some structural resemblance to insertion elements of bacterial genomes. Two of the inserts have counterparts in the tobacco 23S rDNA sequence. The third insert (65 bp) is absent from tobacco and partly accounts for the larger size of the maize gene. Chloroplast 4.5S rDNA has

striking homology (67% in the case of tobacco, 71% in maize) with the 3'
terminal 95 bp of the *E. coli* 23S rDNA (31, 32, 87, 140), but the 23S–4.5S
rDNA spacer sequence clearly has no equivalent in the bacterial gene. Thus the
origin of the 4.5S rRNA gene in chloroplasts may have been caused by the
introduction of an insertion sequence (equivalent to the 23S–4.5S rDNA
spacer) approximately 100 bp from the 3' terminus of a 23S rRNA gene.

Regions which are conserved between *E. coli* and the chloroplast 23S rRNA
genes, either at the primary nucleotide sequence level or at the level of
secondary structures, in many instances correlate with known functional regions (32, 140). They include sites of base modification, the L1 and L24
ribosomal protein binding sites, the 30S–50S ribosome subunit interacting
region, and the site involved in peptidyl transferase activity (puromycin binding and chloramphenicol sensitivity). A very comprehensive analysis (41) of
the secondary structures of 23S rRNAs shows there are over 450 compensating
base changes between maize and *E. coli*. However, it is interesting to note that
the longest stretch of conserved sequence (63 nucleotides) (32, 140) lies in a
region which has not as yet been identified with any specific function. Examination of the sequences surrounding the 5' end of the 23S rDNA and those
surrounding the 3' end of the 4.5S rDNA shows extensive complementarity
(31, 87, 140) and raises the possibility that the potential to assume a cruciform
structure analogous to that formed by *E. coli* precursor rRNA and essential for
processing may also be important for chloroplast rRNA processing.

The *Chlamydomonas reinhardii* 23S rDNA has the distinction of being the
only known chloroplast rRNA gene to contain an intron (114). The 870 bp
intervening sequence is located approximately 270 bp from the 3' end of the
gene in a region which is highly conserved among many different organisms.
Sequence analysis across the junctions with the coding region has shown that
the intron is flanked by 5' CGT oriented as a direct repeat and that the 5 bp
sequence, 5'CGTGA, lies exactly next to one end and also 16 bp away from the
other end, again as a direct repeat (1). Sequences found near one end of the
intron are complementary to portions of the sequence at the other end and may
contribute to the formation of a stem structure in the intron. The existence in
Chlamydomonas chloroplasts of an RNA species of the same size as the intron
has been established by hybridization with a probe derived from and specific to
the intron (114).

The 5' flanking region of the 23S rDNA from *Chlamydomonas,* including
the 282 bp 7S and the 47 bp 3S rDNAs and the first 300 bp of the gene, has been
sequenced (115). Organization of this region is as follows: 7S rDNA-23 bp
spacer-3S rDNA-81 bp spacer-23S rDNA. It is noteworthy that the 7S rDNA,
the 3S rDNA, and the 81 bp spacer are homologous to 5' terminal sequences of
E. coli and other chloroplast 23S rDNAs, the 7S rDNA starting at the equivalent of position 9 of *E. coli* 23S rDNA. The 3S rDNA shows 69%, 74%, and
45% homology with *E. coli,* maize chloroplast, and *Euglena* chloroplast

23S rDNAs respectively. Like the 5' sequences of other prokaryote 23S rRNAs, the 7S and 3S rRNAs of *Chlamydomonas* bear some resemblance to the 5.8S rRNA of cytoplasmic 80S ribosomes. Rochaix & Darlix (115) also noted that the 5' end of the 7S and 3S rRNAs could base-pair with the 23S rRNA sequences which flank the ribosomal intron; they suggest that these small RNAs might act as guide RNAs in the excision of the intron from the 23S rRNA.

To date the only sequence information available for the 23S rDNA of *Euglena* chloroplasts is that for approximately 120 bp at the 5' terminus (100).

5S rDNA

The sequence of 5S rRNA from higher plant chloroplasts is highly conserved (23, 28) and has considerable homology with 5S rRNAs of fern chloroplasts (141), *Anacystis nidulans* (21), and *Prochloron* (88). Although microheterogeneity with respect to the terminal nucleotides is seen in some species (138), it is probably only a reflection of the variability in the processing of the mature RNA from its precursor. In tobacco cpDNA the 103 bp 4.5S rRNA and 121 bp 5S rRNA genes are separated by 256 bp (136). This spacer sequence features a "Pribnow box" and "−35 region" that could function as a signal for the initation of transcription of the 5S rDNA, and 35 bp downstream from the gene there is a twofold rotational symmetry sequence that could signal transcription termination. Furthermore, a potential transcription termination sequence also occurs 15 bp downstream from the end of the 4.5S rDNA. A very similar pattern occurs in maize cpDNA (3). Although there is no direct evidence that these sequences function in transcriptional control in chloroplasts, there is other evidence (56) that suggests the synthesis of 5S rRNA may be separate from the transcription of the 16S–23S–4.5S rRNA cluster.

16S–23S rDNA Spacer

This region has been completely sequenced in maize (71), tobacco (139), and *Euglena* (44, 100) cpDNA and is interesting because it contains genes for two tRNAs (see tRNA section). In *Chlamydomonas* the spacer region immediately adjacent to the 23S rDNA carries the genes for the small 7S and 3S rRNAs (115), but, as discussed above, these appear to be more closely related to sequences within the 5' terminus of other chloroplast 23S rDNAs than to spacer sequences. In higher plant chloroplasts spacer DNA is transcribed as part of the *rrn* operon transcription unit. The primary transcript is subsequently processed to yield mature rRNA and tRNA species and also species corresponding in size to the excised introns of the tRNAs (160).

Despite the difference in size between the 16S–23S spacer rDNAs of *Euglena* and *E. coli,* there are short stretches of homology in the intergenic regions as well as the more extensive homology in the tRNA genes (44). Within the

Euglena rrn operon the spacer sequence shows extensive similarity with parts of the sequence immediately upstream from the 16S rDNA (100). Whether these regions have been conserved because they have functional significance is a matter for conjecture.

GENES FOR TRANSFER RNAs

Mapping of trNA genes on cpDNA

Saturation hybridization of a total chloroplast tRNA fraction to cpDNA indicates that there are between 25 and 45 tRNA cistrons on the chloroplast genome (51, 53, 54, 89, 92, 93). This is in reasonable agreement with the number of chloroplast tRNAs that can be resolved by two-dimensional gel electrophoretic analysis (26, 98). The minimum number of tRNA species required for translation of all codons is 32, and, although the existence of tRNAs for several amino acids [e.g. Asp, Cys, Glu, Gln in spinach, Asp, Cys in *Euglena* (95)] has yet to be demonstrated, there seems to be ample justification for assuming that all the tRNAs involved in chloroplast protein synthesis are coded in the organelle genome.

Hybridization of unfractionated chloroplast tRNA and of individual tRNA species to Southern blots of restricted cpDNA has permitted the construction of tRNA gene maps. Regions hybridizing to tRNAs are dispersed nonrandomly around the circular genome, with some regions carrying a number of cistrons while others carry only one or two (26, 51, 69, 98). The distribution of 22 tRNA genes (coding for 14 different amino acids) on spinach cpDNA (Figure 1) (26) is fairly typical for those species of higher plants so far examined. In *Euglena gracilis* strain Z, seven regions, equivalent to 28% of the cpDNA, show strong hybridization to tRNAs, eight regions (24%) show weak hybridization, while the remainder (48%) does not hybridize at all (98). *Euglena gracilis bacillaris* cpDNA shows a similar pattern (33).

Detailed information on the structure and organization of tRNA genes is now accumulating rapidly as the result of sequencing of cloned restriction fragments of cpDNA. In no case reported so far is the −CCA 3' terminal sequence of the tRNA encoded in the DNA. In this respect chloroplast tRNA genes are different from *E. coli* tRNA genes and resemble those of eukaryote organisms. In most other respects, however, chloroplast tRNAs are more akin to prokaryote than to either eukaryote or mitochondrial tRNAs.

trNA Gene Clusters

The only evidence to date for tight clustering of tRNA genes in cpDNA is that from *Euglena gracilis* strain Z. Sequencing of a 1.6 kb segment of *Eco*Rl fragment G from *Euglena* cpDNA shows the presence of four tRNA genes encompassed within a total of 367 bp (99). The arrangement of the genes is tRNA $_{UAC}^{Val}$-16 bp spacer-tRNA $_{GUU}^{Asn}$-3 bp spacer-tRNA $_{ACG}^{Arg}$-45 bp spacer-tRNA $_{UAG}^{Leu}$. The tRNA^{Leu} gene is of the opposite polarity to that of the other three genes (99). The close proximity and common polarity of three of these genes is consistent with the possibility that they are part of a common transcription unit. Regions 5' to the tRNA^{Val} and to the tRNA^{Leu} genes are AT-rich and sequences similar to those of prokaryote promoters are present (99), but as yet there is no direct evidence about specific transcription start points for these genes.

Another cluster of tRNA genes is located mainly on *Eco*R1 fragment V but extends into *Eco*R1 fragment H of *Euglena* cpDNA. Although the sequence has not been published, the organization is reported to be tRNA^{Tyr} − tRNA^{His} − tRNA^{Met} − tRNA^{Trp} − tRNA^{Glu} − tRNA^{Gly} (54, 98). All of the genes are of the same polarity and they span a total of 575 bp.

Genes for tRNA $_{CAU}^{Ile}$, tRNA $_{GGU}^{Thr}$, tRNA $_{GTG}^{His}$ and tRNA $_{GUU}^{Asn}$

The chloroplast genes for tRNA $_{CAU}^{Ile}$ (66) and tRNA $_{GGU}^{Thr}$(65) of spinach have been sequenced and shown to be colinear with the respective mature tRNA sequences (37, 67). The tRNA^{Ile} gene has the methionine anticodon CAT, but the tRNA reads an isoleucine not a methionine codon, presumably because the C residue in the anticodon is modified and precluded from base pairing with G (37, 66). However, it is interesting to note that the gene shows as much homology (60%) with *E. coli* tRNA$_m^{Met}$ as it does with *E. coli* tRNA^{Ile} (56% or 61% depending on the isoaccepting species). Spinach chloroplast tRNA^{Thr} is 63% and 69% homologous with *E. coli* tRNA$_3^{Thr}$ and tRNA$_4^{Thr}$. Neither gene appears to be part of a tRNA gene cluster. Spinach tRNA $_{CAU}^{Ile}$ maps within the inverted repeat region of the cpDNA. Both copies of the gene have been cloned and their sequences shown to be identical (66).

The sequence of the maize chloroplast tRNA$_{GTG}^{His}$ gene which maps close to the junction of the inverted repeat and the large single copy region (121) is 60% homologous with the *E. coli* gene and 45% homologous with the gene from yeast mitochondria. Analysis of the flanking sequences showed that the 5' end of the tRNA gene overlaps by a few nucleotides the 5' end of a putative protein gene that is coded by the complementary DNA strand and is defined on the one hand by its hybridization to a 1.6 kb chloroplast RNA species and on the other by an open reading frame in the DNA sequence (121). The two genes must be transcribed divergently; this probably involves a shared sequence, but whether their expression is linked is not known. In spinach the tRNA $_{GTG}^{His}$gene lies just outside the inverted repeat region (Figure 1). Its sequence is the same as that of the maize gene except for a C/T base change at position 22 and the addition of a T residue in the variable loop (G. Zurawski, unpublished).

In tobacco the gene for chloroplast tRNA$_{GUU}^{Asn}$ has been identified in the inverted repeat region (68) lying approximately 900 bp distal to the end of the 5S rRNA gene and 850 bp from the junction with the small single-copy region. Its polarity is the reverse of the adjacent rRNA genes, and transcription probably starts approximately 150 bp upstream from the gene where there is a sequence similar to the consensus sequence of *E. coli* promoters. A typical prokaryote-type transcription terminator stem and loop structure occurs immediately after the 3' end (68). Sequence analysis of cloned fragments derived from both of the inverted regions shows that the two copies of the gene are identical and have 71% homology with *E. coli* tRNA^{Asn} (68).

tRNA Genes in the 16S rDNA Leader Sequence

The gene for tRNA$_{GAC}^{Val}$ is approximately 300 bp upstream from the start of the 16S rRNA gene in maize (124) and also in tobacco cpDNA (144). The sequences of the genes from the two species are identical and show 65% homology with the corresponding *E. coli* gene. Both the tRNA^{Val} and the rRNAs are coded by the same DNA strand, and there are reasons for thinking the tRNA gene may be transcribed as part of the rDNA operon. Tohdoh et al (144) have established that there is one strong transcription initiation site on the tobacco rDNA leader sequence and that this site is 20 to 26 bp prior to the start of the tRNA^{Val} gene. It is also just downstream from sequences which resemble the consensus sequences of *E. coli* promoters. RNA polymerase binding experiments have shown that the maize rDNA leader sequence contains three binding sites (124), the middle one of which coincides with the transcription initiation site detected in the tobacco sequence. The 5' leader regions of maize and tobacco rDNA show over 90% sequence homology back to, but not beyond, a point 73 bp preceding the tRNA^{Val} gene. This strong sequence conservation supports the view that the whole region, including the sequence between the tRNA and 16S rRNA genes, is functionally important. Furthermore, this intergenic sequence does not contain any region of strong dyad symmetry such as is commonly associated with *E. coli* transcription terminator regions and as has also been found in similar positions in a number of chloroplast genes. On the other hand, direct analysis of transcripts from maize chloroplasts indicates transcription of the rDNA operon may start from a site between the tRNA^{Val} and 16S rRNA genes (160). Perhaps transcription of the ribosomal RNAs can start either before or after the tRNA^{Val} gene, depending on the developmental stage of the leaf.

Part of the sequence, including that of the tRNA^{Ile} gene, in the 16S–23S rDNA spacer region in *Euglena* shows a significant level of homology (68%) with a 189 bp segment that occurs just upstream from the 16S rDNA (100). The 74 bp corresponding to the tRNA^{Ile} gene, referred to by Orozco et al (100) as a pseudo tRNA gene, has 47 nucleotides in common with the bona fide gene (64% homology) and can be folded into a typical clover-leaf structure. Also present in the 16S rDNA leader sequence, approximately 25 bp downstream from the pseudo tRNA^{Ile} sequence, is a 27 bp region which has some homology with the tRNA^{Ala} gene. When secondary structure interactions are considered, it becomes apparent that the sequence contains the anticodon stem and loop and also the variable loop of a tRNA^{Trp} gene (100). Whether such tRNA-like sequences play a role in controlling the expression of the rDNA operon, or in the processing of its primary transcript, or whether they are simply relics of gene duplication events is not known.

tRNA Genes in the 16S–23S rDNA Spacer Region

It has been known for some time that the spacer region between the 16S and 23S rRNA genes of cpDNA contains genes for tRNAs (26, 89, 97). In all species where the genes have been identified, the tRNAs encoded have been found to be tRNA$_{\text{GAU}}^{\text{Ile}}$ and tRNA$_{\text{UGC}}^{\text{Ala}}$ (44, 69), except in the case of spinach where only the gene for tRNA^{Ile} has as yet been detected (10). Recently the complete sequence of the spacer region in *Euglena* (44, 100), maize (71), and tobacco (139) has been determined. This has provided much interesting information, including establishing that the tRNA genes are transcribed from the same DNA strand as the ribosomal RNAs.

The 16S–23S rDNA spacer region in *Euglena gracilis* strain Z is about 258 bp long (44, 100), and the genes are organized 5' → 3' as follows: 16S rDNA − 87 bp spacer − tRNA$_{\text{GAU}}^{\text{Ile}}$ − 9 bp spacer − tRNA$_{\text{UGC}}^{\text{Ala}}$ − 15 bp spacer − 23S rDNA. This arrangement, including the identity and anticodon specificity of the tRNA genes, is the same as that observed in the spacer region of the *rrnA*, *D* and *X* operons of *E. coli* (100). Each of the three complete rDNA units in the *Euglena* chloroplast genome contains the same tRNA genes, and the organization of the spacer region is the same in each case (100). The sequences of the tRNA^{Ile} and tRNA^{Ala} genes show 77% and 75% homology respectively with those of the corresponding *E. coli* genes. Graf et al (44) noted that part of the 87 bp spacer prior to the tRNA^{Ile} gene has the potential to form a hairpin structure and that the sequence 5'GGTTTTG3' occurs at the 3' end of each of the tRNA genes; they suggest that these features may play a role in the processing of the RNA transcribed from the rDNA operon.

In contrast to the situation in *Euglena,* the 16S–23S rDNA spacer in maize and tobacco (and many other species) exceeds 2 kb in length (4, 5, 71, 139). Sequence analysis of this region has led to the most interesting observation that the tRNA^{Ile} and tRNA^{Ala} genes contain very large intervening sequences (see below). Apart from the size difference, the overall organization of the spacer regions in the two species is the same (see Figure 2). The tRNA coding

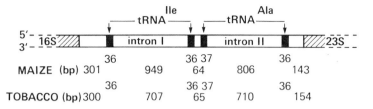

Figure 2. Map of the 16S-23S rDNA spacer region of maize and tobacco cpDNA showing the location of the tRNA genes and introns. Data from (71) and (139)

sequences are identical except for a C/T change in tRNA^{Ala} immediately following the intron splice site (139). The sequences in the "true" spacer regions, i.e. before, between, and following the tRNA genes, are very similar (91% homology) in maize and tobacco (139). The tRNA^{Ile} gene shows 83% and 81% sequence homology, and the tRNA^{Ala} gene 90% and 78% homology with the equivalent genes from *Euglena* chloroplasts and *E. coli* respectively.

Introns in tRNA Genes

As indicated above, the tRNA genes in the 16S–23S rDNA spacer of at least two higher plant chloroplasts contain intervening sequences. In maize the intron in tRNA^{Ile} (Intron I) is 949 bp and that in tRNA^{Ala} (Intron II) is 806 bp (71), whereas in tobacco the corresponding introns are 707 bp and 710 bp (139). Comparison of the introns from maize and tobacco shows that, apart from the minor gaps introduced in order to maximize homology, the sequences are strikingly similar (> 95% homology) (139). Furthermore, the difference in the sizes of the introns between the two species can be seen to be caused by a 229 bp deletion in Intron I and a 103 bp deletion in Intron II in tobacco compared to maize (139). The sequence of Intron I shows considerable similarity to that of Intron II, sufficient in fact to have caused Koch et al (71) and Takawai & Sugiura (139) to suggest that the introns may have had a common origin.

Assignment of the precise location of the splice sites in the tRNA genes was only tentative in that sequence information on the tRNAs themselves had been lacking (71, 139). In the case of maize tRNA^{Ile} it has now been shown by Guillemaut & Weil (52) that the splice site is between the second and third nucleotide beyond the 3' end of the anticodon, not between the first and second nucleotides as suggested by Koch et al (71). The same will almost certainly be true also for the splice site in the tobacco tRNA^{Ile} gene (139).

Of considerable interest is the observation that the sequences of Intron I and II of maize contain open reading frames of 123 and 45 codons respectively (71). The significance of these, however, has now been somewhat cast into doubt following the sequencing of the tobacco introns (139). Intron I in tobacco contains a 71 codon open reading frame and Intron II a 36 codon open reading frame, but the locations of these potential protein coding regions do not correspond to the ones in the maize introns. In view of the overall high level of conservation of the intron sequences, the failure to maintain homology with respect to the open reading frames argues against the latter having a functional significance.

Introns are not confined to the tRNA genes of the spacer region only. The gene for tRNA $^{Leu}_{UAA}$ in maize chloroplasts has been sequenced and shown to contain a 458 bp intron (130) which is spliced into the sequence between the first and second nucleotides of the anticodon. There is no open reading frame in the RNA-like DNA strand of the intron, but Steinmetz et al (130) noted that the complementary DNA strand contains a 60 codon open reading frame. However, there is no evidence that it is transcribed or translated. We have sequenced the genes of five other tRNAs from spinach chloroplasts. Two of them contain introns, three do not (unpublished data).

The presence of introns in chloroplast tRNA genes is somewhat of an anomaly. This property serves as another clear distinguishing characteristic as far as the prokaryote-like nature of chloroplasts is concerned. Introns have not been reported to occur in *E. coli* tRNA genes. However, they do occur in tRNA genes in yeast (147) and *Drosophila* (112), but their size in these organisms is relatively small compared to their size in chloroplasts. And why do the tRNA^{Ile} and tRNA^{Ala} genes in maize and tobacco have introns whereas the equivalent genes in *Euglena* chloroplasts, having extensive sequence similarity, including the same anticodons, and being part of a comparably organized 16S–23S rDNA spacer region, do not? It is not a question of the genes containing introns being nonfunctional. The tRNA for isoleucine with the GAU anticodon from maize chloroplasts has now been purified and its sequence determined and found to be identical to the tRNA^{Ile} gene in the spacer region (52). This tRNA hybridizes to only one restriction fragment of maize cpDNA and that fragment originates in the 16S–23S rDNA spacer region.

It should be noted that as yet only the gene for tRNA^{Ile} has been identified as a constituent of the 16S–23S rDNA spacer region in spinach cpDNA (10) and no sequence data is available. It is important that this point be clarified. The overall size of the spacer region in spinach cpDNA is very similar to that in tobacco and maize cpDNA, and it will be remarkable if the tRNA^{Ala} gene is not also present and if both genes do not have introns.

GENES CODING FOR PROTEINS
Gene Mapping Procedures

The technique of DNA cloning has greatly facilitated the identification and physical mapping of genes on cpDNA. Recombinant plasmids, each carrying a different restriction fragment of the cpDNA, are constructed and individual clones then isolated in large quantities (107). The cpDNA fragment can then be excised from the plasmid and purified for use in the procedures outlined below. In many situations, however, the recombinant plasmid itself may be used directly for the location of protein-coding genes.

TRANSCRIPTION AND TRANSLATION OF CLONED FRAG-MENTS   Probably the most direct means of identifying a protein-coding sequence in a DNA fragment is by transcribing and translating the fragment in vitro. The resultant proteins may then be identified by immunoprecipitation with antibodies to chloroplast proteins, or compared with known proteins by partial proteolytic digestion or two-dimensional gel electrophoresis. This method has been used to map *rbcL* in maize (2), spinach (34, 153), and *Chlamydomonas* (90) and *atpB, atpE,* and the gene for the dicyclohexylcarbodiimide-(DCCD) binding protein *(atpH)* in wheat (60, 61). Genetic nomenclature used here is based in part on that used for bacterial genes; for specification of individual genes, see list of abbreviations (Table 1).

The disadvantage of this method is that if the DNA fragment does not contain the complete gene, then no protein or no full-length protein will be obtained. However, production of a shortened peptide can in itself constitute useful information in that it probably corresponds to the N-terminus of a protein and indicates that the fragment contains the 5' segment of the gene. This then specifies the direction of transcription of the gene.

HYBRID ARREST AND HYBRID RELEASE TRANSLATION OF CHLOROPLAST RNA   These methods involve hybridization of specific DNA fragments to either total or fractionated chloroplast RNA. Because any mRNA which is complementary to the sequences in the DNA fragment will be removed from the RNA population, comparison of the RNA-directed translation products before and after hybridization will reveal any protein whose gene is wholly or partly contained in that DNA. This method is termed "hybrid-arrest translation." If, on the other hand, the DNA-RNA hybrid is melted and the RNA isolated and translated, then it should give only products whose genes are wholly or partly contained in the DNA fragment. This is termed "hybrid-release translation."

These methods have the advantage that the DNA fragment need only contain a portion of the gene sufficient to allow hybridization to the mRNA to yield a full-length polypeptide on translation. The method has been used to localize the gene for a 53,000 M_r polypeptide to a *Eco*RI fragment of *Euglena* cpDNA (118), to map *psbA* (27) and also *atpA, atpB,* and *atpE* (152) on spinach cpDNA.

A refinement of this method has been used to locate precisely *rbcL* and the gene coding for a 2.2 kb mRNA (since shown to be *atpB*) on maize cpDNA (83) and also the genes for a 56,000–M_r (probably *rbcL*) and a 35,000–M_r polypeptide (probably *psbA*) on mustard cpDNA (82). This was achieved by digestion of the DNA-RNA hybrids with S1 nuclease. Where only part of the transcribed region of the gene is contained by the DNA fragment, it is possible, from the size of the S1–resistant duplex region, to estimate the length of that part of the gene present in the particular fragment

GENE PROBES FROM HETEROLOGOUS SPECIES   Because there is a high degree of sequence conservation for plastid genes between different plant species, it is convenient to use an identified gene from one species as a probe to locate the same gene in another species. This approach involves labeling a DNA fragment carrying the characterized gene and using it to probe a southern transfer of restricted total cpDNA or cloned fragments from the second species. By this means the DNA fragment carrying *rbcL* from maize has been used to locate the same gene in species as diverse as spinach (153) and *Euglena* (131). Similarly, an *rbcL* probe from spinach was used to map *rbcL* in tobacco (126), while the locations of *rbcL* and *psbA* on *Atriplex triangularis* and *Cucumis sativa* cpDNAs were determined with DNA probes containing *rbcL* from maize and *psbA* from spinach (101). These same probes, together with probes for *atpA* and *atpB* from spinach, were used to map all four of these genes onto the cpDNA from the fern *Osmunda cinnamonia* (102).

An interesting variation of this technique, and one with considerable potential, is to use cloned *E. coli* genes to locate the equivalent genes on cpDNA from *Chlamydomonas*. Watson & Surzycki (150) found that there was sufficient homology between the protein synthesis elongation factor genes *tufA* and *tufB* of *E. coli* and some *Chlamydomonas* cpDNA fragments to enable them to map the chloroplast elongation factor gene *(tufA)*. Sequences homologous to *E. coli* genes for the α subunit of RNA polymerase and for a number of ribosomal proteins have also been detected in *Chlamydomonas* cpDNA (150). If this approach proves to be generally applicable, then the identification of many other chloroplast genes will be greatly simplified.

Gene for the Large Subunit of RuBP Carboxylase (rbcL)

The evidence establishing that cpDNA contains *rbcL* has been reviewed by Bedbrook & Kolodner (4). Since then the position of this gene on the circular map of cpDNA has been established for a number of plants (Table 1). Chloroplast DNAs from the majority of species studied so far contain inverted repeat regions separated by a large and a small single-copy region (see Figure 1). In all such species, *rbcL* maps in the large single-copy region in approximately the same position relative to the repeated regions (Figure 1). One curious feature of the location of *rbcL* on *Chlamydomonas* cpDNA is that although it maps in the larger of the two single-copy regions, this region, in contrast to the situation in higher plants, is nearer to the 5S rRNA end of the rRNA operon than to the 16S rRNA end. In *Euglena*, *rbcL* has been mapped (131) by hybridizing probes containing internal *rbcL* sequences from both maize and *Chlamydomonas*. It was shown that the coding region in *Euglena* is considerably larger than in maize and that there are two hybridizing regions separated by a short stretch of DNA that does not hybridize to the probes. Because the sequences of the two probes used overlap, it was concluded that *Euglena rbcL* must contain an intervening sequence of about $0.55-1.1$ kb.

A mutant of *Oenothera hookeri* (10) has been described (55) which lacks the large subunit of RuBP carboxylase. When cpDNA from this mutant is transcribed and translated in an in vitro *E. coli* system, no large subunit is synthesised but instead a prominent band of about 30,000 daltons appears, suggesting that the site of the mutation is in *rbcL* about 800 bp downstream from the site of translation initiation (Bottomley & Herrman, unpublished results).

rbcL from maize [(91), updated in (106)] and from spinach (164) has been completely sequenced. Comparison of the two sequences demonstrates a number of interesting features. Within the translated regions the nucleotide sequences are highly conserved (84% homology). Many of the nucleotide changes are silent, usually occurring in the third position of the codon, and thus the overall amino acid homology is about 90%. Conservation of amino acid sequences at the active sites is particularly high. For example, there is complete homology between the 17 amino acid residues around the lysine which has been shown by Lorimer (85, 86) to be responsible for binding the activating CO_2. In the three tryptic peptides which have been shown to lie near the RuBP-binding site (57), there are only two amino acid differences between the maize and spinach sequences, representing over 95% conservation of these regions.

One difference between *rbcL* from spinach and that from maize is the length of the 5' untranslated leader sequences (91, 164). Transcription of spinach *rbcL* is initiated 178–179 bp before the start of translation, whereas in maize this distance is only 63–64 bp. It is possible that sequences in this region are involved in the regulation of transcription, and it is interesting to note that in maize, in contrast to spinach, the large subunit mRNA is differentially expressed in mesophyll and bundle sheath cells. The 3' untranslated sequence has been shown to be 85–88 bp in spinach. Although the length of this region has not been determined in maize, comparison of the sequences from the two species indicates it is about 140 bp long. The occurrence of possible prokaryotic controlling sequences is discussed below. As indicated in Figure 1, the direction of transcription has been determined to be toward the nearer of the two inverted repeat regions (164).

One interesting feature revealed by sequencing *rbcL* of spinach is the presence of a possible N-terminal leader sequence to the LSU protein. In the deduced amino acid sequence of spinach, an alanine residue which is equivalent to the N-terminal alanine of barley and wheat LSU occurs 14 positions after the first methionine. It is argued that translation of spinach LSU starts at the methionine because 13 of these first 14 amino acid residues are conserved between spinach and maize and because there is a 5 bp Shine-Dalgarno (128) sequence (-GGAGG-) just prior to the methionine codon which is complementary to a sequence at the 3' end of chloroplast 16S rRNA. There is no relevant protein sequence data available for spinach LSU, so it is possible that the N-terminus is methionine while in barley and wheat it is alanine. However, it may also be that, although synthesis of spinach LSU commences at the methionine residue, post-translational processing cleaves off the 14 N-terminal amino acids leaving alanine as the N-terminus of the mature protein. Evidence supporting this comes from the observation (81) that LSU synthesized by translation of spinach chloroplast RNA in an *E. coli* cell-free system is 1000–2000 daltons larger than LSU synthesized in isolated organelles.

Gene for the "32,000 dalton" Thylakoid Membrane Protein

In 1973 a polypeptide of about 32,000 daltons was seen as a prominent band among the membrane-associated products from light-driven protein synthesis in isolated pea chloroplasts (8). While a polypeptide of similar M_r has now been reported for a number of plant species (13, 50, 110), the polypeptide itself has not been isolated. Recently it has been shown by Steinback et al (129) that this polypeptide is associated with Photosystem II and is responsible for the binding of herbicides such as 3-(3,4-dichlorophenyl)-1,1,dimethylurea (DCMU) and atrazine. The gene *(psbA)* coding for this protein has been mapped on the cpDNA of several species (Table 1, Figure 1). In all species so far studied, *psbA* is situated in the large single-copy region close to the end of one of the inverted repeat regions (Figure 1). In *Chlamydomonas,* assuming the gene coding for the 32,000 dalton polypeptide D1 (114) is *psbA,* then, as is the case with *rbcL,* it is situated in the single-copy region which is adjacent to the 5S end of the rRNA operon, whereas in other plants it is nearer to the 16S rRNA cistron.

The direction of transcription of *psbA* in spinach is toward the nearer inverted repeat (161) and therefore opposite to that of *rbcL.* The complete nucleotide sequence of *psbA* from both spinach and *Nicotiana debneyi* has been determined (161). The genes from the two species are highly conserved, with over 95% homology between the nucleotide sequences while the deduced amino acid sequences are identical over all 353 residues. The molecular weight of the deduced protein is 38,950 daltons, which is significantly larger than the largest estimate of the in vitro synthesized precursor as determined by SDS-gel electrophoresis. Because of the complete homology of the amino acid sequences, it seems certain that the primary translation product is indeed 38,950 daltons. Whether the discrepancy between the calculated and estimated molecular weights is due to some undetected processing step reducing the size of the protein very rapidly after translation, or to anomalous behavior on gels as has been reported for other hydrophobic proteins (22), is not known. Other features of the protein are the complete absence of lysine residues and the abundance and clustering of hydrophobic amino acids. It has previously been suggested by Edelman & Reisfeld (30), on the basis of radioactive amino acid incorporation, that the 32,000-M_r protein from *Spirodela* contains no lysine. The prokaryotic features of the control of transcription and translation in this gene will be discussed later.

Genes for Subunits of ATP Synthase

Three of the five subunits of the CF_1 part of ATP synthase are known to be synthesized in isolated chloroplasts (94, 96). The positions of the genes (*atpA*, *atpB*, and *atpE* for the α, β, and ε subunits respectively) for these subunits have been mapped on chloroplast DNA from spinach (152). While *atpB* and *E* are located very close to one another, *atpA* is relatively remote, some 40 kb away (Figure 1). The nucleotide sequences of *atpB* and *atpE* have been determined for spinach (162), pea (163), and maize (76). The 5' end of *atpB* in spinach and pea is about 150 bp, that in maize is about 350 bp from the 5' end of *rbcL*, and the genes are transcribed in opposite directions. The most surprising feature of the sequence analysis is a 4 bp overlap of the translated regions of *atpB* and *atpE* in both spinach and maize. The TGA stop codon which terminates translation of *atpB* is preceded by an A which forms the ATG initiation methionine of *atpE*. This means that the two proteins are read in different reading frames. As sequences from either *atpB* or *atpE* of spinach both hybridize to the same large RNA transcript (> 2.4 kb), this also means the two genes are cotranscribed into a dicistronic mRNA. What role, if any, this organization plays in the coordination of expression of the two subunits is not known. However, the overlapping translation stop/start codon is not an essential feature of the two genes because the same genes in pea cpDNA (163) are separated by 22 nucleotides. As with spinach, both *atpB* and *atpE* hybridize to the same pea chloroplast RNA transcript of approximately 2.4 kb, and they must also be transcribed into a dicistronic mRNA.

The β subunits of ATPase from spinach, pea, and maize contain 498, 491, and 498 amino acids respectively and the ε subunits 134, 137, and 137 amino acids respectively. The amino acid homology between the β subunits of ATPase from spinach, pea, and maize lies within the range 86% to 89%. The homology between the ε subunits is slightly less (72–81%). Comparison of the β subunit from spinach and maize with the β subunit from *E. coli* ATPase (119) shows 66% and 67% homology respectively, which is considerably greater than the 26% and 23% homology between the ε subunits of spinach and maize and the ε subunit of *E. coli* ATPase (119).

Two of the three subunits of the CF_0 part of the chloroplast ATP synthase have been shown to be synthesized in isolated chloroplasts, suggesting that they are cpDNA coded. These are subunit I of M_r 15,500 (96) and the DCCD binding subunit III of M_r 8,000 (25). Recently, the position of the gene *(atpH)* for the DCCD binding polypeptide has been mapped on the wheat chloroplast genome and the nucleotide sequence determined (60). This gene is not part of the *atpB* and *atpE* complex but maps approximately 15 kb away. The sequence of the 81 amino acid residues deduced from the nucleotide sequence of *atpH* was identical to that previously obtained for spinach DCCD-binding protein by direct amino acid sequencing (125).

OTHER FEATURES OF cpDNA SEQUENCES

Codon Usage

An inspection of the sequences of *rbcL* from spinach (164), maize (91), and pea (163), *psbA* from spinach and *Nicotiana debneyi* (161), *atpB/E* from spinach (162) and pea (163), and *atpH* from wheat (60) shows that all possible codons are used, including the three stop codons TAA, TAG, and TGA. There is no evidence for any abnormal use of codons as occurs in mitochondrial DNA (6). As might be expected from the high AT content of cpDNA (70), there is a bias toward the use of these two bases, particularly in the third position of the codons. While the first position has 47% AT and the second 57%, the third position contains 69% AT. Because of the redundancy in the code, this bias does not appear to lead to discrimination against the use of any amino acid.

The average GC content of the translated regions of the 10 genes is 42.5%, which is somewhat higher than the overall GC content of 37–39% for cpDNA of higher plants (70). In accordance with this observation, we have noted that sequences lying outside coding regions tend to be AT-rich.

Promoter Sequences

Chloroplasts contain their own DNA-dependent RNA polymerase which is distinct in some respects from bacterial polymerases. However, because *E. coli* extracts will transcribe and translate some chloroplast genes yielding products which are identical with the chloroplast polypeptides (14), it is of interest to compare the sequences of possible promoter sites with those found in prokaryotes. The two regions believed to be important in binding bacterial RNA polymerase are located around 10 and 35 nucleotides upstream from the site of transcription initiation (19, 117). Figure 3 lists the sequences around the putative promoter regions of a number of chloroplast genes, together with the consensus sequence of prokaryotic promoter regions at the −10 and −35 positions. As can be seen, it is possible to identify sequences in all genes which contain the essential features of those from prokaryotes. In the −10 region the one base which has always been found in prokaryotes is the final T of the

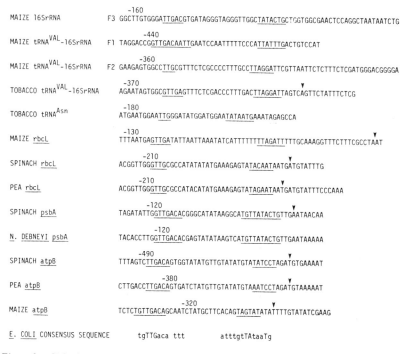

Figure 3. 5' leader sequences of some cpDNA genes showing the comparison with the "−10" and "−35" consensus sequences of some *E. coli* promoter regions (117). The arrows indicate the probable position of transcription initiation. Compiled from (68, 76, 91, 124, 144, 161, 163, 164).

-TATAAT- Pribnow box. This is also present in all cases in the chloroplast genes although, in the case of maize *rbcL,* the Pribnow box is over 20 bp upstream from the transcription initiation site. In the −35 regions the highly conserved prokaryotic sequence -TTG- can also be seen to be present in all chloroplast genes.

Further evidence for the prokaryotic nature of the transcription promoter regions comes from studies on the binding of *E. coli* RNA polymerase to cpDNA fragments. It has been found that this binding is nonrandom and takes place at specific fragments, one of which is near the gene *psbA* (159). In a study of the region near *rbcL* in maize and wheat cpDNA, Koller et al (74) found that *E. coli* polymerase bound specifically to corresponding sites on each DNA. These sites were located outside but near the translated regions of the genes. However, in vitro the *E. coli* enzyme did not distinguish between the coding and noncoding strands so that transcription was bidirectional. Whether this is caused by the heterologous enzyme not recognizing chloroplast control sequences or because some factor is missing under in vitro conditions is not known.

Clearly, it will be necessary to carry out experiments with purified chloroplast RNA polymerase if we are to establish whether *E. coli*-like promoter sequences are functional in chloroplasts.

Transcription Terminator Sequences

The most obvious common feature of prokaryote DNA sequences upstream from transcription termination sites is the presence of complementary sequences capable of forming a short hairpin structure (117). Such a feature can be seen also in *rbcL* from spinach (164), maize (91), and pea (163), as well as in *psbA* from spinach and *N. debneyi* (161). Although the site of transcription termination has not yet been determined in *rbcL* from maize or *psbA* from *N. debneyi*, the homology with the respective spinach genes in the hairpin loop structure is very marked; it is reasonable to assume that termination occurs in a similar position relative to the hairpin structure in these genes.

From these observations, it appears probable that the control of transcription in chloroplast genes is prokaryotic in nature, employing sequences which are strikingly similar in many respects to those employed in prokaryotes.

Translation Initiation

The ribosome-binding sites in bacterial genes have been shown to contain sequences which are complementary to those at the 3' end of 16S rRNA. These "Shine-Dalgarno" (128) sequences are located a short distance upstream from the site of translation initiation. Stormo et al (132) found that of 124 prokaryotic genes, 85% had some combination of the three nucleotides GGA between 5 and 9 bases upstream from the initiation codon. Although *rbcL* from maize (91), spinach (164), and pea (163) all conform to this requirement, no such structures are obvious in the sequences of *atpB* from spinach (162) or pea (163) or of *psbA* from spinach and *N. debneyi* (161). It is of interest that the mRNAs for these genes are translated in vitro in an *E. coli* system, although generally the RuBP carboxylase large subunit is the most prominent band (14). It seems possible that this dominance may be attributable to the presence of the *E. coli*-like Shine-Dalgarno sequences allowing more efficient initiation of translation. If it is assumed that the role of these sequences is to bind the small ribosomal subunit, then it would be expected that the requirements of the chloroplast system would be very similar to that of *E. coli* because the sequences of the 3' end of the 16S rRNA from maize (122), tobacco (145), and *Euglena* (45) chloroplasts are very similar to that of *E. coli*.

The prokaryotic nature of the control of both transcription and translation has been further highlighted by the findings of Gatenby et al (38, 39) that *rbcL* is expressed in vivo in *E. coli* using either the internal promoter of *rbcL* or the external bacteriophage lambda promoter P_L. In the latter case, it was found that the product *rbcL* constituted up to 2% of the total *E. coli* cell protein.

CONCLUSIONS

In this review we have attempted to summarize the information currently available on the location, structure, and organization of genes on cpDNA. We have tabulated those plant species for which cpDNA restriction maps have been published and specified the genes for which approximate map coordinates have been determined. Examination of the primary structure of chloroplast rRNA genes has served to emphasize the similarity of these genes to their bacterial counterparts, the major difference being that the 16S–23S rDNA spacer is much larger in most cpDNAs because of the presence of large intervening sequences within the tRNA genes that are located in the spacer. Likewise, analysis of the primary structure of chloroplast tRNA genes has shown them to be more akin to prokaryote tRNAs than to plant cytoplasmic tRNAs. However, they are different from bacterial tRNA genes in two respects, namely chloroplast tRNA genes do not encode the terminal $-CCA$ sequence of the tRNA and some chloroplast tRNA genes contain introns. The organization of tRNA genes into clusters such as has been demonstrated in *Euglena* chloroplasts may not occur in higher plant chloroplasts. There is no evidence that tRNA genes are arranged as punctuation signals between protein coding genes in chloroplast DNA as they appear to be in some mitochondrial DNAs (6).

Consideration of the primary structure of protein coding genes and their flanking regions has shown that the prokaryote nature of the organelle extends to the nucleotide sequence level. Not only do signal sequences for transcription initiation in cpDNA bear a striking resemblance to the consensus sequence of *E. coli* promoters, but sequences close to the end of the 3' untranslated regions can base pair to form typical prokaryote-like transcription terminator stem-loop structures. Other prokaryote features of which we now have examples in chloroplasts include an overlapping translation stop/start signal and the cotranscription of adjacent genes into a dicistronic mRNA.

Although information on the organization and structure of other chloroplast genes will continue to be of interest in the future, it is clear that we are now in a position to use the significant body of sequence data already accumulated as a basis for designing experiments which probe the mechanisms whereby plastid genes are regulated. The switching-on of genes during plant development, the triggering of the synthesis of specific proteins by light and other factors, the coordination of expression of the genes for the α, β, and ϵ subunits of chloroplast ATPase, the coordinated regulation of nuclear and chloroplast genes coding for different subunits of the one protein, are all problems that can now be tackled. In vitro expression of cloned chloroplast genes in cell-free bacterial systems has proved to be exceedingly useful in the analysis of chloroplast genes. However, for the identification of regulatory sequences within cpDNA which are recognized by chloroplast RNA polymerase and other specific plant regulatory molecules, a cell-free system derived from plastids will have to be employed.

It is clear that in the future the most direct route to the determination of the primary structure of proteins coded by cpDNA will be via the sequencing of the gene. Furthermore, it is also likely that by appropriate genetic constructions, expression of chloroplast genes in bacterial cells will become routine practice. Synthesis of the large subunit of ribulose bisphosphate carboxylase in *E. coli* cells has already been achieved (38, 39). Production of chloroplast proteins in bacterial cultures in sufficient amounts to permit their purification, characterization, and use in the preparation of specific antibodies could provide a means whereby such ephemeral proteins as the "32 kd" thylakoid membrane protein can be studied and also could facilitate the identification of genes whose sequences have been determined but whose products are unknown. If expression of chloroplast genes in bacterial cells can be manipulated to the point where assembly of protein subunits into functional enzymes or structural complexes occurs, then by comparing the products of a range of appropriately modified gene sequences, it should be possible to define those regions of the sequence which are responsible for such properties as transport of proteins across membranes, assembly and role of subunits in a complex, and of course catalytic activity. The potential of recombinant DNA and molecular biology techniques for chloroplast research is tremendous, and the list of important questions begging to be answered is long, so how can the future be anything but productive and exciting?

Literature Cited

1. Allet, B., Rochaix, J. D. 1979. Structure analysis at the ends of the intervening DNA sequences in the chloroplast 23S ribosomal genes of *C. reinhardii*. *Cell* 18:55–60
2. Bedbrook, J. R., Coen, D. M., Beaton, A. R., Bogorad, L., Rich, A. 1979. Location of the single gene for the large subunit of ribulosebisphosphate carboxylase on the maize chloroplast chromosome. *J. Biol. Chem.* 254:905–10
3. Bedbrook, J. R., Dyer, T. A. 1982. Personal communication
4. Bedbrook, J. R., Kolodner, R. 1979. The structure of chloroplast DNA. *Ann. Rev. Plant Physiol.* 30:593–620
5. Bedbrook, J. R., Kolodner, R., Bogorad, L. 1977. *Zea mays* chloroplast ribosomal RNA genes are part of a 22000 base-pair inverted repeat. *Cell* 11:739–49
5a. Bedbrook, J. R., Link, G., Coen, D. M., Bogorad, L., Rich, A. 1978. Maize plastid gene expressed during photoregulated development. *Proc. Natl. Acad. Sci. USA* 75:3060–64

6. Bibb, M. J., Van Etten, R. A., Wright, C. T., Walberg, M. W., Clayton, D. A. 1981. Sequence and gene organization of mouse mitochondrial DNA. *Cell* 26:167–80
7. Bisaro, D., Siegel, A. 1980. Sequence homology between chloroplast DNAs from several higher plants. *Plant Physiol.* 65:234–37
8. Blair, G. E., Ellis, R. J. 1973. Protein synthesis in chloroplasts. I. Light-driven synthesis of the large subunit of Fraction I protein by isolated pea chloroplasts. *Biochim. Biophys Acta* 319:223–34
9. Bohnert, H. J., Crouse, E. J., Schmitt, J. M. 1982. Organization and expression of plastid genomes. In *Encyclopedia of Plant Physiology: Nucleic Acids and Proteins in Plants*, ed. D. Boulter, B. Parthier, 14B:475–530. Berlin: Springer
10. Bohnert, H. J., Driesel, A. J., Crouse, E. J., Gordon, K., Herrmann, R. G. 1979. Presence of a transfer RNA gene in the spacer sequence between the 16S and 23S

rRNA genes of spinach chloroplast DNA. *FEBS Lett.* 103:52–56

11. Borbely, G., Simoncsits, A. 1981. 3' terminal conserved loops of 16S rRNAs from the cyanobacterium *Synechococcus* AN PCC 6301 and maize chloroplast differ only in two bases. *Biochem. Biophys. Res. Commun.* 101:846–52

12. Bottomley, W., Bohnert, H. J. 1982. The biosynthesis of chloroplast proteins. See Ref. 9, pp. 531–96

13. Bottomley, W., Spencer, D., Whitfeld, P. R. 1974. Protein synthesis in isolated spinach chloroplasts: Comparison of light-driven and ATP-driven synthesis. *Arch. Biochem. Biophys.* 164:106–17

14. Bottomley, W., Whitfeld, P. R. 1979. Cell-free transcription and translation of total spinach chloroplast DNA. *Eur. J. Biochem.* 93:31–39

15. Bovenberg, W. A., Kool, A. J., Nijkamp, H. J. J. 1981. Isolation, characterization and restriction endonuclease mapping of the *Petunia hybrida* chloroplast DNA. *Nucleic Acids Res.* 9:503–17

16. Bowman, C. M., Dyer, T. A. 1979. 4.5S ribonucleic acid, a novel ribosome component in the chloroplasts of flowering plants. *Biochem. J.* 183:605–13

17. Bowman, C. M., Koller, B., Delius, H., Dyer, T. A. 1981. A physical map of wheat chloroplast DNA showing the location of the structural genes for the ribosomal RNAs and the large subunit of ribulose 1,5-bisphosphate carboxylase. *Mol. Gen. Genet.* 183:93–101

18. Buetow, D. E. 1983. Molecular biology of chloroplasts. In *Photosynthesis: Carbon Assimilation and Plant Productivity*, ed. Govindjee, Vol. 4. New York: Academic. In press

19. Bujard, H. 1980. The interaction of E. *coli* RNA polymerase with promoters. *Trends Biochem. Sci.* 5:274–78

20. Chu, N. M., Oishi, K. K., Tewari, K. K. 1981. Physical mapping of the pea chloroplast DNA and localization of the ribosomal RNA genes. *Plasmid* 6:279–92

21. Corry, M. J., Payne, P. I., Dyer, T. A. 1974. The nucleotide sequence of 5S rRNA from the blue-green alga *Anacystis nidulans*. *FEBS Lett.* 46:63–66

22. Darley-Usmar, M. V., Fuller, S. D. 1981. M_r-Values of mature subunits I and III of beef heart cytochrome *c* oxidase in relationship to nucleotide sequences of their genes. *FEBS Lett.* 135:164–66

23. Delihas, N., Anderson, J., Sprouse, H. M., Dudock, B. 1981. The nucleotide sequence of the chloroplast 5S ribosomal RNA from spinach. *Nucleic Acids Res.* 9:2801–5

24. Delius, H., Koller, B. 1980. Sequence homologies between *Escherichia coli* and chloroplast ribosomal DNA as seen by heteroduplex analysis. *J. Mol. Biol.* 142:247–61

25. Doherty, A., Gray, J. C. 1980. Synthesis of a dicyclohexylcarbodiimide-binding proteolipid by isolated pea chloroplasts. *Eur. J. Biochem.* 108:131–36

26. Driesel, A. J., Crouse, E. J., Gordon, K. Bohnert, H. J., Herrmann, R. G., et al. 1979. Fractionation and identification of spinach chloroplast transfer RNAs and mapping of their genes on the restriction map of chloroplast DNA. *Gene* 6:285–306

27. Driesel, A. J., Speirs, J., Bohnert, H. -J. 1980. Spinach chloroplast mRNA for a 32,000 dalton polypeptide. Size and localization on the physical map of the chloroplast DNA. *Biochim. Biophys. Acta* 610:297–310

28. Dyer, T. A., Bowman, C. M. 1979. Nucleotide sequences of chloroplast 5S ribosomal ribonucleic acid in flowering plants. *Biochem. J.* 183:595–604

29. Edelman, M. 1981. Nucleic acids of chloroplasts and mitochondria. In *The Biochemistry of Plants;* Vol. 6: *Proteins and Nucleic Acids*, ed. A. Marcus, pp. 249–301. New York: Academic

30. Edelman, M., Reisfeld, A. 1980. Synthesis, processing and functional probing of P-32,000, the major membrane protein translated within the chloroplast. In *Genome Organization and Expression in Plants*, ed. C. J. Leaver, pp. 353–62. New York/London: Plenum

31. Edwards, K., Bedbrook, J., Dyer, T., Kössel, H. 1981. 4.5S rRNA from *Zea mays* chloroplasts shows structural homology with the 3' end of prokaryotic 23S rRNA. *Biochem. Int.* 2:533–38

32. Edwards, K., Kössel, H. 1981. The rRNA operon from *Zea mays* chloroplasts: nucleotide sequence of 23S rDNA and its homology with *E. coli* 23S rDNA. *Nucleic Acids Res.* 9:2853–69

33. El-Gewely, R., Lomax, M. I., Lau, E. T., Helling, R. B., Farmerie, W., Barnett, W. E. 1981. A map of specific cleavage sites and tRNA genes in the chloroplast genome of *Euglena gracilis*. *Mol. Gen. Genet.* 181:296–305

34. Erion, J. L., Tarnowski, J., Weissbach, H., Brot, N. 1981. Cloning, mapping, and *in vitro* transcription-translation of the gene for the large subunit of ribulose-1,5-bisphosphate carboxylase from spinach chloroplasts. *Proc. Natl. Acad. Sci. USA* 78:3459–63

35. Fluhr, R., Edelman, M. 1981. Conserva-

tion of sequence arrangement among higher plant chloroplast DNAs: molecular cross hybridization among the Solanaceae and between *Nicotiana* and *Spinacia*. *Nucleic Acids Res.* 9:6841–53

36. Fluhr, R., Edelman, M. 1981. Physical mapping of *Nicotiana tabacum* chloroplast DNA. *Mol. Gen. Genet.* 181:484–90

37. Francis, M. A., Dudock, B. S. 1982. Nucleotide sequence of a spinach chloroplast isoleucine tRNA. *J. Biol. Chem.* 257:11195–98

38. Gatenby, A. A., Castleton, J. A. 1982. Amplification of maize ribulose bisphosphate carboxylase large subunit synthesis in *E. coli* by transcriptional fusion with the lambda *N* operon. *Mol. Gen. Genet.* 185:424–29

39. Gatenby, A. A., Castleton, J. A., Saul, M. W. 1981. Expression in *E. coli* of maize and wheat chloroplast genes for large subunit of ribulose bisphosphate carboxylase. *Nature* 291:117–21

40. Gillham, N. W., Boynton, J. E., Harris, E. H. 1982. Evolution of plastid DNA. In *DNA and Evolution: Natural Selection and Genome Size,* ed. T. Cavalier-Smith. New York: Wiley. In press

41. Glotz, C., Zwieb, C., Brimacombe, R., Edwards, K., Kössel, H. 1981. Secondary structure of the large subunit ribosomal RNA from *Escherichia coli, Zea mays* chloroplasts, and human and mouse mitochondrial ribosomes. *Nucleic Acids Res.* 9:3287–306

42. Gordon, K. H. J., Crouse, E. J., Bohnert, H.-J., Herrmann, R. G. 1981. Restriction endonuclease cleavage site map of chloroplast DNA from *Oenothera parviflora (Euoenothera* Plastome IV). *Theor. Appl. Genet.* 59:281–96

43. Gordon, K. H. J., Crouse, E. J., Bohnert, H.-J., Herrmann, R. G. 1981. Physical mapping of differences in chloroplast DNA of the five wild-type plastomes in *Oenothera* subsection *Euoenothera. Theor. Appl. Genet.* 61:373-84

44. Graf, L., Kössel, H., Stutz, E. 1980. Sequencing of 16S–23S spacer in a ribosomal RNA operon of *Euglena gracilis* chloroplast DNA reveals two tRNA genes. *Nature* 286:908–10

45. Graf, L., Roux, E., Stutz, E., Kössel, H. 1982. Nucleotide sequence of a *Euglena gracilis* chloroplast gene coding for the 16S rRNA: Homologies to *E. coli* and *Zea mays* chloroplast 16S rRNA. *Nucleic Acids Res.* 10:6369–81

46. Gray, M. W., Doolittle, W. F. 1982. Has the endosymbiont hypothesis been proven? *Microbiol. Rev.* 46:1–42

47. Gray, P. W., Hallick, R. B. 1977. Restriction endonuclease map of *Euglena gracilis* chloroplast DNA. *Biochemistry* 16:1665–71

48. Gray, P. W., Hallick, R. B. 1978. Physical mapping of the *Euglena gracilis* chloroplast DNA and ribosomal RNA gene region. *Biochemistry* 17:284–89

49. Gray, P. W., Hallick, R. B. 1979. Isolation of *Euglena gracilis* chloroplast 5S ribosomal RNA and mapping the 5S rRNA gene on chloroplast DNA. *Biochemistry* 18:1820–25

50. Grebanier, A. E., Coen, D. M., Rich, A., Bogorad, L. 1978. Membrane proteins synthesized but not processed by isolated maize chloroplasts. *J. Cell Biol.* 78:734–46

51. Groot, G. S. P., van Harten-Loosbroek, N. 1981. Physical mapping of 4S RNA genes on chloroplast DNA of *Spirodela oligorhiza. Curr. Genet.* 4:187–90

52. Guillemaut, P., Weil, J.-H. 1982. The nucleotide sequence of the maize and spinach chloroplast isoleucine transfer RNA encoded in the 16S to 23S rDNA spacer. *Nucleic Acids Res.* 10:1653–59

53. Haff, L. A., Bogorad, L. 1976. Hybridization of maize chloroplast DNA with transfer ribonucleic acids. *Biochemistry* 15:4105–9

54. Hallick, R. B. 1983. Chloroplast DNA. In *The Biology of Euglena,* ed. D. E. Buetow, Vol. 4. New York: Academic. In press

55. Hallier, U. W., Schmitt, J. M., Heber, U., Chaianova, S. S., Volodarsky, A. D. 1978. Ribulose 1,5-bisphosphate carboxylase-deficient plastome mutants of *Oenothera. Biochim. Biophys. Acta* 504:67–83

56. Hartley, M. R. 1979. The synthesis and origin of chloroplast low molecular-weight ribosomal ribonucleic acid in spinach. *Eur. J. Biochem.* 96:311–20

57. Hartman, F. C., Norton, I. L., Stringer, C. D., Schloss, J. V. 1978. Attempts to apply affinity labelling techniques to ribulose bisphosphate carboxylase/oxygenase. In *Photosynthetic Carbon Assimilation,* ed. H. W. Siegelman, G. Hind, pp. 245–69. New York/London: Plenum

58. Helling, R. B., El-Gewely, M. R., Lomax, M. I., Baumgartner, J. E., Schwartzbach, S. D., Barnett, W. E. 1979. Organization of the chloroplast ribosomal RNA genes of *Euglena gracilis bacillaris. Mol. Gen. Genet.* 174:1–10

59. Herrmann, R. G., Possingham, J. V.

1980. Plastid DNA—The plastome. In *Chloroplasts*, ed. J. Reinert, pp. 45–96. Berlin: Springer

60. Howe, C. J., Auffret, A. D., Doherty, A., Bowman, C. M., Dyer, T. A., Gray, J. C. 1983. Location and nucleotide sequence of the gene for the proton-translocating subunit of wheat chloroplast ATP synthase. *Proc. Natl. Acad. Sci. USA.* In press

61. Howe, C. J., Bowman, C. M., Dyer, T. A., Gray, J. C. 1982. Localization of wheat chloroplast genes for the beta and epsilon subunits of ATP synthase. *Mol. Gen. Genet.* 186:525–30

62. Jenni, B., Stutz, E. 1978. Physical mapping of the ribosomal DNA region of *Euglena gracilis* chloroplast DNA. *Eur. J. Biochem.* 88:127–34

63. Jenni, B., Stutz, E. 1979. Analysis of *Euglena gracilis* chloroplast DNA. Mapping of a DNA sequence complementary to 16S rRNA outside of the three rRNA gene sets. *FEBS Lett.* 102:95–99

64. Jurgenson, J. E., Bourque, D. P. 1980. Mapping of rRNA genes in an inverted repeat in *Nicotiana tabacum* chloroplast DNA. *Nucleic Acids Res.* 8:3505–16

65. Kashdan, M. A., Dudock, B. S. 1982. Structure of a spinach chloroplast threonine tRNA gene. *J. Biol. Chem.* 257:1114–16

66. Kashdan, M. A., Dudock, B. S. 1982. The gene for a spinach chloroplast isoleucine tRNA has a methionine anticodon. *J. Biol. Chem.* 257:11191–94

67. Kashdan, M. A., Pirtle, R. M., Pirtle, I. L., Calagan, J. L., Vreman, H.-J., Dudock, B. S. 1980. Nucleotide sequence of a spinach chloroplast threonine tRNA. *J. Biol. Chem.* 255:8831–35

68. Kato, A., Shimada, H., Kusuda, M., Sugiura, M. 1981. The nucleotide sequence of two tRNA^Asn genes for tobacco chloroplasts. *Nucleic Acids Res.* 9:5601–7

69. Keller, M., Burkard, G., Bohnert, H. J., Mubumbila, M., Gordon, K., et al. 1980. Transfer RNA genes associated with the 16S and 23S rRNA genes of *Euglena* chloroplast DNA. *Biochem. Biophys. Res. Commun.* 95:47–54

70. Kirk, J. T. O., Tilney-Bassett, R. A. E. 1978. *The Plastids. Their chemistry, structure, growth and inheritance.* New York/Amsterdam: Elsevier/North Holland. 960 pp. 2nd ed.

71. Koch, W., Edwards, K., Kössel, H. 1981. Sequencing of the 16S–23S spacer in a ribosomal RNA operon of *Zea mays* chloroplast DNA reveals two split tRNA genes. *Cell* 25:203–13

72. Koller, B., Delius, H. 1980. *Vicia faba* chloroplast DNA has only one set of ribosomal RNA genes as shown by partial denaturation mapping and R-loop analyses. *Mol. Gen. Genet.* 178:261–69

73. Koller, B., Delius, H. 1982. Electron microscopic analysis of the extra 16S rRNA gene and its neighbourhood in chloroplast DNA from *Euglena gracilis* strain Z. *FEBS Lett.* 139:86–92

74. Koller, B., Delius, H., Dyer, T. A. 1982. The organization of the chloroplast DNA in wheat and maize in the region containing the LS gene. *Eur. J. Biochem.* 122:17–23

75. Kolodner, R., Tewari, K. K. 1979. Inverted repeats in chloroplast DNA from higher plants. *Proc. Natl. Acad. Sci. USA* 76:41–45

76. Krebbers, E. T., Lorrinua, I. M., McIntosh, L., Bogorad, L. 1982. The maize chloroplast genes for the β and ε subunits of the photosynthetic coupling factor CF_1 are fused. *Nucleic Acids Res.* 10:4985–5002

77. Kumar, A., Cocking, E. C., Bovenberg, W. A., Kool, A. J. 1982. Restriction endonuclease analysis of chloroplast DNA in interspecific somatic hybrids of *Petunia*. *Theor. Appl. Genet.* 62:377–83

78. Kusuda, J., Shinozaki, K., Takaiwa, F., Sugiura, M. 1980. Characterization of the cloned ribosomal DNA of tobacco chloroplasts. *Mol. Gen. Genet.* 178:1–7

79. Lamppa, G. K., Bendich, A. J. 1979. Chloroplast DNA sequence homologies among vascular plants. *Plant Physiol.* 63:660–68

80. Lamppa, G. K., Bendich, A. J. 1981. Fine scale interspersion of conserved sequences in the pea and corn chloroplast genomes. *Mol. Gen. Genet.* 182:310–20

81. Langridge, P. 1981. Synthesis of the large subunit of ribulose bisphosphate carboxylase may involve a precursor. *FEBS Lett.* 123:85–89

82. Link, G. 1981. Cloning and mapping of the chloroplast DNA sequences for two messenger RNAs from mustard (*Sinapis alba* L.). *Nucleic Acids Res.* 9:3681–94

83. Link, G., Bogorad, L. 1980. Sizes, locations, and directions of transcription of two genes on a cloned maize chloroplast DNA sequence. *Proc. Natl. Acad. Sci. USA* 77:1832–36

84. Link, G., Chambers, S. E., Thompson, J. A., Falk, H. 1981. Size and physical organization of chloroplast DNA from mustard (*Sinapis alba* L.) *Mol. Gen. Genet.* 181:454–57

85. Lorimer, G. H. 1981. Ribulose bisphosphate carboxylase: Amino acid sequence

of a peptide bearing the activator carbon dioxide. *Biochemistry* 20:1236–40

86. Lorimer, G. H., Miziorko, M. 1980. Carbonate formation on the ε–amino group of a lysyl residue as the basis for the activation of ribulosebisphosphate carboxylase by CO_2 and Mg^{2+}. *Biochemistry* 19:5321–28

87. Machatt, M. A., Ebel, J. P., Branlant, C. 1981. The 3' terminal region of bacterial 23S ribosomal RNA: structure and homology with the 3' terminal region of eukaryotic 28S rRNA and with chloroplast 4.5S rRNA. *Nucleic Acids Res.* 9:1533–49

88. MacKay, R. M., Salgado, D., Bonen, L., Stackebrandt, E., Doolittle, W. F. 1982. The 5S ribosomal RNAs of *Paracoccus denitrificans* and *Prochloron*. *Nucleic Acids Res.* 10:2963–70

89. Malnoe, P., Rochaix, J.-D. 1978. Localization of 4S RNA genes on the chloroplast genome of *Chlamydomonas reinhardii*. *Mol. Gen. Genet.* 166:269–75

90. Malnoe, P., Rochaix, J.-D., Chua, N.-H., Spahr, P.-F. 1979. Characterization of the gene and messenger RNA of the large subunit of ribulose 1,5-diphosphate carboxylase in *Chlamydomonas reinhardii*. *J. Mol. Biol.* 133:417–34

91. McIntosh, L., Poulsen, C., Bogorad, L. 1980. Chloroplast gene sequence for the large subunit of ribulose bisphosphate carboxylase of maize. *Nature* 288:556–60

92. Meeker, R., Tewari, K. K. 1980. Transfer ribonucleic acid genes in the chloroplast deoxyribonucleic acid of pea leaves. *Biochemistry* 19:5973–81

93. Meeker, R., Tewari, K. K. 1982. Divergence of tRNA genes in chloroplast DNA of higher plants. *Biochim. Biophys. Acta* 696:66–75

94. Mendiola-Morgenthaler, L. R., Morgenthaler, J. J., Price, C. A. 1976. Synthesis of coupling factor CF_1 protein by isolated spinach chloroplasts. *FEBS Lett.* 62:96–100

95. Mubumbila, M., Burkard, G., Keller, M., Steinmetz, A., Crouse, E., Weil, J.-H. 1980. Hybridization of bean, spinach, maize and *Euglena* chloroplast transfer RNAs with homologous and heterologous chloroplast DNAs. *Biochim. Biophys. Acta* 609:31–39

96. Nelson, N., Nelson, H., Schatz, G. 1980. Biosynthesis and assembly of the proton translocating adenosine triphosphate complex from chloroplasts. *Proc. Natl. Acad. Sci. USA* 77:1361–64

97. Orozco, E. M. Jr., Gray, P. W., Hallick,

R. B. 1980. *Euglena gracilis* chloroplast ribosomal RNA transcription units. I. The location of transfer RNA, 5S, 16S and 23S ribosomal RNA genes. *J. Biol. Chem.* 255:10991–96

98. Orozco, E. M. Jr., Hallick, R. B. 1982. *Euglena gracilis* chloroplast transfer RNA transcription units. I Physical map of the transfer RNA gene loci. *J. Biol. Chem.* 257:3258–64

99. Orozco, E. M. Jr., Hallick, R. B. 1982. *Euglena gracilis* chloroplast transfer RNA transcription units. II Nucleotide sequence analysis of a tRNA^Val-tRNA^Asn-tRNA^Arg-tRNA^Leu gene cluster. *J. Biol. Chem.* 257:3265–75

100. Orozco, E. M. Jr., Rushlow, K. E., Dodd, J. R., Hallick, R. B. 1980. *Euglena gracilis* chloroplast ribosomal RNA transcription units. II. Nucleotide sequence homology between the 16S–23S ribosomal RNA spacer and the 16S ribosomal RNA leader regions. *J. Biol. Chem.* 255:10997–11003

101. Palmer, J. D. 1982. Physical and gene mapping of chloroplast DNA from *Atriplex triangularis* and *Cucumis sativa*. *Nucleic Acids Res.* 10:1593–605

102. Palmer, J. D., Stein, D. B. 1982. Chloroplast DNA from the fern *Osmunda cinnamonia:* Physical organization, gene localization and comparison to angiosperm chloroplast DNA. *Curr. Genet.* In press

103. Palmer, J. D., Thompson, W. F. 1981. Rearrangements in the chloroplast genomes of mung bean and pea. *Proc. Natl. Acad. Sci. USA* 78:5533–37

104. Palmer, J. D., Thompson, W. F. 1982. Chloroplast DNA rearrangements are more frequent when a large inverted repeat sequence is lost. *Cell* 29:537–50

105. Palmer, J. D., Zamir, D. 1982. Chloroplast DNA evolution and phylogenetic relationships in *Lycopersicon*. *Proc. Natl. Acad. Sci. USA* 79:5006–10

106. Poulsen, C. 1981. Comments on the structure and function of the large subunit of the enzyme ribulose bisphosphate carboxylase-oxygenase. *Carlsberg Res. Commun.* 46:259–78

107. Rawson, J. R. Y., Andrews, W. H. 1982. Cloning of chloroplast DNA in Bacterial Plasmid Vectors. In *Methods in Chloroplast Molecular Biology*, ed. M. Edelman, R. B. Hallick, N.-H. Chua, pp. 493–506. Amsterdam: Elsevier/North Holland. In press

108. Rawson, J. R. Y., Clegg, M. T., Thomas, K., Rinehart, C., Wood, B. 1981. A restriction map of the ribosomal RNA genes and the short single-copy DNA

sequence of the pearl millet chloroplast genome. *Gene* 16:11–19

109. Rawson, J. R. Y., Kushner, S. R., Vapnek, D., Alton, N. K., Boerma, C. L. 1978. Chloroplast ribosomal RNA genes in *Euglena gracilis* exist as three clustered tandem repeats. *Gene* 3:191–209

110. Reisfeld, A., Gressel, J., Jakob, K. M., Edelman, M. 1978. Characterization of the 32,000 dalton membrane protein. 1. Early synthesis during photoinduced plastid development of *Spirodela*. *Photochem. Photobiol.* 27:161–65

111. Rhodes, P. R., Zhu, Y. S., Kung, S. D. 1981. *Nicotiana* chloroplast genome I. Chloroplast DNA diversity. *Mol. Gen. Genet.* 182:106–11

112. Robinson, R. R., Davidson, N. 1981. Analysis of a *Drosophila* tRNA gene cluster: two tRNA^{Leu} genes contain intervening sequences. *Cell* 23:251–59

113. Rochaix, J. D. 1978. Restriction endonuclease map of the chloroplast DNA of *Chlamydomonas reinhardii*. *J. Mol. Biol.* 126:597–617

114. Rochaix, J. E. 1981. Organization, function and expression of the chloroplast DNA of *Chlamydomonas reinhardii*. *Experientia* 37:323–32

115. Rochaix, J. D., Darlix, J. L. 1982. Composite structure of the chloroplast 23S rRNA genes of *Chlamydomonas reinhardii*: evolutionary and functional implications. *J. Mol. Biol.* 159:383–95

116. Rochaix, J. D., Malnoe, P. 1978. Anatomy of the chloroplast ribosomal DNA of *Chlamydomonas reinhardii*. *Cell* 15:661–70

117. Rosenberg, M., Court, D. 1979. Regulatory sequences involved in the promotion and termination of RNA transcription. *Ann. Rev. Genet.* 13:319–53

118. Rutti, B., Keller, M., Ortiz, W., Stutz, E. 1981. Analysis of *Euglena gracilis* chloroplast DNA. The DNA fragment *Eco*R1.N carries genetic information for a 53,000 M_r polypeptide. *FEBS Lett.* 134:15–19

119. Saraste, M., Gay, N. J., Eberle, A., Runswick, M. J., Walker, J. E. 1981. The ATP operon: nucleotide sequence of the genes for the γ, β, and ε subunits of *Escherichia coli* ATP synthase. *Nucleic Acids Res.* 9:5287–96

120. Schmitt, J. M., Bohnert, H.-J., Gordon, K. H. J., Herrmann, R., Bernardi, G., Crouse, E. J. 1981. Compositional heterogeneity of the chloroplast DNAs from *Euglena gracilis* and *Spinacia oleracea*. *Eur. J. Biochem.* 117:375–82

121. Schwarz, Z., Jolly, S. O., Steinmetz, A.,

A., Bogorad, L. 1981. Overlapping divergent genes in the maize chloroplast chromosome and *in vitro* transcription of the gene for tRNA^{His}. *Proc. Natl. Acad. Sci. USA* 78:3423–27

122. Schwarz, Z., Kössel, H. 1979. Sequencing of the 3'-terminal region of a 16S rRNA gene from *Zea mays* chloroplast reveals homology with *E. coli* 16S rRNA. *Nature* 279:520–22

123. Schwarz, Z., Kössel, H. 1980. The primary structure of 16S rDNA from *Zea mays* chloroplast is homologous to *E. coli* 16S rRNA. *Nature* 283:739–42

124. Schwarz, Z., Kössel, H., Schwarz, E., Bogorad, L. 1981. A gene coding for tRNA^{Val} is located near the 5'-terminus of 16S rRNA gene in *Zea mays* chloroplast genome. *Proc. Natl. Acad. Sci. USA* 78:4748–52

125. Sebald, W., Wachter, E. 1980. Amino acid sequence of the proteolipid subunit of the ATP synthase from spinach chloroplasts. *FEBS Lett.* 122:307–11

126. Seyer, P., Kowallik, K. V., Herrmann, R. G. 1981. A physical map of *Nicotiana tabacum* plastid DNA including the location of structural genes for ribosomal RNAs and the large subunit of ribulose bisphosphate carboxylase/oxygenase. *Curr. Genet.* 3:189–204

127. Shen, G. F., Chen, K., Wu, M., Kung, S. D. 1982. *Nicotiana* chloroplast genome IV *N. accuminata* has larger inverted repeats and genome size. *Mol. Gen. Genet.* 187:61–66

128. Shine, J., Dalgarno, L. 1975. Determinant of cistron specificity in bacterial ribosomes. *Nature* 254:34–38

129. Steinback, K. E., McIntosh, L., Bogorad, L., Arntzen, C. J. 1981. Identification of the triazine receptor protein as a chloroplast gene product. *Proc. Natl. Acad. Sci. USA* 78:7463–67

130. Steinmetz, A., Gubbins, E. J., Bogorad, L. 1982. The anticodon of the maize chloroplast gene for tRNA$^{Leu}_{UAA}$ is split by a large intron. *Nucleic Acids Res.* 10:3027–37

131. Stiegler, G. L., Matthews, H. M., Bingham, S. E., Hallick, R. B. 1982. The gene for the large subunit of ribulose-1,5-bisphosphate carboxylase in *Euglena gracilis* chloroplast DNA: Location, polarity, cloning, and evidence for an intervening sequence. *Nucleic Acids Res.* 10:3427–44

132. Stormo, G. D., Schneider, T. D., Gold, L. M. 1982. Characterization of translational initiation sites in *E. coli*. *Nucleic Acids Res.* 10:2971–96

133. Sugiura, M., Kusada, J. 1979. Moleuclar

cloning of tobacco chloroplast ribosomal RNA genes. *Mol. Gen. Genet.* 172:137–41

134. Takaiwa, F., Kusuda, M., Sugiura, M. 1982. The nucleotide sequence of chloroplast 4.5S rRNA from a fern, *Dryopteris acuminata. Nucleic Acids Res.* 10:2257–60

135. Takaiwa, F., Sugiura, M. 1980. Cloning and characterization of 4.5S and 5S RNA genes in tobacco chloroplasts. *Gene* 10:95–103

136. Takaiwa, F., Sugiura, M. 1980. Nucleotide sequence of the 4.5S and 5S ribosomal RNA genes from tobacco chloroplasts. *Mol. Gen. Genet.* 180:1–4

137. Takaiwa, F., Sugiura, M. 1980. The nucleotide sequence of 4.5S ribosomal RNA from tobacco chloroplasts. *Nucleic Acids Res.* 8:4125–29

138. Takaiwa, F., Sugiura, M. 1981. Heterogeneity of 5S RNA species in tobacco chloroplasts. *Mol. Gen. Genet.* 182:385–89

139. Takaiwa, F., Sugiura, M. 1982. Nucleotide sequence of the 16S–23S spacer region in an rRNA gene cluster from tobacco chloroplast DNA. *Nucleic Acids Res.* 10:2665–76

140. Takaiwa, F., Sugiura, M. 1982. The complete nucleotide sequence of a 23S rRNA gene from tobacco chloroplasts. *Eur. J. Biochem.* 124:13–19

141. Takaiwa, F., Sugiura, M. 1982. The nucleotide sequence of chloroplast 5S ribosomal RNA from a fern *Dryopteris acuminata. Nucleic Acids Res.* 10:5369–73

142. Tewari, K. K. 1979. Chloroplast DNA: Structure, transcription and replication. In *Nucleic Acids in Plants*, ed. T. C. Hall, J. W. Davies, 1:41–108. Boca Raton, Fla: CRC Press

143. Thomas, J. R., Tewari, K. K. 1974. Conservation of 70S ribosomal RNA genes in the chloroplast DNAs of higher plants. *Proc. Natl. Acad. Sci. USA* 71:3147–51

144. Tohdoh, N., Shinozaki, K., Sugiura, M. 1981. Sequence of a putative promoter region for the rRNA genes of tobacco chloroplast DNA. *Nucleic Acids Res.* 9:5399–406

145. Tohdoh, N., Sugiura, M. 1982. The complete nucleotide sequence of a 16S ribosomal RNA gene from tobacco chloroplasts. *Gene* 17:213–18

146. Tomioka, N., Shinozaki, K., Sugiura, M. 1981. Molecular cloning and characterization of ribosomal RNA genes from a blue-green alga, *Anacystis nidulans. Mol. Gen. Genet.* 184:359-63

147. Valenzuela, P., Venegas, A., Weinberg, F., Bishop, R., Rutter, W. J. 1978. Structure of yeast phenylalanine-tRNA genes: an intervening DNA segment within the region coding for the tRNA. *Proc. Natl. Acad. Sci. USA* 75:190–94

148. Van Charldorp, R., Van Knippenberg, P. H. 1982. Sequence, modified nucleotides and secondary structure at the 3′end of small ribosomal subunit RNA. *Nucleic Acids Res.* 10:1149–58

149. van Ee, J. H., Vos, Y. J., Planta, R. J. 1981. Physical map of chloroplast DNA of *Spirodela oligorrhiza;* analysis by the restriction endonucleases *PstI, XhoI* and *SacI. Gene* 12:191–200

150. Watson, J. C., Surzycki, S. J. 1982. Extensive sequence homology in the DNA coding for elongation factor Tu from *Escherichia coli* and the *Chlamydomonas reinhardii* chloroplast. *Proc. Natl. Acad. Sci. USA* 79:2264–67

151. Weil, J. H., Parthier, B. 1982. Transfer RNA and aminoacyl-tRNA synthetases in plants. In *Encyclopedia of Plant Physiol: Nucleic Acids and Proteins in Plants I*, ed. D. Boulter, B. Parthier, 14A:65–112. Berlin: Springer

152. Westhoff, P., Nelson, N., Bünemann, H., Herrmann, R. G. 1981. Localization of genes for coupling factor subunits on the spinach plastid chromosome. *Curr. Genet.* 4:109–20

153. Whitfield, P. R., Bottomley, W. 1980. Mapping of the gene for the large subunit of ribulose bisphosphate carboxylase on spinach chloroplast DNA. *Biochem. Int.* 1:172–78

154. Whitfeld, P. R., Herrmann, R. G., Bottomley, W. 1978. Mapping of the ribosomal RNA genes on spinach chloroplast DNA. *Nucleic Acids Res.* 5:1741–51

155. Whitfeld, P. R., Leaver, C. J., Bottomley, W., Atchison, B. A. 1978. Low molecular weight (4.5S) ribonucleic acid in higher-plant chloroplast ribosomes. *Biochem. J.* 175:1103–12

156. Wildeman, A. G., Nazar, R. N. 1980. Nucleotide sequence of wheat chloroplastid 4.5S ribonucleic acid. Sequence homologies in 4.5S RNA species. *J. Biol. Chem.* 255:1896–900

156a. Willey, D. L., Huttly, A. K., Phillips, A. L., Gray, J. C. 1983. Localization of gene for cytochrome *f* in pea chloroplast DNA. *Mol. Gen. Genet.* In press

157. Wurtz, E. A., Buetow, D. E. 1981. Intraspecific variation in the structural organization and redundancy of chloroplast ribosomal DNA cistrons in *Euglena gracilis. Curr. Genet.* 3:181–87

158. Young, R. A., Steitz, J. A. 1978. Complementary sequences 1700 nucleotides

apart form a ribonuclease III cleavage site in *Escherichia coli* ribosomal precursor RNA. *Proc. Natl. Acad. Sci. USA* 75:3593–97

159. Zech, M., Hartley, M. R., Bohnert, H. J. 1981. Binding sites of *E. coli* DNA-dependent RNA polymerase on spinach chloroplast DNA. *Curr. Genet.* 4:37–46

160. Zenke, G., Edwards, K., Langridge, P., Kössel, H. 1982. The rRNA operon from maize chloroplasts: analysis of *in vivo* transcription products in relation to its structure. In *Cell Function and Differentiation*, ed. G. Akoyunoglou. In press

161. Zurawski, G., Bohnert, H.-J., Whitfeld, P. R., Bottomley, W. 1982. Nucleotide sequence of the gene for the 32,000-M_r thylakoid membrane protein from *Spinacia oleracea* and *Nicotiana debneyi* predicts a totally conserved primary

translation product of M_r 38,950. *Proc. Natl. Acad. Sci. USA* 79:7699–7703

162. Zurawski, G., Bottomley, W., Whitfeld, P. R. 1982. Structure of the genes for the β and ε subunits of spinach chloroplast ATPase indicates a dicistronic mRNA and an overlapping translation stop/start signal. *Proc. Natl. Acad. Sci. USA* 79:6260–64

163. Zurawski, G., Bottomley, W., Whitfeld, P. R. 1983. Molecular analysis of adjacent transcriptional units from pea chloroplast DNA: *rbcL*, the gene for the large subunit of ribulose bisphosphate carboxylase and *atpBE*, the genes for the β and ε subunits of ATPase. In press

164. Zurawski, G., Perot, B., Bottomley, W., Whitfeld, P. R. 1981. The structure of the gene for the large subunit of ribulose 1,5-bisphosphate carboxylase from spinach chloroplast DNA. *Nucleic Acids Res.* 9:3251–70

Ann. Rev. Plant Physiol. 1983. 34:311–26

REGULATION OF ION TRANSPORT

Anthony D. M. Glass

Department of Botany, University of British Columbia, 2075 Wesbrook Mall, Vancouver, B.C. Canada V6T 2B1

CONTENTS

INTRODUCTION[1]

The fluxes of inorganic ions across plant membranes have been the subject of intensive investigations for almost a hundred years. In the main, these investigations have sought to clarify the mechanisms responsible for these fluxes; in particular, the nature of the driving forces. By contrast, the regulation of ion transport has received comparatively little attention. In this review the term regulation of ion transport will be reserved for intrinsic processes which serve to maintain rates of transport or internal outputs of transport (turgor, volume, or

---

[1]Abbreviations used: [ ], concentration with subscripts i = internal (tissue), o = outside, c = cytoplasmic, v = vacuolar; Ψ, water potential; Ψ_π, osmotic potential; Ψ_p, pressure potential (turgor); Ψ_m, matric potential; FCCP, (*p*-trifluoromethoxy) carbonyl cyanide phenylhydrazone; $\Delta\bar{\mu}$, electrochemical potential difference with subscripts referring to specific ions; Pi, inorganic phosphate.

311

0066–4294/83/0601–0311$02.00

concentration) at prescribed levels. Therefore, regulations generally, but not exclusively, maintain the status quo, buffering the cell against environmental vagaries as well as the perturbations caused by growth. However, regulations may also guide the cell or organism from one physiological or developmental state to another (50).

Inorganic ions which are accumulated in plant cells exert effects which fall into two broad categories. First are the effects which arise from a lowering of the free energy of aqueous solutions of these ions (of particular importance here are the phenomena of osmosis and turgor). Second are effects related to their nutritional roles. There exists, nevertheless, considerable flexibility in the choice of particular solutes for osmotic rather than nutritional function (54). The low-K^+, high-sugar ("low-salt") roots of Hoagland and Broyer probably reflect this capacity to substitute an organic solute for the osmotic function normally performed by K^+ (62). The topic of regulation of ion transport, therefore, inevitably overlaps extensively with aspects of mineral nutrition and osmoregulation. Although previous issues of this series have not included a review specifically devoted to the regulation of ion transport, various aspects of this topic are included in reviews on mineral nutrition (13), osmoregulation (83), and regulation of pH (75). In addition, the reviews by Pitman & Cram (61, 62), Cram (16), and Raven (65) represent outstanding contributions to understanding in this area.

Under conditions of adequate availability of inorganic ions, [ ]$_i$ approach upper limits (set points or "desired" levels) whose values must presumably reflect the requirements for both osmotic and nutritional functions. Our knowledge of the mechanisms involved in the maintenance of these set points is rudimentary. We are aware that certain inorganic ions, notably sodium and potassium chlorides, and diverse organic solutes make prominent contributions to Ψ_{Π_i} and hence to Ψ_p (16, 39, 46). We are also aware, from extensive comparative studies, that adjustments in the concentrations of these solutes are commonly a consequence of osmotic perturbation (39, 41, 77, 82). Similarly, there is extensive documentation of the increased capacity for ion uptake associated with low external ion availability (16). Nevertheless, beyond these strong indications of the existence of the corresponding regulatory systems, we are woefully ignorant of the details of the mechanisms responsible.

In order to describe fully the regulation of ion transport within a particular system, one should ideally have knowledge of (a) the mechanisms of active and passive fluxes to and from the compartment, (b) the cellular and whole plant functions of these fluxes, (c) the means whereby perturbations of particular cell outputs are sensed and the identity of the signals which serve to alter existing fluxes, and (d) the manner in which the transport system is adjusted by these signals to achieve regulation. In this review, points c and d will be emphasized, principally at the cellular level.

OSMOTIC EFFECTS

Wall-less cells

Perhaps the most fundamental consequence of the evolution of a semiperme-able membrane that enclosed charged macromolecules was the associated need for volume regulation (68). This arises from the inherent tendency for mobile ions to distribute themselves according to the Gibbs–Donnan equilibrium. The resulting ionic asymmetry and consequent osmotic flow of water would cause swelling and lysis were it not for counteractive forces. The extrusion of a hypo-osmotic fluid by means of contractile vacuoles represents one means of dealing with hypertonicity (40). Rothstein (68) has suggested that the universal tendency to extrude Na^+, thus generating a sodium-poor, potassium-rich cytoplasm, may have arisen originally from the need to extrude the predomi-nant external ion, namely Na^+, in order to regulate volume. Through a continuous extrusion of Na^+ against its electrochemical potential gradient, osmotic equilibrium is maintained between the cell's interior and the bathing medium (53).

Wall-less cells are generally considered to be unable to sustain internal hydrostatic pressures (Ψ_p) significantly in excess of those of the medium (16). Following osmotic shock there is rapid swelling or shrinkage in accord with the Boyle-Van't Hoff relation and subsequent volume readjustment at constant pressure and osmotic potential (Ψ_{π_i}) as total internal solute content is changed (16, 49, 83). It has been suggested therefore that the cell output which is monitored during these changes must be volume rather than Ψ_p or Ψ_{π_i} (16). Alternatively, plasma membrane deformation may serve as the appropriate feedback signal (47, 82). Whichever cell parameter is sensed, there is little detailed information regarding the mechanism responsible for transducing this signal to restore volume (e.g. 9). During volume recovery in red blood cells following osmotic perturbation, large dissipative ion fluxes occur down their respective electrochemical potential gradients (10, 38); K^+ and Cl^- efflux during regulatory volume decrease (RVD) and Na^+ and Cl^- influx during regulatory volume increase (RVI). Sensitivity of the sodium flux to the inhibi-tor amiloride, which is not inhibitory to steady-state fluxes, indicates that fluxes associated with volume regulation are different from steady-state fluxes. Furthermore, these fluxes are electrically silent, Na^+ and K^+ in exchange for H^+, and Cl^- in exchange for HCO_3^- (10). The means for transducing the volume changes into signals which initiate additional transport pathways are unknown. However, it has been suggested that elevation of $[Ca^{2+}]_i$ associated with swelling may trigger increased K^+ permeability leading to RVD (38).

In *Dunaliella parva*, glycerol accumulation has been demonstrated to be strongly correlated with $[NaCl]_o$ (9). Similar findings have been documented for mannitol in *Platymonas subcordiformis* (48). Among other wall-less cells,

Poteriochromonas malhamensis may regulate volume by synthesis or degradation of isofloridoside following osmotic perturbation (47). The correlations between external salinity and internal glycerol concentration, together with the documented sensitivity of isolated enzymes to inhibition by NaCl, have been advanced as an argument against salt accumulation as a means of osmoregulation in *Dunaliella* (9). However, despite the obvious importance of these organic solutes in maintaining osmotic equilibrium, it may require considerable time to increase solute concentration biosynthetically. In *Platymonas*, for example, mannitol accounted for 20% of Ψ_{π_i} 20 min after hyperosmotic shock (48). By 120 min this percentage had increased to 45%. Kirst (48) has suggested, therefore, that inorganic ions (K^+, Na^+, and Cl^-) may serve to "bridge the concentration gap" as the immediate response to osmotic perturbation. Gimmler et al (30) have questioned the exclusive role of glycerol synthesis in volume regulation in *Dunaliella* since volume recovery following hyperosmotic shock was essentially unaffected by the inhibitor FCCP at concentrations which were inhibitory to glycerol synthesis.

The argument based upon in vitro effects of salt upon enzyme activity is not a strong one. In vivo the enzyme may be protected from harmful effects of salt (71). Recent work suggests that $[Na^+]_i$ in *Dunaliella* is strongly correlated with $[NaCl]_o$ (31, 32). It is also clear from isotopic studies in *D. parva* (33) that substantial influxes of Na^+ and Cl^- are maintained at steady state. Following hyperosmotic shock, increased Na^+ and Cl^- influxes doubled cell $[Na^+]$ and $[Cl^-]$ within 2 min. The increase in $[Na^+]$ was insensitive to temperature (33), consistent with the observed capacity to restore cell volume in presence of FCCP noted by Gimmler (30). Subsequently, a temperature-sensitive extrusion of these ions restored [ ]$_i$ to former levels within 2 hours at 25°C. During RVD in media of constant osmolarity but varying $[K^+]$, the rate of volume restoration varied inversely with $[K^+]_o$ in *D. maritima* (67). The authors concluded that RVD was achieved via increased permeability to K^+ leading to increased K^+ efflux. By contrast, RVI, following hyperosmotic treatment, was independent of $[K^+]_o$. Thus, particularly in the initial response, inorganic ion fluxes may play a significant role in volume regulation in these algae. The apparent use of existing $\Delta\bar{\mu}$, as in red blood cells, provides for rapid and energetically cheap ionic readjustment (7).

An interesting role for inorganic phosphate in controlling the enzyme pathway leading either to glycerol or starch synthesis has been proposed for *Dunaliella* (29). Following hyperosmotic shock $[Pi]_i$ increases, directing carbon flow to glycerol rather than starch through the inhibitory action of Pi on ADPG-pyrophosphorylase. In *Poteriochromonas*, activation of the enzyme isofloridoside-phosphate synthase following cell shrinkage appears to depend upon a Ca^{2+} regulated cascade of reactions operating via membrane participation (47). In concluding this discussion on wall-less cells, it should be emphasized that volume regulation is not a phenomenon restricted to wall-less cells, but represents a continuing process of adjustment associated with growth and cell division in all cells. Dainty, for example, has stressed the problems associated with the maintainance of osmotic equilibrium between cytoplasm and vacuole (20). During cell elongation, in order to control vacuole enlargement, there is an obvious need to regulate solute fluxes between these compartments. Failure to do so would be to invite swelling or shrinking of the cytoplasm and its organelles (including the vacuole) which would almost certainly lead to impaired metabolic function. In *Platymonas*, for example, one of the earliest metabolic responses to hypo-osmotic stress is the severe inhibition of photosynthesis associated with chloroplast swelling during the first seconds of stress (49).

Walled Cells

By enclosing the protoplast within a cell wall, the problem of Donnan swelling by a hypertonic cytoplasm is resolved. Water influx associated with $\Delta\Psi$ now generates increased turgor (Ψ_p) because of the high volumetric elastic modulus of most plant cells (83). This means of resolving $\Delta\Psi$ is not without problems. Under the constraints of such a strait jacket, the threshold (critical) turgor ($\Psi_{P_{th}}$) required to achieve cell expansion is extremely high in most walled plant cells. Values in the range of 0.6–0.8 MPa have been reported for various crop plants (8). This is 10^4 times larger than the values of Ψ_p (20 to 30 Pa) which can be tolerated by naked cells such as red blood cells (64). Considering the number of environmental factors which might reduce Ψ_p below $\Psi_{P_{th}}$ on a daily or seasonal basis and the critical dependence of cell expansion and numerous biochemical processes upon Ψ_p (41), it would be surprising if this cell output were not regulated.

Perhaps the strongest evidence for the regulation of turgor in plant cells has come from the extensively studied giant-celled marine algae such as *Valonia*, *Codium*, and *Halicystis* (7, 46, 82). These organisms maintain an almost constant Ψ_p over a wide range of Ψ_{π_o}. The details of such regulations, particularly in the case of *Valonia*, have been reviewed frequently in recent years (39, 46, 82, 83), and hence only a brief outline of these processes will be given here. In these algae a large proportion ($\sim 95\%$) of vacuolar osmolality is accounted for by inorganic ions, particularly Na^+, K^+, and Cl^- (39). Moreover, the osmotic adjustments required to maintain Ψ_p in the face of declining external Ψ_{π_o} are the result of active K^+ transport in *Valonia* and active Cl^- transport in *Codium* and *Halicystis*. Nevertheless, passive fluxes of Cl^-, acting as counterion for K^+, are extremely important in the osmotic adjustment to increased salinity in *Valonia* (39). In this organism, there is good evidence that Ψ_p itself is the primary signal responsible for initiating turgor regulation (39, 82). Pressure changes of the sort which cause adjustments of ion fluxes are too small to act as driving forces for ion movements, and moreover, since proteins appear to be quite resistant to changes of pressure in the normal biological range (55), it is evident that changes associated with osmotic stress would require amplification. Ultrastructural evidence for a close appression of the membrane against the cell wall in turgid *Pelvetia* embryos has been advanced (7). Chemical or physical modifications of the membrane, associated with Ψ_p changes, could alter the thickness of the membrane resulting in altered membrane activity (7). Various generalized negative feedback models, based upon electrical analogs, have been proposed to describe regulation of the cell's ion concentration or turgor (see 7, 16). While such generalized models may provide a convenient language for the description of the processes of regulation, they may also prematurely channel our perception of regulation, and they have been criticized as overly simplistic and of limited value in advancing knowledge of regulation (82). In their place, Zimmermann et al (82, 83) have advanced a more quantitative electromechanical model for turgor detection based upon mechanical compression of the membrane associated with turgor changes. Although based principally on measurements using *Valonia*, the model may have application to turgor sensing in cells which regulate via biochemical interconversions and also to volume regulation in wall-less cells. The basic tenet of the model is that the sensing membrane may be compressed through both electrical forces associated with membrane potential and by mechanical forces (Ψ_p in this context). Evidence supporting a dependence of membrane compression upon Ψ_p is adduced indirectly by measurements of the dielectrical breakdown voltages which are shown to be a function of membrane thickness. Estimates of the transverse elastic compressive modulus obtained from the relationship between breakdown voltages and Ψ_p indicate that the modulus is sufficiently low to sense rather moderate changes of Ψ_p. The net effect of changes in $\Psi\pi_o$ in *Valonia* is to bring about flux changes which eventually restore original Ψ_p. By applying the vacuolar perfusion technique to *V. ventricosa*, Gutknecht (39) was able to demonstrate that K^+ influx, and to a lesser extent efflux were directly influenced by the administered pressure. Subsequent studies using the pressure probe revealed that in *V. utricularis* K^+ influx decreases with increased Ψ_p up to values of about 0.2 MPa, beyond which influx becomes insensitive to Ψ_p (82). Over the same range of Ψ_p, efflux increased severalfold. Clearly, in this organism the most obvious gap in present understanding lies in the link between pressure sensing and the transport processes.

It is clear from the extensive plant literature dealing with what is conveniently referred to as osmoregulation (but may in fact be turgor, concentration, or volume regulation) that plants from a diverse range of habitats maintain turgor at a more or less constant value. At the same time, however, it is equally apparent that the sorts of mechanisms which have been developed on the basis of studies with particular giant algal cells may not apply universally throughout the plant kingdom.

In giant-celled marine algae, turgor appears to be regulated directly, via compensatory changes of ion fluxes, in response to environmental or experimental perturbations. By contrast, in other plants, there appear to be direct responses to external ion concentration rather than to turgor per se. Thus in *Porphyra purpurea*, where KCl forms the major fraction of Ψ_{π_i} at high salinity, influxes of K^+ and Cl^- failed to respond to turgor manipulation via added mannitol (66). The author concluded that the primary signal for the adjustment of $[ion]_i$ was probably $[K^+]_o$. In *Codium decorticatum* also, although $[NaCl]_i$ was adjusted in response to decreasing Ψ_{π_o} (independently of salinity), $[K^+]_i$ was determined by $[K^+]_o$ (6). *Chaetomorpha linum* represents yet another example where turgor changes in response to $[K^+]_o$ (84) or may remain constant as Ψ_{π_o} is changed at constant $[K^+]_o$. In *Chara corallina* fluxes of Cl^-, K^+, and HCO_3^- were found to be insensitive to turgor, as was Ψ_{π_i} (69). In media of constant Ψ_π, the regulation of Ψ_π may effectively be equivalent to Ψ_p regulation (7). Moreover, since in marine environments, where Ψ_p changes must be almost exclusively linked to changes of $[ion]_o$, Ψ_p might be regulated directly through effects of Ψ_p on transport or biochemical reactions, or indirectly by effects of $[ion]_o$ on these processes.

The process of Ψ_p maintenance in higher plants is considerably more complicated and far from clear. Osmotic adjustment to salinity in halophytes by means of salt accumulation has long been recognized (82). More recently, the maintenance of turgor by osmotic adjustment following drought conditions has been emphasized (41, 77). Salt tolerance and related aspects of osmoregulation in halophytes and nonhalophytes have been reviewed extensively in this series (28, 37) and elsewhere (80), and these sources should be consulted for further details. It is clear that adjustment to the reduced Ψ_π of saline soils in halophytes is achieved by the accumulation of inorganic ions (particularly Na^+ and Cl^-) to considerably higher levels, particularly in leaf tissue, than are found in nonhalophytes. This capacity may be constitutive in halophytes and must presumably reflect inherent differences in set points since these differences are found even when halophytes and nonhalophytes are grown under identical conditions in media of low ion content (28). An important aspect of the adjustment to salinity is the need for subcellular compartmentation of the absorbed salt within the vacuole and the associated need for osmotic adjustment of the cytoplasm by "compatible" solutes (28, 80). The spatial separation of absorbing organs (roots) from other sites which must undergo osmotic adjustment greatly complicates the integration of turgor regulation. Because leaves are dependent upon ion translocation (as well as photosynthetic activity) to provide the solutes for the maintainance of Ψ_p, the translocation step represents a likely site for regulation via signals originating at the shoot. Though a participatory role for plant growth substances seems likely, there is limited information regarding the manner in which they might mediate control in vivo (61, 62). In grapevine leaves (23), the time course of responses to NaCl treatment revealed increased ABA levels to be among the earliest changes. Concentrations of reducing sugars increased rapidly in the first days after salt treatment and subsequently declined. Inorganic ion concentrations, particularly Na^+, K^+, and Cl^-, increased throughout the duration of the experiment. Osmotic adjustment via increased concentrations of K^+, Cl^-, sugars, and amino acids in response to drought conditions has been documented for leaves of many crop species (77). Nevertheless, details are lacking regarding the sequence of events and factors responsible for controlling the levels of solutes resulting from imposed stress. Although increased ABA synthesis associated with loss of turgor represents one of the earliest responses to dehydration (58), the nature of the impact of this hormone (in vivo) upon the absorption and translocation of ions is still not clear.

Attempts to document direct relationships between Ψ_p and ion transport in roots of higher plants have, with few exceptions, failed to demonstrate the anticipated responses. In beet and carrot slices, increased ion uptake in response to nonpermeating osmotica such as mannitol has been reported (18, 25, 76). Generally the responses have been expressed after some hours (18, 76), although Enoch & Glinka (25) reported stimulated K^+ uptake and reduced efflux in carrot slices at 30 min after mannitol application. By contrast, in roots of corn, wheat, and barley, similar treatments led to inhibition of ion transport (14, 27, 57). Considering that water stress associated with drought is usually experienced very gradually under natural conditions, it may be naive to anticipate immediate compensatory changes of ion transport rates in response to reduced turgor. Perhaps greater emphasis should be placed upon long-term adjustments of ion transport. Finally, it should be appreciated that attempts to investigate effects of turgor loss on ion transport by exposure of tissues to nonpenetrating solutes fail to simulate drought conditions which are associated with reduced Ψ_m rather than Ψ_π (52).

NUTRITIONAL EFFECTS

Introduction

Consideration of the widely used influx isotherms for ion uptake (26), particularly as they apply to "low-salt" roots, might erroneously lead to the conclusion that declining [ ]$_o$ would drive the tissue from set point toward deficiency. However, long-term studies reveal an essential constancy of [ ]$_i$, independent of [ ]$_o$, which indicates accommodation on ion fluxes to diminishing availability (1, 79). The accommodation of uptake is even clearer in short-term studies of ion uptake as a function of previous nutrient provision. For example, although net NO_3^- uptake by barley in the concentration range from 10-500 μM demonstrated typical concentration-dependent curves, with V_{max} inversely related to prior NO_3^- provision, uptake at concentrations corresponding to those of the growth medium (from 100–500 μM) were constant at $\sim$ 1.9 μmol $g^{-1}h^{-1}$ (21). Nevertheless, there are finite limits to which such adjustments can be made. In the latter study, uptake rates at growth concentrations of 10 and 60 μM NO_3^- were 0.55 and 1.52 μmol $g^{-1}h^{-1}$ respectively. Clearly, at concentrations where uptake is insufficient to sustain set point, growth rates must decline. Even in the long-term studies cited above (e.g. 79), $[K^+]_i$ and growth rates were somewhat reduced by declining $[K^+]_o$. Nevertheless, considering that $[K^+]_i$ and growth rates fell by only 25% despite a 1000-fold decline of $[K^+]_o$, substantial accommodation of transport rates must have occurred.

Mechanisms

The negative correlations between uptake and $[ion]_i$ have been documented in a large number of cases (e.g. 16). Since influx to the vacuole is commonly more strongly reduced than plasmalemma influx by ion accumulation (3, 14, 19), negative feedback between $[ion]_v$ and fluxes to the vacuole have been proposed (19, 59). Reductions of plasmalemma influx are also associated with ion accumulation (11, 14, 34, 51, 60, 69). A strong case for the maintenance of $[ion]_c$ within rather narrow limits, common to most eukaryote cells, has been advanced by Wyn Jones et al (80). It is difficult to envisage how this could be achieved without feedback from $[ion]_c$, and the most likely targets for this feedback would be tonoplast and plasmalemma transporters. Under conditions of nutrient deprivation, it is reasonable to presume that $[\ ]_c$ is maintained at the expense of $[\ ]_v$. Considering relative vacuolar and cytoplasmic volumes, this could be achieved with only small changes of $[\ ]_v$.

Recently Pitman et al (63) have provided estimates of $[\ ]_c$ and $[\ ]_v$ using electron probe X-ray microanalysis. In $CaSO_4$ – grown plants, K^+ appeared to be 3–4 times more concentrated in the cytoplasm than in the vacuole. However, their data also demonstrated that growth in 2.5 mM K^+ generally increased $[K^+]_c$ to $\sim$3x the values for $CaSO_4$ – plants. It seems likely, therefore, that plasmalemma influx may be sensitive to feedback from $[\ ]_c$ and possibly also from $[\ ]_v$. In *Lemna* (81) the time courses of readjustment of K^+ influx following step-up or step-down experiments were interpreted to arise from alterations of $[K^+]_c$, thus pointing to the cytoplasm as the origin of the regulatory signal. In *Chara*, plasmalemma Cl^- influx is extremely sensitive to $[Cl^-]_c$ (69). Generally, plasmalemma efflux appears not to be important in the mechanisms of regulation. In studies of phosphate uptake in *Spirodela* (e.g. 5), efflux rates were reported to be $\sim$10% of accumulation rates and did not increase significantly with growth at higher $[Pi]_o$. The close correspondence between K^+ influx and net uptake in barley (45) and *Lemna* (81) led to the conclusion that efflux of this ion was not significant in regulation. In apparent conflict with the above observations, other workers have reported constant values of K^+ and Cl^- influx and increasing values of efflux during salt accumulation (42). Recent studies of NO_3^-, uptake by barley using $^{36}ClO_3^-$ as tracer for NO_3^-, together with continuous short-term measurements of net NO_3^- uptake, point to efflux as a determining factor in short-term adjustments of uptake to perturbations in $[NO_3^-]_o$ (22). Plasmalemma NO_3^- ($^{36}ClO_3^-$) influx, by contrast, was independent of $[NO_3^-]_i$. The simplest type of system which might achieve these effects is based upon passive adjustments of influx or efflux to the changing $\Delta\bar{\mu}$ for the particular ion. In this form of control, referred to as "pump and leak," [ ]$_i$ at equilibrium is determined by [ ]$_o$ rather than by some set point (2). As a consequence it may be inappropriate to consider this as regulation per se. During Cl^- accumulation by carrot slices, $\Delta\bar{\mu}_{Cl_{io}^-}$ increased from 17kJ mol^{-1} in non loaded tissue to 23 kJ mol^{-1} in KCl-loaded tissue (15). The author concluded that the increased gradient opposing Cl^- entry was probably insufficient to account for the tenfold reduction of influx. The situation is perhaps clearer for barley roots in which $\Delta\bar{\mu}_{K^+_{io}}$ declined from 5.84 to 4.11 kJ mol^{-1} as influx declined from 3.05 to 1.61 μmol g^{-1}h^{-1} (36).

The observed adjustments of influx according to ion availability might have arisen either through changes in the number of transporters or through direct control of their activity. There is every reason to assume that plants are capable of regulating ion transport through induction and repression of carrier synthesis as in the well-documented bacterial and fungal systems (4, 56). In green plants, however, evidence for such mechanisms is indirect and equivocal. The failure to "induce" increased transport capacity under "inducing" conditions (i.e. when a particular nutrient is withheld) in the presence of inhibitors of protein synthesis is consistent with, but hardly strong evidence for, induction of carrier synthesis (e.g. 43, 78). This is particularly so when the inhibitors have potent side effects. Even were side effects absent, the linkage between protein synthesis and "induction" of carrier synthesis is not necessarily direct. In *Neurospora crassa* there is both kinetic and genetic evidence for the existence of two phosphate transporters (56). The high affinity system appears to be a derepressible system which leads to increased V_{max}, but constant K_m, following growth at low $[Pi]_o$.

There is kinetic evidence for two phosphate transport systems in sterile-grown barley (11). However, growth at low $[Pi]_o$ for 8 days led to reduced K_m values for the high affinity system at constant V_{max}. K_m values for K^+ influx into barley roots increased and V_{max} values decreased within 3 h of pretreatment of $CaSO_4$– grown roots with KCl (34). Similarly, growth of barley at constant

$[K^+]_o$ (from 5–100 μM) for 18 days generated plants with K_m values ranging from .041 to 0.139 mM and V_{max} values from 11.72 to 0.78 μmol g^{-1}h^{-1} (73). With a knowledge of the relationships between K_m, and V_{max} and root $[K^+]_i$, influx can be predicted at particular $[K^+]_o$ and $[K^+]_i$ from a modified form of Michaelis-Menten equation (72). Comparison of predictions using terms derived during rapid K^+-loading of $CaSO_4$–grown roots with those obtained under steady-state conditions reveals differences which suggest long-term and short-term adjustment of fluxes in response to $[K^+]_i$ (73). Substantial differences in kinetic constants at corresponding $[K^+]_i$ have been established both between and within species (24, 73). Under conditions in which set points are not realized, these differences may have significant bearing upon nutrient acquisition and growth rates.

Evidence for control of the activity of existing transporters is also indirect and based upon the time course of adjustments. Reduction of K^+ influx associated with increased $[K^+]_i$ was evident within 15–30 min (35). Similarly, pretreatment in 15 μM phosphate reduced ^{32}Pi influx into Pi-starved roots within 1 hr (51). Such reductions might operate via allosteric effects of the accumulated ion upon the transporter (34, 44).

In *Chara,* Cl^- influx is extremely sensitive to $[Cl^-]_c$ and $[H^+]_c$. Kinetic analyses suggest that Cl^- is the first ion to bind externally to the $1Cl^-:2H^+$ symporter and the first to dissociate internally (70). Reduction of influx associated with high $[Cl^-]_c$ in this organism is thought to result from intrinsic kinetic properties of the transporter. Stated simply, the dissociation of Cl^- at the cytoplasmic phase is reduced because of high $[Cl^-]_c$. A somewhat similar model has been proposed for the regulation of ion influx in barley roots (2). In the latter, flux reduction is achieved by the interplay between diffusion into a membrane cavity and transport into or out of the cavity by means of a tetrameric transporter. This model is an elaborate form of "pump and leak" system. The arguments advanced earlier regarding driving forces for increased efflux make such a model unlikely.

Temperature Acclimation

A situation which may be analogous to the accommodation of uptake following nutrient deprivation is the adjustment of ion fluxes associated with root/shoot temperature differentials (12). Considering the high Q_{10} values for ion transport, it is likely that shoot growth might be severely limited by the root's capacity for uptake and translocation when roots are cooled below shoot temperature. However, rates of ion uptake and translocation by cooled roots are increased, over a period of days, to values comparable to those of controls maintained at higher temperatures (12). Clarkson (12) has proposed that this compensation is obtained through increased numbers of ion transporters rather than control of their activity. These adjustments to temperature differentials should perhaps be viewed as a particular case of a general low-temperature response rather than a specific adjustment to root/shoot temperature differentials because Cl^- influx in *Chara* displays similar temperature acclimation (69).

A Common Mechanism for Regulation?

Specificity in regulating the uptake of a particular ion might be expected to require that uptake be sensitive only to feedback from [ ]$_i$ of that ion or its metabolic derivatives and insensitive to the concentrations of other ions. Indeed, the kinetic model for regulation of Cl^- influx in *Chara* (70) is based upon the presumption that both influx and the transinhibition occur at the same locus with equivalent specificity. However, Cl^- influx into barley and carrot is reduced by accumulation of NO_3^- as well as Cl^- (14). Net nitrate uptake is also sensitive to $[Cl^-]_i$ as well as to $[NO_3^-]_i$ (74). If, as is believed, NO_3^- and Cl^- do not compete for uptake (74), then feedback must be exerted at a site other than the uptake binding site (14), unlike the situation proposed for *Chara* (69, 70). Pretreatment of barley roots with Rb^+ has been demonstrated to reduce Na^+ influx (2) although the reverse was not true. Cram (17) has proposed that the internal NO_3^-/Cl^- interaction may be part of a general regulatory system because, in carrot slices, loading with KCl reduced uptake of several apparently unrelated solutes. The feature which these solutes are suggested to possess in common is their dependence upon the proton motive force (17). The obvious drawback of this attractive hypothesis is the apparent failure to provide for specificity in the regulation of uptake of diverse ions, unless feedback be "fine tuned" by the specific ions.

CONCLUSIONS

The maintenance of prescribed set points for turgor or ion concentration in the face of environmental perturbations demands adjustments of rates of ion transport and/or biochemical reactions which constitute components of the more general process of homeostasis. What is most impressive about plant responses to these perturbations is the physiological plasticity which is evident within a single genotype and the diversity of mechanisms exhibited by different species. We should be cautious in presuming that models which apply to one group of organisms necessarily apply universally, or that mechanisms based on a particular ion will apply to all others. In a previous review in this series (13), it was suggested that the regulation of ion transport will not be fully understood until the transport mechanisms can be described in molecular rather than kinetic terms. Biochemical information regarding the transport process must certainly represent the minimum requirement for a complete description of transport regulation. It is to be hoped that the recent improvements in techniques for tissue and protoplast culture may be applied to the quest for ion transport mutants, which may give valuable impetus to the goal of biochemical characterization of transport mechanisms.

ACKNOWLEDGMENTS

I am grateful to my colleagues Drs. M. Y. Siddiqi and C. E. Deane-Drummond for the many discussions of topics included in this review. The continuing financial support of the Natural Sciences and Engineering Council of Canada is also gratefully acknowledged.

Literature Cited

1. Asher, C. J., Loneragan, J. F. 1967. Response of plants to phosphate concentration in solution cultures: I. Growth and phosphorus content. *Soil Sci.* 103:225–33

2. Bange, G. G. J. 1978. Ionic equilibration of the symplast of low-salt barley roots. *Acta Bot. Neerl.* 27:183–98

3. Bange, G. G. J. 1979. Comparison of compartmental analysis for Rb^+ and Na^+ in low- and high-salt barley roots. *Physiol. Plant.* 46:179–86

4. Beever, R. E., Burns, D. J. W. 1980. Phosphorous uptake, storage and utilization by fungi. *Adv. Bot. Res.* 8:127–219

5. Bieleski, R. L., Johnson, P. N. 1972. The external location of phosphatase activity in phosphorus deficient *Spirodela oligorrhiza. Aust. J. Biol. Sci.* 25:707–20

6. Bisson, M. A., Gutknecht, J. 1975. Osmotic regulation in the marine alga *Codium decorticatum. J. Membrane Biol.* 24:183–200

7. Bisson, M. A., Gutknecht, J. 1980. Osmotic regulation in algae. In *Plant Membrane Transport: Current Conceptual Issues,* ed. R. M. Spanswick, W. J. Lucas, J. Dainty, pp. 131–42. Amsterdam/New York/Oxford: Elsevier/North Holland

8. Boyer, J. S. 1970. Leaf enlargement and metabolic rates in corn, soybean, and sunflower at various leaf water potentials. *Plant Physiol.* 46:233–35

9. Brown, A. D., Borowitzka, L. J 1979. Halotolerance of *Dunaliella.* In *Biochemistry and Physiology of Protoza,* ed. M. Levandowsky, S. H. Hutner, 1:139–90. New York: Academic

10. Cala, P. M. 1970. Volume regulation by *Amphiuma* red blood cells: the membrane potential and its implications regarding the nature of the ion flux pathways. *J. Gen. Physiol.* 76:683–708

11. Cartwright, B. 1972. The effect of phosphate deficiency on the kinetics of phosphate absorption by sterile excised barley roots and some factors affecting the ion uptake efficiency of roots. *Soil Sci. Plant Anal.* 4:313–22

12. Clarkson, D. T. 1976. The influence of temperature on the exudation of xylem sap from detached root systems of rye *(Secale cereale)* and barley *(Hordeum vulgare). Planta* 132:297–304

13. Clarkson, D. T., Hanson, J. B. 1980. The mineral nutrition of higher plants. *Ann. Rev. Plant Physiol.* 31:239–98

14. Cram, W. J. 1973. Internal factors regulation nitrate and chloride influx in plant cells. *J. Exp. Bot.* 24:328–41

15. Cram, W. J. 1975. Relationships between chloride transport and electrical potential differences in carrot root cells. *Aust. J. Plant Physiol.* 2:301–10

16. Cram, W. J. 1976 Negative feedback regulation of transport in cells. The maintenance of turgor, volume and nutrient supply. In *Encyclopedia of Plant Physiology,* New ser, ed. U. Luttge, M. G. Pitman, 2A:284–316. Berlin/Heidelberg/NewYork: Springer

17. Cram, W. J. 1980. A common feature of the uptake of solutes by root parenchyma cells. *Aust. J. Plant Physiol.* 7:41–49

18. Cram, W. J. 1980. Chloride accumulation as a homeostatic system: negative feedback signals for concentration and turgor maintenance in a glycophyte and a halophyte. *Aust. J. Plant Physiol.* 7:237–49

19. Cram, W. J., Laties, G. G. 1971. The use of short-term and quasi steady influx in estimating plasmalemma and tonoplast influx in barley root cells at various external and internal salt concentrations. *Aust. J. Biol. Sci.* 24:633–46

20. Dainty, J. 1968. The structure and possible function of the vacuole. In *Plant Cell*

Organelles, ed. J. B. Pridham, pp. 40–46. London/New York: Academic

21. Deane-Drummond, C. E. 1982. Mechanisms for nitrate uptake into barley (*Hordem vulgare* L. cv. Fergus) seedlings grown at controlled nitrate concentration in the nutrient medium. *Plant Sci. Lett.* 24:72–89

22. Deane-Drummond, C. E., Glass, A. D. M. 1983. Short-term studies of nitrate uptake into barley plants using ion specific electrodes and $^{36}C10_3^-$. I. Regulation of net uptake by NO_3^- efflux. *Plant Physiol.* In press

23. Downtown, W. J. S., Loveys, B. R. 1981. Abscisic acid content and osmotic relations of salt-stressed grapevine leaves. *Aust. J. Plant Physiol.* 8:443–52

24. Dunlop, J., Glass, A. D. M., Tomkins, B. J. 1979. The regulation of potassium uptake by ryegrass and white clover roots in relation to their competition for potassium. *New Phytol.* 83:365–70

25. Enoch, S., Glinka, Z. 1981. Changes in potassium fluxes in cells of carrot storage tissue related to turgor pressure. *Physiol. Plant.* 53:548–52

26. Epstein, E. 1976. Kinetics of ion transport and the carrier concept. In *Encyclopedia of Plant Physiology*, New ser. ed. U. Luttge, M. G. Pitman, 2B:70–94. Berlin/Heidelberg/New York: Springer

27. Erlandsson, G. 1975. Rapid effects on ion and water uptake by changes of water potential in young wheat plants. *Physiol. Plant.* 35:256–62

28. Flowers, T. J., Troke, P. F., Yeo, A. R. 1977. The mechanism of salt tolerance in halophytes. *Ann. Rev. Plant Physiol.* 28:89–121

29. Gimmler, H., Moller, E. M. 1981. Salinity-dependent regulation of starch and glycerol metabolism in *Dunaliella parva*. *Plant Cell Environ.* 4:367–75

30. Gimmler, H., Schirling, R., Tobler, U. 1977. Cation permeability of plasmalemma of halotolerant alga *Dunnialiella parva*. 1. Cation induced osmotic volume changes. *Z. Pflanzenphysiol.* 83:145–58

31. Gimmler, H., Schirling, R., Tobler, U. 1978. Cation permeability of plasmalemma of halotolerant alga *Dunaliella parva*. II. Cation content and glycerol concentration of the cells as dependent upon external NaCl concentration. *Z. Pflanzenphysiol.* 87:435–44

32. Ginzburg, M. 1981. Measurement of ion concentrations and fluxes in *Dunaliella parva*. *J. Exp. Bot.* 32:321–32

33. Ginzburg, M. 1981. Measurement of ion concentration in *Dunaliella parva* subject to osmotic shock. *J. Exp. Bot.* 32:333–40

34. Glass, A. D. M. 1976. Regulation of potassium absorption in barley roots: an allosteric model. *Plant Physiol.* 58:33–37

35. Glass, A. D. M. 1977. Regulation of K^+ influx in barley roots: evidence for direct control by internal K^+. *Aust. J. Plant Physiol.* 4:313–18

36. Glass, A. D. M., Dunlop, J. 1979. The regulation of K^+ influx in excised barley roots: relationships between K^+ influx and electrochemical potential differences. *Planta* 145:395–97

37. Greenway, H., Munns, R. 1980. Mechanisms of salt tolerance in nonhalophytes. *Ann. Rev. Plant Physiol.* 31:149–90

38. Grinstein, S., DuPre, A., Rothstein, A. 1982. Volume regulation by human lymphocytes. Role of Calcium. *J. Gen. Physiol.* 79:844–68

39. Gutknecht, J., Bisson, M. A. 1977. Ion transport and osmotic regulation in giant algal cells. In *Water Relations in Membrane Transport in Plants and Animals*, ed. A. M. Jungreis, T. K. Hodges, A. Kleinzeller, S. G. Schultz, pp. 3–14. New York: Academic

40. Heywood, P. 1978. Osmoregulation of the alga *Vacuolaria virescens*. Structure of the contractile vacuole and nature of its association with the golgi apparatus. *J. Cell Sci.* 31:213–24

41. Hsiao, T. C., Acevedo, E., Fereres, E., Henderson, D. W. 1976. Stress metabolism, water stress, growth and osmotic adjustment. *Philos. Trans. R. Soc. London Ser. B.* 273:479–500

42. Jackson, P. C., Edwards, D. G. 1966. Cation effects on chloride fluxes and accumulation levels in barley roots. *J. Gen. Physiol.* 50:225–41

43. JeanJean, R. 1973. The relationship between the rate of phosphate absorption and protein synthesis during phosphate starvation in *Chorella pyrenoidosa*. *FEBS Lett.* 32:149–51

44. Jensen, P., Pettersson, S. 1978. Allosteric regulation of potassium uptake in plant roots. *Physiol. Plant.* 30:24–29

45. Johansen, C. Edwards, D.G., Loneragan, J.F. 1970. Potassium fluxes during potassium absorption by intact barley roots of increasing potassium content. *Plant Physiol.* 45:601–14

46. Kauss, H. 1978. Osmotic regulation in algae. *Prog. Phytochem.* 5:1–27

47. Kauss, H. 1981. Sensing volume changes by *Poteriochromonas* involves a Ca^{2+}- regulated system which controls

activation of isofloridoside-phosphate synthase. *Plant Physiol.* 68:420–24

48. Kirst, G. O. 1977. Coordination of the ionic relations and mannitol concentrations in the euryhaline alga, *Platymonas subcordiformis* (Hazen) after osmotic shocks. *Planta* 135:69–75

49. Kirst, G. O., Kramer, D. 1981. Cytological evidence for cytoplasmic volume control in *Platymonas subcordiformis* after osmotic stress. *Plant Cell Environ.* 4:455–62

50. Kondo, T. 1982. Persistence of the potassium uptake rhythm in the presence of exogenous sucrose in *Lemna gibba* G3. *Plant Cell Physiol.* 23:467–72

51. Lefebvre, D. D., Glass, A. D. M. 1982. Regulation of phosphate influx in barley roots: Effects of phosphate deprivation and reduction of influx with provision of orthophosphate. *Physiol. Plant.* 54:199–206

52. Lucas, W. J., Alexander, J. M. 1981. Influence of turgor pressure manipulation on plasmalemma transport of HCO_3^- and OH^- in *Chara corallina*. *Plant Physiol.* 68:553–59

53. McKnight, A. D. C., Leaf, A. 1977. Regulation of cellular volume. *Physiol. Rev.* 57:510–73

54. Mott, R. L., Steward, F. C. 1972. Solute accumulation in plant cells. 1. Reciprocal relations between electrolytes and non-electrolytes. *Ann. Bot.* 36:621–39

55. Muller, K., Ludemann, H. D., Jaenicke, R. 1981. Pressure-dependent deactivation and reactivation of dimeric enzymes. *Naturwissenschaften.* 68:524–25

56. Pardee, A. B., Palmer, L. M. 1973. Regulation of transport systems: a means of controlling metabolic rates. In *Rate Control of Biological Processes*, ed. D. D. Davies, pp. 133–44. S.E.B. Symp. 27. Cambridge: Cambridge Univ. Press

57. Parrondo, R. T., Smith, R. C., Lazurick, K. 1975. Rubidium absorption by corn root tissue after a brief period of water stress and during recovery. *Physiol. Plant.* 35:34–38

58. Pierce, M., Raschke, K. 1981. Synthesis and metabolism of abscisic acid in detached leaves of *Phaseolus vulgaris* after loss and recovery of turgor. *Planta* 153:156–65

59. Pitman, M. G. 1969. Simulation of Cl^- uptake by low-salt barley roots as a test of models of salt uptake. *Plant Physiol.* 44:1417–27

60. Pitman, M. G., Courtice, A. C., Lee, B. 1968. Comparison of potassium and sodium uptake by barley roots at high

and low salt status. *Aust. J. Biol. Sci.* 21:871–81

61. Pitman, M. G., Cram, W. J. 1973. Regulation of inorganic ion transport in plants. In *Ion Transport in Plants*, ed. W. P. Anderson, pp. 465–81. London/New York: Academic

62. Pitman, M. G., Cram, W. J. 1977. Regulation of ion content in whole plants. In *Integration of Activity in the Higher Plant*, ed. D. H. Jennings, pp. 391–424. S.E.B. Symp. 31. Cambridge: Cambridge Univ. Press

63. Pitman, M. G., Lauchli, A., Stelzer, R. 1981. Ion distribution in roots of barley seedlings measured by electron probe X-ray microanalysis. *Plant Physiol.* 68:673–79

64. Rand, R. P., Burton, A. C. 1964. Mechanical properties of the red cell membrane. *Biophys. J.* 4:115–35

65. Raven, J. A. 1977. Regulation of solute transport at the cell level. See Ref. 62, pp. 73–99

66. Reed, R. H., Collins, J. C., Russell, G. 1981. The effects of salinity upon ion content and ion transport of the marine red alga *Porphyra purpurea* (Roth) C. Ag.. *J. Ext. Bot.* 32:347–67

67. Riisgard, H. U., Nielsen, K. N., Sogaard-Jensen, B. 1980. Further studies on volume regulation and effects of copper in relation to pH and EDTA in the naked marine flagellate *Duniella maritima*. *Mar. Biol.* 56:267–76

68. Rothstein, A. 1964. Membrane function and physiological activity of microorganisms. In *The Cellular Functions of Membrane Transport*, ed. J. F. Hoffman, pp. 23–39. New Jersey: Prentice Hall

69. Sanders, D. 1981. Physiological control of chloride transport in *Chara corallina*: I. Effects of low temperature, cell turgor pressure, and anions. *Plant Physiol.* 67:1113–18

70. Sanders, D., Hansen, U-P. 1980. Mechanisms of Cl^- transport at the plasmamembrane of *Chara corallina*: II. Transinhibition and the determination of binding order from a reaction kinetic model. *J. Membr. Biol.* 58:139-53

71. Schobert, B. 1977. Is there an osmotic regulatory mechanism in algae and higher plants. *J. Theor. Biol.* 68:17-26

72. Siddiqi, M. Y., Glass, A. D. M. 1982. Simultaneous consideration of tissue and substrate potassium concentration in K^+ uptake kinetics: A model. *Plant Physiol.* 69:283–85

73. Siddiqi, M. Y., Glass, A. D. M. 1983. Studies of the growth and mineral nutrition of barley varieties. II. Potassium up-

take and its regulation. *Can. J. Bot.* In press

74. Smith, F. A. 1973. The internal control of nitrate uptake into excised barley roots with differing salt contents. *New Phytol.* 72:769–82

75. Smith, F. A., Raven, J. A. 1979. Intracellular pH and its regulation. *Ann. Rev. Plant Physiol.* 30:289–311

76. Sutcliffe, J. F. 1954. The absorption of potassium ions by plasmolysed cells. *J. Exp. Bot.* 5:215–31

77. Turner, N. C., Jones, M.M. 1980. Turgor maintenance by osmotic adjustment: a review and evaluation. In *Adaptation of Plants to Water and High Temperature Stress*, ed. N. C. Turner, P. J. Kramer, pp. 87–104. New York: Wiley

78. Ullrich, W. R., Schmitt, H. D., Arntz, E. 1981. Regulation of nitrate uptake in green algae and duckweeds. Effects of starvation and induction. In *Biology of Inorganic Nitrogen and Sulfur*, ed. H. Bothe, A. Trebst, pp. 244–51. Berlin/Heidelberg/New York: Springer

79. Williams, D. E. 1961. The absorption of potassium as influenced by its concentration in the nutrient medium. *Plant Soil.* 15:387–99

80. Wyn Jones, R. G., Brady, C. J., Speirs, J. 1979. Ionic and osmotic regulation in plants. In *Recent Advances in the Biochemistry of Cereals*, ed. D. L. Laidman, R. G. Wyn Jones, pp. 63–103. London: Academic

81. Young, M., Sims, A. P. 1972. The potassium relations of *Lemna minor* L.: I. Potassium uptake and plant growth. *J. Exp. Bot.* 23:958–69

82. Zimmermann, U. 1977. Cell turgor regulation and pressure-mediated transport processes. See Ref. 62, pp. 117–54

83. Zimmermann, U. 1978. Physics of turgor- and osmoregulation. *Ann. Rev. Plant Physiol.* 29:121-48

84. Zimmermann, U., Steudle, E. 1971. Effects of potassium concentration and osmotic pressure of sea water on the cell-turgor pressure of *Chaetomorpha linum*. *Mar. Biol.* 11:132–37

Ann. Rev. Plant Physiol. 1983. 34:327–46

HERITABLE VARIATION IN PLANT CELL CULTURE

Frederick Meins, Jr.

Friedrich Miescher-Institut, CH-4002 Basel, Switzerland

CONTENTS

INTRODUCTION

Plant tissues and cells in culture undergo variation. This variation is of particular interest because it may provide hints as to the basic mechanisms underlying genotypic and phenotypic stability in normal development. Variation is also of considerable practical importance in the application of plant tissue culture as a technique for crop improvement. When used for plant propagation or as a system for genetic manipulation, untoward variation must be minimized. On the other hand, spontaneous and induced genetic changes arising in culture may provide novel forms of variation for use by the plant breeder. Both applications of tissue culture depend on understanding heritable cellular variation and on learning how to regulate it.

Students of plant cell and tissue culture are faced with the same fundamental

327

0066–4294/83/0601–0327$02.00

problems encountered in earlier studies of bacterial variation (50), namely, deducing events at the cell level from the behavior of populations and differentiating between physiological adaptations, random mutations, and directed heritable changes. This review concentrates on the experimental approaches available for dealing with these problems and, in particular, the means of distinguishing between two major sources of heritable, cellular variation: mutation and epigenetic change. For more complete surveys of the types of variation encountered in culture, the reader is referred to several recent reviews (23, 52, 53, 60, 105, 109) and the bibliography of Thomas (123).

TERMINOLOGY

Variation is defined here in the narrow sense as phenotypic changes that are: (*a*) *stable,* i.e. they persist in the absence of the event that induced the change (74); and (*b*) *heritable,* i.e. the new phenotype is transmitted to daughter cells when they divide (89). Changes in phenotype that persist only so long as the cells or tissues are maintained in a new environment are referred to as physiological responses (51).

Genetic variation is used to describe heritable variation that is sexually transmitted to progeny of plants regenerated from cultured cells. The term *mutant* is reserved for the special case in which a trait is transmitted meiotically according to well-established laws of inheritance (23, 59, 60). When the nature of the heritable change is not known, the term *variant* is used.

In the plant tissue culture literature, *clone* is used variously to denote plants propagated vegetatively from the same parent plant, plants regenerated from the same tissue-culture line, and cells descended from a common parental cell. Here, clone is defined operationally as the cells descended from the same cultured cell, and is not meant to imply that sister cells within a clone are genetically or phenotypically identical. As recommended by Chaleff (23), plants derived from cell and tissue culture are called the *R generation.* Progeny obtained by self-fertilization of the regenerated plants are designated the R_1 generation, and so forth for each successive sexual generation.

SOURCES OF VARIATION
Qualitative Traits

SELECTION   The major source of phenotypic diversity in culture is likely to be selection. Cultures are often initiated from complex tissues consisting of cells in different phases of the cell cycle (128), exhibiting different developmental states (41), with different ploidy levels (29, 94), and—in rare cases—even with different numbers of nuclei (39). The relative abundance of these cell types in cultured tissues and cell suspensions is determined by three main factors (69) : *(a)* the relative rates at which the cell types proliferate; *(b)* cell-cell interactions; and *(c)* the heritable interconversion of cell types.

Because it is unlikely that any two cell types have exactly the same proliferation rate, cultured tissues derived from a complex explant will tend to change in cellular constitution. This is particularly striking during the first few transfer generations in which tissues commonly change in friability, color, and growth rate (41). Even after prolonged culture, tissue lines and cell suspensions often remain highly heterogeneous in cellular constitution (126). Part of this heterogeneity could be accounted for by the stabilizing effects of cell-cell interactions.

In complex populations, one species can interact with another in such a way that the growth rates are balanced and a steady state or even a stable cyclical variation in abundances is established (55). With regard to tissue culture, this means that, in principle, a histogenetic equilibrium can be established even when the growth rates of the constituent cell types cultured separately differ. For example, cytokinin habituated cells of tobacco produce sufficient quantities of growth factors in culture to support the proliferation of neighboring nonhabituated cells in the same tissue (121). Thus, nonhabituated cells with an absolute cytokinin requirement can persist at low frequencies in habituated tissues serially propagated for many transfers on media without added cytokinin (68, 72).

Observations of this type lead to two important conclusions. First, because patterns of cellular interaction can persist when tissues are subcultured, stability of a trait at the tissue level does not necessarily indicate that the trait is inherited by individual cells. Heritability of a trait must be established by cloning cells and showing that the clones express the same phenotype as the parent cell population (117). Second, variation detected after prolonged culture may still reflect selection of cell types present in the initial tissue explant. Even after histogenetic equilibrium has been reached, local fluctuations in the abundance of different cell types within a tissue could give rise to sectors that retain their special character when subcultured separately.

HERITABLE CHANGES   The most direct way to show that a heritable trait arises in culture is to begin with a cloned line that does not exhibit the new trait and then to clone a second time after the trait is detected. Although heritable variation is often inferred from the persistence of a new trait on subculturing, the number of cases in which selection of preexistent cell types has been excluded by clonal analysis is small, e.g. variation in anthocyanin pigmentation in carrot cultures (35), alkaloid content in *Catharanthus roseus* cultures (135), and growth factor requirement of tobacco cultures (68, 72, 106).

RATE OF VARIATION   The basic parameter used to express growth in culture is cell number. A similar parameter is needed for the quantitative description of variation. When dealing with mutation in populations of microorganisms, mutation rate is used for this purpose (50). By analogy it is appropriate to define variation rate p as the probability of a heritable event per cell generation:

$$p = 0.693 \, M/(N_t\text{-}N_o)$$

where M is the number of heritable events that have occurred in the time required for the number of cells to increase from N_o to N_t.

It must be emphasized that in bulk cultures the frequency of variants in a population of dividing cells does not provide an estimate of variation rate. There are two reasons for this. First, the frequency of variant cells will depend on when the heritable event occurred. Second, if the variant and wild type cells differ in proliferation rate, there will be selection for the more rapidly growing cell type. Depending upon the conditions of culture and when the assay is performed, two cell populations can give markedly different frequencies of variant cells even when the variation rate is the same.

When the proliferation rates of the different cell types are known, the variation rate may be calculated from an analysis of the way the frequency of variant cells changes with time (69). Unfortunately, this information is difficult to obtain and a direct approach is not usually feasible. A powerful indirect method suitable for plant tissue culture was proposed by Luria & Delbrück (56) to measure mutation rate in populations of bacteria. The mathematical basis for the method is described in detail by Lea & Coulson (54), and more readable descriptions may be found in books by Hayes (50) and Stent (114). In brief, numerous cultures with equal numbers of cells are scored for the presence of the variant phenotype. If the number of heritable events is distributed at random among the cultures, then it follows from the Poisson law that the average number of heritable events per culture M is related to the fraction of cultures *without* variants P(0) by:

$$M = -\ln P(O)$$
with standard deviation;
$$S.D. = \sqrt{\frac{e^M - 1}{x}}$$

for x replicate cultures (54). The beauty of this method lies in the fact that M is calculated from the incidence of cultures without variant cells; hence, the estimate of rate is not affected by cell selection or interactions between parental and variant cell types.

The net rate of variation in response to an experimental treatment can be estimated directly provided that the heritable events occur in physically separated cells and that clones arising from different cells are distinguishable. These conditions are satisfied when single cells plated in agar or suspensions of protoplasts are treated with putative mutagens and treated and control cultures are assayed for variant clones. In principle, variant clones could arise either from variant cells present in the population before the experiment was begun or from heritable events that occurred after the cells were physically isolated. In control cultures, these sources of variation cannot be distinguished, and the frequency of variant clones does not provide an estimate of the spontaneous variation (or "mutation") rate. On the other hand, because the frequency of variant clones obtained with treated cultures is the sum of the frequency of variants in the control cultures plus the incidence of heritable events resulting from the experimental treatment, the net frequency—treated minus control—equals the frequency of induced heritable events per cell, which is equivalent to variation rate expressed on a per cell generation basis.

Two precautions should be observed in calculating variation rates by this method. First, the frequency of variant clones is only meaningful when expressed on a per viable cell basis. When large numbers of clones are obtained under nonselective conditions, this frequency is the fraction of clones that show the variant phenotype. When a selective agent is added it may not be possible to distinguish cells that did not survive the cloning procedure and mutagen treatment from wild-type cells that did not survive treatment with the selective agent. If this is the case, then additional control and treated cultures without the selective agent are needed to correct the frequencies for differences in cell viability. Second, because of sampling error, it is essential to show that the frequency in treated cultures is significantly higher than the frequency in control cultures (27). A suitable test is the chi square test of significance for binomial proportions corrected, if necessary, for continuity (108). There are few reports in the literature in which variation rates have been measured and sufficient information is provided to establish that the experimental treatment significantly increases the incidence of variants. Representative estimates of variation rates and, for comparison, somatic mutation rates measured *in planta* are listed in Table 1.

Quantitative Traits

Sometimes different clones isolated from the same tissue line differ quantitatively in some character, e.g. pigmentation, alkaloid content, or requirement for specific growth factors and nutrients (3, 10, 31, 35, 40, 106, 120–122, 135). If the rate of variation is slow relative to the time required to clone and assay the clones, then the quantitative differences among clones will primarily

Table 1 Rates of somatic variation

Types of variation	Species	Phenotypic change	Variation rate[a]	References
Bulk culture	*Nicotiana tabacum*	to cytokinin autotrophy	4.8×10^{-3}[b, c]	77
		to cytokinin requiring	8.2×10^{-3}[b, c]	73, 75
		to temperature sensitivity	2.0×10^{-7}[d]	62
Induced by mutagens in protoplast suspensions	*Nicotiana tabacum* (androgenic haploids)	to valine resistance	1.3×10^{-5}[c]	13
	Nicotiana plumbaginifolia (androgenic haploids)	to chlorate resistance	4.5×10^{-4}[c]	64
Somatic mutation *in planta*	*Nicotiana*	to purple petal color (stable)	5.0×10^{-6}	99
		to speckled petal color (mutable)	8.7×10^{-4}	99
	Tropaeolum majus	to yellow petal color	8.2×10^{-6}	111
	Tradescantia	to pink petal color	2.8×10^{-4}	112

[a] Expressed as heritable events per cell or per cell generation.
[b] Estimated by the Luria & Delbruck (56) method.
[c] Difference in frequency of variants in control and treated cultures significant at 5% level. Calculated from data provided in the reference cited.
[d] Based on the assumption that the growth rate and cloning efficiency of the two phenotypes are equal.

reflect quantitative differences in the cells from which the clones were derived. If the variation rate is rapid, however, as cells in the clones proliferate they may occasionally change phenotype, and hence different clones from the same parent tissue will contain different proportions of variant and wild-type cells. Therefore, clones assayed for phenotype at the tissue level may differ quantitatively even when the difference between the variant and wild-type cells is qualitative.

These two sources of quantitative variation can be distinguished by cloning cells from sister clones that differ widely in their expression of the variant phenotype (72). If individual cells differ quantitatively, then subclones should exhibit roughly the same phenotype as the clone from which they were derived. On the other hand, if the clones consist of mixtures of wild-type and variant cells, then subclones from clones of high as well as low phenotypic expression should give a bimodal distribution: one peak with high expression and a second peak with low expression. Moreover, the different cloned lines should only differ in the relative abundance of subclones in the two peaks.

In practice, this test does not provide a clear-cut distinction between the two sources of variation. For example, the distributions of subclones from sister clones of tobacco tissue exhibiting different degrees of cytokinin habituation are not bimodal (72). Some of the subclones exhibit the same degree of habituation as the parent clone, indicating that the trait is stable; but most subclones exhibit a phenotype intermediate between the lowest and highest values obtained with the parental distribution of clones. Similar results have been obtained in a detailed clonal analysis of anthocyanin accumulation by cultured wild carrot cells (35). The best interpretation of these findings appears to be that individual cells differ quantitatively in phenotype and also shift to higher and lower states of phenotype expression at rates that are rapid relative to the time required for cloning.

The analysis of quantitative variation is further complicated by the observation that different cloned lines from the same tissue may also differ in stability. Some cloned lines retain their degree of expression of a particular trait for many transfer generations, whereas other lines may increase or decrease in expression at different rates (79, 120, 135). This suggests that quantitative differences in rate of variation as well as in phenotypic expression of a specific trait can be inherited at the cellular level.

THE NATURE OF HERITABLE VARIATION

Distinguishing Between Mutation and Epigenetic Change

In principle, cellular variation could result from genetic mutation, epigenetic change, or a combination of both processes. Genetic mutations involve random alterations in genetic constitution such as point mutations, deletions, duplications, and rearrangements of the genetic material. Epigenetic changes are more difficult to define because their underlying mechanism is as yet unknown. The concept has its origins in developmental biology. Aristotle (2) proposed that organisms arise from a germ by *epigenesis*, the progressive formation of new structures (129, 130). It is believed that epigenesis, with few exceptions, involves selective gene expression rather than a sorting out of genetic determinants (83). Nevertheless, it was recognized from studies of protozoa and cultured animal cells that some states of determination and differentiation are inherited by individual cells (9, 19, 28, 42). This posed the fundamental problem of how cells thought to be genetically equivalent can inherit different characters.

To resolve this paradox, Nanney (90) proposed that there are two systems of inheritance: genetic systems concerned with the transmission of developmental potentialities between sexual generations of organisms, and epigenetic systems concerned with the somatic transmission of patterns of gene expression. Thus, the term epigenetic change has come to denote heritable, cellular alterations that do not result from permanent changes in the cell genome (49). These changes differ from rare, random mutations in several important ways (67, 90). First, the heritable alteration is directed, i.e. it occurs regularly in response to specific inducers. Under inductive conditions the rate of epigenetic variation is high, greater than 10^{-3} per cell generation. In contrast, well-characterized gametic mutations in plants usually exhibit spontaneous rates of less than 10^{-5} (112). Second, the variant phenotype, although stable, is potentially reversible, and the reversal process is directed and occurs at high rates. Third, the range of phenotypes generated by epigenetic changes is limited by the genetic potentiality of the cell. Finally, by definition, epigenetic changes are not transmitted meiotically.

With the exception of meiotic transmissibility, none of the criteria cited above, when applied individually, provides an absolute test for epigenetic inheritance. For example, rapid mutation can result from the action of mutator genes (95), and the rate of epigenetic variation is likely to be low under noninductive conditions. Paramutation is both directed and potentially reversible (15), and regulatory mutations could effect the expression of normally silent genes. What is needed is a functional test in which each of the distinguishing criteria can be evaluated.

THE PLANT REGENERATION TEST   When variant cells remain totipotent, mutations can be distinguished from epigenetic changes by regenerating plants from variant clones and assaying tissues from the R generation and the progeny from the appropriate genetic crosses for the variant trait. It is important that the genetic tests be extended to several sexual generations because there may be nonconcordance of male and female gametes of the R generation (65), and certain new traits that arise in culture are only lost in the R_2 generation (24). With this precaution, the demonstration that a trait is inherited in crosses according to well-established laws of inheritance provides unequivocal evidence for mutation.

Failure to detect the variant trait in R generation plants is more difficult to interpret. There are several examples in which a variant phenotype is expressed in culture but not in the plant (24, 132). A further complication is that cultures established from different plant parts may exhibit different stable phenotypes that persist for many transfer generations. Certain cultures of maize-endosperm tissue retain the capacity to produce endosperm-specific zein polypeptides (K. Shimamoto, personal communication). Cultures from different organs of maize (47, 84, 113), English ivy (5, 6, 116), and *Citrus* species (26) exhibit different competences for organogenesis. Callus cultures of *Ruta graveolens* produce the same essential-oil components as the plant part from which they were derived (87); and cloned lines of leaf, stem-cortex, and pith tissues of tobacco exhibit different cytokinin requirements (76). Therefore, failure to detect the variant trait, even in cultured material, could result from tissue-specific expression of the trait. The critical test is to assay cultured tissues from the regenerated plant that are of the same tissue type as the tissue from which the variant was originally isolated.

In some cases, the variant trait is regularly lost in the R generation (10, 81), whereas in other cases the trait is lost in plants from some tissue lines, but not from others (25). The key point is that the regular loss of a variant trait, while consistent with an epigenetic mechanism, does not rule out mutation. Plants are usually regenerated from uncloned tissue lines or from cloned lines many cell generations after the clone was isolated. It may be argued, therefore, that variant cells are no longer organogenetic and that the regenerated plants arise from a few wild-type cells generated by low rates of back mutation (80). To distinguish between back mutation and directed reversal, the rate of plant regeneration must be measured on a per cell generation basis. If the rate obtained is high, as is the case for cytokinin-habituated cells of tobacco (73, 75), then it is reasonable to conclude that the plant-regeneration process regularly induces reversion or that the plants develop from revertant cells that arise at high rates in culture.

OTHER TESTS   Often the plant regeneration test is not feasible because either tissues have lost their capacity for organogenesis or the regenerated plants are not fertile (125). Several alternative ways of distinguishing between mutation and epigenetic change have been proposed. Point mutations may lead to alterations in the primary amino-acid sequence of polypeptides, and the demonstration that such alterations are associated with the variation process provides good evidence for mutation (107). More commonly, a change in the activity or the regulatory properties of an enzyme is observed from which a specific structural alteration is inferred. Changes of this type are not reliable criteria for mutation. Patterns of isozymes sometimes vary with tissue type, stage of development, or physiological state (100). Changes in the properties of an enzyme could simply reflect an alteration in the pattern of gene expression, which may or may not have a mutational basis. For example, 5-methyltryptophan-resistant variants of tetraploid potato have been isolated that exhibit anthranilate synthetase activity far less sensitive to tryptophan inhibition than wild-type activity (21). The cultured tissues contain two electrophoretic forms of the enzymes, one tryptophan sensitive, the other tryptophan insensitive. Most of the activity exhibited by wild-type tissues is of the sensitive type, whereas most of the activity of the variant tissue is of the insensitive type. Therefore, the heritable change to 5-methyltryptophan resistance appears to involve a shift in the relative abundance of two forms of the same enzyme. Whether this shift results from mutation in a regulatory gene or from an epigenetic change is not known.

Somatic sectoring of regenerated plants in which the sector tissues exhibit the revertant phenotype has also been suggested as evidence for mutation (63). This does not seem justified. Somatic sectors can arise by somatic mutation (22, 36, 37), but they can also arise in chimeric plants consisting of a mixture of genetically different cells (115). Because it is not known whether or not regenerated plants arise from single cultured cells (16), the sectors may simply reflect heterogeneity of the tissue from which the plants were derived.

There are several reports, reviewed elsewhere (52, 60), showing that mutagens can increase the frequency of variation in culture. In at least two cases, valine resistance (13) and chlorate resistance (85; A. Müller, personal communication) in tobacco, it has been shown that the variants recovered after treatment were true mutants. Although an increased rate of variation in response to mutagen treatment is consistent with mutation, it does not necessarily rule out other less well-defined genetic modifications, epigenetic changes, or transient physiological responses. Certain epigenetic changes can mimic single-gene mutations in *Drosophila* and tobacco (45, 71). Mutagens also appear to induce both epigenetic changes and transient forms of drug resistance in cultured plant cells (27, 61).

When genetic tests are not feasible, the best alternative for characterizing a heritable change would appear to be an examination of the rates of spontaneous and induced variation in recently cloned populations of cells. If the rates of spontaneous variation in both the forward and back directions are very low, i.e. less than 10^{-5}, and both rates increase dramatically when cells are treated with specific inducers that are not obvious mutagens, then it is likely that the heritable change has an epigenetic basis.

Epigenetic Changes

Of the many reports of mutation in culture, few have been verified with breeding experiments (23, 52, 60, 105, 125). There are even fewer well-documented cases of epigenetic change. The one studied in detail is cytokinin habituation (70). The heritable conversion of cultured tobacco cells to the cytokinin habituated phenotype is a gradual, progressive process (72) which, unlike mutation, is strongly influenced by the physiological and developmental state of the cells (76, 77). The induction process is directed and occurs at high rates, greater than 10^{-3}, in response to cytokinin or 35°C treatment (68, 77). Once established, the habituated state is extremely stable; reversion only occurs at high rates when cloned lines are induced to form plants (73, 75). Finally, cytokinin habituation involves the expression of a preexistent potentiality of the plant cell (76) which appears to be specified by a single nuclear gene (71, 78,

82). The best interpretation of these results appears to be that cytokinin habituation results from epigenetic changes in the expression of a gene that is normally silent in cultured pith cells of tobacco. Tissues sometimes lose their requirement for both auxin and cytokinin, and this fully habituated phenotype is also inherited at the cellular level (57). The evidence that this change is epigenetic is less complete than for cytokinin habituation; for example, the phenotype is regularly lost during plant regeneration (58, 98), and the fully habituated state can be induced in culture by treating cells with auxin (119).

The type of plant regeneration obtained with *Citrus grandis* cultures depends upon the origin of the tissue (26). When placed on an inductive medium containing zeatin, callus cultures from leaf and stem form shoot-buds, whereas callus cultures from reproductive parts, e.g. ovary tissue, form embryoids. Stem-derived callus occasionally gives rise to variants which on induction form embryoids, indicating that the tissues have undergone a stable change in developmental competence. What is needed to establish that this change in competence has an epigenetic basis are experiments with cloned lines in which tissues from regenerated plants are returned to culture and assayed for shoot-bud and embryoid formation.

The urease activity of tobacco suspension cultures undergoes slow, cellular changes in response to the nitrogen source in the medium (110). Cultures grown on nitrate gradually increase in urease levels when grown on urea-containing medium. The induced state, once established, is inherited by individual cells and different clones differ in level of induction. This change appears to be metastable, i.e. the urease activity of induced lines gradually declines when cells are serially cultured on nitrate-containing medium. Nevertheless, these lines are more rapidly induced when treated with urea a second time, indicating that the cells "remember" their previous urea treatment. The frequency of high-urease cells as a function of time of induction was compared with mathematical models in which different variation rates were assumed. If high-urease cells arise continuously under inductive conditions, then the frequencies are consistent with an induction rate of 8×10^{-5} per cell generation. The results were also consistent with the hypothesis that high-urease cells only arise during the transition from nitrate to urea-containing medium and at a rate of 2×10^{-3} per cell generation. The important point is that the rates observed are 10^{-3} to 10^{-4} greater than the frequency of other biochemical variants isolated from the same line using resistance to amino acids or amino-acid analogs as a selection method. In the absence of breeding experiments, these findings provide fairly strong evidence that the variation in urease activity results from epigenetic changes.

There are hints that variation in chilling resistance has an epigenetic basis. Dix & Street (34) have isolated cell lines of *Nicotiana sylvestris* that exhibit different degrees of enhanced chilling resistance. The resistant state was expressed, although at a reduced level, in tissue from the few regenerated plants tested. None of the tissue from R_1 generation plants, however, exhibited enhanced resistance, suggesting that this trait is not transmitted meiotically (33).

Chromosomal Variation

Plant cells cultured from a wide range of species exhibit a high incidence of polyploidy, aneuploidy, and chromosomal rearrangement (8, 30, 118). Of 55 species studied in detail, only 11 retained the same chromosome number in culture as in the parent plant (8); and even when chromosome number is conserved, there may be a high incidence of chromosomal rearrangement (86).

Several factors influence the extent and type of chromosomal variation in culture. There is a strong genetic effect. Different varieties of the same species, e.g. cultivars of celery (17), maize (4), and oat (66), vary dramatically in chromosomal stability; and there is evidence that a simple dominant trait generated in culture induces karyotypic instability in tobacco (91). The tendency to vary also depends on the epigenetic state of the cultured material. Tissues with a high degree of tissue organization, such as shoot meristems, usually give rise to plants with normal chromosome numbers, whereas callus cultures from the same plant may not (30). Finally, chromosomal constitution depends on culture conditions, e.g. the concentration and type of hormones in the medium, and whether cells are grown in suspension or callus culture (20, 93, 97, 133). It is not known, however, whether these differences result from changes in the rate of variation or selection for preexistent karyotypes in the cell population.

Karyotypic changes appear to be an important factor in the loss of organogenetic capacity observed in long-term cultures. There is a strong negative correlation between the incidence of aneuploidy and the capacity of cultured tissues to regenerate plants (20, 133), and there tends to be selection for more normal karyotypes in the plant regeneration process (97, 127).

Morphological variation of regenerated tobacco plants is correlated with aneuploidy (134), and the changes in growth habit, flower and leaf morphology, and fertility observed are similar to those found in monosomic types of *Nicotiana tabacum* (20) and trisomic types of the closely related species, *N. sylvestris* (46). On the other hand, there are also well-documented examples of morphological variation in the R generation of dihaploid tobacco (18), tetraploid potato (104), and diploid alfalfa (96) in which chromosome number is normal. Clonal analysis of cytokinin-habituated cells in tobacco has shown that variation in chromosome number is not correlated with degree of habituation nor the potential of habituated cells to return to the normal state (11). Even when a chromosomal aberration is associated with a specific phenotype, the two traits are not necessarily causally related. Cells in plants regenerated from

Crepis capillaris tissue that are auxin-habituated retain their habituation specific karyotype but lose their habituated character (98). The most reliable test for linkage between a particular karyotype and a variant phenotype would appear to be the demonstration that the two traits segregate in the same way in crosses of R generation plants.

Other Forms of Variation

Variation that has features of both epigenetic changes and genetic mutation has now been described in at least ten different genera (53). This strange form of variation occurs spontaneously at high frequency and in some cases is transmitted in sexual crosses of R generation plants. Unexpectedly, variation affecting many different characters is often expressed in the R generation derived from a single, highly inbred or vegetatively propagated plant. The traits affected are often complex and probably multigenic such as growth, vegetative anatomy, and disease resistance.

Plants regenerated from different leaf-protoplast derived lines of "Russet Burbank," a cultivar of tetraploid potato, vary dramatically in growth habit, maturity date, photoperiod requirements for flowering, tuber characteristics, and resistance to early and late blight (104). Of 65 lines tested, each from a different protoplast-derived clone, there was significant variation in 22 of 35 characters examined, and the lines differed from the parent cultivar by up to 17 characters (101). The tendency to vary also appears to depend on the genotype of the parent cultivar. Plants regenerated from a diploid variety of potatoes are uniform in phenotype (131), whereas plants regenerated from a tetraploid variety using the same culture procedure are highly variable (124).

The important question is whether this variation arises in culture or reflects the diversity of somatic cells in the parent plant. In experiments with the "Maris Bard" cultivar of tetraploid potato, more than one plant was regenerated from each clone (124). The R generation exhibited wide interclonal as well as intraclonal variation. In contrast, replicate plants derived by shoot culture from most regenerates were uniform. Therefore, at least some of the observed variation was generated by the culture procedure itself.

In the case of the "Russet Burbank" cultivar, different variant clones have the same chromosome number, although there is evidence to suggest that some of the clones have undergone chromosomal rearrangement (102, 103). Unfortunately, this cultivar exhibits low fertility, and as yet it has not been possible to distinguish between epigenetic and genetic mechanisms or to establish the relationship between chromosomal changes and phenotypic variation by breeding tests.

There are several examples from other plant species in which wide variation clearly has a genetic basis. Plants regenerated from callus cultures from a single, diploid alfalfa plant show extensive morphological variation (96). Good sexual heritability of at least three different variant phenotypes has been obtained in back crosses with wild-type plants (E.T. Bingham, personal communication). High sexual heritability is exhibited by several traits in dihaploid tobacco plants obtained by regenerating plants from anther cultures and doubling the chromosome complement by colchicine treatment (1, 88, 92). For example, using plants regenerated from a single, inbred plant, Arcia et al (1) found that the R_1 and R_2 generations differed significantly in yield, plant height, number of leaves, and alkaloid content when compared with conventional F_1 and F_2 generations obtained from the same inbred line. A second cycle of anther culture generated further genetic variability, even when the colchicine treatment was omitted (18). Similar increases in variability in each of five anther-culture cycles have been reported for *Nicotiana sylvestris* (32). Therefore, it appears that genetic variability results from the anther-culture procedure and not from residual heterozygosity in the plants from which the cultures were derived.

The nature of culture-induced genetic variation has been studied using tobacco plants heterozygous in two nuclear loci that affect chloroplast differentiation. Barbier & Dulieu (7, 37a) regenerated plants directly from leaf explants and from plated cells derived from cloned leaf callus after up to four transfers in culture. They found significant variation after a single transfer, which tended to increase in subsequent transfers. Genetic analysis showed that the variation can be accounted for by single deletions or reversions and double events, such as two reversions or one deletion and one reversion. The frequencies of single and double events were not statistically independent.

Cultures of maize capable of plant regeneration exhibit far fewer gross, chromosomal abnormalities than organogenetic callus cultures of tobacco (38, 48, 65). Nevertheless, there is extensive genetic variablity in the R generation. Male-fertile plants resistant to the toxin produced by *Drechslera* (=*Helminthosporium*) *maydis* race T have been regenerated from scutellar cultures of maize carrying Texas male-sterile cytoplasm treated with sublethal concentrations of the toxin (44). This trait, which is inherited cytoplasmically, also occurs spontaneously and at high frequency in nonselected cultures (14). Restriction endonuclease analysis shows that the mitochondrial DNAs from different lines selected for toxin resistance differ from each other and from the mitochondrial DNA isolated from seed-grown normal and Texas male-sterile plants (43). There is also evidence for variation in the nuclear genome. Of 51 plants regenerated from scutellar cultures of A188xW22 R-nj R-nj, 9 segregated for mutant traits in the R_1 generation, and of the normal R_1 plants, 8 segregated for mutant traits in the R_2 generation (65). A wide variety of mutants affecting the kernel and seedling were also recovered in regenerates of W64A and S65 maize lines (38). The incidence of mutants exhibiting Mendelian ratios in the R_2 generation was high, 0.8 to 1.6 per plant, whereas no mutants were recovered from the seed-grown parent lines. The mutations obtained with different regenerants from the same culture were usually different, suggesting that the extensive genetic variation observed arose in culture. However, because cultures were not of clonal origin, and some of the regenerated plants exhibit genetic mosaicism (38, 65), the possibility that this variation is inherent in somatic tissues of the parent plants cannot be completely ruled out.

CONCLUDING REMARKS

There is no satisfactory substitute for using cloned materials, measuring variation rates, regenerating plants, and performing genetic crosses in establishing the nature of heritable variation. The few cases studied in detail show that heritable changes arising in culture can affect qualitative as well as quantitative traits and can result from epigenetic changes, mutations in specific gene loci, chromosomal rearrangements, and less well-defined genetic modifications. Because the number of examples is still small, it is not possible to assess the relative contribution of these different mechanisms.

Cultured plant cells exhibit an extremely high incidence of spontaneous, heritable variation which sometimes persists in the R generation and is transmitted in sexual crosses. The incidence, type, and stability of variation depends on the genotype of the parent plant and the epigenetic state of the cell.

There are several lines of evidence to support the hypothesis that the tissue culture procedure itself induces a profound cellular destabilization. Many different new traits are often recovered from the same population of cultured cells. Mutations and epigenetic changes in different independent traits can even arise in the same cell (24), and double mutations affecting two different genes are sometimes nonindependent events (7). This suggests that diverse heritable changes with different underlying mechanisms can be generated by a single, primary, destabilizing event. According to this interpretation, the wide spectrum of variants arising from cultured material reflects destabilization, followed by the action of selection and secondary heritable changes in the cell population.

Destabilization has several important practical consequences. It is commonly assumed that different plants regenerated from the same tissue and sister cells in the same clone are genetically and epigenetically identical. In general, this is probably not the case. Even when specific mutants are selected for in culture it is likely that the regenerated plants will vary in other traits as well. There is growing interest in using tissue culture methods to modify cells genetically by protoplast fusion, organelle transfer, and DNA transformation (12). Because of the extremely broad range of phenotypes that can be generated in culture by spontaneous variation, it is essential to use only the most stringent criteria in distinguishing intrinsic variation from new phenotypes resulting from the transfer of genetic information. Finally, destabilization may provide novel forms of variation affecting plant morphology, disease resistance, and yield that are not available using more traditional methods. Variation is a challenging problem that is likely to provide insights into genetic and epigenetic homeostasis and is of critical importance in the application of tissue culture for crop improvement. Little is known about the genetic and epigenetic influences on the variation process, the prospects of controlling the stability of cultured cells, or the mechanisms underlying destabilization. What is needed are more detailed studies of the variation process itself, using totipotent plant materials bearing several heterozygous genetic markers that permit the analysis of phenotypes at the single cell, tissue, and whole plant level.

ACKNOWLEDGMENTS

I am grateful to members of the plant groups at the Friedrich Miescher-Institut for their helpful comments, and I thank my colleagues who kindly supplied unpublished results and copies of papers in press.

Literature Cited

1. Arcia, M. A., Wernsman, E. A., Burk, L. G. 1978. Performance of anther-derived dihaploids and their conventionally inbred parents as lines, in F_1 hybrids, and in F_2 generations. *Crop Sci.* 17:413–18

2. Aristotle. 1941. On the generation of animals. In *The Basic Work of Aristotle*, ed. R. McKeon, pp. 665–80. New York: Random House

3. Arya, H. C., Hildebrandt, A. C., Riker, A. J. 1962. Growth in tissue culture of single-cell clones from grape stem and *Phylloxera* gall. *Plant Physiol.* 37:387–92

4. Balzan, R. 1978. Karyotype instability in tissue cultures derived from the mesocotyl of *Zea mays* seedling. *Caryologia* 31:75–87

5. Banks, M. S. 1979. Plant regeneration from callus from two growth phases of English ivy, *Hedera helix* L. *Z. Pflanzenphysiol.* 92:349–53

6. Banks, M. S., Christensen, M. R., Hackett, W. P. 1979. Callus and shoot formation in organ and tissue cultures of *Hedera helix* L., English ivy. *Planta* 145:205–7

7. Barbier, M., Dulieu, H. L. 1980. Effets génétiques observès sur des plantes de tabac régénéréés à partir de cotylédons par culture *in vitro*. *Ann. Amélior. Plantes* 30:321–44

8. Bayliss, M. W. 1980. Chromosomal variation in plant tissues in culture. *Int. Rev. Cytol. Suppl.* 11A:113–44

9. Beale, G. H. 1958. The role of the cytoplasm in antigen determination in *Paramecium aurelia*. *Proc. R. Soc. London Ser. B* 148:308–14

10. Binns, A., Meins, F. Jr. 1973. Evidence that habituation of tobacco pith cells for cell division-promoting factors is heritable and potentially reversible. *Proc. Natl. Acad. Sci. USA* 70:2660–62

11. Binns, A., Meins, F. Jr. 1980. Chromosome number and the degree of cytokinin habituation of cultured tobacco cells. *Protoplasma* 103:179–87

12. Bottino, P. J. 1975. The potential of genetic manipulation in plant cell cultures for plant breeding. *Radiat. Bot.* 15:1–16

13. Bourgin, J.-P. 1978. Valine-resistant plants from *in vitro* selected tobacco cells. *Mol. Gen. Genet.* 161:225–30

14. Brettell, R. I. S., Thomas, E., Ingram, D. S. 1980. Reversion of Texas male-sterile cytoplasm maize in culture to give fertile, T-toxin resistant plants. *Theor. Appl. Genet.* 58:55–58

15. Brink, R. A. 1973. Paramutation. *Ann. Rev. Genet.* 7:129–52

16. Broertjes, C., Keen, A. 1980. Adventitious shoots: Do they develop from one cell? *Euphytica* 29:73–87

17. Browers, M. A., Orton, T. S. 1982. A factorial study of chromosomal variabil-

ity in callus cultures of celery (*Apium graveolens*). *Plant Sci. Lett.* 26:65–73

18. Brown, J. S., Wernsman, E. A., Schnell, R. J. II. 1983. Effect of a second cycle of anther culture on flue-cured lines of *Nicotiana tabacum* L. *Crop Sci.* In press

19. Cahn, R. D., Cahn, M. B. 1966. Heritability of cellular differentiation in retinal pigment cells *in vitro*. *Proc. Natl. Acad. Sci. USA* 55:106–14

20. Cameron, D. R. 1959. The monosomics of *Nicotiana tabacum*. *Tob. Sci.* 3:164–66

21. Carlson, J. E., Widholm, J. M. 1978. Separation of two forms of anthranilate synthetase from 5-methyltryptophan-susceptible and resistant cultured *Solanum tuberosum* cells. *Physiol. Plant.* 44:251–55

22. Carlson, P. S. 1974. Mitotic crossing-over in a higher plant. *Genet. Res.* 24:109–12

23. Chaleff, R. S. 1981. *Genetics of Higher Plants. Applications of Cell Culture.* Cambridge: Univ. Press

24. Chaleff, R. S., Keil, R. L. 1981. Genetic and physiological variability among cultured cells and regenerated plants of *Nicotiana tabacum*. *Mol. Gen. Genet.* 181:254–58

25. Chaleff, R. S., Parsons, M. F. 1978. Direct selection *in vitro* for herbicide-resistant mutants of *Nicotiana tabacum*. *Proc. Natl. Acad. Sci. USA* 75:5104–7

26. Chaturvedi, H. C., Mitra, G. C. 1975. A shift in morphogenetic pattern in *Citrus* callus tissue during prolonged culture. *Ann. Bot.* 39:683–87

27. Christianson, M. L., Chiscon, M. O. 1978. Use of haploid plants as bioassays for mutagens. *Environ. Health Perspect.* 27:77–83

28. Coon, H. G. 1966. Clonal stability and phenotypic expression of chick cartilage cells *in vitro*. *Proc. Natl. Acad. Sci. USA* 55:66–73

29. D'Amato, F. 1964. Endopolyploidy as a factor in plant tissue development. *Caryologia* 17:41–52

30. D'Amato, F. 1977. Cytogenetics of differentiation in tissue and cell cultures. In *Applied and Fundamental Aspects of Plant Cell, Tissue, and Organ Culture,* ed. J. Reinert, Y. P. S. Bajaj, pp. 343–57. Berlin-Heidelberg-New York: Springer

31. Davey, M. R., Fowler, M. W., Street, H. E. 1971. Cell clones contrasted in growth, morphology and pigmentation isolated from a callus culture of *Atropa belladonna* var. *lutea. Phytochemistry* 10:2559–75

32. De Paepe, R., Bleton, D., Gnangbe, F.

1981. Basis and extent of genetic variability among doubled haploid plants obtained by pollen culture in *Nicotiana sylvestris. Theor. Appl. Genet.* 59:177–84

33. Dix, P. J. 1977. Chilling resistance is not transmitted sexually in plants regenerated from *Nicotiana sylvestris* cell lines. *Z. Pflanzenphysiol.* 84:223–26

34. Dix, P. J., Street, H. E. 1976. Selection of plant cell lines with enhanced chilling resistance. *Ann. Bot.* 40:903–10

35. Dougall, D. K., Johnson, J. M., Whitten, G. H. 1980. A clonal analysis of anthocyanin accumulation by cell cultures of wild carrot. *Planta* 149:292–97

36. Dulieu, H. 1974. Somatic variations on a yellow mutant in *Nicotiana tabacum* L. (a_1^+/a_1 a_2^+/a_2) I. Non-reciprocal genetic events occurring in leaf cells. *Mutation Res.* 25:289–304

37. Dulieu, H. L. 1975. Somatic variations on a yellow mutant in *Nicotiana tabacum* L. (a_1^+/a_1 a_2^+/a_2). II. Reciprocal genetic events occurring in leaf cells. *Mutat. Res.* 28:69–77

37a. Dulieu, H. L., Barbier, M. 1982. High frequencies of genetic variant plants regenerated from cotyledons of tobacco. In *Variability in Plants Regenerated from Tissue Culture,* ed. E. Earle, Y. Demarly, pp. 211–27. New York: Praeger

38. Edallo, S., Zucchinali, C., Perenzin, M., Salamini, F. 1981. Chromosomal variation and frequency of spontaneous mutation associated with *in vitro* culture and plant regeneration in maize. *Maydica* 26:39–56

39. Esau, K. 1938. The multinucleate condition in fibers of tobacco. *Hilgardia* 11:427–34

40. Fosket, D. E. 1981. Stability of the cytokinin requirement in Paul's Scarlet Rose cells in culture. *In Vitro* 17:322–30

41. Gautheret, R. J. 1966. Factors affecting differentiation of plant tissues grown *in vitro*. In *Cell Differentiation and Morphogenesis,* pp. 55–95. Amsterdam: North-Holland

42. Gehring, W. 1968. The stability of the determined state in cultures of imaginal disks in *Drosophila*. In *The Stability of the Differentiated State,* ed. H. Ursprung, pp. 134–54. Berlin: Springer

43. Gengenbach, B. G., Connelly, J. A., Pring, D. R., Conde, M. F. 1981. Mitochondrial DNA variation in maize plants regenerated during tissue culture selection. *Theor. Appl. Genet.* 59:161–67

44. Gengenbach, B. G., Green, C. E., Donovan, C. M. 1977. Inheritance of selected pathotoxin resistance in maize plants re-

generated from cell cultures. *Proc. Natl. Acad. Sci. USA* 74:5113–17

45. Goldschmidt, R. 1935. Gen und Ausseneigenschaft (Untersuchungen an *Drosophila*) I und II. *Z. Indukt. Abstamm. Vererbungsl.* 69:38–131

46. Goodspeed, T. H., Avery, P. 1939. Trisomics and other types in *Nicotiana sylvestris. J. Genet.* 38:381–458

47. Green, C. E., Phillips, R. L. 1975. Plant regeneration from tissue cultures of maize. *Crop Sci.* 15:417–21

48. Green, C. E., Phillips, R. L., Wang, A. S. 1977. Cytological analysis of plants regenerated from maize tissue cultures. *Maize Genet. Coop. Newsl.* 51:53–54

49. Harris, M. 1964. *Cell Culture and Somatic Variation.* New York: Holt, Rinehart & Winston

50. Hayes, W. 1968. *The Genetics of Bacteria and their Viruses,* pp. 179–200. New York: Wiley. 2nd ed.

51. Holtzer, H., Weintraub, H., Mayne, R., Mochan, B. 1972. The cell cycle, cell lineage, and cell differentiation. *Curr. Top. Dev. Biol.* 7:229–56

52. King, P. J. 1983. Biochemical mutants of higher plants via somatic cell culture. *Biol. Rev. Cambridge Philos. Soc.* In press

53. Larkin, P. J., Scowcroft, W. R. 1981. Somaclonal variation—a novel source of variability from cell cultures. *Theor. Appl. Genet.* 60:197–214

54. Lea, D. E., Coulson, C. A. 1949. The distribution of the numbers of mutants in bacterial populations. *J. Genet.* 49:264–85

55. Lotka, A. J. 1956. *Elements of Mathematical Biology.* New York: Dover

56. Luria, S. E., Delbrück, M. 1943. Mutation of bacteria from virus sensitivity to virus resistance. *Genetics* 28:491–511

57. Lutz, A. 1971. Aptitudes morphogénétiques des cultures de tissus d'origine unicellulaire. *Colloq. Int. CNRS* No. 193:163–68

58. Lutz, A., Belin, C. 1974. Analyse des aptitudes organogénétiques de clones d'origine unicellulaire issus d'une souche anergiée de tabac, en fonction de concentrations variées en acide indolylacétique et en kinétine. *C.R. Acad. Sci. Ser. D* 279:1531–33

59. Maliga, P. 1976. Isolation of mutants from cultured plant cells. In *Cell Genetics in Higher Plants,* ed. D. Dudits, G. L. Farkas, P. Maliga, pp. 59–76. Budapest: Hungarian Acad. Sci.

60. Maliga, P. 1980. Isolation, characterization, and utilization of mutant cell lines in higher plants. *Int. Rev. Cytol. Suppl.* 11A:225–50

61. Maliga, P., Lázár, G., Sváb, Z., Nagy, F. 1976. Transient cycloheximide resistance in a tobacco cell line. *Mol. Gen. Genet.* 149:267–71.

62. Malmberg, R. L. 1979. Temperature-sensitive variants of *Nicotiana tabacum* isolated from somatic cell culture. *Genetics* 92:215–21

63. Malmberg, R. L. 1980. Biochemical, cellular and developmental characterization of a temperature-sensitive mutant of *Nicotiana tabacum* and its second site revertant. *Cell* 22:603–9

64. Marton, L., Dung, T. M., Mendel, R. F., Maliga, P. 1982. Nitrate reductase deficient cell lines from haploid protoplast cultures of *Nicotiana Plumbaginifolia. Mol. Gen. Genet.* 186:301–4

65. McCoy, T. J., Phillips, R. L. 1982. Chromosomal stability in maize (*Zea mays* L.). *Can. J. Genet. Cytol.* In press

66. McCoy, T. J., Phillips, R. L., Rines, H. W. 1981. Cytogenetic analysis of plants regenerated from oat (*Avena sativa*) tissue cultures: High frequency of partial chromosome loss. *Can. J. Genet. Cytol.* 24:37–50

67. Meins, F. Jr. 1972. Stability of the tumor phenotype in crown gall tumors of tobacco. *Prog. Exp. Tumor Res.* 15:93–109

68. Meins, F. Jr. 1974. Mechanisms underlying the persistence of tumour autonomy in crown-gall disease. In *Tissue Culture and Plant Science 1974,* ed. H. E. Street, pp. 233–64. London/New York: Academic

69. Meins, F. Jr. 1975. Cell division and the determination phase of cytodifferentiation in plants. In *Cell Cycle and Cell Differentiation,* ed. J. Reinert, H. Holtzer, pp. 151–75. Berlin-Heidelberg-New York: Springer

70. Meins, F. Jr. 1982. Habituation of cultured plant cells. In *Molecular Biology of Plant Tumors,* ed. G. Kahl, J. Schell, pp. 3–31. New York: Academic

71. Meins, F. Jr. 1982. The nature of the cellular, heritable change in cytokinin habituation. See Ref. 37a, pp. 202–10

72. Meins, F. Jr., Binns, A. 1977. Epigenetic variation of cultured somatic cells: Evidence for gradual changes in the requirement for factors promoting cell division. *Proc. Natl. Acad. Sci. USA* 74:2928–32

73. Meins, F. Jr., Binns, A. N. 1978. Epigenetic clonal variation in the requirement of plant cells for cytokinins. In *The Clonal Basis of Development,* ed. S. Subtelny, I. M. Sussex, pp. 185–201. New York: Academic

74. Meins, F. Jr., Binns, A. N. 1979. Cell

determination in plant development. *BioScience* 29:221–25

75. Meins, F. Jr., Binns, A. N. 1982. Rapid reversion of cell-division-factor habituated cells in culture. *Differentiation* 23:10–12

76. Meins, F. Jr., Lutz, J. 1979. Tissue-specific variation in the cytokinin habituation of cultured tobacco cells. *Differentiation* 15:1–6

77. Meins, F. Jr., Lutz, J. 1980. The induction of cytokinin habituation in primary pith explants of tobacco. *Planta* 149:402–7

78. Meins, F. Jr., Lutz, J. 1980. Epigenetic changes in tobacco cell culture: studies of cytokinin habituation. In *Genetic Improvement of Crops: Emergent Techniques*, ed. I. Rubenstein, B. Gengenbach, R. L. Phillips, C. E. Green, pp. 220–36. Minneapolis: Univ. Minn.

79. Meins, F. Jr., Lutz, J., Binns, A. N. 1980. Variation in the competence of tobacco pith cells for cytokinin- habituation in culture. *Differentiation* 16:71–75

80. Melchers, G. 1971. Transformation or habituation to autotrophy and tumor growth and recovery. *Colloq. Int. CNRS.* No. 193:229–34

81. Mok, M. C., Gabelman, W. H., Skoog, F. 1976. Carotenoid synthesis in tissue cultures of *Daucus carota* L. *J. Am. Soc. Hort. Sci.* 101:442–49

82. Mok, M. C., Mok, D. W. S., Armstrong, D. J., Rabakoarihanta, A., Kim, S.-G. 1980. Cytokinin autonomy in tissue cultures of *Phaseolus*: A genotype-specific and heritable trait. *Genetics* 94:675–86

83. Morgan, T. H. 1934. *Embryology and Genetics*. New York: Columbia Univ.

84. Mott, R. L., Cure, W. W. 1978. Anatomy of maize tissue cultures. *Physiol. Plant.* 43:139–45

85. Müller, A. J., Grafe, R. 1978. Isolation and characterization of cell lines of *Nicotiana tabacum* lacking nitrate reductase. *Mol. Gen. Genet.* 161:67–76

86. Murata, M., Orton, T. J. 1983. Chromosome structural changes in cultured celery cells. *In Vitro*. In press

87. Nagel, M., Reinhard, E. 1975. Das ätherische Oel der Callus-Kulturen von *Ruta graveolens* L. I. Die Zusammensetzung des Oels. *Planta Med.* 27:151–58

88. Nakamura, A., Yamada, T., Kadotani, N., Itagaki, R., Oka, M. 1974. Studies on the haploid method of breeding in tobacco. *Soc. Adv. Breed. Res. Asia Oceania* 6:107–31

89. Nanney, D. L. 1957. The role of the cytoplasm in heredity. In *The Chemical Basis of Heredity*, ed. W. D. McElroy, B. Glass, p.134–66. Baltimore: Johns Hopkins Univ.

90. Nanney, D. L. 1958. Epigenetic control systems. *Proc. Natl. Acad. Sci. USA* 44:712–17

91. Ogura, H. 1978. Genetic control of chromosomal chimerism found in a regenerate from tobacco callus. *Jpn. J. Genet.* 53:77–90

92. Oinuma, T., Yoshida, T. 1974. Genetic variation among doubled haploid lines of burley tobacco varieties. *Jpn. J. Breed.* 24:211–16

93. Orton, T. J. 1980. Chromosomal variability in tissue cultures and regenerated plants of *Hordeum*. *Theor. Appl. Genet.* 56:101–12

94. Partanen, C. R. 1965. On the chromosomal basis for cellular differentiation. *Am. J. Bot.* 52:204–9

95. Potrykus, I. 1970. Mutation und Rückmutation extrachromosomal vererbter Plastidenmerkmale von *Petunia*. *Z. Pflanzenzücht.* 63:24–40

96. Reisch, B., Bingham, E. T. 1981. Plants from ethionine-resistant alfalfa tissue cultures: Variation in growth and morphological characteristics. *Crop Sci.* 21:783–88

97. Sacristán, M. D., Melchers, G. 1969. The caryological analysis of plants regenerated from tumorous and other callus cultures of tobacco. *Mol. Gen. Genet.* 105:317–33

98. Sacristán, M. D., Wendt-Gallitelli, M. F. 1970. Transformation to auxin-autotrophy and its reversibility in a mutant line of *Crepis capillaris* callus culture. *Mol. Gen. Genet.* 110:355–60

99. Sand, S. A., Sparrow, A. H., Smith, H. H. 1960. Chronic gamma irradiation effects on the mutable V and stable R loci in a clone of *Nicotiana*. *Genetics* 45:289–308

100. Scandalios, J. G. 1974. Isozymes in development and differentiation. *Ann. Rev. Plant Physiol.* 25:225–58

101. Secor, G. A., Shepard, J. F. 1981. Variability of protoplast-derived potato clones. *Crop Sci.* 21:102–5

102. Shepard, J. F. 1981. Protoplasts as sources of disease resistance in plants. *Ann. Rev. Phytopathol.* 19:145–66

103. Shepard, J. F. 1982. The regeneration of potato plants from leaf-cell protoplasts. *Sci. Am.* 246(5):112–21

104. Shepard, J. F., Bidney, D., Shahin, E. 1980. Potato protoplasts in crop improvement. *Science* 208:17–24

105. Siegemund, F. 1981. Selektion von Resistenzmutanten in pflanzlichen Zellkulturen—eine Uebersicht. *Biol. Zentralbl.* 100:155–66

106. Sievert, R. C., Hildebrandt, A. C. 1965. Variation within single cell clones of

tobacco tissue cultures. *Am. J. Bot.* 52:742–50

107. Siminovitch, L. 1976. On the nature of hereditable variation in cultured somatic cells. *Cell* 7:1–11

108. Simpson, G. G., Roe, A., Lewontin, R. C. 1960. *Quantitative Zoology*, revised. New York/Chicago/San Francisco/Atlanta: Harcourt, Brace & World

109. Skirvin, R. M. 1978. Natural and induced variation in tissue culture. *Euphytica* 27:241–66

110. Skokut, T. A., Filner, P. 1980. Slow adaptive changes in urease levels of tobacco cells cultured on urea and other nitrogen sources. *Plant Physiol.* 65:995–1003

111. Sparrow, A. H., Baetcke, K. P., Shaver, D. L., Pond, V. 1968. The relationship of mutation rate per roentgen to DNA content per chromosome and to interphase chromosome volume. *Genetics* 59:65–78

112. Sparrow, A. H., Sparrow, R. C. 1976. Spontaneous somatic mutation frequencies for flower color in several *Tradescantia* species and hybrids. *Environ. Exp. Bot.* 16:23–43

113. Springer, W. D., Green, C. E., Kohn, K. A. 1979. A histological examination of tissue culture initiation from immature embryos of maize. *Protoplasma* 101:269–81

114. Stent, G. S. 1971. *Molecular Genetics. An Introductory Narrative.* San Francisco: Freeman

115. Stewart, R. N. 1978. Ontogeny of the primary body in chimeral forms of higher plants. See Ref. 73, pp. 131–60

116. Stoutemyer, V. T., Britt, O. K. 1965. The behavior of tissue cultures from English and Algerian ivy in different growth phases. *Am. J. Bot.* 52:805–10

117. Street, H. E. 1977. Single-cell clones—derivation and selection. In *Plant Tissue and Cell Culture*, ed. H. E. Street, pp. 207-22. Oxford/London/Edinburgh/Melbourne: Blackwell. 2nd ed.

118. Sunderland, N. 1977. Nuclear cytology. In *Plant Cell and Tissue Culture*, ed. H. E. Street, 2:177–206. Berkeley: Univ. Calif. Press

119. Syōno, K., Furuya, T. 1974. Induction of auxin- nonrequiring tobacco calluses and its reversal by treatments with auxins. *Plant Cell Physiol.* 15:7–17

120. Tabata, M., Hiraoka, N. 1976. Variation of alkaloid production in *Nicotiana rustica* callus cultures. *Physiol. Plant.* 38:19–23

121. Tandeau de Marsac, N., Jouanneau, J.-P. 1972. Variation de l'exigence en cytokinine de lignées clonales de cellules de tabac. *Physiol. Vég.* 10:369–80

122. Tandeau de Marsac, N., Péaud-Lenöel, M. C. 1972. Cultures photosynthétiques de lignées clonales de cellules de tabac. *C. R. Acad. Sci. Ser. D* 274:1800–2

123. Thomas, B. 1981. Bibliography of mutant isolation from plant cell cultures. *Plant Mol. Biol. Newsl.* 2:77–89

124. Thomas, E., Bright, S. W. J., Franklin, J., Lancaster, V. A., Miflin, B. J., Gibson, R. 1982. Variation amongst protoplast-derived potato plants. *Theor. Appl. Genet.* 62:65–68

125. Thomas, E., King, P. J., Potrykus, I. 1979. Improvement of crop plants via single cells *in vitro*—an assessment. *Z. Pflanzenzücht.* 82:1–30

126. Thorpe, T. A. 1980. Organogenesis *in vitro:* Structural, physiological, and biochemical aspects. *Int. Rev. Cytol. Suppl.* 11A:71–111

127. Torrey, J. G. 1967. Morphogenesis in relation to chromosome constitution in long-term plant tissue culture. *Physiol. Plant* 20:265–75

128. Van't Hof, J. 1973. The regulation of cell division in higher plants. *Brookhaven Symp. Biol.* 25:152–65

129. Waddington, C. H. 1956. *Principles of Embryology.* London: Allen & Unwin

130. Wardlaw, C. W. 1970. Enigmas of epigenesis. In *Cellular Differentiation in Plants and Other Essays*, pp. 117–35. Manchester: Manchester Univ. Press

131. Wenzel, G., Schieder, O., Przewozny, T., Sopory, S.K., Melchers, G. 1979. Comparison of single cell culture-derived *Solanum tuberosum* L. plants and a model for their application in plant breeding programs. *Theor. Appl. Genet.* 55:49–55

132. Widholm, J. 1980. Differential expression of amino acid biosynthetic control isozymes in plants and cultured cells. *In Plant Cell Cultures: Results and Perspectives*, ed. F. Sala, B. Parisi, R. Cella, O. Ciferri, pp. 157–59. Amsterdam: Elsevier/North Holland

133. Yeoman, M. M., Forche, E. 1980. Cell proliferation and growth in callus cultures. *Int. Rev. Cytol. Suppl.* 11A:1–24

134. Zagorska, N. A., Shamina, Z. B., Butenko, R. G. 1974. The relationship of morphogenetic potency of tobacco tissue culture and its cytogenetic features. *Biol. Plant.* 16:262–74

135. Zenk, M. H., El-Shagi, H., Arens, H., Stöckigt, J., Weiler, E. W., Deus, B. 1977. Formation of the indole alkaloids serpentine and ajmalicine in cell suspension cultures of *Catharanthus roseus.* In *Plant Tissue Culture and its Biotechnological Applications,* ed. W. Barz, E. Reinhard, M.H. Zenk, pp. 27–43. Hiedelberg/Berlin/New York: Springer

Ann. Rev. Plant Physiol. 1983. 34:347–87

PHLOEM LOADING OF SUCROSE

Robert T. Giaquinta

E. I. du Pont de Nemours and Company, Central Research and Development Department, Experimental Station, Wilmington, Delaware 19801

CONTENTS

INTRODUCTION

The partitioning of assimilates between their sites of production in photosynthesizing leaves and their sites of utilization in harvestable regions is, unquestionably, a major determinant of crop yield (48). The potential of regulating the translocation of assimilates by either plant growth regulators or genetics promises substantial opportunities for increasing the yield of most major agronomic crops. Because partitioning is highly integrated and orchestrated throughout plant growth and development, control can be exerted potentially at several target sites in both source and sink regions.

347

0066–4294/83/0601–0347$02.00

In source leaves, for example, the availability of sucrose for export to the harvestable regions is governed in part by: (a) the rate of photosynthesis; (b) the extent of partitioning of fixed carbon between transport and nontranport carbohydrates such as sucrose and starch; (c) the rate of sucrose synthesis; (d) compartmentation of sucrose into "nontransport" and "transport" pools; (e) intercellular transfer of assimilates to the phloem region; and (f) the transport and accumulation of sucrose into the phloem (phloem loading). The reader is referred to several recent reviews on these topics in relation to assimilate partitioning (29, 30, 32, 42, 93, 104).

I plan to focus my remarks on two aspects of export leaf physiology: the cellular pathway of assimilate transfer to the phloem and the mechanism of sucrose accumulation during phloem loading. "Discovering the rules" associated with these processes has relevance not only to the regulation of partitioning in relation to yield, but also to increasing the uptake and systemic mobility of xenobiotics such as crop protectants.

PATHWAY OF SUGAR TRANSFER

There is little doubt that sucrose is synthesized in mature leaves by the enzyme sucrose phosphate synthetase. This enzyme is located exclusively in the mesophyll cytoplasm. Because the rate of sucrose synthesis and, perhaps more relevantly, its movement to the phloem are among the principal determinants of export, it is important to consider, in a quantitative manner, the structural and metabolic basis for sugar transfer from the mesophyll cells to the region of the phloem. Pathways of transport that warrant consideration are those of assimilate transfer: (a) between mesophyll cells; (b) from the mesophyll to the vein boundary; and (c) from the vein boundary into the phloem sieve tubes.

Transfer to the Phloem Region

Basically, solute movement toward the phloem can occur through the symplasm via plasmodesmata by either volume flow of solution or by solute diffusion in the absence of solvent movement. Alternatively, intercellular transfer can proceed via the apoplast by crossing the cell membrane-wall-membrane boundaries. After considering the solute permeabilities of biological membranes, solute diffusion parameters through plasmodesmata, plasmodesmata structure, and cytoplasm viscosity, Tyree (111) concluded that theoretical considerations alone dictate that cell-to-cell transfer of solutes must occur mainly by diffusion through plasmodesmata, even though plasmodesmata themselves represent a high resistance to flow.

Although detailed quantitative studies on the distribution and frequency of plasmodesmata between various types of leaf cells are lacking, there are a few notable studies (unfortunately in different systems) which provide some insight into the intercellular transfer of solutes. In an excellent review on plasmodesmata, Gunning (50) pointed out that Haberlandt's "Principles of Expeditious Translocation" dealt with the favorable arrangement of interconnecting cells within a leaf in relation to the channeling of assimilates to the veins. Although intuitively it is believed that assimilates are transferred toward the phloem within the symplasm, little definitive data exist on this subject. For example, plasmodesmata connections, which typically (but not exclusively) are formed at the time of cell division during leaf differentiation, are prevalent throughout the mesophyll. In the palisade mesophyll, they are more abundant on the periclinal walls than on the anticlinal walls which usually face the intercellular air spaces. In the spongy layer where intercellular air spaces are abundant, symplastic continuity is achieved by groups of branched plasmodesmata in the areas of limited wall contact (28). Estimates of solute transfer through these plasmodesmata, however, are generally lacking or are rough estimates at best. For instance, Brinckman & Lüttge (see 50), studying electrical coupling between mesophyll in *Oenotherma,* calculated that plasmodesmata occupied 0.38% of the cell wall of the mesophyll junctions with a frequency of 300 μm^2 wall. By knowing the magnitude of assimilate export, sugar flux through the plasmodesmata was estimated to be 3.8×10^{-21} mol.sec^{-1} per plasmodesma. Assuming a transport sugar concentration of 5% within the mesophyll and an unrestricted plasmodesmata length and radius of 0.5 μm and 20 nm, respectively, this flux translates to a pore velocity of 2 μsec^{-1} and a mass flow through the plasmodesmata of four volume changes per second (50). Without measurements of the plasmodesmata frequency and distribution in the other cell types it is not possible to determine whether these performance parameters of assimilate production and export are consistent with a specific cell-to-cell transfer route all the way to phloem.

If symplastic transport occurs, we would predict that leaf cells would retain [14C] assimilates during their transfer to the veins. Studies by Kaiser et al (66) show that isolated protoplasts from leaves of *Papaver* and *Spinacia* show very little leakage of ^{14}C-assimilates after [14]C022 fixation. These authors also showed that[14]C022-derived assimilates were retained within the leaf symplasm during intercellular transfer in intact leaves. Although these studies provide support for symplastic transport, they do not exclude the necessity for sugar entry into a limited region of the apoplast adjacent to the phloem prior to loading.

In contrast to the paucity of quantitative structural and transport data on assimilate movement in the mesophyll, detailed studies do exist for assimilate transfer across the mesophyll to vein boundary in wheat leaves (73). Wheat represents a special situation because all longitudinal veins in this C-3 monocot are enclosed within a suberized cell layer called the mestome sheath. Thus, photosynthate must traverse this presumably apoplastic barrier enroute to the phloem. Quantitative structural analyses are consistent with a symplastic route of sugar transfer through the mestome sheath. The total surface area of the inner tangential walls of the mestome sheath cells which abut the vascular region is 2.6×10^{-1} .cm^{-2}. Given a photosynthetic rate for wheat of 15 mg (CH$_2$0) dm^{-2}.hr$^-$1 (4.2×10^{-8} g sec$^-$1.cm^{-2} leaf) and a sucrose export rate of 2×10^{-8} g.sec^{-1} cm$^-$2 (58.5 pmol.sec^{-1} .cm^{-2} leaf), the sugar flux rate across the inner tangential wall was estimated to be 2.3×10^{-2} pmol.sec^{-1}. cm^{-2} sheath (73). Because the inner tangential wall contains 2×10^8 plasmodesmata/cm^2 leaf (1.5% of the wall area), the flux per plasmodesma is 2.9×10^{-7} *pmol.sec*$^{-1}$. Based on the above, sugar flux across the mestome sheath by diffusion through plasmodesmata would necessitate a sucrose concentration gradient across the sheath of 50μg cm^{-3} (0.146 mM). This concentration, which is entirely consistent with in vivo sugar concentrations, can be achieved by 50 chloroplasts photosynthesizing for 2 min (73). These studies indicate that assimilates pass through the mestome sheath via a symplastic route by diffusion down a concentration gradient. Again, it is probable that assimilates, after passing through the mesotome sheath, enter the apoplast prior to loading into the phloem. It also should be pointed out that wheat leaves contain both a transverse and a lateral network of phloem. It is possible that functional differences exist between these tissues.

A symplastic pathway is also indicated for assimilate transport between the mesophyll and bundle sheath in C$_4$ plants. In corn, for example, extensive plasmodesmatal connections exist among the mesophyll cells and between individual mesophyll and bundle sheath cells (19,20). In the latter case the plasmodesmata frequency in the interfacing walls in the C$_4$ *Salsola ali* is 14×10^8 cm^{-2} cell wall. This calculates to about 0.1% of the cell wall being occupied by plasmodesmata pores (92). In corn leaves, the outer tangential and radial walls of the bundle sheath contain a continuous suberin lamella whereas the inner tangential walls contain suberin mainly at the sites of plasmodesmata aggregation (19). This further indicates that movement of assimilates through the bundle sheath is restricted largely to a symplastic pathway, while transpiration water movement occurs almost exclusively via a cell wall or apoplastic route.

It is also important to note that in the leaves in certain C-3 dicot legumes (soybean, mung bean, and winged bean) a unique cell layer called the paraveinal mesophyll (PVM) exists in the center of the leaves at the level of the phloem and spans the interveinal space (23,26,27). Its position in the leaf strongly suggests that assimilates produced in the mesophyll are transferred to and through this layer enroute to the phloem. Thus, in species with large interveinal distances, the paraveinal mesophyll represents a specialized cellular network for channeling assimilates to the vascular system. Although no quantitative studies exist on plasmodesmata distribution and frequency within the paraveinal mesophyll (PVM) in relation to solute flux, the observation that abundant plasmodesmata exists between the mesophyll-PVM, PVM-PVM, and PVM-bundle sheath suggests a symplastic transport route through this network (26). Even though quantitative structural studies on assimilate movement to the veins are fragmentary and limited to special cell types, the studies cited above indicate that assimilates most likely move to the phloem via a symplastic route. Comprehensive studies in a given leaf are sorely needed on the extent of wall contact between various cell types, the distribution and frequency of plasmodesmata connections between these cell types, and the distribution of suberized regions in order to fully elucidate and appreciate the pathway(s) of assimilate transfer to the veins in various species.

Transfer to the Loading Sites

The pathway of assimilate transfer into the sieve element-companion cell complex of the phloem from the surrounding mesophyll has generally been discussed in terms of two extreme and mutually exclusive routes—the apoplast versus the symplast. It is worthwhile to consider the body of evidence which deals with this issue and also to address whether the operation of one pathway necessarily excludes the operation of the other.

EVIDENCE FOR AN APOPLASTIC STEP   Several lines of structural, physiological, and biochemical evidence along with theoretical considerations support the view that assimilates at some point exit the symplasm and enter the apoplast prior to loading. However, it is important to point out that the majority of studies have been limited to a few species, and thus a generalization to other organisms is not completely warranted. Also, the view that loading via the apoplast is the major route of transfer in a given species does not exclude the parallel operation of a symplastic pathway for transfer of other solutes into the phloem. The following lines of evidence have been interpreted as support for sucrose entry into the apoplast prior to loading.

EXISTENCE OF A CONCENTRATION STEP   Evidence from both plasmolysis and high-resolution tissue autoradiographic studies leaves little doubt that the phloem has a substantially higher sugar concentration than the surrounding mesophyll. Incipient plasmolysis studies coupled with freeze substitution electron microscopy in sugar beet leaves (33) showed that a marked concentration of solutes occurred at the interface between sieve element-companion cell complexes and the adjacent mesophyll and phloem parenchyma. The osmotic pressure of the sieve element-companion cell complex was approximately 30 bars compared to the osmotic pressure of 8 and 13 bars for the adjacent phloem parenchyma and mesophyll cells, respectively. In corn, plasmolysis studies also showed a high solute concentration in the sieve element-companion cell complex compared to other cell types (except for the bundle sheath which, because of its suberized walls, did not plasmolize even in 1.5 M sucrose (20). That the concentration differences among cell types in corn were not as great as those found for sugar beet could reflect the fact that sucrose, which readily accumulates in various cells, was used as the osmoticum rather than relatively nonpermeable mannitol.

Similarly, high-resolution tissue autoradiographic studies, following either fixation of $^{14}CO_2$ or the accumulation of exogenous $[^{14}C]$ sucrose in sugar beet (25,91), bean (45), Vicia (13), and maize (91), show that sucrose accumulates to a significantly higher level in the phloem compared to the surrounding mesophyll (see Figure 1). In sugar beet, Geiger (29) has estimated that the sieve element-companion cell complex contains 80% of the total leaf sucrose (45 $\mu g \cdot cm^{-2}$ leaf blade) and has a sucrose concentration of 0.8M. It should be pointed out, however, that accumulation in the phloem per se does not address whether the sugar entered the phloem directly from the apoplast or indirectly via a symplastic pathway from the mesophyll.

STRUCTURAL CONSIDERATIONS   There is little doubt that assimilates are concentrated within the phloem, so we must ask whether there is an adequate symplastic continuity between the mesophyll and phloem to account for this accumulation. Although quantitative data on the distrubution and frequency of plasmodesmata are not available, data exist for several species showing a general paucity of direct plasmodesmata connections between mesophyll cells and adjacent companion cells and sieve elements. In sugar beet, plasmodesmata connections between the mesophyll and companion cells or sieve elements are rare or absent (33). Similarly, plasmodesmata are scarce between mesophyll and type A transfer cells[1] in Vicia and Tussilago [(less than 2% of the plasmodesmata found in the veins were between the transfer cells and bundle sheath (50,52)]. A similar situation exists in corn, in that plasmodesmata connections were scarce or rare between the bundle sheath and the companion cells or between bundle sheath and thin-walled sieve tubes (19).

The studies cited above for sugar beet, Vicia, and corn show that there is very little direct symplastic continuity between the assimilate-producing mesophyll or bundle sheath cells and the adjacent companion cells and sieve tubes of the phloem. In other plants, however, such as Cucurbita pepo and Fraxinus, plasmodesmata connections can be quite abundant between the sheath and companion cells (28,108,109).

The fact that some species lack direct connections between the mesophyll and companion cells does not preclude an indirect symplastic transfer of assimilate from the mesophyll or bundle sheath to the phloem via the vascular

[1]See (50) for classification of transfer cell types.

parenchyma cells. In fact, plasmodesmata connections are present in the mesophyll/phloem parenchyma/companion cell/sieve element transfer route in sugar beet (33) and in the bundle sheath/phloem parenchyma/type A and B transfer cell route in *Vicia* and *Tussilago* (52). Similarly in corn (19,20), symplastic continuity exists between the bundle sheath/vascular parenchyma/ thick-walled sieve tubes. Thus, the qualitative presence or absence of plasmodesmata in the absence of solute flux data cannot be used to distinguish unequivocally between the two pathways.

In this regard, it is also important to point out that different pathways of loading can occur for different solutes, and that the degree of apoplastic and symplastic transport may differ in plants having different leaf morphology. For example, data which will be presented later in this review are consistent with an accumulation of sucrose in the companion cell-sieve element complex in sugar beet, corn, and *Vicia* from the apoplast by a membrane transport process. As mentioned above, symplastic continuity exists between the bundle sheath/ vascular parenchyma/thick-walled sieve tubes in corn. These thick-walled sieve tubes, however, do not appear to be involved with either photosynthate transport or sugar storage. Instead, these sieve tubes may function in the retrieval of solutes from xylem (19) or in the exchange of solutes between the xylem and phloem. Similarly, type A transfer cells are viewed as modified companion cells which function in the collection and passage of photosynthate, whereas type B transfer cells appear to be modified phloem parenchyma which function in the retrieval and recycling of intraveinal solutes (50). Pate and co-workers (93) have demonstrated substantial phloem-phloem, xylem-xylem, and xylem-phloem exchange of solutes in lupin. In species like sugar beet, where there are no distinguishing structural characteristics between the phloem parenchyma or companion cells, it is entirely feasible that the symplastic continuity which exists between phloem and nonphloem cells is necessary for either retrieval or recycling of solutes in the vascular bundle.

Plasmodesmata may also function in the transfer of other metabolites (nonsucrose) from the mesophyll to the phloem for translocation. In this latter case an important distinction is made between the active accumulation of sucrose into the companion cell-sieve element complex in order to produce the driving force for translocation and the entry of other phloem mobile metabolites for transport within the plant. It is also possible that assimilates are transported symplastically to the phloem parenchyma for temporary storage during periods of reduced sink demand and are later remobilized to the phloem. This retrieval or metering could occur by either a symplastic or apoplastic pathway.

SUGAR ENTRY IN THE APOPLAST   There are several lines of evidence which considered collectively are consistent with, but do not necessarily prove, the view that sugars enter the apoplast prior to loading. Geiger and co-workers

(29, 34, 100) have shown that 20 mM sucrose fed to the apoplast of an abraded sugar beet leaf was readily available for loading and produced translocation rates equal to those produced by leaves photosynthesizing in $^{14}CO_2$. In other experiments, these workers attempted to disrupt symplastic continuity of the mesophyll by plasmolyzing the mesophyll with 0.8 M mannitol prior to $^{14}CO_2$ or [^{14}C] sucrose feeding. Although the mesophyll cytoplasm could not be completely disrupted by plasmolysis, photosynthesis, and thus $^{14}CO_2$-derived translocation rates, were completely inhibited by the osmoticum. That addition of 20 mM [^{14}C] sucrose to the free space restored translocation to the rate produced by photosynthesis prior to the mannitol treatment indicated that phloem loading of sucrose can occur from the free space. However, because symplastic continuity was not completely disrupted by plasmolysis, it is difficult to exclude the possibility that sucrose was accumulated in the mesophyll and then transported via a symplastic route into the phloem. It is also likely that exogenous [^{14}C]-sucrose entered the phloem by both direct transfer from the apoplast and indirect transfer after accumulation in the mesophyll. The results obtained from experiments in which $^{14}CO_2$-derived assimilates were trapped in the free space appear less subject to these possibilities (described below).

Kursanov & Brovchenko (74) have reported that up to 20% of the total leaf sugars, mainly hexoses, were present in the apoplast of sugar beet leaves. They proposed that the apoplast represents the major route of assimilate transfer to the veins. Although this value seems quite high, data from isotopic trapping experiments and measurements of sucrose turnover within the apoplast are consistent with the free space being the route for sugar entry into the phloem. Geiger and co-workers (29,34) employed a technique that would intercept and trap a portion of the $^{14}CO_2$-derived assimilates that would be transferred to the phloem if passage occurred via the free space. These experiments showed that increasing the rates of leaf photosynthesis and translocation by increasing light intensity resulted in a concomitant increase in the interception of ^{14}C label within the sucrose trapping solution which perfused the apoplast. Similarly, application of 4 mM ATP, which enhanced translocation of assimilates by 75% without affecting photosynthesis, resulted in a 66% increase in the rate of ^{14}C entry in the free space trapping solution. In other experiments, Fondy & Geiger (25) identified and quantitated the major ^{14}C-labeled compounds in the free space of sugar beet leaves during steady state photosynthesis. The concentrations of free space sucrose and glucose were calculated to be 70 and 200 μM, respectively (assuming a total free space volume of 3.4×10^{-3} cm^{-3} cm^{-2} leaf) after a 2 hr photosynthesis period in $^{14}CO_2$. Although the static glucose concentration was unexpectedly high, its specific activity was low. In contrast, the specific activity of free space sucrose approached the activity of the supplied $^{14}CO_2$, indicating that free space sucrose and not glucose was the precursor for the exported sucrose. A similar conclusion was reached in corn. In mature corn leaves, ^{14}C sucrose (0.25%) was the only sugar present in the xylem exudate following exposure to [14]CO_2, and its concentration increased with increasing photosynthetic light (60).

NONPERMEABLE PROBES   Additional evidence for an apoplastic step prior to loading is provided from studies employing nonpermeable chemical modifiers. Giaquinta (35,40) showed that the relatively nonpermeable sulfhydryl group modifier, p-chloromercuribenzene sulfonic acid (PCMBS), markedly and reversibly inhibited the uptake and phloem loading of exogenously supplied ^{14}C] sucrose in sugar beet leaves. Treatment with PCMBS did not result in short-term inhibition of ^{14}C] glucose uptake, photosynthesis, or respiration, suggesting its site of inhibition was at the membrane level, possibly the sucrose carrier itself (see later sections). The observation that PCMBS supplied to the apoplast of sugar beet leaves during photosynthesis in [14]CO_2 markedly inhibited assimilate translocation indicated that ^{14}C] sucrose entered the apoplast prior to its uptake into the phloem. This interpretation is based on the assumption that the nonpenetrating PCMBS has no effect on intracellular transport or plasmodesmata function.

APOPLASTIC VERSUS SYMPLASTIC PATHWAY   It is important to reiterate that entry into the apoplast may not be a universal mechanism for loading. Based on free space sugar analysis following [14]CO_2 fixation, and the lack of direct loading of exogenous ^{14}C] stachyose, the principal transport sugar in *Cucurbita*, Madore & Webb (82) concluded that the symplastic pathway is the major, if not sole, pathway for loading in *Cucurbita*. These authors addressed several potential pitfalls in experiments which had been interpreted to indicate an apoplastic route of loading. While the various types of experiments supporting an apoplastic step, particularly those using exogenous sucrose, are open to alternate interpretations, conclusions based on "static" measurements of sugars, as was the case for the studies on *Cucurbita*, do not address the issue of whether sugars traverse the apoplast enroute to the phloem.

Although an entirely symplastic pathway remains a possibility for sugar loading, a symplastic pathway has to be reconciled with several features of the loading system. First, phloem loading is a selective process. Although selectivity during loading can be achieved readily by a membrane transport process, it is difficult to reconcile selectivity with symplastic transfer unless some as yet unknown discriminating mechanism exists at the level of the plasmodesmata. Second, there is no disagreement that a marked concentration of sugar occurs within the phloem. This concentration mechanism is the driving force for long-distance transport. Unless plasmodesmata contain an active pumping mechanism or act as "one-way valves," or are part of an endoplasmic reticu-lum-mediated pumping mechanism, active transport of sugars across the phloem membrane is the simplest explanation for accumulation. Third, the presence of plasmodesmata among mesophyll and phloem cells is not a con-vincing argument for a symplastic route of loading. As mentioned above, these plasmodesmata most likely were formed at the time of cell division during leaf differentiation. Because sugar unloading in some sinks can occur via a sym-plastic pathway (46,104), it is not possible to determine whether these plas-modesmata are actually functioning in assimilate loading or whether they represent pathways which were used for unloading during early leaf develop-ment. Fourth, it should be recongnized that the companion cells and sieve tubes of the phloem undergo both a structural and biochemical differentiation which is temporally correlated with the acquisition of export capacity. For example, Fellows & Geiger (21) found that the sink-to-source transition in sugar beet leaves was accompanied by the ability of the phloem to accumulate sugars to above a threshold value in osmotic pressure to cause export by mass flow. Also, phloem transfer cells first begin to develop their extensive wall ingrowths (i.e. amplification of membrane surface area) at the onset of export, and the ingrowths increase in parallel with export during leaf maturation (51). This membrane amplification along with the acquisition of ATPase activity on the phloem membrane at the onset of export (11) suggest the development of a membrane transport system for loading from the apoplast (43). Finally, the recent demonstrations of depolarization of sieve tube membrane potential (116) and proton cotransport during active sugar loading (59) are consistent with active sucrose transport across the phloem membranes. Based on the above, one is forced to conclude that loading via the apoplast remains a more than distinct possibility, and proponents of an entirely symplastic pathway of loading must reconcile their data and interpretations with the above discussion.

Nature of Sugar Efflux into the Apoplast

The above studies provide evidence that sucrose exits from the symplasm and enters the free space enroute to the phloem. We know very little about the actual site or mechanism of sugar release.

In terms of mechanism, the magnitude of sucrose efflux into the apoplast appears more consistent with some type of facilitated transfer into the apoplast rather than a diffusion process. Assuming a mass transfer rate of sucrose in sugar beet leaves of 1.2 nmol•min^{-1}•cm^{-2} leaf, and a membrane surface of the sieve element-companion cell complex of 0.88 cm^2•cm^{-2} lamina, the flux across the sieve element-companion cell membranes would be 1.36×10^3 pmol min^{-1}•cm^{-2} membrane. Doman & Geiger (18) calculated that the efflux rate of sucrose, if it occurs uniformly across all the mesophyll, would be 120 pmol sucrose min^{-1}•cm^{-2} mesophyll membrane (based on a mesophyll membrane area of 10 cm^2•cm $^{-1}$54^2). Because this flux rate is about 10^6 greater than the values reported for passive efflux of sucrose across cell membranes of intact cells (18), a facilitated transfer mechanism is implied.

With respect to the site(s) of efflux, because *(a)* there is evidence that sugars move through the mesophyll within the symplasm (66), and *(b)* sucrose release over the entire mesophyll and subsequent movement toward the vein would be greatly impeded by the transpiration stream occuring in the opposite direction, it is highly likely that sucrose is released in close proximity to the sieve element-companion cell complex. Likely candidates for the sites of release are either the phloem parenchyma (which are symplastically connected to the mesophyll and have a lower solute content than the mesophyll) or the mesophyll cells adjacent to the companion cells (18). The latter possibility, however, would dictate that the plasma membranes of the mesophyll cells abutting the phloem are specialized to facilitate sucrose efflux.

A localized entry of sucrose into the apoplast proximal to the phloem would result in even greater efflux rates than noted above by Doman & Geiger (18), because their rates were based on a uniform efflux of sugar over the entire mesophyll. Thus the enhanced efflux would provide further evidence for a facilitated rather than passive or diffusional efflux of sugars. The localized efflux into the apoplast of the phloem region would also achieve a higher local sucrose concentration at the sites of loading (18). This would explain the apparent discrepancy between the low apoplast sucrose concentration [70μM calculated on the basis of entire free space volume (25)] and the K^m of 15–25 mM for sucrose loading and translocation (18,36,100). The actual sucrose concentration at the phloem membranes is not known.

It should also be pointed out that a *localized* sucrose entry in the apoplast of intraveinal regions may not be opposed by transpirational water movement. Gunning (50) reviewed the earlier literature on the morphology of minor veins in relation to the movement of transpirationally fed dyes and surmised that the aqueous environment of the apoplast in the central tissues of the leaf may be a static "backwater" because it is bypassed by the transpirational stream.

The energy requirement for the sugar efflux step is open to speculation. Brovchenko (6,7) suggests that sugar efflux from the mesophyll into the free space is dependent both on ATP from photophosphorylation and oxidative metabolism. Addition of ATP caused both an increase in sugar efflux from the mesophyll and a reabsorption of sugar (mainly glucose) into the veins. These results support those by Sovonick et al (100), which show that ATP increases both sucrose turnover in the free space and translocation without affecting leaf photosynthesis. Sovonick et al (100) also showed that 4 mM dinitrophenol (DNP) inhibited the translocation of exogenously supplied ^{14}C sucrose from sugar beet source leaves by 80%. Interestingly, 4 mM ATP restored the DNP inhibited rate to approximately 80% of the control rate. The stimulation of sucrose efflux and sucrose loading by ATP, however, could be occuring by different mechanisms. Several studies on solute transport show that the stimulation of solute transport by ATP is caused by ion chelation (42). In many of these studies the stimulation by ATP could be mimicked by EDTA, suggesting that ATP was acting as a divalent metal chelator rather than as an energy source (81). The stimulation of loading by ATP, particularly in the presence of DNP (100), however, suggests that ATP is most probably active as an energy source.

In regard to sucrose efflux, K^+ or Na^+ at 12.5 mM supplied to either bean or sugar beet leaf slices caused a rapid and selective efflux of sucrose into the surrounding medium (54). The cation-induced efflux was prevented by adding Ca^{+2} to the medium, suggesting that sucrose efflux may be responding to either the K^+ or Ca^{+2} concentrations of the free space. This is supported by the recent findings of Doman & Geiger (18) which show that foliar applications of 5 to 30 mM KCl stimulated translocation derived from $^{14}CO_2$ by 15% without affecting leaf photosynthesis. The K^+ stimulation of translocation was accompanied by a 50% increase in the rate of ^{14}C entering the free space of the leaf. Similar conclusions were reached for the effects of K^+ on sugar loading in willow (95). These results indicate that the increase in loading and export is caused by a K^+-induced increase in sucrose entry into the free space and not by a stimulation of loading per se.

The mechanism of sucrose efflux from either the phloem parenchyma or mesophyll is not known, but the K^+ effects may be consistent with a sucrose/cation symport mechanism. For example, isolated mesophyll protoplasts from tobacco and wheat leaves release both sugar and K^+ into the external medium (62,63). The sugar release was inhibited 70% by 50 mM KCl but was unaffected by similar concentrations of NaCl, or by $CaCl_2$ and $MgCl_2$ at 2-5 mM. These authors interpret their results in terms of a sucrose/K^+ cotransport mechanism for sugar release. Caution is warranted, however, in drawing comparisons between isolated protoplasts (which are subject to alterations in membrane permeability) and the sugar efflux step occuring in intact leaves.

The increased efflux into the apoplast adjacent to the phloem may by coupled to the phloem loading process itself (32). As discussed below, phloem loading appears to occur by a sucrose/H^+ cotransport mechanism with K^+ movement occurring in response to the membrane potential. Below an apoplast pH of 5.5, Van Bel & Van Erven (112) found that K^+ efflux was coupled to sucrose and proton uptake. If this occurred during phloem loading in vivo, then the loading process could increase the K^+ in the free space, which, in turn, could increase sucrose efflux from the cells in the vicinity of the phloem.

Because the availability of sucrose for loading appears to be an important control point in export, elucidating the factors which regulate the efflux of sucrose from the mesophyll or phloem parenchyma promises to shed light on an important target site for regulating translocation.

CHARACTERISTICS OF PHLOEM LOADING

Structural Correlates of Sugar Loading

The structure of the vascular system of dicot and monocot leaves is highly specialized for the high rates of solute flux demanded by phloem loading and translocation (29, 42). The veins of dicotyledenous leaves typically undergo a repeated series of branching and anastomosing into smaller diameter veins. The phloem in this interconnecting minor vein network represents the primary loading sites for sucrose (Figure 1A). This network of minor vein phloem is quite extensive and represents an efficient collecting system for photosynthate which is synthesized in the surrounding mesophyll. For example, sugar beet leaves contain 70 cm of minor vein•cm^{-2} leaf blade and a 33 μm length of minor vein can collect photosynthate from 29 mesophyll cells. Furthermore, assimilates produced in the mesophyll of sugar beet leaves need to transit only 2 mesophyll cell diameters (approximately 73 μm) before they are loaded into the minor vein phloem (29).

The minor veins in a dicot leaf are usually comprised of a single xylem element, vascular parenchyma cells, and one or two sieve elements which are surrounded by two to four companion cells (Fig. 1B). The organelle- and cytoplasm-rich companion cells, which are twice the diameter of the sieve elements in exporting leaves, figure prominently in solute loading into the sieve elements. For instance, quantitative microautoradiographic studies following $^{14}CO_2$ fixation in soybean leaves demonstrated the similarities in the kinetics of [^{14}C] sucrose labeling within the companion cells and kinetics of ^{14}C export (24). This, along with the high sucrose concentration of the sieve element-companion cell complex (0.3 to 0.8M), indicates that the sieve element-companion cell complex represents the primary site of sucrose entry into the translocation system.

The companion cell-sieve element complex provides an expanded membrane surface which is capable of maintaining high flux rates necessary for loading. For example, although the sieve element-companion cell complex (se-cc complex) represents only approximately 0.6% of the total volume of a sugar beet leaf, its surface area equals 0.88 cm^2•cm^2 leaf blade, 75% of which represents the surface area occupied by the companion cells. Based on the measured translocation rate of 2.8 nmol sucrose•min^{-1}•cm^{-2} leaf blade, the sucrose flux rate through the se-cc complex was calculated at 3.2 nmol sucrose•min^{-1}•cm^{-2} membrane (100). In studies where 10 mM [^{14}C] sucrose was fed to sugar beet leaves, Fondy & Geiger (25) calculated a flux rate of sucrose across the se-cc cell membrane of 16 pmol sec^{-1}•cm^{-2} membrane, based on estimated surface area of 4.7 × 10 μm^2 se-cc membrane•cm^{-2} leaf.

In some species, the companion cells are modified into transfer cells which display numerous cell wall ingrowths, greatly amplifying the membrane sur-

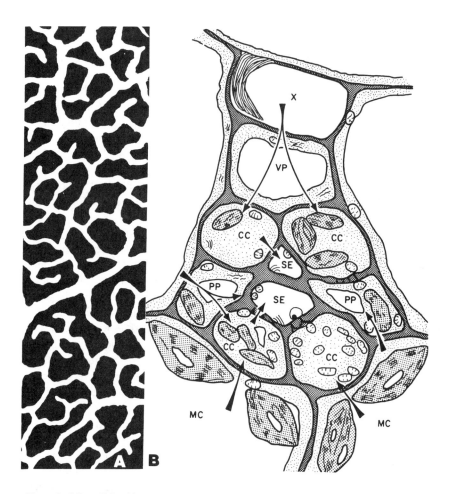

Figure 1   Minor Vein phloem.

(A)   High resolution autoradiograph of *Phaseolus vulgaris* leaf tissue showing accumulation of [^{14}C] sucrose into the minor vein network (^{14}C-label denoted by white); after (45).

(B)   Tracing of an electron micrograph of a cross-section of minor vein from a tobacco leaf. Arrows illustrate possible assimilate entry routes into the sieve element-companion cell complex. X = Xylem; VP = vascular parenchyma; CC = companion cell; SE = sieve element; PP = phloem parenchyma; MC = mesophyl cell. (micrograph by R.T. Giaquinta.)

face area. In absolute terms, Gunning et al (52) have determined that a 58 μm length of vein in *Vicia faba* has 3650 μm^3 of companion cell wall ingrowths which collectively generate 30340 μm^2 of membrane surface over and above the 9789 μm^2 that would have occurred if the vein lacked ingrowths. Additionally, a 1mm length of vein in *Vicia* has a transfer cell membrane surface area of 0.69 mm^2. This translates to 4.84 cm^2 transfer cell membrane·cm^{-2} leaf blade, assuming a value of 70 cm vein·cm^{-2} blade (29). Using the measured export rate of 1.3 μg sucrose·min^{-1}·cm^{-2} for sugar beet (29), the membrane surface area of 4.84 cm^2·cm^{-2} leaf in *Vicia* results in a transmembrane flux of sucrose of 4.9 ng cm^{-2}sec^{-1} or 14.3 pmol·cm^{-2}sec^{-1} (52). This approximates the range of solute fluxes (1-10 pmol cm^{-2}sec^{-1}) found in plant cells (42, 52). Therefore, the structural correlates of sucrose loading across the phloem membranes are consistent with the observed rates of assimilate export from leaves.

The increased size and membrane surface area of companion cells relative to their associated sieve elements suggest that the companion cells are the primary sites of sucrose loading. The presence of numerous complex plasmodesmata (branched on the companion cell side) between the cell wall of the companion cells and sieve elements, along with the similarity in osmotic pressure (28-30 bars) in these two cell types, suggest that sucrose moves from the companion cells to the phloem by either diffusion or volume flow. There is structural support for this view. In *Tussilago* and *Vicia* minor veins, approximately 40% of the intraveinal plasmodesmata interconnect the companion cells and sieve elements (50, 52). Because the average contact between each sieve element and type A transfer cell wall was 3.5 μm^2 per length of vein, a typical minor vein with four sieve elements would have a total area of wall contact of 0.1 cm^2·cm^{-2} (assuming 70 cm vein cm^{-2} leaf). This wall area would have 5.95 × 10^7 channels (occupying 0.87% of the wall) on the transfer cell face and 2.14 × 10^7 channels (0.65% of the wall) on the sieve element face (52) to mediate sugar transfer. Assuming sugar transfer between the transfer cell and sieve element occurs exclusively via plasmodesmata, and that the sucrose concentration in the companion cells is the same as the sieve tube exudate (15%), Gunning (50) calculates that the flux through the plasmodesmata would be 7 volume changes sec^{-1} (2.4 μm sec^{-1} in nonoccluded channels on the sieve element side). Less than a 0.001 bar pressure drop would be required to drive such a flux. If the plasmodesmata channels were constricted by the presence of a desmotubule, the resulting volume change of 250 sec^{-1} (90 μm sec^{-1} velocity) could be achieved by a pressure drop of 0.26 bar.

Assuming that water enters the minor vein of the phloem in the same ratio that it is present in the sieve tube sap, then the water flow across the sieve element-companion cell-complex in sugar beet approximates 1.5 × 10^{-7} mol·min^{-1}·10^{-2} membrane. Thus, active loading of sucrose and the resultant water flux across the phloem membranes appears to be sufficient to account for the pressure differences and solute flux rates necessary to drive translocation (29).

As alluded to above, some species have companion cells which are modified into transfer cells having cell wall ingrowths. No detailed comparative studies exist on phloem loading in plants which have companion cells versus transfer cells. There is suggestive evidence that loading may differ in plants which differ in their degree of membrane surface amplification. Cytochemical localization studies show that ATPase is present on the plasmalemma of both the companion cells and sieve elements in species where the companion cells lack cell wall ingrowths. Interestingly, species that have transfer cells (with wall ingrowths) contain ATPase only on the transfer cell membrane and not on the associated sieve element (4,11). If ATPase activity is related to the loading mechanism (see below), then loading into both the companion cell and sieve element may compensate for the reduced membrane surface area in the absence of cell wall ingrowths (42). Other ways to compensate for the reduced membrane surface area in species lacking transfer cells may include an increased number of smaller companion cells per unit length of vein, or a more favorable geometrical relationship between the phloem and assimilate-producing mesophyll (50). There is great need for detailed quantitative studies on the structural correlates of solute fluxes into the phloem, both within a given species and within different plants.

Selective Loading of Sugars

The types of sugars found in the translocation stream are the result, in part, of the selectivity of the loading process (29, 42). Although my remarks will center on the specificity of the carrier mechanism located in the phloem membranes, which is responsible for establishing the driving force for translocation, it is important to recognize that solutes can also enter the phloem by nonmediated processes. Most exogenous or endogenous solutes that can penetrate the phloem membranes (or which in some species may enter the phloem via a symplastic route) will be transported by mass flow in the translocation stream. The diversity of compounds found in the sieve tubes is not surprising when one considers that (a) the sieve tubes represent the major conduits for nutrient transport within the plant, and (b) mechanisms or pathways other than active transport across the phloem membranes can influence the types of solutes found in the phloem sap. For example, control of the solute composition of the phloem could occur by: (a) selectivity in the efflux of sugars from the mesophyll or phloem parenchyma; (b) sequestering of solutes into nontransport and transport compartments in regions in close proximity to the phloem; and (c) biochemical transformation within the phloem cells themselves which would convert nontransportable solutes into transportable ones (29, 42).

Comprehensive compilations of phloem mobile sugars show that the nonreducing raffinose series of sugars (sucrose, raffinose, stachyose, and verbascose) as well as sugar alcohols (mannitol and sorbitol), represent the major, if not sole, carbohydrates that are transported in higher plants. Reducing sugars are notably absent from the translocation stream. Assuming that sugars are loaded into the phloem from apoplast, the selectivity of loading is ultimately based on carrier recognition of the structural and chemical characteristics of the transport sugar. Sucrose, which is the principal transport sugar in most crops, is composed of α-D-glucopyranose and β-D-fructofuranose joined through two anomeric carbons via a 1-2 linkage. The furanose configuration of the fructose portion of the molecule confers a substantial free energy of hydrolysis (Δ G= -6600 cal•mol^{-1} at pH 6.6, compared to about –7000 cal•mol^{-1} for ATP). This high energy feature may somehow relate to the "advantage" and thus "selection" of sucrose as the major translocated sugar in plants (42).

The rare β-fructoside feature of sucrose is also found in the other members of the raffinose series. Trehalose [α-D-glucopyranosyl-(1-1-α-D-glucopyranoside)] is the only other nonreducing dissacharide found in nature, and it represents the principal transport sugar in insect hemolymph and fungi. Neither trehalose nor raffinose, however, compete with the putative sucrose carrier involved in loading into developing *Ricinus* cotyledons (68). Other sugars such as melibiose, maltose, and galactose also do not appear to compete with sucrose uptake (59, 68). Additionally, stachyose is not loaded into plants like sugar beet which translocate mainly sucrose (25). This can be taken as evidence that the carrier for loading is not simply keying on the sucrosyl portion of the raffinose series.

With regard to the molecular requirements of a sugar for loading, it is also noteworthy that the substitution of an H for an OH in the C-2 glucose moiety of sucrose to form 2-deoxysucrose does not reduce phloem mobility (94). Similarly, the raffinose series has galactose moieties added to the glucose portion of the sucrose molecule. Because modifications of the glucose moiety do not affect transport, it appears important to determine whether substitutions on the unique fructofuranoside portion of the molecule affect transport. In this regard, it would be interesting to determine whether a trisaccharide comprised of glucose-fructose-fructose is transported.

As outlined above, the specificity of phloem loading can be determined by the selectivity of the process presenting the sugar to the loading sites or by selective mechanisms operating at the phloem membranes. Studies by Hendrix (56, 57) on *Cucurbita* show that selectivity resides in the processes which control the availability of the sugar presented to the phloem. In *Cucurbita*, stachyose is the major sugar formed by photosynthesis, and it represents the principal sugar in the translocation stream. The observation that exogenously supplied [^{14}C] sucrose is loaded and translocated in *Cucurbita* suggests that phloem loading is not the system that specifies which sugar is transported. This brings up the possibility that sucrose may be transferred to and accumulated within the phloem where it is then synthesized to stachyose prior to export. Based on labeling patterns of the glucose and galactose moieties of the translocated stachyose following [^{14}C] sucrose feeding to the leaves, Hendrix (56) concluded that stachyose was synthesized in the mesophyll. This conclusion is supported to some extent by the recent findings of Madore & Webb (83) which show that isolated mesophyll cells from *Cucurbita* are capable of a limited synthesis of stachyose from [^{14}C] bicarbonate. Unfortunately, these studies are difficult to interpret because the presence of osmoticum appears to inhibit stachyose (but not sucrose) synthesis severely; therefore the evidence for the mesophyll-only site of stachyose synthesis remains equivocal.

Selectivity of loading in sugar beet leaves appears to reside in the membranes of the sieve element-companion cell complex. Fondy & Geiger (25) showed by microdensitrometry of autoradiographs that of the several sugars supplied to the apoplast of a translocating sugar beet leaf, only sucrose was accumulated into the phloem against a concentration gradient. It is noteworthy, however, that 10 mM [^{14}C] sucrose appears to enter and turn over within both the mesophyll and minor veins. Assuming that all the ^{14}C in the minor vein existed as sucrose, the calculated sucrose concentration in the vein (480 mM) was 43-fold greater than that of the supplied sugar. That the 10 mM equilibrium concentration of the phloem was exceeded soon after the sucrose was supplied supported an active accumulation of sucrose into the phloem. Uptake of fructose, mannitol, or stachyose occurred into the mesophyll of sugar beet leaves rather than the phloem, where they were converted to sucrose prior to transport to the phloem. Madore & Webb (82) also found that free space stachyose was not loaded directly into the phloem in *Cucurbita* leaves but instead entered the mesophyll where it was converted to sucrose. However, they interpret their data in terms of a symplastic pathway for loading.

Sucrose hydrolysis in the free space is a prerequisite for sugar transport in many plant tissues (42). Kursanov & Brovchenko (74) have long proposed that hexoses are the principal sugars that are loaded into the phloem. Giaquinta (36,38), however, showed that asymmetrically labeled sucrose, [^{14}C] (fructosyl) sucrose, retained its asymmetry during accumulation and translocation in sugar beet. These results were taken as evidence for the presence of a sucrose specific carrier mechanism.

Because so little is known about the structural requirements of sugars which are necessary for loading into the phloem, the design of specific sucrose derivatives or affinity labels promises to provide considerable insight into the molecular nature of the sucrosyl carrier. This approach would be equally valuable in species which translocate either sugar alcohols or the raffinose series. In red blood cells, for example, glucose derivatives such as 6-0-n-propyl-D-glucose and n-propyl-β-D-glucopyranoside, which provide steric hinderance at C1 and C6 positions, have been used to show that the glucose carrier, which spans the erythrocyte membrane, binds the C1 end of the glucose molecule at the external side of the membrane and the C6 end on the cytoplasmic side. Moreover, these sugar derivatives show that this orientation is maintained during translocation across the membrane (3).

Sucrose/Proton Cotransport Mechanism

The coupling of nonelectrolyte transport with cation gradients has received considerable attention and wide support in bacteria, algae, yeast, fungi, and animal cells. Although sugar and amino acid transport in animal cells is linked with transmembrane Na^+ gradients, Mitchell's chemiosmotic hypothesis of energy transduction emphasized the important and ubiquitous role of protons in both ATP formation and solute transport (101). Over the last 5 years there has been an increasing body of evidence supporting the existence of an active sucrose/proton cotransport mechanism in higher plants. Several of these studies have proposed that sucrose loading into the phloem occurs by such a cotransport system (2, 13, 15, 36, 37, 40, 41, 59, 64, 65, 68, 70, 84, 85).

The driving force for phloem loading of sucrose via a proton cotransport mechanism is viewed as the electrochemical potential gradient of protons that is established by an active primary proton pump—possibly a proton-translocating membrane ATPase. The energy in this proton gradient is coupled to the secondary active transport of sucrose. This proposal leads to several predictions and testable hypotheses in relation to phloem loading which are addressed below.

PHLOEM CHARACTERISTICS   The characteristics of the phloem are consistent with the hypothesis that sucrose loading is driven by an ATPase-generated proton gradient. Chemical composition studies (42) show that, along with its high sucrose concentration (0.8 M), the phloem has a relatively low proton concentration (pH 7.5 to 8.5) and high potassium concentration (100-200 mM). Although the pH of the apoplast at the actual sites of loading is not known, if we assume its pH approximates that of the total apoplast pH (pH 5.5 to 6), then a substantial proton gradient of 1.5 to 3 pH units exists across the phloem membrane. A proton electrochemical gradient of this magnitude is sufficient to drive both oxidative- and photophosphorylation as well as solute accumulation (17).

Cytochemical localization studies (11) also show that the phloem membranes contain ATPase activity which develops at the onset of export capacity. Thus, the presence of a substantial transmembrane proton gradient and its presumed generating mechanism (ATPase) can be taken as circumstantial evidence consistent with (but certainly not proof of) a cotransport system.

DEPENDENCE OF LOADING AND TRANSLOCATION ON A PROTON GRADIENT   If a proton gradient is required for sugar transport, then increasing the alkalinity of the apoplast should decrease the cosubstrate concentration and magnitude of the proton gradient, and in turn decrease sucrose loading and subsequent translocation. Although this strategy of altering cosubstrate concentration has been employed for Na^+–dependent sugar transport in animal cells, it should be recognized that, unlike Na^+ additions, it is very difficult to differentiate between the effects of external proton concentration on the proton gradient and the pH dependence of the carrier (or any biological) process (101). Given this proviso, sucrose uptake at low external sucrose concentrations is markedly inhibited by alkaline pH and, conversely, stimulated by acidic pH (15, 36, 84). The pH dependence of sucrose uptake was also reflected in sucrose translocation in intact sugar beet plants (36); translocation (measured as ^{14}C arrival in a developing sink leaf) of [^{14}C] sucrose at pH 5 from a sugar beet source leaf was decreased approximately 40% upon changing the pH of the sucrose feeding solution to pH 8. Also, tissue autoradiographs showed that sucrose loading into the minor vein phloem is most intense at acid rather than alkaline pH values (15).

In *Chlorella,* Komor & Tanner (71) found pH–dependent K_m values for hexose uptake. This was interpreted in terms of the existence of two conformational states of the carrier: a high affinity, protonated form at acid pH, and a lower affinity, unprotonated form at more alkaline pH. Similarly, Giaquinta (36) and Delrot & Bonnemain (15) found that the K_m for sucrose uptake in sugar beet and *Vicia* leaves increased with increasing alkalinity. The K_m for protons for sucrose uptake in both *Beta* and *Vicia* at low sucrose concentrations was near 0.01 μM, suggesting that the carrier is half maximally protonated at an apoplast pH of 8. It should also be emphasized that determining the effects of proton concentration on sucrose uptake is complicated because of the simultaneous presence of two sucrose uptake systems (see *Kinetics*).

CORRELATION BETWEEN LOADING, H^+ FLUX, AND ATPASE ACTIVITY   As discussed above, the nonpermeable sulfhydryl group modifier PCMBS, (p-chloromercuribenzenesulfonate) was shown to inhibit sucrose uptake and translocation of $^{14}CO_2$-derived assimilates from intact sugar beet leaves (35). This provided support for an apoplastic step of sucrose loading into the phloem. That PCMBS did not inhibit photosynthesis, respiration, or hexose accumulation (35, 40) indicated that its effects were limited to the plasmalemma, possibly at the sucrose carrier. Importantly, the PCMBS inhibition of sucrose uptake could be reversed completely by dithioerythritol (35). Because PCMBS inhibits sucrose uptake and translocation, this modifier has been used to provide further insights into the loading process. The inhibition of sucrose uptake in sugar beet leaves by PCMBS was accompanied by an inhibition of active proton efflux (40) and an inhibition of the cytochemical localization of phloem membrane ATPase activity (R. T. Giaquinta, unpublished results). PCMBS was also shown to inhibit basal and K^+-stimulated ATPase activity of cell membranes isolated from corn roots (43). These observations were taken as evidence for two sites of PCMBS inhibition: one at the sucrose carrier itself, the other at the energy transducing system, presumably the ATPase-mediated proton pump in the phloem membrane.

Membrane potential measurements during sucrose uptake in developing soybean cotyledons (77) clearly demonstrate two sites of inhibition by PCMBS. Lichtner & Spanswick (77) showed that sucrose addition to soybean cotyledons caused an immediate and transient depolarization of the membrane potential consistent with an electrogenic sucrose transport system. Although addition of PCMBS depolarized the membrane potential to the predicted diffusion potential, the membrane potential was completely restored when PCMBS was removed by washing the tissue with PCMBS-free buffer. Interestingly, sucrose addition to the PCMBS-treated tissue (after restoration of the membrane potential) did not induce a membrane depolarization, indicating that PCMBS was inhibiting the sucrosyl carrier itself. The observations by Giaquinta (41) that PCMBS (added directly to the H^+ efflux assay) completely inhibited active proton efflux and sucrose uptake, whereas PCMBS pretreatment (followed by washing in buffer) substantially reversed the inhibition of proton efflux, but not sucrose uptake, are also consistent with PCMBS acting at two sites: the carrier and the membrane potential. The depolarization of the membrane potential by PCMBS could result from either a reversible inhibition of the ATPase or an increase in the passive permeability of the cell membranes to ions (77), or both. The observations that PCMBS and PCMB inhibit *(a)* the cytochemical localization of ATPase in the phloem membranes in sugar beet (R. T. Giaquinta, unpublished) and pea (11), respectively, and *(b)* in vitro ATPase (43, 55) support the former, whereas the PCMBS-induced increase in cation permeability in animal and algal cells is consistent with the latter alternative (75, 77). PCMBS and other mercurials have been shown to inhibit K^+ uptake into root cells without inhibiting cellular respiration (55, 78).

Although PCMBS has been used in the study of sucrose uptake, its effects on the membrane are not well understood (75) and may depend on specific assay conditions. For example, Delrot et al (16) found that PCMBS inhibited sucrose uptake but not proton or potassium transport in *Vicia faba* leaves. The reasons for the discrepancy between these results and the studies cited above are not apparent and further work seems warranted.

While PCMBS has been used to show a correlation between inhibition of sucrose uptake, proton flux, and ATPase activity, the stimulation of these processes by fusicoccin indicates that the converse is also true (14, 40, 85). Fusicoccin, a diterpene glucoside, catalyzes an active H^+/K^+ exchange in plant cells presumably by stimulating a proton pump (86). The stimulation of sucrose uptake by fusicoccin (14, 40, 85) is complex in that the magnitude appears to depend on sucrose concentration, buffer, pH, osmoticum (12), and possibly potassium status of the plant (102). Tissue autoradiographs also show that fusicoccin enhances sucrose accumulation into the minor veins (14).

SUCROSE-DEPENDENT PROTON TRANSPORT   The experiments noted above provide circumstantial evidence for a sucrose/proton cotransport mechanism for phloem loading. As Tanner (101) aptly points out, the most critical evidence for establishing a sucrose/proton system is the demonstration of a sugar-induced net uptake of protons from the medium. More rigorously, the sugar-dependent proton influx should show the same sugar specificity and concentration dependence as the sugar uptake system. It is also important that the net proton uptake be transient, because reestablishement of the proton gradient is a prerequisite for transport (101).

The detailed studies by Komor and colleagues (68-70, 87) and Hutchings (65) on germinating *Ricinus* cotyledons provide the strongest evidence for a sucrose/proton cotransport in higher plants. Because these cotyledons are intimately involved in the absorption of endosperm-derived sucrose, this uptake system is presumed to be analogous to phloem loading in mature leaves (70). These workers have shown that the addition of sucrose to *Ricinus* cotyledons caused an alkalinization (0.1 to 0.2 pH unit) of the external medium which continued for 15-30 min before returning to the original pH value (65, 69). This was consistent with the coupled entry of both sucrose and protons followed by an active proton efflux to reestablish the proton gradient. Importantly, sugars such as methylglucoside, fructose, glucose, and raffinose, which do not compete with the sucrose carrier, did not cause the alkalinization, indicating that the proton cotransport system is specific for sucrose (65, 68). Also, increasing concentrations of sucrose up to 30-50 mM resulted in increased net proton influx. The K_m for the sucrose-induced net proton influx was 5 mM (65, 68) compared to about 25 mM for sucrose uptake. The discrepancy in K_m for proton and sucrose uptake is thought to reflect the presence of diffusion barriers within the tissue (68).

Tissue diffusion barriers also appear to explain the low proton/sucrose stoichiometry (0.3 proton per sucrose) found in *Ricinus*. That the proton/sucrose stoichiometry decreases with increasing sucrose concentration in both *Ricinus* cotyledons (65) and *Vicia* leaves (13) indicates the presence of a second sucrose transport system which is not linked to proton cotransport (see *Kinetics*).

The transient influx of protons during sucrose uptake into *Ricinus* cotyledons is accompanied by a simultaneous efflux of K^+. This K^+ efflux is not mechanistically coupled to sucrose uptake (either by a H^+/K^+ exchanger or by an electrogenic K^+ antiport), but instead, it occurs passively in response to the change in membrane potential caused by an electrogenic sucrose/H^+ influx (8).

The studies on *Ricinus* cotyledons provide compelling evidence for a sucrose proton cotransport system in higher plant tissues. Although autoradiographic studies show that [^{14}C] sucrose accumulates in the minor veins of this tissue, it appears to do so only after ^{14}C-labeling of the nonphloem tissues of the cotyledon (70, 87). While there may be technical reasons for the apparent duality in loading sites, the studies on *Ricinus* do not allow us to state unequivocally that sucrose uptake in this system is occurring solely into the minor vein phloem. The majority of sucrose uptake could occur into non-phloem cells with subsequent transfer to the phloem. It would be interesting to determine whether isolated cells or protoplasts from the cotyledon mesophyll (i.e. nonphloem) contain a sucrose/proton cotransport system. In this regard, Delrot (13) proposed that two sugar cotransport systems exist in *Vicia* leaves: *(a)* a sucrose/H^+ cotransport into the phloem; and *(b)* a hexose/H^+ cotransport into mesophyll. It should be recalled, however, that exogenously supplied sucrose does enter the mesophyll (25, 53). Similar to *Ricinus* cotyledons, addition of 20 mM sucrose to *Vicia* leaf fragments in solutions below pH 5 caused a transient (20 to 40 min) alkalinization (0.05 to 0.1 pH unit) of the medium. PCMBS prevented the sucrose-induced proton transport (13).

The most convincing evidence for a sucrose/proton cotransport mechanism of phloem loading in photosynthesizing leaves comes from Heyser's studies on corn leaves (59). In these studies, the longitudinal vascular bundles of a segment of a mature corn leaf are perfused with 5 mM KCl while continually measuring the pH of the xylem perfusate with a surface microelectrode. This system has several advantages in the study of phloem loading (59): *(a)* unlike developing cotyledons, it represents a mature, photosynthesizing leaf; *(b)* solutions can be introduced into one end of the vascular system and retrieved at the distal end of the segment; *(c)* the vascular bundles are surrounded by a suberized lamella which restricts the movement of the xylem perfusate to the mesophyll; and *(d)* the sieve element-companion cell complex is symplastically isolated from the vascular parenchyma, so phloem loading occurs exclusively from the apoplast. Perfusion of 25 mM sucrose through the vascular system caused a marked and energy-dependent alkalinization (0.6 to 0.9 pH unit) of the perfusate within 10 min (Figure 2). The pH rise was transient and returned to the orginal pH value within 30 to 60 min. Importantly, the alkalinization was specific for sucrose because mannitol, glucose, fructose, galactose, and polyethyleneglycol, all at 25 mM, failed to elicit the pH transient. That the addition of 25 mM sucrose to media containing these sugars resulted in alkalinization (Figure 2) indicated that the sucrose/H$^+$ cotransport system was fully functional in the presence of these sugars and that no competition existed. That Heyser (59) also found a close correlation between the time course of [^{14}C] sucrose uptake and proton influx provides further evidence for a sucrose/H$^+$ cotransport system in phloem loading. However, one cannot determine whether the sugar/proton transport system is operating at the thin- or thick-walled sieve tubes which are present in the vascular bundle of corn leaves.

SUCROSE-DEPENDENT MEMBRANE DEPOLARIZATIONS   Aside from the sucrose-induced alkalinizations in the vascular bundles of corn leaves (59), all the studies cited above are complicated to varying degrees because of the high nonphloem/phloem ratio of tissues that were used. The elegant (and painstaking) experiments of Wright & Fisher (116) show that sucrose entry into the sieve tubes appears to be accompanied by a depolarization of the sieve tube membrane potential. In these studies, microelectrodes were inserted into the exuding phloem sap from severed stylets of aphids which had been feeding on willow bark strips (115, 116). The extent of depolarization of the sieve tube

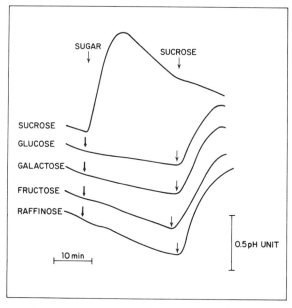

Figure 2 Sucrose-dependent proton transport in corn leaves. Note that only the addition of sucrose results in the alkalinization of the external medium. See text for details; drawn from Heyser (59).

membrane potential was dependent on sucrose concentration and displayed saturation kinetics with a K_m ranging from 2–50 mM. While addition of mannitol, sorbitol, 1–O-methylglucose, ribose, arabinose, and L-glucose gave only marginal depolarizations, fructose, D-glucose, 3–0-methylglucose, and raffinose gave 50 to 70% of the response of sucrose. Unexpectedly, mannose was equally effective as sucrose. To my knowledge, mannose has not been tested in the sucrose-specific transport system of *Ricinus* cotyledons. It would be interesting to determine whether a similar response is observed. The efficacy of hexoses, particularly 3–0-methylglucose (a sugar not thought to be loaded), in eliciting depolarizations of the sieve tube membrane potential is unexpected and somewhat disturbing. However, Lichtner & Spanswick (77) found that addition of glucose and fructose to developing soybean cotyledons also caused a similar membrane depolarization. In agreement with the results of Lichtner & Spanswick (77), the sieve tube membrane potential was depolarized by PCMBS and could be restored after removal of PCMBS. Sucrose additions following restoration of the membrane potential failed to elicit a membrane depolarization (J. P. Wright and D. B. Fisher, personal communication). These results are in accord with those of Lichtner & Spanswick (77) described above.

Wright & Fisher (116) also observed that a light-to-dark transition caused an immediate hyperpolarization (15 to 20 mV) of the sieve tube membrane potential which reached a maximum in 5 min and which slowly returned to the original potential. Reillumination caused a rapid depolarization followed by a slow repolarization. The significance of this to loading is not evident, but Guy et al (53) have recently proposed that light and pH synergistically convert a sugar carrier in mesophyll protoplasts to a fully activated form. The sieve tube membrane may contain a similar system, or the electrical measurements of the sieve tubes are somehow also monitoring the electrical events associated with sugar uptake into the surrounding cells.

ENERGETICS OF UPTAKE   The active accumulation of sucrose against a concentration gradient clearly requires the expenditure of metabolic energy, most likely ATP. The translocation performance parameters in a sugar beet leaf (29, 100) suggest that the ATP requirement for phloem loading could be met by metabolizing only 1.4% of the amount of sucrose that is earmarked for translocation. Stated another way, only 0.3% of the total ATP derived from photosynthate is needed to supply the ATP requirement for loading (assuming 1 ATP/sucrose loaded).

Active transport of sugar in algae and bacteria is accompanied by an increase in respiration rate. The sugar-induced increase in respiration presumably represents the regeneration of cellular ATP which has been utilized to power the transport process. In *Ricinus* cotyledons (which have a potent sucrose transport system), the addition of sucrose also caused an increase in respiration rate [20% above the basal rate (68-70)]. Although, the magnitude of the enhanced respiration was dependent on the concentration of the supplied sucrose, the increase in respiration was not caused by internal sucrose acting as a substrate for respiration (68). Also, the K_m value for the sucrose-induced stimulation of respiration (25 mM) was identical to the K_m for sucrose uptake. Given the measured stoichiometry of 4.2 to 5.5 sucrose molecules accumulated/extra oxygen respired, and assuming a P/O ratio of 3, the ATP requirement for sucrose uptake was calculated to be 1.1 to 1.4 ATP/sucrose (69, 70).

If loading occurs by a sucrose/proton cotransport system as the above data indicate, then the driving force is not a direct coupling between ATP and the sucrose carrier per se, but instead it is the electrochemical potential difference of protons ($\Delta\mu \, H^+$) across the cell membrane, or proton motive force (pmf): $pmf = \Delta\psi - \frac{2.3 \, RT}{F} \Delta pH,$ where R is the gas constant, T is absolute temperature, F is the faraday constant, and $\Delta\psi$ is the membrane potential.

Using the distribution of dimethyloxazolidinedione (DMO) and tetraphenylphosphonium ion (TPP^+) to measure the pH gradient and electrical potential, respectively, Komor (68) concluded that the energy contained in a pH gradient of 1.5 pH units and an "average" membrane potential of -150 mV [sieve tube membrane in willow $= -155$ mV (116)] was more than sufficient to meet the energy required for the hundredfold accumulation of sucrose that was measured in *Ricinus* cotyledons.

MODEL FOR PHLOEM LOADING OF SUCROSE   The schematic in Figure 3 represents a speculative model for sucrose loading across the phloem membranes. It is envisioned that sucrose in the free space interacts with a sucrose-specific carrier protein having SH-sensitive groups exposed at the external membrane surface. Although virtually nothing is known about how an electrochemical gradient interacts with *any* energy transducing system, two models (Figure 3a, b) which are based on neutral solute uptake across bacterial and fungal membranes (96) can be discussed. In model A, the uncharged carrier on the external membrane surface binds with sucrose and proton(s). The binding of cosubstrates may occur sequentially (as depicted) or independently as has been suggested for galactoside transport in *Escherichia coli* (114). The resulting charged ternary complex (carrier-sucrose-H^+) is driven across the membrane in response to the inside negative potential. After the substrate dissociates at the inner membrane surface the carrier returns to the external surface. In model B, the carrier is viewed as being negatively charged. The binding of sucrose and protons (the number of protons determined by the valency of the carrier z) results in a neutral ternary complex which translocates across the membrane. The uncomplexed, negatively charged carrier is returned to the external membrane surface in reponse to the membrane potential.

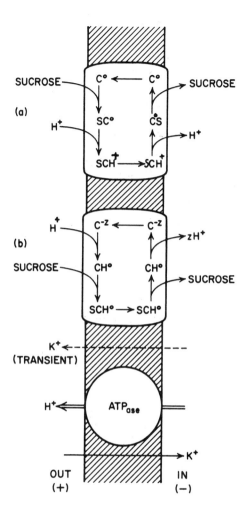

Figure 3   Schematic model for sucrose loading across phloem membranes. See text for details.

The manner in which these events are accomplished obviously is not known. A current view of the facilitated transport of glucose across red blood cell membranes is that glucose binds to a transmembrane protein with its C-1 end at the external side of the membrane and its C-6 end at the cytoplasm side (3). The "alternating conformation" model proposes that the carrier at any one time contains a single sugar binding site which exists in either an outward- or cytoplasmic-facing form, and that transport of the bound sugar results from a conformational change between these two forms.

With regard to membrane conformational changes, protonation of membrane polyelectrolytes by either intra- or transmembrane proton gradients has been proposed to cause conformational changes in energy transducing systems. (17, 44, 47). Likewise, electrochemical gradients and proton concentration have been shown to cause vertical and lateral displacements of proteins within bacterial membranes (1, 9). One can only speculate whether any of these examples are relevant to sucrose loading into the phloem.

As Figure 3 depicts, the driving force for sucrose loading is the electrochemical potential gradient of protons which is established by an active, vectorial, proton-translocating ATPase. The "downhill" movement of protons is thought to be coupled to the secondary active transport of sucrose. The transient K^+ efflux associated with proton influx results from charge compensation during electrogenic sucrose uptake (8). The net accumulation of K^+ into the phloem can occur by a similar mechanism or via a H^+/K^+ exchanging ATPase.

Collectively, the studies cited in this section strongly suggest that sucrose loading into the phloem occurs by a sucrose/proton cotransport mechanism.

Kinetics—Revisited

Kinetic analyses of sucrose uptake from the free space are consistent with the participation of a carrier-mediated process during phloem loading. A note of caution is warranted, however, because we should recognize that given the heterogeneity in cell types present in a leaf and the multiple diffusion barriers within the tissue, it is exceedingly difficult to interpret transport kinetics solely in terms of phloem loading.

Studies on [^{14}C] sucrose uptake into abraded surfaces of sugar beet leaves in relation to export have demonstrated a biphasic concentration dependence for uptake and translocation. Sucrose uptake in these studies was interpreted in terms of two saturable transport systems (100). The high affinity system (System I) which operated at external sucrose below 50-100 mM displayed a K_j (K_m) and J_{max} (V_{max}) for sucrose of 16 mM and 70 µg C (490 nmol)• min^{-1}• dm^{-2}, whereas the K_j and J_{max} for the low affinity system (System II) observed at high sucrose (100-400 mM) were 620 mM and 330 µg C•min^{-1}•dm^{-2}, respectively (100). Interestingly, both systems produced loading into the translocation stream. This indicated that sucrose uptake at high external sucrose concentrations was not simply reflecting transport into a terminal storage compartment such as the vacuoles of the mesophyll or phloem parenchyma. Additionally, sucrose uptake at the higher concentrations did not appear to be simple diffusions because (a) the high concentration sucrose concentration of the phloem (0.8M) results in a gradient in the opposite direction, and (b) low temperature and anoxia inhibit this sucrose uptake by 60-80% (43, 100).

We now recognize that exogenously supplied sucrose can enter both the mesophyll and the phloem (25, 53), so it is unlikely that kinetic analyses reflect solely the transport mechanism operating at the sieve element-companion cell complex. Several studies show that the biphasic concentration dependence of sucrose uptake may actually be composed of a saturable uptake system which operates at low sucrose concentrations and a linear (rather than a saturable) component which predominates at high external sucrose concentrations. In germinating *Ricinus* cotyledons, Komor (69) observed a linear "diffusion-like" uptake process superimposed on a saturable carrier-mediated transport system (K_m = 25 mM). Similarly, sucrose uptake into developing soybean cotyledons showed a nonsaturable linear component which dominates total sucrose uptake into the cotyledons at high external sucrose concentrations (76). It is interesting to note that earlier studies by Kriedemann & Beevers (72) demonstrated that sucrose uptake into germinating *Ricinus* cotyledons which were attached to the embryonic axis of the intact seedling showed a biphasic response to increasing sucrose concentration. Up to 100 mM sucrose, the uptake rate showed satura- tion kinetics. Between 100 and 500 mM sucrose, sucrose uptake was linearly dependent on sucrose concentration. Interestingly, cotyledons which had their embryonic axis removed showed a similar concentration dependence for suc- rose as did attached cotyledons up to 100 mM sucrose. However, the cotyledons minus the axis lacked the linear component of sucrose uptake at higher sucrose concentrations. That sucrose concentrations between 100-500 mM failed to increase the uptake rate into cotyledons which lacked a strong translocation sink, argues against the view that the linear phase is simply diffusion in a terminal storage pool. Kriedemann & Beevers (72) suggest that the low concentration, saturable system, represents sucrose uptake and utiliza- tion into the cotyledons, whereas uptake at high sucrose concentrations occurs directly into the phloem. This would be consistent with the observation that sucrose which is taken up into sugar beet leaves at high external sucrose concentrations is readily available for translocation (100). In this regard, isolated vascular bundles from sugar beet petioles, which contain both phloem and nonphloem cells, also show a saturable and linear component for sucrose uptake with increasing sucrose concentration (110). The saturable component, operating at sucrose concentrations below 25 mM, displays a K_m for sucrose of 12.5 mM, is inhibited by dinitrophenol and phlorizin, and transports sucrose against a concentration gradient (110). The linear component was suggested to occur by carrier-mediated facilitated diffusion and to function in vivo in the uptake or retrieval of sucrose into the vascular parenchyma after its release from the sieve tubes (a region of high sucrose concentration). It should be recognized that at high external concentrations sucrose could be entering the phloem directly as proposed or indirectly via nonphloem cells. In fact, recent autoradiographic studies on sucrose uptake into *Ricinus* cotyledons (70, 87) suggest that sucrose at concentration ≤ 30 mM first enters the mesophyll of the cotyledon prior to transport to and accumulation within the phloem.

Detailed studies by Maynard & Lucas (88, 89) on sucrose uptake into sugar beet leaf discs also show that total sucrose uptake can be resolved into two transport systems: one showing Michaelis-Menten saturation kinetics, the other a linear component which dominates at high external sucrose concentrations. The saturable uptake process dominating at the lower sucrose concentrations is thought to reflect the active sucrose/H^+ transport mechanism envisioned for phloem loading and sucrose uptake, while the in vivo significance (if any) of the linear component remains to be established. In fact, the linear phase of uptake is found in non phloem tissues (89), and it may represent a phase of transport that is only present during the feeding of exogenous sugar. Maynard & Lucas (89) suggest that this linear component of sugar transport may represent a general retrieval system present in all plant cells. The biphasic concentration dependence of translocation in sugar beet in response to exogenous sucrose (100) may result from sucrose entry into the phloem via an apoplastic and symplastic route (mesophyll→phloem parenchyma→phloem).

The saturable and linear components apparently can be separated on the basis of their differential sensitivity to proton concentration, PCMBS, anoxia, and competing sugars. For example, Giaquinta (36) observed that the rate of sucrose uptake in sugar beet leaves was markedly pH dependent. This pH dependence, however, was influenced by the external sucrose concentration. At low sucrose concentrations (0.1 to 5 mM) there was greater than a 50% inhibition of sucrose uptake at pH 8 compared to pH 5, whereas at high sucrose concentrations (50-100 mM) the pH dependence of uptake was marginal. Similarly, Delrot & Bonnemain (15) observed that sucrose uptake into *Vicia* leaves was inhibited by alkaline pH at sucrose concentrations less than 5 mM, while sucrose uptake was insensitive to pH at sucrose concentrations exceeding 20 mM. Sucrose uptake into germinating *Ricinus* cotyledons was also sensitive to pH at 5 mM sucrose (72), but not at a sucrose concentration of 30 mM (68). Likewise, sucrose uptake into developing soybean cotyledons at high sucrose concentrations was insensitive to pH between pH 5 and 8, whereas at 1 mM, sucrose uptake is reduced by alkaline pH (76,103). In contrast to these studies, Maynard & Lucas (103) observed that sucrose uptake into sugar beet leaves was inhibited by alkaline pH even at 200 mM external sucrose concentration. The reason for the discrepancy between this study and the ones mentioned previously is not readily apparent, although assay conditions or the nutrient status of the plant material may be involved. Thus, although it appears that sucrose uptake at high sucrose concentrations is not markedly sensitive to pH (and thus may not reflect a sugar/proton cotransport system), this still needs to be firmly established.

As mentioned in the preceding section, the nonpenetrating sulfhydryl reagent PCMBS markedly inhibits sucrose uptake, proton fluxes, phloem membrane ATPase activity, and sieve tube membrane electrical potential. That PCMBS markedly inhibits the saturable component with only marginal inhibition of the linear component in sugar beet leaves (89, R.T. Giaquinta, unpublished data) and in developing soybean embryos (76,103) supports the contention that the saturable component operating at low external sucrose concentrations reflects the active sucrose/proton cotransport system. This view is further strengthened by the observation that maltose at high concentrations (100 and 400 mM) competitively inhibits the linear portion component of uptake (89), even though this dissacharide does not compete with the sucrose/proton cotransport system (59, 68).

Interestingly, Maynard & Lucas (89) observed that the saturable system was insensitive to anoxia, whereas the linear component was inhibited by anaerobiosis. The lack of marked inhibition of the saturable component by anoxia is an interesting and unexpected result, given the known energy dependence of loading along with other studies which show that loading and transport are oxygen sensitive (100). It should be noted that in vivo loading of $^{11}CO_2$-derived assimilates in C_4 plants (as inferred from the rate of ^{11}C export from a photosynthesizing leaf) was also insensitive to anoxia, whereas loading and export in C-3 plants were markedly inhibited by anoxia (105). The reason for the apparent differences in oxygen sensitivity in C_3 versus C_4 plants is not known. One wonders about the oxygen permeability into various suberized cell types and possibly the ATP content of the different cell types. It may be relevant that sugar loading into phloem in C_4 plants, unlike C_3 plants, has been suggested to occur down a concentration gradient (80,106). Lush & Evans (80) calculated that loading down a concentration gradient was feasible as long as the speed of translocation exceeded 200 $cm \cdot hr^{-1}$. Troughton and co-workers (106), using ^{11}C techniques, have shown that translocation velocities in corn can equal and exceed this value of 200 $cm \cdot hr^{-1}$. More interestingly, they also showed a biphasic relationship between translocation velocity and endogenous leaf sucrose concentration. These authors suggested that at sucrose concentrations below 8% (dry weight basis), loading may occur by an active accumulation of sucrose into the phloem, whereas at high leaf sucrose concentrations (perhaps in the bundle sheath) loading may occur down a concentration gradient. The sucrose-dependent alkalizations found by Heyser (59) in corn leaves are consistent with an active sucrose/H^+ cotransport mechanism from the apoplast at a free space sucrose concentration of 25 mM. Loading at higher sucrose concentrations probably occurs by a facilitated transport mechanism rather than by diffusion via a symplastic pathway, because the companion cell-thin wall sieve tube complex is essentially symplastically isolated from both the vascular parenchyma and bundle sheath (19,59).

Although it is tempting to speculate about the significance of transport kinetics in relation to loading, not enough information is present at this time to couple these transport kinetics mechanistically with a particular model for sucrose loading or perhaps even, to phloem loading itself.

Control Mechanisms

Control of sucrose export from leaves can be exerted at the level of *(a)* production and partitioning of photosynthate; *(b)* intra- and inter-cellular compartmentation of assimilates; *(c)* localized entry of sucrose into the apoplast near the loading sites; and *(d)* the active loading mechanism on the sieve element-companion cell complex. Although I plan to limit my remarks to possible control mechanisms that may operate at the loading sites, it is important to stress that translocation within the plant is a well-integrated and orchestrated process. Thus, when addressing the topic of regulation no single site can be treated independently from the entire system.

It should also be recognized that, based on the kinetics of sucrose uptake, the capacity of the phloem membranes to load in vivo concentrations of sucrose from the free space may not be rate-limiting. However, because loading is viewed to occur from the apoplast, the loading system would seem to play an important transducing role for mediating changes in export in relation to changes in sink demand.

Our understanding of the loading process is too meager to provide anything more than sheer speculation as to the nature of control mechanisms operating at the phloem membranes. As stated previously (32), this speculative approach (in some cases based on transport phenomena in other systems) may provide some testable hypotheses which will spur research into this uncharted area of phloem physiology.

Given the characteristics of the phloem itself, it is not too surprising that internal sucrose concentration, turgor changes, and water fluxes have been proposed as likely candidates for mediating phloem loading, especially in response to changes in sink demand. High sucrose concentrations within the phloem have the potential of inhibiting the further accumulation of sucrose through "transinhibition." In "transinhibition" high internal concentrations of solutes are presumed, through mass action, to prevent the dissociation of the hypothetical carrier-substrate complex on the inside surface of the plasmalemma (Figure 3). This sequestering of the putative carrier-substrate complex at the inner surface of the membrane results in less free carrier at the outside surface for additional substrate binding and transport. High internal concentrations of solutes such as amino acids, sugars, and ions have been reported to alter the further entry of these solutes in bacteria, algae, yeast, and animal cells (see 42, 68, and references therein). Similarly, high internal sucrose concentrations (achieved by preloading of tissues) have been shown to inhibit the uptake of [^{14}C] sucrose in castor bean cotyledons (68) and sugar beet source leaves (42). Conversely, low internal sucrose concentrations stimulated sucrose transport into these tissues. Although these results can be interpreted as control of phloem loading by internal sucrose concentration, the relevance of these studies to in vivo loading still remains to be established.

Internal solute concentrations have also been proposed to exert allosteric control of transport carriers. Based on transport kinetics, Glass & Dunlop (49) and Hodges (61) have suggested models for K^+ transport in plant roots in which the binding of K^+ to allosteric sites on the carrier complex induces a conformational change which reduces the affinity of the carrier toward additional substrate (K^+). Hodges (61) found that K^+ uptake and K^+-stimulated, Mg^{+2}-dependent plasmamembrane ATPase activity displayed kinetics which were consistent with a carrier showing negative cooperativity toward K^+. Similarly shaped Eadie-Hofstee plots were also found for sucrose uptake in sugar beet source leaves (43). However, interpretation of transport kinetics in a heterogeneous tissue in terms of enzyme theory is, at best, exceedingly difficult. It is interesting to note, however, that recent (but unconfirmed) kinetic studies on a plasmamembrane ATPase in sugar beet roots (79) purport to show that sucrose allosterically affects in vitro ATPase activity in the presence of K^+ or Na^+. Sucrose in the presence of the substrate MgATP and Na^+ was suggested to activate ATPase activity. Lindberg (79) cites studies in animal tissues which also purport to show an allosteric interference between actively transported monosaccharides and membrane Na^+/K^+ ATPase activity. The effect of sugar in the above studies was dependent on the Na/K ratio. It is also noteworthy that monosaccharide transport in animal cells is believed to occur via a sugar/Na^+ cotransport. One can only speculate as to whether the proton-translocating ATPase proposed for phloem membranes also represents a cation-dependent, sucrose-sensitive system.

In contrast to internal sucrose concentration, other studies indicate that phloem loading responds to changes in sieve tube turgor (97-99). It is envisioned that an increase in sink demand would decrease phloem turgor, and this decrease would be rapidly propagated along the sieve tube pathway to the loading sites. The decrease in cell turgor would in turn cause compensatory changes in sucrose and K^+ loading. Smith & Milburn (97) found that the decrease in phloem sucrose concentration in the exudate of *Ricinus* plants (which were placed in continual darkness) was accompanied by an increase in phloem K^+ levels and partial maintenance of phloem turgor. That phloem turgor was maintained at low internal sucrose concentrations by an increase in K^+ loading was interpreted as an osmoregulatory response during periods of reduced sucrose availability. In related experiments (98, 99), the rates of sucrose and K^+ loading were able to respond to a reduction in sieve tube turgor caused by stem incision which created an artificially high sink demand. The reduction of the phloem pressure potential to near zero at the site of incision was thought to be rapidly propagated along the phloem to the loading sites where the decrease in solute potential caused a compensatory increase in loading. Thus, changes in sink demand in the intact plant may regulate phloem loading through changes in turgor at the loading sites. It is interesting to call attention to the previously mentioned studies of Kriedemann & Beevers (72), which showed that removal of a strong translocation sink (embryonic axis) from developing *Ricinus* cotyledons resulted in the elimination of the linear (but not the saturable) component of sucrose uptake. This system may be extremely useful in studying the effects of sink demand on loading. For instance, it would be interesting to know whether the internal sucrose concentration of the cotyledons increased after sink removal or whether [^{14}C] sucrose uptake into the veins (as measured by tissue autoradiography) was less in the detached cotyledons compared to cotyledons which contained the embryonic axis. In this regard, preliminary results from my laboratory show that an increase in sink demand in sugar beet plants is accompanied by an increase in sucrose uptake into source leaf tissue.

Because lateral water flow and sucrose loading into the phloem appear to be coupled (90), it is also possible that the increase in loading in response to a decrease in turgor results from an increase in the rate of free space water movement into the sieve tubes (31). Given the paucity of data and our lack of understanding of the loading mechanism in vivo, it is not possible to rule out unequivocally any of the above osmotic mechanisms in relation to control of loading.

Changes in the difference of hydrostatic pressure across a membrane are probably too small to be a significant force for solute transfer (see 42), for any pressure or turgor difference must be transduced and amplified at the membrane level. Based on the data emerging on the mechanism of phloem loading, the membrane potential (possibly via the ATPase) seems a likely candidate for this transducing mechanism. In this regard, pressure-sensing mechanisms have been suggested in plant cell membranes (10). Turgor changes have also been linked to membrane conformational changes as well as changes in membrane potential (32). It remains to be established whether turgor-induced changes in membrane potential are in any way related to phloem loading. Also, as mentioned in a previous section, sucrose loading may be operating in tandem with sucrose efflux from the mesophyll or phloem parenchyma. This should not be overlooked as a control or feedback mechanism.

Elucidating the control mechanisms associated with phloem loading promises to be an exciting area for further research. The role of plant hormones and ions in relation to sucrose efflux from the mesophyll or phloem parenchyma and sucrose loading into the phloem deserve serious consideration and study.

Development of Loading Capacity

Several studies have documented that developing sink leaves begin to export photosynthate at the time they acheive 30-50% of their maximal size (21, 107-109). The import-to-export transition in dicotyledon leaves begins at the leaf tip and develops basipetally through the leaf blade. Although structural development of the minor vein phloem coincides with the commencement of assimilate export in certain species, structural maturation of the phloem alone does not appear to govern export (109). Tissue autoradiographic studies by Fellows & Geiger (21) in developing sugar beet leaves show that the minor vein phloem of sink leaves was not able to accumulate exogenously supplied [^{14}C] sucrose. In the sugar beet leaves, export is thought to be related to the ability of the minor veins to accumulate sugar above a threshold solute concentration necessary to cause mass flow from the leaf.

Biochemical differentiation of the phloem membranes also coincides with the acquisition of loading and export capacity. The increase in membrane surface area resulting from the development of wall ingrowths in transfer cells was correlated with acquisition and development of export in *Pisum* leaves (51). Transfer wall ingrowths failed to develop in either dark-grown leaves or leaves of albino mutants of peas, or in nonchlorophyll-containing regions of variegated leaves. The increase in membrane surface area of the transfer cells at the onset of export is also accompanied by the appearance of membrane ATPase activity (11). Because phloem membranes have an unusual arrangement of freeze-etch particles on the outward facing side of the inner transfer cell membrane (5), it would be interesting to determine whether distinct changes in particle appearance or arrangement are also correlated with the acquisition of membrane ATPase and sucrose-loading capacity. If so, this may indicate that loading commences with the insertion of a sucrose transport carrier complex in the phloem membrane. This possibility warrants consideration, because studies in *Chlorella* show that the induction of glucose transport (after a few hours in the presence of the "inducer" glucose) is accompanied by the synthesis and appearance of a 30 kilodalton, SH-containing intrinsic membrane protein (22). This is interesting because recent studies have demonstrated that application of 2% glucose to *Pisum* leaves induced transfer cell wall ingrowths in the dark (58). The onset of export is accompanied by the appearance of sucrose phosphate synthetase and sucrose synthesis (39, 42), so it is very tempting to speculate that the "inducer" for the development of the phloem membrane carrier system is the sucrose concentration in the free space (43). Clearly, this is a testable hypothesis.

A similar development of transport function was seen in germinating *Ricinus* cotyledon (68), a system which apppears to closely approximate phloem loading in leaves. The ability of a *Rincinus* cotyledon to accumulate exogenous sucrose develops within the first few days of germination and reaches a maximum activity (on a fresh weight basis) on the fourth day. Transport capacity dramatically decreases by day 6. The rapid onset and decline of transport capacity is consistent with a net systhesis and degradation of the sucrose transport system (68). This view is supported by studies on the inducible hexose system in *Chlorella*, which show that the decline in glucose uptake and disappearance of a specific membrane protein occurs with a half time of 4 hours after the removal of the "inducer" (glucose). This argues for turnover of the sugar carrier protein rather than simply inactivation of the carrier (22).

If, as is the case for sugar uptake in bacteria, algae (22), and germinating *Ricinus* cotyledons (68), the sucrose carrier on the phloem membrane is turning over, and its presence is dependent on the continual presence of "inducer", then phloem loading in a senescing leaf becomes an interesting system to study. Because leaf senescence is accompanied by a decline in photosynthesis (i.e. sucrose production) and a degradation of leaf protein into amino acids for export (113), it would be interesting to determine whether the ability to load and translocate sucrose is differentially impaired relative to amino acids. Again, this is readily testable. Studies on the development (and degradation) of the phloem transport system promise to provide important insights into the nature and regulation of the loading process.

Summary

In this review I have addressed the structural and metabolic correlates of the transfer and accumulation of sucrose within the phloem. Although several lines of evidence are consistent with a localized entry of sugars into the apoplast prior to active loading into the phloem by a sucrose/proton cotransport mechanism, the data have been obtained from only a few species, and our understanding of the mechanisms and controls is far from complete. Throughout the review I have also attempted to identify areas where research is needed and have suggested some specific lines of experimentation, with the hope that this will stimulate additional research.

Clearly, we are in the early stages of "discovering the rules" of assimilate transport and partitioning. The challenge to current and future students of translocation is to devise research strategies and methodologies that will enable one to probe relevant mechanistic processes and to extend these approaches and findings to intact, translocating systems. Only then can we understand and appreciate the intricacies and coordination involved in translocation.

ACKNOWLEDGMENTS

I gratefully acknowledge the assistance of T. Sparre and N. L. Sadler in preparing this manuscript. I also thank Dr. J. H. Thorne for reading the manuscript.

Literature Cited

1. Amar, A., Rottem, S., Razin, S. 1978. Disposition of membrane proteins as affected by changes in the electrochemical gradient across mycoplasma membranes. *Biochem. Biophys. Res. Commun.* 84:306–12
2. Baker, D. A. 1978. Proton co-transport of organic solutes by plant cells. *New Phytol.* 81:485–97
3. Baldwin, S. A., Lienhard, G. E. 1981 Glucose transport across plasmamembranes: facilitated diffusion systems. *Trends Biochem. Sci.* 6:208–11
4. Bentwood, B. J., Cronshaw, J. 1978. Cytochemical localization of adenosine triphosphatase in the phloem of *Pisum sativum* and its relation to the function of transfer cells. *Planta* 140:111–20
5. Briarty, L. G. 1973. Repeating particles associated with membranes of transfer cells. *Planta* 113:373–77
6. Brovchenko, M. I. 1976. Energy dependence of assimilate evacuation into the apoplast and loading of conducting system terminals of the leaf *Fiziol.Rast.* 24:327–34
7. Brovchenko, M. I., Slobodlskaya, G. A., Chmora, S. N., Lipatova, T. F. 1976. Effects of CO_2 and O_2 on photosynthesis and photosynthesis-linked escape of assimilates into the void space of the leaf in sugar beet. *Sov. Plant Physiol.* 233:1232–40
8. Cho. B.H., Komor, E. 1980. The role of potassium in charge compensation for sucrose-proton-symport by cotyledons of *Ricinus communis*. *Plant Sci. Lett.* 17:425–35
9. Copps, T. P., Chelack, W. S., Petkau, A. 1976. Variation in distribution of membrane particles in *Acholeplasma laidlawii B* with pH. *J. Ultrastruct. Res.* 55:1–3
10. Coster, H. G. L., Steudle, E., Zimmermann, U. 1976. Turgor pressure sensing in plant cell membranes. *Plant Physiol.* 58:636–43
11. Cronshaw, J. 1981. Phloem structure and function. *Ann. Rev. Plant Physiol.* 32:465–84
12. Delrot, S. 1981. *Etude des méchanismes de l'absorption des glucides par les tissus foliaires et de leur accumulation dans les Nervures.* PhD thesis. Univ. Poitiers, France
13. Delrot, S. 1981. Proton fluxes associated with sugar uptake in *Vicia faba* leaf tissues. *Plant Physiol.* 68:706–11
14. Delrot, S., Bonnemain, J. L. 1978. Etude du mecanisme de l'accumulation

des products de la photosynthese dans les nervures. *C.R. Acad. Sci.* 287:125–30
15. Delrot, S., Bonnemain, J. L. 1981. Involvement of protons as a substrate for the sucrose carrier during phloem loading in *Vicia faba* leaves. *Plant Physiol.* 67:560–64
16. Delrot, S., Despeghel, J. P., Bonnemain, J. L. 1980. Phloem loading in *Vicia faba* leaves: Effect of N-ethylmaleimide and parachloromercuribenzenesulfonic acid on H^+ extrusion, K^+ and sucrose uptake. *Planta* 149:144–48
17. Dilley, R. A., Giaquinta, R. T. 1975. H^+ ion transport and energy transduction in chloroplasts. *Curr. Top. Memb. Transp.* 7:49–107. New York/San Francisco/London: Academic
18. Doman, D. C., Geiger, D. R. 1979. Effect of exogenously supplied foliar potassium on phloem loading in *Beta vulgaris* L. *Plant Physiol.* 64:528:33
19. Evert, R. F., Eschrich, W., Heyser, W. 1977. Distribution and structure of the plasmodesmata in mesophyll and bundle-sheath cells of *Zea mays* L. *Planta* 136:77–89
20. Evert, R. F., Eschrich, W., Heyser, W. 1978. Leaf structure in relation to solute transport and phloem loading in *Zea mays* L. *Planta* 138:279–94
21. Fellows, R. J., Geiger, D. R. 1974. Structural and physiological changes in sugar beet leaves during sink to source conversion. *Plant Physiol.* 54:877–85
22. Fenzl, F., Decker, M., Haass, D., Tanner, W. 1977. Characterization and partial purification of an inducible protein related to hexose proton co-transport of *Chlorella vulgaris*. *Eur. J. Biochem.* 72:509–14
23. Fisher, D. B. 1967. An unusual layer of cells in the mesophyll of the soybean leaf. *Bot. Gaz* 128:215–18
24. Fisher, D. B., Housley, T. L., Christy, A. L. 1978. Source pool kinetics for ^{14}C-photosynthate translocation in morning glory and soybean. *Plant Physiol.* 61:291–95
25. Fondy, B. R., Geiger, D. R. 1977. Sugar selectivity and other characteristics of phloem loading in *Beta vulgaris* L. *Plant Physiol.* 59:953–60
26. Franceschi, V. R., Giaquinta, R. T. 1983. Paraveinal mesophyll of soybean leaves in relation to assimilate transfer and compartmentation. I. Ultrastructure

and histochemistry during vegetative development. *Planta.* 157: In press

27. Francheschi, V. R., Giaquinta, R. T. 1983. Paraveinal mesophyll of soybean leaves in relation to assimilate transfer and comparmentation. II. Structural, metabolic and compartmental changes during reproductive growth. *Planta.* 157: In press

28. Gamalei, Yu. V., Pakhomova, M. V. 1980. Distribution of plasmodesmata and parenchyma transport of assimilates in the leaves of several dicots. *Fiziol. Rast.* 28:901–12

29. Geiger, D. R. 1975. Phloem loading. In *Transport in Plants I,* ed. M. H. Zimmermann, J. A. Milburn, pp. 395–450. New York: Springer. 535pp

30. Geiger, D. R. 1979. Control of partitioning and export of carbon in leaves in higher plants. *Bot. Gaz* 140:241–48

31. Geiger, D. R., Fondy, B. R. 1980. Response of phloem loading and export to rapid changes in sink demand. *Ber. Dtsch. Bot. Ges.* 93:177–86

32. Geiger, D. R., Giaquinta, R. T. 1982. Translocation of photosynthates. In *Photosynthesis: CO₂ Assimilation and Plant Productivity,* ed. Govindjee, pp. 345–86. New York: Academic.

33. Geiger, D. R., Giaquinta, R. T., Sovonick, S. A., Fellows, R. J. 1973. Solute distribution in sugar beet leaves in relation to phloem loading and translocation. *Plant Physiol.* 52:585–89

34. Geiger, D. R., Sovonick, S. A., Shock, T. L., Fellows, R. J. 1974. Role of free space in translocation in sugar beet. *Plant Physiol.* 54:892–98

35. Giaquinta, R. T. 1976. Evidence for phloem loading from the apoplast: Chemical modification of membrane sulfhydryl groups. *Plant Physiol.* 57:872–75

36. Giaquinta, R. T. 1977. Phloem loading of sucrose. pH Dependence and selectivity. *Plant Physiol.* 59:750–53

37. Giaquinta, R. T. 1977. Possible role of pH gradient and membrane ATPase in the loading of sucrose into the sieve tubes. *Nature* 267:369–70

38. Giaquinta, R. T. 1977. Sucrose hydrolysis in relation to phloem translocation in *Beta vulgaris. Plant Physiol.* 60:339–43

39. Giaquinta, R. T. 1978. Source and sink leaf metabolism in relation to phloem translocation. *Plant Physiol.* 61:380–85

40. Giaquinta, R. T. 1979. Phloem loading of sucrose: involvement of membrane ATPase and proton transport. *Plant Physiol.* 63:744–48

41. Giaquinta, R. T. 1980. Sucrose/proton

cotransport during phloem loading and its possible control by internal sucrose concentration. In *Plant Membrane Transport: Current Conceptual Issues,* ed. R. M. Spanswick, W. J. Lucas, J. Dainty, pp. 273–82. Amsterdam: North-Holland, Elsevier. 670 pp.

42. Giaquinta, R. T. 1980. Translocation of sucrose and oligosaccharides. In *The Biochemistry of Plants,* 3:271–320. New York: Academic

43. Giaquinta, R. T. 1980. Mechanism and control of phloem loading of sucrose. *Ber. Dtsch. Bot. Ges.* 93:187–201

44. Giaquinta, R. T., Dilley, R. A. 1977. Chemical modification of chloroplast membranes. In *Encyclopedia of Plant Physiology,* ed. A. Trebst, M. Avron, 5:297–303. Berlin/Heidelberg: Springer

45. Giaquinta, R. T., Geiger, D. R. 1977. Mechanism of cyanide inhibition of phloem translocation. *Plant Physiol.* 59:178–80

46. Giaquinta, R. T., Lin, W., Sadler, N. L., Franceschi, V. R. 1983. Pathway of phloem unloading of sucrose in corn roots. *Plant Physiol.* In press

47. Giaquinta, R. T., Ort, D. R., Dilley, R. A. 1975. The possible relationship between a membrane conformational change and photosystem II dependent hydrogen ion accumulation and ATP synthesis. *Biochemistry* 14:4392–96

48. Gifford, R. M., Evans, L. T. 1981. Photosynthesis, carbon partitioning, and yield. *Ann. Rev. Plant Physiol* 32:485–509

49. Glass, A. D. M., Dunlop, J. 1979. The regulation of K⁺ influx in excised barley roots. *Planta* 145:395–97

50. Gunning, B. E. S. 1976. The role of plasmodesmata in short distance transport to and from the phloem. In *Intercellular Communication in Plants: Studies on Plasmodesmata,* ed. B. E. S. Gunning, A. W. Robards, pp. 203–27. Berlin: Springer

51. Gunning, B. E. S., Pate, J. S. 1974. Transfer cells In *Dynamic Aspects of Plant Ultrastructure,* ed. A. W. Robards, pp. 441–80. London: McGraw-Hill

52. Gunning, B. E.S., Pate, J. S., Minchin, F. R., Marks, I. 1974. Quantitative aspects of transfer cell structure in relation to vein loading in leaves and solute transport in legume nodules. *Symp. Soc. Exp. Biol.* 28:87–126

53. Guy, M., Reinhold, L., Rahat, M., Seiden, A. 1981. Protonation and light synergistically convert plasmalemma sugar carrier system in mesophyll proto-

plasts to its fully activated form. *Plant Physiol.* 67:1146–50

54. Hawker, J. S., Marschner, H., Downton, W. J. S. 1974. Effects of sodium and potassium on starch synthesis in leaves. *Aust. J. Plant Physiol.* 1:491–501

55. Hendrix D. L., Higinbotham, N. 1974. Heavy metals and sulfhydryl reagents as probes of ion uptake in pea stem. In *Membrane Transport in Plants,* ed. U. Zimmermann, J. Dainty, pp. 412–17. New York: Springer

56. Hendrix, J. E. 1973. Translocation of sucrose by squash plants. *Plant Physiol.* 52:688–89

57. Hendrix, J. E. 1977. Phloem loading in squash. *Plant Physiol.* 60:567–69

58. Henry, Y., Steer, M. W. 1980. A reexamination of the induction of phloem transfer cell development in pea leaves *(Pisum sativum)*. *Plant Cell Environ.* 3:377–80

59. Heyser, W. 1980. Phloem loading in the maize leaf. *Ber. Dtsch. Bot. Ges.* 93:221–28

60. Heyser, W., Evert, R. F., Fritz, E., Eschrich, W. 1978. Sucrose in the free space of translocations maize leaf bundles. *Plant Physiol.* 62:491–94

61. Hodges, T. K. 1973. Ion absorption by plant roots. *Adv. Agron.* 25:163–207

62. Huber, S. C., Moreland, D. E. 1980. Translocation: efflux of sugars across the plasmalemma of mesophyll protoplasts. *Plant Physiol.* 65:560–62

63. Huber, S. C., Moreland, D. E. 1981. Co-transport of potassium and sugars across the plasmalemma of mesophyll protoplasts. *Plant Physiol.* 67:163–69

64. Humphreys, T. E. 1980. Sugar proton cotransport and phloem loading. *What's New in Plant Physiol.* 11:9–12

65. Hutchings, V. M. 1978. Sucrose and proton cotransport in *Ricinus* cotyledons. I. H$^+$ influx associated with sucrose uptake. *Planta* 138:229–35

66. Kaiser, W. M., Paul, J. S., Bassham, J. A. 1979. Release of photosynthates from mesophyll cells *in vitro* and *in vivo*. *Plant Physiol.* 94:377–85

67. Kholodova, V., Sokolova, S., Turkina, M., Meshcherjakov, A. 1976. Transport and accumulation of di- and monosaccharides in sugar beet tissues. *Wiss. Z. Humboldt-Univ. Berl. Math. Naturwiss. Reihe* 25:127–32

68. Komor, E. 1977. Sucrose uptake by cotyledons of *Ricinus communis L.*: characteristics, mechanism, and regulation. *Planta* 137:119–31

69. Komor, E., Rotter, M., Tanner, W. 1977. A proton-cotransport system in a

higher plant: sucrose transport in *Ricinus communis*. *Plant Sci. Lett.* 9:153–62

70. Komor, E., Rotter, M., Waldhauser, J., Martin, E., Cho, B. H. 1980. Sucrose proton symport for phloem loading in the *Ricinus* seedling. *Ber. Dtsch. Bot. Ges.* 93:211–19

71. Komor, E., Tanner, W. 1975. Simulation of a high- and low-affinity sugar uptake system in *Chlorella* by a pH-dependent change in the K_m of the uptake system. *Planta* 123:195–98

72. Kriedemann, P., Beevers, H. 1967. Sugar uptake and translocation in the castor bean seedling. I. Characteristics of transfer in intact and excised seedling. *Plant Physiol.* 42:161–73

73. Kuo, J., O'Brian, T. P., Canny, M. J. 1974. Pit-field distribution, plasmodesmatal frequency, and assimilate flux in the mestone sheath cells of wheat leaves. *Planta* 121:97–118

74. Kursanov, A. L., Brovchenko, M. I. 1970. Sugars in the free space of leaf plates: their origin and possible involvement in transport. *Can. J. Bot.* 48:1243–50

75. Lichtner, F. T., Lucas, W. J., Spanswick, R. M. 1981. Effects of sulfhydryl reagents in the biophysical properties of the plasmalemma of *Chara corallina*. *Plant Physiol.* 68:899–904

76. Lichtner, F. T., Spanswick, R. M. 1981. Sucrose uptake by developing soybean cotyledons. *Plant Physiol.* 68:693–98

77. Lichtner, F. T., Spanswick, R. M. 1981. Electrogenic sucrose transport in developing soybean cotyledons. *Plant Physiol.* 67:869–74

78. Lin, W. 1980. Corn root protoplasts: isolation and general characterization of ion uptake properties. *Plant Physiol.* 66:550–54

79. Lindberg, S. 1982. Sucrose and ouabain effects on the kinetic properties of a membrane-bound (Na$^+$K$^+$Mf^{2+}) AtPase in sugar beet roots. *Plant Physiol.* 54:455–60

80. Lush, W. M., Evans, L. T. 1974. Translocation of photosynthetic assimilate from grass leaves, as influenced by environment and species. *Aust. J. Plant Physiol.* 1:417–31

81. Lüttge, U., Schöch, E. V., Ball, E. 1974. Can externally applied ATP supply energy to active ion uptake mechanisms of intact plant cells? *Aust. J. Plant Physiol.* 1:211–20

82. Madore, M., Webb, J. A. 1981. Leaf free space analysis and vein loading in *Cucurbita pepo*. *Can. J. Bot.* 59:2550–57

83. Madore, M., Webb, J.A. 1982. Stachyose systnesis in isolated mesophyll cells of *Cucurbita pepo*. *Can J. Bot.* 60:126–30

84. Malek, F., Baker, D. A. 1977. Proton co-transport of sugars in phloem loading. *Planta* 135:297–99

85. Malek, F., Baker, D. A. 1978. Effect of fusicoccin on proton co-transport of sugars in the phloem loading of *Ricinus communis L. Plant Sci. Lett.* 11:233–39

86. Marrè, E. 1979. Fusicoccin: a tool in plant physiology. *Ann. Rev. Plant Physiol.* 30:273–88

87. Martin, E. Komor, E. 1980. Role of phloem in sucrose transport by *Ricinus* cotyledons. *Planta* 148:367–73

88. Maynard, J. W., Lucas, W. J. 1982. A reanalysis of the two-component phloem loading system in *Beta vulgaris. Plant Physiol.* 69:734–39

89. Maynard, J. W., Lucas, W. J. 1982. Sucrose and glucose uptake into *Beta vulgaris* leaf tissues: A case for general (Apoplastic) retrieval systems.*Plant Physiol.* 70:1436–43

90. Minchin, P. E. H., Thorpe, M. R. 1982. Evidence for a flow of water into sieve tubes associated with phloem loading. *J. Exp. Bot.* 33:233–40

91. Moss, D. N., Rasmussen, H. P. 1969. Cellular localization of CO_2 fixation and translocation of metabolites. *Plant Physiol.* 44:1063–68

92. Olesen, P. 1975. Plasmodesmata between mesophyll and bundle sheath cells in relation to the exchange of C_4-acids. *Planta* 123:199–202

93. Pate, J. S. 1980. Transport and partitioning of nitrogenous solutes. *Ann. Rev. Plant Physiol.* 31:313–40

94. Pavlinova, O. A., Göring, H., Turkina, M. V., Ehwald, R. 1978. Use of 2-deoxy-d-glucose for study of biosynthesis and transport of oligosaccharides in plants. *Fiziol. Rast.* 25:213–21

95. Peel, A. J., Rogers, S. 1982. Stimulation of sugar loading into sieve elements of willow by potassium and sodium salts. *Planta* 154:94–96

96. Rottenberg, H. 1976. The driving force for proton(s) metabolites cotransport in bacterial cells. *FEBS Lett.* 66:159–63

97. Smith, J. A. C., Milburn, J. A. 1980. Osmoregulation and the control of phloem-sap composition in *Ricinus communis L. Planta* 148:28–34

98. Smith, J. A. C., Milburn, J. A. 1980. Phloem transport, solute flux and the kinetics of sap exudation in *Ricinus communis L. Planta* 148:35–41

99. Smith, J. A. C., Milburn, J. A. 1980.

100. Sovonick, S. A., Gieger, D. R., Fellows, R. J. 1974. Evidence for active phloem loading in the minor veins of sugar beet. *Plant Physiol.* 54:886–91

101. Tanner, W. 1980. Proton sugar cotransport in lower and higher plants. *Ber. Dtsch. Bot. Ges.* 93:167–76

102. Thompson, R. G., Dale, J. E. 1981. Export of ^{14}C and ^{11}C labelled assimilate from wheat and maize leaves: effects of parachloromercurobenzylsulphonic acid and fusicoccin and of potassium deficiency. *Can. J. Bot.* 59:2439–44

103. Thorne, J. H. 1982. Characterization of the active sucrose transport system of immature soybean embryos. *Plant Physiol* 70:953–58

104. Thorne, J. H., Giaquinta, R. T. 1982. Pathways and mechanisms associated with carbohydrate translocation in Plants. In *Physiology and Biochemistry of Storage Carbohydrates in plants*, ed. D. H. Lewis, pp xxx–xxx. *Symp. Soc. Exp. Bot.* London: Cambridge. In press

105. Thorpe, M. R., Minchin, P. E. H., Dye, E. A. 1979. Oxygen effects on phloem loading. *Plant Sci. Lett.* 15:345–50

106. Troughton, J. H., Currie, B. G. 1977. Relations between light level, sucrose concentration, and translocation of carbon 11 in *Zea mays* leaves. *Plant Physiol.* 59:808–20

107. Turgeon, R., Webb, J. A. 1973. Leaf development and phloem transport in *Cucurbita pepo:* transition from import to export. *Planta* 113:179–91

108. Turgeon, R., Webb, J. A. 1976. Leaf development and phloem transport in *Cucurbita pepo:* maturation of the minor veins. *Planta* 129:265–69

109. Turgeon, R., Webb, J. A., Every, R. F. 1975. Ultrastructure of minor veins in *Cucurbita pepo* leaves. *Protoplasma* 83:217–32

110. Turkina, M. V., Sokolova, S. V. 1972. Membrane transport of sucrose in plant tissue. *Fiziol. Rast.* 19:912–19

111. Tyree, M. T. 1970. The symplast concept. A general theory of symplastic transport according to the thermodynamics of irreversible processes. *J. Theor. Biol.* 26:181–214

112. Van Bel, A. J. E., Van Erven A. J. 1979. A model for proton and potassium cotransport during the uptake of glutamine and sucrose by tomato internode disks. *Planta* 145:77–82

113. Wittenbach, V. A., Ackerson, R. C., Giaquinta, R. T. 1980. Changes in

photosynthesis, ribulosebisphosphate carboxylase, proteolytic activity, and ultrastructure of soybean leaves during senescence. *Crop Sci* 20:225–31

114. Wright, J. K., Riede, I., Overath, P. 1981. Lactose carrier protein of *Escherichi coli:* Interaction with galactosides and protons. *Biochemistry* 20:6404–15

115. Wright, J. P., Fisher, D. B. 1980. Direct measurement of sieve tube turgor pressure using severed aphid stylets. *Plant Physiol.* 65:1133–35

116. Wright, J. P., Fisher, D. B. 1981. Measurement of the sieve tube membrane potential. *Plant Physiol.* 67:845–48

Ann. Rev. Plant Physiol. 1983. 34:389-417

DEVELOPMENTAL MUTANTS IN SOME ANNUAL SEED PLANTS

G. A. Marx

Department of Seed and Vegetable Sciences, New York State Agricultural Experiment Station, Cornell University, Geneva, New York 14456

CONTENTS

INTRODUCTION

Development, the orderly, coordinated train of events that attend the change from a single-celled zygote to a multicelled adult, is at once an enduring source of fascination and a surpassing scientific challenge. Its facets in one way or another touch upon most biological disciplines, and, partly because of its complexity and primacy, literature on the subject is highly fragmented. A prodigious literature notwithstanding, few unifying theories are available for testing. There may be merit, therefore, in viewing development from the perspective of the whole plant, because this may afford insights that might not otherwise be apparent at lower levels of organization. Inevitably, however, reductionism is forced upon us. But the manner in which the whole is separated into its constituent parts may be of some consequence in the way problems are viewed or thinking is channeled. Sinnott (166) distinguished several common characteristics of an organized system: polarity, differential gradients, symmetry, and spirality. All have physiological and architectural implications and all span disciplinary boundaries.

Nature, too, through the action of spontaneous and induced mutations, offers a way of subdividing plant form and function within a temporal framework. Mutants may be likened unto natural experiments, in effect both "asking" and "answering" questions about growth and development. Departures from normal may be localized or general, depending on the time and place of gene action; or they may be adaptively significant or trivial, depending on how well or to what degree the plant accomodates the change. Mutants may act at the beginning or at the end of a canalized process. That these deviations may provide a powerful complement to traditional methods of investigating regulation and development is the underlying thesis of this review. A recent book (178) elaborates upon this thesis more fully.

Discussion will be limited to mutants (oligogenes) with qualitative, discrete effects. Oligogenes are individually manipulable and hence analyzable in Mendelian terms. These properties impart genetic definition, experimental control, and precision not possible with genes whose action is quantitative.

Genetic variation as the basis for the vast diversity of plant life is as much the sine qua non of botany as it is of genetics, but for many specialties in botany the focus is at or above the species level, and the variation is rarely, if ever, treated in genetic terms. Thus to adopt an approach in which mutants play a central role would be for some to embrace an unfamiliar notion. With the use of mutants the emphasis shifts from the more common practice of comparing members of different species to one of comparing the norm with deviations from the norm within species. Moreover, well-characterized qualitative mutants are available in large numbers in only a handful of species, nearly all of which are annual, diploid, economic plants. Economic plants are often considered as too plastic for taxonomic treatment. Also, for some, the word mutant, unlike the word gene, connotes a change that departs so fundamentally from the norm as to be totally unrepresentative.

If rigorously defined genotypes are to assume a larger role in plant experimentation, then their attributes must become widely known and, equally important, they must be readily available. Isogenic lines differing (presumably) at one locus are especially valuable for evaluating allelic effects without the confounding influence of other genetic differences. In mammals, mutants and defined inbred stocks are utilized worldwide in impressive numbers for biomedical research (47, 72, 164). To spur even greater acceptance, Festing (35) has documented in detail their uses and properties. *Drosophila*, too, is important in

modern developmental studies, largely because of its extensive genetic literature (4, 178). In some of the well-studied plant species the number of simply inherited
mutants exceeds 500, and the diversity of action is such that it cannot be
summarized readily. Virtually every plant part is subject to modification. The
variation can best be comprehended by consulting the multitude of original
descriptions. Many of the descriptions are contained in specialized genetic
newsletters, and in some instances these are the sole source of information.
Published summaries of gene descriptions (8, 12, 13, 53, 83, 115–117, 126,
141, 152, 154, 159, 167, 176, 198) afford access to much of the literature. The
broad range of variation in maize is exceptionally well portrayed in (26, 116).

Temporal and site-specificity is a prominent characteristic of the variation,
but mutants also exhibit a wide difference in magnitude of effect. A mutant of
barley that drastically reduces the number of roots (90) exemplifies a class of
mutants that produces large or even profound changes. Such mutants are
particularly notable because they draw attention to several developmentally
relevant issues. The manner in which a single gene can bring about radical
changes, quite apart from what that change may be, is for example a matter of
fundamental import. From a broader perspective it is of interest to consider how
the plant accommodates, compensates for, or otherwise adjusts to large
changes in structure or function, for in many instances such changes are not
deleterious to the plant. Moreover, because the genes have large effects, the
overall character and habit of the plant can be readily and substantially modified by experimentally manipulating relatively few genes.

Mutants frequently exhibit other features important to development because
they may reveal how apparently unrelated characters may indeed be related.
Pleiotropy (multiple effects) is especially common. The *wex* mutant of *Pisum,*
for example, alters pollen morphology and also causes many pollen grains to
remain in the tetrad stage (96). In addition, the foliage color is changed and
plant vigor is reduced. It is not at all clear in this example what the primary
target of gene action is, if indeed one can be specified. Different genes may also
combine or interact to affect a character or function. In maize, normal yellow
pollen is converted to white by the combined action of two independently
inherited genes (25). The change renders the pollen nonfunctional, giving
support to the belief that synthesis of pigment is prerequisite for the pollen to
function. In this example, two genes are involved in the process, but resolving
power and definition increases as more genes are added to a system that
controls a given process.

For the most part, the mutants treated in the following subsections have been
examined beyond the stage of genetical characterization, but they represent a
small part of the total array of known variants. The examples chosen betray the
author's predilections. Organelle and biochemical mutants are excluded from

the discussion. However, a number of the mutants considered here are often subsumed in the terms biochemical (114, 147) and physiological (181), but the distinctions are arbitrary and the terms are scarcely preemptive.

MUTANTS AFFECTING DIFFERENT SYSTEMS

Shoot Elongation

Some of the many mutants that influence plant stature do so indirectly, i.e. through general impairment of the plant's physiology. Here, however, discussion is limited to mutants that control stem elongation, seemingly without interfering with the coordinated and balanced development of the plant. For convenience, geneticists use descriptive names to distinguish among simply inherited differences in stature, e.g. gigas, tall, dwarf, compactum, brachytic, nana, etc (33, 66, 67, 94, 95, 123). These names connote qualitative differences in height, but they do not reflect absolute height classes. Genes that influence internode length are strongly influenced by background genotype and by environmental conditions.

As with other mutants, gene action may be site specific, as is evident in a mutant of cucumber (153) which causes a striking increase in the length of the hypocotyl without much influence on the ensuing internodes, or a variant of rice (155) which expressly lengthens the uppermost internode of the plant. Such genes are useful when considered individually, but they may contribute even more as members of an interacting group of genes that affects internode length within a species. If the factors contributing to a gene system are known and understood, then complex interactions may serve more to clarify than to confuse. The well-established interacting system of length factors in the garden pea illustrates these interactions (98) and provides background for a discussion of contemporary work to follow.

Alternative alleles at a single locus, *Le,* determine the classical tall (*Le*) vs dwarf (*le*) difference in peas. The tall and dwarf classes are rightly described as discrete but only in terms of what the locus controls, namely internode length. Final height, however, is compounded of both internode length and internode number. Other sets of genes, mainly those governing flowering and photoperiodism (110), influence plant height indirectly by controlling the number of vegetative and reproductive internodes that a given plant produces. These genes have a confounding influence if final height is sole measure of stature.

But like *Le,* other genes of peas influence internode length directly. The actions of two independent loci, *La* and *Cry,* are particularly well known (93, 98). *La* has two known alleles whereas *Cry* has three. Combinations of these, interacting with the alleles at the *Le-le* locus, produce a range of internode length phenotypes. They may, for example, produce discrete subclasses within the dwarf (*le*) class itself, and certain combinations yield internode length phenotypes that exceed those produced by *Le* (tall). The combination *le, la cry*^{c} yields a phenotype known as cryptodwarf whereas *le, la cry*^{s} is known as slender, the length of the initial internodes of both phenotypes surpassing those of tall (*Le*) plants. At the other end of the scale, the mutant *nana* (*na*) is capable of reducing sharply the internode length in *Le* as well as *le* plants (193). *Nana* plants, although only a few centimeters in height, are still "tall" (*Le*) or "dwarf" (*le*) with respect to the genotype at the *Le* locus (111). These relationships dictate that for proper interpretation all the genes in the interacting system must be considered together. Focusing on only one locus in a system may actually prove misleading. The genes within the system may differentially promote or inhibit stem length and the phenotype reflects the balance between these tendencies. Physiological studies involving dwarf or tall cultivars of peas rarely take into consideration the genotype of the plant beyond the *Le-le* difference. This in part might contribute to contradictory findings among different experimenters.

The oft-cited pioneer studies of maize mutants by Phinney and associates (127, 128) indicated that different dwarf mutants influence growth by interfering with different steps in the gibberellin pathway. Recent summaries may be found in (21, 40, 46). Comparisons of dwarf and normal plants for GA content have, in most species studied, showed GA to be reduced in amount or absent in the dwarf forms (46).

Genetic information has been exploited in recent physiological studies of shoot elongation in maize, peas, and barley. In maize GA_1 was found to be the only bioactive GA in the pathway leading to normal growth, and other GAs in the pathway are active only as a result of their metabolism to GA_1 (128a, 172). Evidence suggests that mutant d_1 probably controls the step $GA_{20} \rightarrow GA_1$, and d_5 controls a step prior to GA_{20}. This appears to be fulfillment of the original hypothesis that nonallelic genes concerned with the same growth substance control different steps in the reaction sequence leading to a product necessary for normal growth (127).

Potts et al (134) succeeded in demonstrating a large and near-qualitative difference in the gibberellin-like content of shoots of light grown tall (*Le*) and dwarf (*le*) peas, a difference not found previously by others. Choosing specific tissues and the proper stage of plant development contributed to the success. In this species, *Le* is the normal allele and *le* the mutant. The GA-like compound found in *Le* plants was tentatively identified by GCMS/MIN as GA_1. The authors concluded that *Le* controls the conversion of GA_{20} to the highly active GA_1 by promoting 3 ß-hydroxylation; *le* cannot effect this conversion. Thus, a pattern similar to that in maize has begun to emerge.

Introduction of an extreme dwarf called *nana* (*na*) extends the phenotypic range of internode length in the pea system. Little or no GA-like substances were recovered in the developing shoots of *na* plants, but exogenously applied

GA$_3$ evoked dramatic stem elongation (133). The *na* mutant was first isolated in an *Le* background, and it was subsequently demonstrated that *na* is epistatic to both *Le* and *le* (111, 146), i.e. both talls and dwarfs are extremely short in the presence of *na*. Slender plants (*le la cry*^{s}) which, as mentioned, have longer internodes than normal tall (*Le*) plants, do not show a major qualitative increase in GA-like substances over the dwarfs (*le la cry*) (133), nor do they respond to AMO 1618 (93, 133), a compound that inhibits gibberellin synthesis. Since little or no native GA-like substances were recovered from either the shortest (*nana*) or the longest (slender) phenotypes in the internode length spectrum, Potts el at (133) hypothesized that *la* and *cry*^{s} influence some process at or beyond the perception of the gibberellin signal.

If an allele of a given gene can be identified because it produces a discernible effect on a process, and if a number of alleles at different loci are known to affect the same process, then it follows that the allelic status of all the genes in the system bears importantly on the outcome of the entire process. It is thus in interpreting the internode length system of genes in peas. For example, since *na* markedly reduces the stature of the dwarf internode length in most genotypes, including combinations involving *Le,* the impression may be gained that *na* has an effect that overrides (is epistatic to) all other loci in the system. However, when *na* is present in combination with *la cry*^{c} and *la cry*^{s}, the effect of *na* itself is overridden (100, 133). Describing experimental plants in terms of phenotypic class obviously lacks the required definition.

Systems similar to that described above are likely to exist in other species. Slender plants similar to those found in peas have been described in *Phaseolus* (88) and barley (37). Successful exploitation of such systems relies on a well-established fund of genetic information.

Three light-insensitive barley mutants affecting shoot elongation were investigated in the context of the Jacob-Monod model, with a regulatory gene and a producer gene triggered by light and dark periods (33). In this system a mutation in the regulatory function results in the *gigas* (giant) phenotype which is insensitive to GA. A mutation in the producer gene is a lethal dwarf which is unable to produce GA and which is highly sensitive to GA. The third gene controls the utilization of the hormone. It is epistatic to the *gigas* expression. The plants are viable and completely insensitive to GA. Exogenously supplied GA produces several predicted mimics. Clearly, parallels exist between this and the pea system.

Foliar Architecture

Although leaves are determinate organs of relatively simple structure, extraordinary variation in leaf size, shape, arrangement, and specialization contributes in large measure to the total diversity of plant life. Remarkable diversity in leaf morphology also exists within species where changes may be slight or extreme, confined to the leaf itself or parts thereof, or extend to other tissues and organs as well. In some instances, deviation from normal morphology may be so extreme as to seemingly transcend the boundaries of the species in which the mutant was found and to reflect characteristics of other species or even other genera (14). Yet no definite relationship is evident between the magnitude or severity of the change in leaf morphology and the impact that that change may have on the plants' physiological well-being. Certainly not all deviations from the norm are harmful.

Of the mutants that exert a range of pleiotropic effects, the *lanceolate* leaf mutant of tomato has drawn the most attention, having been studied from the biochemical, physiological, and morphogenetic (19, 20, 104) point of view. Normal tomato plants (*la/la*) have odd-pinnately compound leaves. Heterozygous (*La/+*) plants have simple, entire leaves, small flowers and fruits, weak apical dominance, and other abnormalities. Homozygous dominant plants (*La/La*) die without producing seeds but, curiously, this class is composed of three distinct phenotypes which appear regularly in the progeny of selfed heterozygotes, even in highly inbred lines. Whereas *lanceolate* justifiably is classed as a foliage mutant, its action is so early and so profound that it apparently affects many vital processes, and from this standpoint the mutant may as well be classed as a genetic abnormality. Indeed, two of the phenotypes in the *La/La* class of plants lack an organized shoot (19).

A range of severe abnormalities was ascribed to two independently inherited, temperature sensitive, dominant mutants in *Phaseolus*. Allelic dosage accounted for differences in expression (163). The behavior of a single dominant lanceolate leaf mutant (*Lan*) in common bean is very similar to the *Lac* mutant of tomato (6). It too is incapable of reproduction in the homozygous dominant condition, and it also shows irregularities in phenotypic expression.

The leaf blades in a barley mutant, *Lfl,* are reduced to a tiny appendage (183). Like *Lac* of tomato, *Lfl* is incompletely dominant, but in this case the heterozygote, although distinguishable from the two homozygotes, is nearly normal. Homozygous dominant plants succumb under ordinary growing conditions, but under special environmental conditions some plants produce one culm bearing a few seeds. Thus, this barley mutant has features in common with *Lac* in tomato, but its effects are less extreme.

Less extreme examples also occur in *Pisum* (11) and *Phaseolus* (149), where the normal compound leaves are converted to simple leaves by the action in each case of a single recessive factor pair, *unifoliata* (*uni*). Heterozygotes are normal. The alteration in leaf morphology invariably is accompanied by a radical change in the reproductive organs and sterility, but otherwise the plants grow and develop reasonably well. Thus, although the *unifoliata* mutants exert powerful effects, their effects are less pervasive than the *lanceolate* mutant of tomato. Inasmuch as reproductive appendages are construed as modified leaves, mutants such as *unifoliata* may serve to point up this homology. This relationship is suggested especially in the *crispa* (*cri*) mutant of *Pisum* where the leaflets are folded and the margins rolled under, the stipules are narrow and pointed, and the first leaf internode is markedly foreshortened (11). Pollen fertility of this mutant is high, but because the pods, like the leaves, exhibit morphological irregularities the seed set is somewhat impaired in comparison with the normal phenotype. Mutants of this type are candidates for experiments such as those done with maize (129) and other species (156, 157) where the normal phenotype is restored in part or in full after being treated with certain compounds. It would be equally instructive to resolve pleiotropic action into separate functions such as was done in the recessive *laciniata* (*lac*) of *Pisum* (14), a mutant which, like the *Lac* mutant of tomato, is sterile. Fertility was restored in the pea mutant after repeated mutagenic treatment of heterozygous (normal) plants.

Mutants that influence leaf morphology with little or no apparent influence on other tissues form a class which includes changes in overall size (reflecting differences in cell number and/or size) or shape (unequal growth rate in different dimensions) of leaves. Sinnott (166) considers some of the abundant evidence for the existence of genes for size, form, and shape. The outstanding challenge is to learn how genes control the relative distribution of growth.

Where the alterations in external form involve changes in the parts of leaves, it may be the base, margin, or apex, or one or the other leaf surface or both, the venation, etc, that is affected. One component of a compound leaf may be affected with little or no apparent effect on other component parts. Regardless of the nature of the change, the specificity with respect to site of action may be an important way to subdivide the leaf into zones or compartments. The presence of marginal meristems in plant leaves may complicate matters or, on the other hand, may actually facilitate analysis. Serrated vs entire leaf margins, for example, is a simply inherited difference that presumably is manifestation of gene action operating at the level of the marginal meristem.

Since leaves obviously participate in the source-sink relationships of plants, the distribution of assimilates between the vegetative and reproductive portions of the plant is of great practical concern. The aim frequently is to tip the balance in favor of reproductive growth. A classic approach to this question is to remove parts surgically (e.g. 50). An alternative approach makes use of mutants with altered leaf morphology.

Leaf surface area in cotton is sharply reduced by mutants that cause deeply cleft and narrow lobed leaves (30). In contrast, the potato leaf (*c*) mutant of tomato decreases the leaf segmentation (148). In *Pisum*, the pinnately compound leaf of normal plants is composed of two large foliaceous stipules, one or more pairs of subterminal leaflets, and terminal tendrils (98). Each component may be modified by the action of each of three independently inherited genes. Stipules are markedly reduced by *st* and the tendrils are converted to leaflets by *tl*. A third gene, *afila* (*af*), converts the leaflets of normal plants to tendrils. Together these three loci generate eight different phenotypes. Vegetative apices of plants of various of these phenotypes were examined sterologically (105). The multiple tendrils of *af* leaves were found to be formed from numerous secondary branches on the leaflet primordia. An interaction between *af* and *tl* produces a unique foliar morphology: clusters of small leaflets. The genotype *af, st, Tl,* produces a leaf composed mainly of tendrils because *af* converts leaflets to tendrils and *st* reduces stipule size. The latter phenotype has been dubbed "leafless" although no parts are lost, merely radically modified.

The physiological behavior of modified leaf types of peas and of cotton have been examined in relation to population density, photosynthate partitioning, and water use efficiency. The morphological changes alter virtually all the interrelationships among physical and biological factors that influence plant growth and development (55–57, 59, 68, 80, 81, 124, 130, 140, 169, 191). For example, photosynthetic structures below the reproductive portion of the plant are exposed to light, soil temperature is increased, evaporation is changed, gas exchange is affected, etc. Any change in population density invokes the need for further compensatory adjustments.

The *af, st, tl* genes of *Pisum* were participants along with several other genes that affect leaf morphology in a study that revealed additional unexpected gene interactions (97). Leaflets and stipules of otherwise normal plants (*Af, St, Tl*) become undulated or sinuate in the presence of *sil*. Substituting *af* for *Af*, as mentioned, converts the leaflets into tendrils (without affecting the stipules), but when *sil* and *af* are present together, the stipule tips become deeply incised, and adventitious tendrils emerge from a cleft in the stipule tip. The morphology of the adventitious tendrils themselves varies according to the allelic status of the *Tl* locus. The introduction of specific wax genes into the system elicits further unpredicted interactions which imply a homology between the dorsiventral lamina of the leaflet and the anterior-posterior aspect of the stipule. The behavior also implies an interrelatedness among component parts of the leaf that appeared to be virtually independent and autonomous when *af, st, tl* were considered alone, i.e. without *sil*. However these findings are interpreted, it is clear that the leaf can be modified and molded to extreme degrees, and many of these changes can be wrought without affecting reproductive morphology.

Other foliage mutants of *Pisum* are known and these have stimulated conjecture concerning the evolution of leaf form (11, 15, 44). Though untested and unproved, these ideas help to visualize individual steps in the developmental path leading from a simple (*unifoliata*) condition to compound leaf. There is enough architectual variation within the species to give credence to the concepts. Combinations of genes often show morphological traits which are not present in either parent, just as in the case of the *af tl* combination.

Pigmentation

Pigments formed as products of secondary metabolism are not usually a part of discussions of plant development; they are rather more in the province of phytochemistry, perhaps in part because the physiological role of these compounds is still unclear (69). Yet the temporal, site, qualitative, and distributional aspects of pigment formation may be regarded as important markers in the developmental process, if only as an accompaniment. Early geneticists found the wide range of flower colors displayed by plants to be well suited to Mendelian analysis, and a sizable literature on the subject had already accumulated by 1920. These studies also provided the material basis for some of the first contributions to what has come to be known as biochemical genetics. Individual genes have been shown to produce differences in anthocyanin structure via hydroxylation, methylation, glycosylation, acylation, etc. Genetic studies have also assisted in establishing biosynthetic pathways (73, 74, 82). By feeding pigment precursors to flowers of mutant *Antirrhinum* plants, anthocyandin is formed only if the precursor occupies a position after the genetic block in the biochemical pathway (51, 52, 179).

In some plants a truly imposing body of information concerning the control of pigmentation is at hand. This is perhaps best illustrated in maize, but a comparable situation obtains in several other species as well. Most of the classical principles of genetics, i.e. dominance, pleiotropy, epistasis, etc, are known to operate within these overall systems. Flower color is but a part of a much larger fabric. Pigment may form at different stages and at different sites throughout ontogeny via a system of genes that may exercise control in a highly specific manner. For example, Styles et al (180) named 17 main tissues and origins of maize in which anthocyanin can occur. These represent stages of the life cycle from the appearance of coleoptile, through various stages of vegetative and reproductive development, to the formation of mature kernels. Within a given tissue differences exist in the rate and time (onset and cessation) the genes exert their action (173, 174, 180). In maize and *Pisum* about 30 genes are known to influence flavonoid synthesis (12, 26). Nearly all these genes have been mapped as to chromosome and position in the chromosome. In *Phaseolus* 9 genes are involved in the pigmentation and patterning of the seeds alone, interactions of which produce an ornate array of hues and patterns (135–137).

One locus in this group, the *C* locus, is considered a complex locus with numerous "alleles" and with a variety of effects, not only on the seeds but on other plant parts as well. The manifold effects of this locus were detailed by Prakken (137), and the locus appears to offer a model for studies of gene regulation. The *Cr* locus in *Pisum* also exhibits properties of a compound locus because the two biochemical functions ascribed to this gene (B-ring methylation and 3-rhamnosylation) are inseparable by genetic analysis, but a temporal difference in these two functions is evident (174).

Despite the magnitude and intricacy of these and other systems, they remain for the most part an untapped potential, at least with respect to regulation and development. There are, however, a few examples that serve to indicate the utility of the systems. Prominent among these is the discovery by McClintock (92) of controlling (transposable) elements the ramifications of which are so significant as to occupy a major portion of modern genetics (36).

The pigment genes of maize were also used to investigate the developmental differences in action of the *R* and *B* alleles and to test the validity of a theory concerning gene regulation in higher plants (24, 179). These studies examined the nature and extent of the tissue specificity of alleles at these loci. A high degree of stability in tissue specificty was demonstrated as measured by the absence of spontaneous mutational change of function from one tissue type to another in 10^{13} replications of three different parental alleles (24).

The preceding studies reveal the differences in the onset of pigment formation in different tissues, including the kernel. These temporal differences are also evident in *Pisum*. The monopodial growth habit and sequential flowering and fruiting habit of this plant result in progressive maturation of the fruits and seeds from the first formed to the last reproductive node. Advancing maturity of seeds within a pod and differences in maturity of seeds borne on successive nodes may be measured in a variety of ways (99). The onset of expression of the seed gene *Pl* correlates well with chemical and physical measures of maturity, its expression evidently marking the advent of physiological maturity (102). An analogous situation exists in sorghum and maize where the formation of a pigmented (black) layer also is associated with physiological maturity (192). The *Pl* gene of peas is only the first of a series of seed-pigmenting genes that come into expression in regular and consistent order as seed maturity advances. As in maize, pigment formation occurs over a period of two or three days, and this provides a quantitative dimension to the changes that take place.

The so-called brown-midrib mutants of maize and sorghum control the formation of brown pigment in various parts of the plant but particularly in the midrib of the leaves (39, 54, 132). These genes reduce the lignin content of stalks and leaves and as a result have implications on plant rigidity and on digestibility when the plant is used for animal feed.

Wettstein-Knowles studied the distribution and synthesis anthocyanins produced in the vegetative organs of tomato (194, 195). The 12 loci studied showed, as in other species, marked tissue-specificity. She further characterized the quantitative changes of some of the mutants in relation to changes in time and growth environment. One mutant, *atv*, showed the property of producing anthocyanin at the expense of growth, even under conditions unfavorable for anthocyanin synthesis (32, 195).

The principal pigment of the mature tomato fruit is lycopene, a product of the carotenoid biosynthetic path. The process of fruit ripening in tomatoes is a classical platform for developmental studies. Changes and interactions of many kinds occur over time, changes that can be viewed as a series of separate steps because various aspects of ripening can be prevented or modified (49). These steps are triggered by ethylene. Mutants that block certain steps or otherwise modify the process have aided in establishing the biosynthetic pathway (82).

Results obtained by Grierson et al (49) show that the problem of the regulation of specific genes during ripening is open to direct biochemical attack, and that alterations in enzyme activity seem to involve both an RNA synthesis and enzyme synthesis. Furthermore, since the conversion of a mature green fruit to a ripe fruit involves a transition from one physiological state to another, ripening may be considered a change in physiological age (49). *Nr, nor,* and *rin* are three mutants that block or retard ripening (182). Because these genes show incomplete dominance, ripening in heterozygous plants is slowed in comparison with normal plants but not prevented as in mutant plants.

Mutations affecting chlorophyll and carotenoids are the most commonly occurring mutations in plants. Some have been shown to influence discrete steps in photosystem I and others in photosystem II. This review will not attempt to treat this vast and involved subject. Still, the specificity of these mutants provides numerous opportunities for developmental studies, particularly those mutants that come into expression during ontogeny. In the *alt* mutant of *Pisum* the onset of chlorosis evidently marks the transition from growth that is dependent upon seed reserves and growth that is autonomous, since the seedlings are normal green until they reach the 5–6 node stage of development (1). Ordinarily the two phases are regarded as aspects of one continuous process, but it is possible that responses or effects measured in the seedling stage may be misleading when used to represent the whole plant. Mutants such as *alt* also may prove useful in studies of seed development. In this case the time or developmental stage at which the seed reserves are exhausted may be a measure of the conditions under which the seed on the heterozygous mother plant was produced. Exposing the mutant seedling to different environmental variables may hasten or prolong the utilization of the seed reserves, including the chemical responsible for the chlorosis itself.

Flowering, Photoperiodism, and Senescence

The shift from the vegetative to the reproductive state, marked by flowering, constitutes one of the most arresting developmental events in plant biology and one in which a genetic approach has served admirably. As a result, a number of similarities have begun to emerge with respect to response to environmental cues and to the genetic mechanisms responsible for those responses (2, 109, 110, 171, 190). Whereas flowering is a topic subject to regular review, less attention has been given to the phenomena that may attend the flowering process. These may be manifest before, during, and/or after anthesis, all as a part of the overall reproductive behavior of plants (118, 171). Schwabe (158) lists more than 25 such modifications, some of which have been examined at the anatomical and histological level (150). Because all are related in one way or another to the advent of flowering, and because it is of interest to show that the same genes that control flowering may also control related phenomena, brief attention must be paid to flowering itself as a framework for discussion.

The sheer diversity in flowering behavior and a complex interaction among internal and external cues have thwarted efforts to discover a satisfactory, much less a common, basis of regulation. Still, combined physiological-genetical studies have succeeded in identifying components of the control mechanism in some plants. Murfet (110) has reviewed the progress in *Pisum*, a system that well illustrates the many possible interactions.

Peas exhibit an array of differences in time and node of flower bud initiation, flower formation, and duration of the reproductive period. Analyses carried out under defined conditions in a systematic manner reveal distinct responses and response classes which can be ascribed to actions and interactions among four major loci: *Lf, Sn, Hr, E*. These distinctions, however, are not based on flowering time alone, but on several aspects of reproductive behavior, including duration. The genes or gene combinations may exert their influence in a physiological sense by controlling flowering time, or in a morphological sense by controlling flowering node. Moreover, because growth temperature affects flowering time (via its influence on growth rate) more than it affects flowering node, the two criteria of flowering, which ordinarily are closely correlated, may not always be so. Differences in flowering time and node are evident both in long days (LD) and in short days (SD), but complete separation of the phenotypic classes requires cultivation and comparisons under both day-lengths. Day-neutral, quantitive, and near-qualitative flowering responses are evident under SD. SD sets the stage for higher order interactions. High temperature, in combination with SD, actually delays flowering time in photoperiodically sensitive genotypes, whereas under LD, high temperatures hasten flowering. *Sn* and *Hr* are mainly responsible for the differential response to photoperiod, but *Lf* and *E* interact to modify the type or degree of response. Because the genes act in concert and are interdependent to the extent that any one gene cannot be properly evaluated by itself, i.e. without reference to the status of the remaining genes, the four genes are best regarded as a gene system. This in turn implies that the physiological mechanism underlying the behavioral differences comprises positive, negative, and counterbalancing effects, a conclusion which is consistent with the growing belief that flowering is a response to hormonal balance (40, 110). Once understood, these relationships can be used to make exacting predictions of flowering behavior and to provide a firm basis for investigation at the physiological level.

This background points up the elements in a known genetic system that involves flowering but extends beyond that to other reproductive-related phenomena, including maturation, partitioning of photosynthates, and senescence.

One combination of genes within the *Lf, Sn, Hr, E* system yields a genotype known as G2. The behavioral properties displayed by the G2 line are relevant to whole plant and apical senescence because the line responds in a singular manner to differences in daylength (138). Under LD, the G2 line flowers early (at a low node), then senesces after producing an ordinary complement of fruits and seeds. G2 plants also flower early under SD, but under these conditions flower and pod formation continue for a protracted period of time. Hence, apical senescence is not an inevitable consequence of flowering and fruiting as in ordinary pea lines. Yet fruit production is seen to play some role in apical senescence because removal of flowers and young fruits prevents senescence in LD. These facts have led to the conclusion (138, 139) that photoperiodic control of senescence in G2 peas involves: (*a*) a promotory growth factor, probably a gibberellin, which is produced in the leaves under SD and is translocated to the shoot tip; and (*b*) a senescence factor associated with developing fruits. This interpretation was, however, challenged by Noodén & Leopold (119), who have suggested that continued apical growth under SD may result from a photoperiod-dependent modification in the number of fruits during a given time period. They contended that the reduced fruit load present at any one time may sufficiently attentuate the drain on the plants' resources to allow sustained apical growth. Grounds for this alternative come from the work of Ingram & Browning (71), who showed that, although a much greater number of fruits is produced in SD, fruit growth of G2 peas is more rapid in LD. It is also predicated on a common observation that maturing pods beyond a certain age no longer compete for assimilates. Another view (143) is that the growth factor produced in SD affects the shoot apex indirectly by reducing fruit growth and thereby reduces the senescence-producing effect of the fruits.

These alternative hypotheses emphasize the indirect effect of fruit and seed growth and thus invoke the larger question of source-sink relationships, all within a temporal framework. Collective growth rates of the fruits thus bear importantly on the question of senescence of G2 peas since it affects the intensity of demand at a given time. Clarification on this point came from studies showing that even though seed growth is needed for senescence to take place in G2 peas, the rate of seed growth per se could not account for the observed differences in apical behavior (43). In other studies with other genotypes (63, 64), a marked differential response to growth temperature and radiation intensity was demonstrated in pea cultivars with a genetic capacity to produce more than two flowers per raceme and those regularly producing fewer than two flowers per node. Up to 13 flowers per raceme were developed on a multiflowered cultivar.

The question reduces to whether the variety of behavioral effects manifested in G2 is the consequence of direct action on the apical meristem and everything else follows from that or whether the effects are indirect, resulting from differences in nutrient demand and assimilate distribution. Pertinent to this question is the fact that the G2 line derives its behavioral characteristics from a combination of genes, namely *lf Sn Hr E,* (110). This raises the question as to the individual contribution of each gene in the system to the whole. Gene *Sn* with some justification is considered by Murfet and Reid (142, 144) to be pivotal, and *Hr* to be a modifier of *Sn.* Barber (5) showed that the effect of *Sn* was not confined to an increase in the flowering node; it simultaneously conferred the ability to respond to photoperiod and vernalization, delayed the appearance of the first leaf with four leaflets, and decreased the internode length. *Sn* also shows some activity and influences reproductive behavior under LD (112). Flowering is strongly delayed under SD when *Hr* is combined with *Sn,* but the *Hr Sn* combination in the presence of recessive *lf* and dominant *E* represents the G2 genotype which flowers early. Thus, all genes play a part in the phenotypic expression of G2, and an allelic change in any one of the loci causes a change in the phenotype.

Phaseolus may be regarded either as a short-day plant (SDP) or as day neutral, depending on the criterion of measurement. Photoperiod exerts no control over initiation of flower buds in beans (120, 122), so by this fundamental criterion the species is day-neutral. Photoperiod does, however, affect postinitiation development (7). Under LD the first formed flower buds are inhibited from further development and this inhibition has the effect of delaying anthesis, so by this more restricted defintion *Phaseolus* is an SDP. Experiments with different bean genotypes grown under different environments have identified rather clear-cut response reactions to factorial combinations of temperature and daylength (2, 28, 122, 190). These reactions have similarities with those identified in *Pisum.* Wallace & Enriquez (190) listed 17 characteristics which they used to compare a photoperiod-sensitive cultivar with an insensitive one, but the differences were ascribed to only two loci (28, 103, 122). Despite striking differences in vegetative and reproductive morphology, the two cultivars give similar yields, indicating that the different genes involved regulate postflowering activities which become expressed as differences in distribution of photosynthates.

Further evidence that the *Lf, Sn, Hr, E* gene system in peas affects more than flowering and photoperiodism is revealed by its interaction with the recently discovered mutant *veg* (45). Homozygous *veg* plants failed to flower and steadfastly remained vegetative even after determined measures were taken to encourage flowering (145). After prolonged vegetative growth, *veg* plants eventually die, presumably as a result of a failure of the root system. The mutant is maintained as a heterozygote. Reid & Murfet (145) introduced *veg* into several combinations of the *Lf, Sn, Hr, E* gene system and observed that although flowering itself was prevented in all cases, the reproductive-related events associated with the presence of *Sn* and *Hr* occurred at about the same developmental stage as they would if *veg* were not present. Specific evidence that the signal from the flowering stimulus had reached the apex was shown by a slightly more open apical bud, followed later by the production of numerous lateral branches. Responses of this type typically occur in plants without *veg* but from which the flowers have been removed. These observations point to the conclusion that the developmental events that accompany flowering may be part of one overall signal and not merely incidental to or dependent on the flowering process. Past evidence has been equivocal on this point because the two processes have been confounded. Thus, *veg* may serve to gain further insight into events at the apex as well as to study aspects of growth and development of the pea plant, including apical and whole plant senescence, without the confounding influence of flowering and fruiting (145).

Reproductive Morphology and Sex Expression

The advent of flowering sets off all the events that culminate in fruits and seeds, and thus exposes to view another profusion of structural and functional adaptations and bears witness to the genetic potential for change. But within this extraordinary diversity certain common features are evident. The flowers of higher seed plants are determinate structures bearing successive whorls of specialized appendages: perianth (calyx and corolla), androecium, and gynoecium, the differentiation of which, as Heslop-Harrison (60) suggested, "may be regarded simply as an example of morphogenesis and studied as such." A wealth of mutants affects these parts, individually or collectively. Modifications may take the form of degeneration of tissues, arrestment of development, partial or complete fusion of parts, etc (89, 156, 157, 165). The stamens of the tomato mutant *stamenless-2* (*sl-2*) are abnormal and nonfunctional, but normalcy was restored upon treatment with GA_3 (156). Time-course studies revealed that as GA_3 application was delayed, its restorative effect was progressively diminished, a result that was interpreted in the light of the concepts of determination and canalization (157).

Mutants conferring male sterility have received the most attention at the cytological level, the mutant typically being compared with a normal counterpart (48, 177). Their relative frequency is indicated by the recent paper announcing the availability of over 300 such mutant strains in barley (62).

Whereas male and female sterility and other floral anomalies can and do affect sex expression, the term ordinarily is used in separate context, more nearly at the level of function. Because of its considerable infraspecific variation and because of its value in combined genetic and physiological studies, the cucumber (*Cucumis sativus L.*) admirably fulfills the role of a model plant with respect to the regulation of sex expression. Frankel & Galun (38) give a useful overview of the subject, from which much of the following was drawn.

Interpretation is aided by first considering each of several different levels of morphological variation within the plant: (*a*) the individual flower, (*b*) the flower(s) at a given node or leaf axil, and (*c*) the distribution of flowers along the shoot axis. Although the embryonal floral bud is bisexual, the mature flower will be female, male, or hermaphrodite depending on whether the pistil, stamen, or both develop into a functional organ. Some leaf axils bear more than one flower and such flowers may be of the same or of different sex. Finally, the pattern or sequence in which the individual flowers occur along the elongating shoot establishes the overall sexual status of a given plant. These patterns are the basis for a series of reasonably well defined but modifiable phenotypic classes. Most cultivars are monoecious, i.e. the plants bear both male and female flowers. Among the other classes are: androecious, andromonoecious, gynoecious, and hermaphrodite. These phenotypes result from an interacting system of major genes at two or three separate loci. Modifying genes and environmental conditions play a decided role in influencing the distinctness of the phenotypic classes, and overlapping is a feature of the scheme. Nevertheless, under defined genetic and environmental conditions definite distributional patterns exist. Typically, the first few nodes of normal monoecious cultivars contain male flowers exclusively; this is followed by a mixture of male and female flowers, with an acropetal tendency toward femaleness. The genotype determines the deviation from this normal pattern, the extremes being the production of all female or all male flowers. These genetically determined differences are the basis for a theory concerning the evolution of the cucumber (86) and for schemes to facilitate breeding work (87, 113).

Sex expression can also be experimentally manipulated by exogenously supplied growth regulators (21, 22). The compounds serve, as do the genes, as decisive switches and provide a relationship among genetics, physiology, and development. There is abundant evidence to show that gibberellins increase male tendency, possibly through suppression of female flowers rather than through promotion of male flowers. Female flowers are converted to male when monoecious plants are treated with GA. GA also induces the production of male flowers in all female (gynoecious) plants. This permits all female plants to be self-pollinated and thus to be maintained as inbred lines. Gibberellins show specificity of action, GA_{4+7} being much more effective in promoting male flower production than GA_3. Auxins, on the other hand, encourage production of female flowers, but less so than another growth regulator, ethylene. Auxin was later found to play an indirect role in sex expression by evoking increased production of ethylene by the plants (22).

In two dioecious plants, hemp (*Cannabis sativa*) and spinach (*Spinacia oleracea*), sex expression is also under genetic control, and in these species exogenously supplied gibberellin also enhances maleness and auxin and ethylene enhance femaleness (21). Experiments in which the role of leaves and roots were examined separately suggested that gibberellins which are synthesized in the leaves cause male sex expression and cytokinins which are synthesized in the roots cause female sex expression (16).

The phenomenon of parthenocarpy, the formation of fruit without fertilization of the ovules, is another available way to relate physiology to gene action. Mutants that cause parthenocarpy (130, 131) have been used to advantage in comparing seeded and nonseeded forms with respect to endogenous levels (9) and to their response to exogenously supplied growth substances, morphactin being a compound of special interest (151). Once again the balance between plant growth substances and the timing of action appear to be features of developmental behavior.

Seeds

Molecular biological techniques have revealed considerable genetic polymorphism for seed constituents, especially with respect to the storage protein fractions. Although biochemical mutants are not unrelated to the present topic and although the field of research is highly active, attention here is confined to mutants with morphological effects. The *dek* mutants of maize (161) and the embryo lethal mutants of *Arabidopsis thaliana* (106, 107), although they may be classed as genetic disorders, represent a promising means of analyzing plant embryo development because the mutants characteristically interrupt development at different stages. Development is not appreciably impaired, but altered, in a series of endosperm mutants in maize which modify the form and texture such that the kernel (caryopsis) becomes variously and peculiarly deformed (26). The kernel phenotypes of over 30 identified genes were roughly classified as: extremely collapsed, shrunken, defective, etched and pitted, sugary and glassy, opaque, floury, viviparous, and germless.

These morphological differences are characteristically accompanied by changes in chemical composition (17, 18, 29, 41, 42, 79). Emphasis on the chemical aspect accounts for this group of variants frequently being treated as biochemical mutants (114). The chief properties of a few mutants serve to indicate the nature of the chemical changes involved. The *sugary* (*su*) gene is key in distinguishing sweet corn from a number of other types, for it substantially increases kernel sugar and phytoglycogen content. A specific modifier, sugary enchancer (*en*), acts to increase sugar content even more (34).

Another mutant, *shrunken-2* (*sh-2*) codes for still higher levels of sugar, but *sh-2* kernels, unlike *su*, are virtually devoid of phytoglycogen content (170). Some mutants in the group are of interest from the standpoint of industrial utilization, whereas others are important to animal and human nutrition. One mutant, *opaque-2* (*op-2*), is widely known for its capacity to increase lysine content (29, 79). The discovery of high lysine barley mutants has also stimulated considerable interest (121, 160, 162, 185–187). But whatever considerations have motivated studies of these mutants, the mutants have come to be characterized in impressive detail. Since they all may influence the integrated events that are a part of the maturation process, they may rightly be viewed as developmental mutants.

The classical round vs wrinkled seed of the garden pea is another gross morphological difference with biochemical implications. Wrinkled seed may occur as the result of a recessive gene pair at not just one but either of two loci, *r* and *rb*, and although the seed shape is phenotypically similar, the underlying chemical composition is distinctly different (85). Morphology of the starch granules and the ratio of amylose/amylopectin content were among the differences noted. More recently the cell development and starch granule formation were traced in three genotypes: *R Rb*, *r Rb*, and *R rb* (16). Genotypic differences were noted but for all genotypes distinct gradients in cell development from the center to the periphery of the cotyledons and toward the cotyledon-hypocotyl axis were observed, indicating that the cotyledon, like the storage tissues of other species, is a heterogeneous population of cells of physiologically different ages. Although defined genotypes were not used in another study of seed development in peas, the inclusion of six diverse strains allowed seed development to be divided into three phases of high growth rate separated by two phases of low growth rate (58). The seed was suggested to be composed of two sinks, the testa and the embryo, each competing for the reserves of the endosperm. It was suggested further that each of the three tissues interact in ways determined by the genotype.

Another familiar seed characteristic of the garden pea is the yellow (*I*) vs green (*ii*) cotyledon. Remarkably, scarcely anything is known about this classical difference. That the alleles at the locus effect a developmental difference in cotyledon color between the genotypes is apparent only when the seeds are mature or nearing maturity; prior to that the seeds of both genotypes are green and indistinguishable because both produce chlorophyll. Thus, in the *ii* genotype chlorophyll pigment is retained to the dry seed stage, whereas in *I* plants the chlorophyll is degraded or synthesis is inhibited during seed maturation and thus represents a developmental phenomenon accessible to analysis. Moreover, in terms of gene action, it should be noted that the pollen from *I* parents has an immediate effect on cotyledon color of *ii* plants when the flowers of *ii* plants are fertilized by pollen from *I* plants. This phenomenon also occurs with respect to the round vs wrinkled difference in peas and, like the *I-i* difference, has not been exploited in developmental studies. A related phenomenon, gene dosage, has been explored extensively in maize where the endosperm is not absorbed during kernel formation and where this tissue constitutes a substantial portion of the whole kernel (26).

Epicuticular Wax

Glossy, bloomless, waxless, and eceriferum are descriptive names commonly applied to mutants that control the formation, distribution, or composition of epicuticular wax (10, 65, 125, 199). Epidermal cells are known to be the major sites of synthesis for these relatively nonpolar compounds, the function of which is presumed to be protective (84, 184). Wax composition varies widely, not only among species but also among plants of a single species and among parts of a single plant (184, 196).

Significant progress has been made in recent years toward the understanding of wax chemistry and biosynthesis, in large part because of technological advances such as various types of chromotography, tracers, wax inhibitors, and others (3, 108, 184, 196). Genetics has made an important additional contribution, especially in exposing various blocks in the biosynthetic pathway (196, 197). In barley, a species which is the object of intensive research, a large system of genes containing at least 70 separate loci has been found to control epicuticular wax formation (91, 196, 197). These genes control a broad range of expression in amount, composition, structure, and distribution of wax within and between genotypes. Within this system, similar phenotypes may be specified by different genes. The nature of the wax coating depends on its physical and chemical properties which in turn is a manifestation of differences in gene action (65, 196). Long chain fatty acid synthesis involves the elongation of fatty acid chains, either de novo or by elongating preformed chains. Inasmuch as fatty acid synthesis is an involved process and numerous variations in the process are known, detailed knowledge of mutant action provides a means of understanding the parallel and sequential elongation systems that take part in wax lipid classes. They show, for example, that during certain steps in the process, enzymes or subunits thereof are shared (196).

Wax structure is highly variable; the coating may appear as dense or sparse platelets, rods, or tubes. In some cases the wax structure has been related with wax chemistry (184, 196).

Tissues

Coe (23) enumerates the favorable properties of the maize aleurone layer as an experimental tissue: its bulk and visibility, its extended field display that is formed by systematic developmental processes, its active biosynthetic state, and its phasic constitution. The planes of sequential cell division in this tissue have been ascertained and have proved to be systematic and generally consistent. These properties have allowed mutational changes and clonal proliferation to be followed in considerable detail.

Genetical sophistication in maize has also been instrumental in tracing the destinies of cells derived from the shoot meristem, from its initiation in the embryo through the differentiation of the leaves and stem, to the ear and tassel (27, 189). By extending a method used by Steffensen (175), Coe and his associates (27, 189) have analyzed maize organogenesis by examining the clones from genetically marked cells and by constructing a fate map of the shoot apex. A fate map is defined as a prediction of the role each cell in the apical meristem is likely to play in producing the plant body (189). Specifically, the technique relies on treating the seed of genotypes which are heterozygous for specific dominant anthocyanin genes with X rays to eliminate the dominant allele (and thus the pigment) and then examining the clones in various parts of the plant (27, 76, 189). The results provided estimates of the number of embryonal cells contributing to various parts of the plant body, from the tassel on down. The meristems were found to be compartmented, both in early embryogenesis and in late embryogenesis, to form groups of organs or parts of organs (27, 76, 189). The reproductive parts in the tassel were set off from a subset of cells that form the vegetative body sections. Extension of this line of investigation disclosed more details concerning the variability of the parameters and led to the observation that proliferation of individual cell lineages is highly variable and that spatial location rather than lineal descent is the deciding factor in determining the final fate of a cell (77).

The *Argenteum* (Arg) mutant of *Pisum* (101) has a demonstrated value in physiological and biochemical studies where a single epidermal cell layer is desired. *Arg* plants have a gray-green, silvery cast, shown to result from the presence of extensive air space between the leaf epidermis and the subjacent cell layers (61). Preliminary ultrastructural studies suggest a weakening of the middle lamella, but whatever its basis, the adaxial and abaxial epiderme may be detached from the leaves in relatively large sheets with little contamination of the parenchyma tissue. This property has been exploited in studies of stomatal physiology where initial experiments indicate that stomatal responses to various treatments parallel that observed in nonmutant plants (31, 75). *Arg* was also used recently in a study designed to determine the distribution of secondary metabolites and their biosynthetic enzymes (70). Differences in enzyme activity were noted between the epidermal layer and parenchyma.

Abnormal stomatal development was found to be associated (as a pleiotropic effect) with one of the wax-deficient mutants of barley (200). Double and triple stomatal complexes, extra cells, and abnormal subsidiary cells were among the irregularities noted. A sequential pattern of development was postulated.

Numerous pustules occur in the pods of pea plants dominant for *Np* (*neoplas-*

tic pod). Gene expression is highly influenced by the lighting environment, particularly light quality, so *Np* and *np* plants may be phenotypically alike under some conditions (i.e. no pustule formation). The mature neoplasm is highly polyploid and growth apparently stems from induced mitotic activity of the subsidiary cells which surround the stomata (168). Subsidiary cells occur only on the pods, as do the neoplasms. Later studies have shown that tumor development is apparently dependent on hormones produced by developing seeds because killing the seeds impedes tumor development (78).

Concluding Remarks

It may be axiomatic that the wondrous diversity of plant life is a manifestation of genetic variation, but, except for a few species, that variation is poorly characterized in genetical terms, if at all. In those few exceptional species, however, numerous mutants with clear, discrete, and widely differing expressions have been identified and characterized in some detail. Heritable deviations from normal plant structure and function are projected here as a way to understand the normal condition rather than as idle anomalies or as signs of degeneration. Indeed, Waddington (188, p.13) reminds us of Bateson's (1894) observation: "each organ has its well defined and characteristic morphology, and when, in an aberrant individual, it varies from the norm, it tends to appear in some other almost equally definite shape." To understand the means by which a plant makes adjustments to change is to understand much of the mystery of development. Moreover, many mutants confer adaptive advantages.

Mutants have wide-ranging yet specific action, and they may, in effect, dissect the whole plant and its parts and functions into smaller units. Their actions may expose continuities and relationships not otherwise evident in the normal plant. They may be looked upon as natural experiments, the results of which we would do well to heed. Because mutants with similar action occur in different species, it may be profitable to look for common control mechanisms.

Genetical characterization lends needed definition to a mutant and enhances its value as a research tool. Delineating a mutant's expression, determining its response to environmental variation, fixing its position on the chromosome, assessing its interactive effects with other genes, and developing isogenic lines all contribute to a mutant's usefulness.

Species that contain a broad range of defined variation have the important advantage of offering not just one but groups of genes that mediate the same plant process. Because there is a growing belief that certain developmental processes are regulated through hormonal balances, individual genes in these groups may participate by controlling either promotory or inhibitory effects. Mutants exerting powerful pleiotropic effects may provide another avenue of inquiry either by inducing new mutations within a mutant line or by identifying chemical compounds that differentially affect parts of the phenotypic syndrome.

As genetic information accumulates, it may become possible to select mutants for particular research purposes just as one chooses a chemical from the laboratory shelf.

ACKNOWLEDGMENTS

The indispensable technical assistance of Mmes. Antinelli, Marx, and Porterfield is gratefully acknowledged.

Literature Cited

1. Acree, T., Marx, G. A. 1972. Evidence for a transmissible substance affecting pigment synthesis in *Pisum. Experientia* 28:1505–6
2. Aitken, Y. 1974. *Flowering Time, Climate and Genotype.* Carlton, Vic: Melbourne Univ. Press. 193 pp.
3. Avato, P., Mikkelsen, J. D., Wettstein-Knowles, P. von. 1980. Effect of inhibitors on synthesis of fatty acyl chains present in waxes on developing maize leaves. *Carlsberg Res. Commun.* 45:329–47
4. Baker, W. K. 1978. A genetic framework for *Drosophilia* development. *Ann. Rev. Genet.* 12:451–70
5. Barber, H. N. 1959. Physiological genetics of *Pisum* II. The genetics of photoperiodism and vernalisation. *Heredity* 13:33–60
6. Bassett, M. J. 1981. Inheritance of a lanceolate leaf mutation in the common bean. *J. Hered.* 72:431–32
7. Bentley, B., Morgan, C. B., Morgan, D. G., Saad, F. A. 1975. Plant growth substances and effects of photoperiod on flower bud development in *Phaseolus vulgaris. Nature* 256:121–22
8. Bernard, R. L., Weiss, M. G. 1973. Qualitative genetics. In *Soybeans: Improvement, Production, and Uses,* ed. B. E. Caldwell, pp. 117–54. Madison: Am. Soc. Agron. 681 pp.
9. Beyer, E. M. Jr., Quebedeaux, B. 1974. Parthenocarpy in cucumber: mechanism of action of auxin transport inhibitors. *J. Am. Soc. Hortic. Sci.* 99:385–90
10. Bianchi, G., Avato, P., Salamini, F. 1979. Glossy mutants of maize IX. Chemistry of glossy-4, glossy-8, glossy-15, and glossy-18 surface waxes. *Heredity* 42:391–95
11. Blixt, S. 1967. Linkage studies in *Pisum.* VII. The manifestation of the genes *cri* and *coch* and the double-recessive in *Pisum. Agri Hort. Genet.* 25:131–44
12. Blixt, S. 1972. Mutation genetics in *Pisum. Agri Hort. Genet.* 30:1–293
13. Blixt, S. 1974. The pea. In *Handbook of Genetics,* ed. R. C. King, pp. 181–221. New York: Plenum. 631 pp.
14. Blixt, S. 1976. A crossing programme with mutants in peas. *IAEA Bull.* AG-25/3:21–36
15. Blixt, S. 1978. Some genes of importance for the evolution of the pea in cultivation (and a short presentation of the Weibullshom-PGA pea collection). *Proc. Conf. Broadening Genet. Base Crops, Wageningen,* pp. 195–202
16. Boyer, C. D. 1981. Starch granule formation in developing seeds of *Pisum sativum* L.: Effect of genotype. *Am. J. Bot.* 68:69–65
17. Boyer, C. D., Daniels, R. R., Shannon, J. C. 1977. Starch granule (amyloplast) development in endosperm of several *Zea mays* L. genotypes affecting kernel polysaccharides. *Am. J. Bot.* 64:50–56
18. Boyer, C. D., Garwood, D. L., Shannon, J. C. 1976. Interaction at the amylose extender and waxy mutants of *Zea mays. Plant Physiol.* 57:102
19. Caruso, J. L. 1968. Morphogenetic aspects of a leafless mutant in tomato. I. General patterns in development. *Am. J. Bot.* 55:1169–76
20. Caruso, J. L. 1971. Isozymes of the lanceolate mutant in tomato. *Am. J. Bot.* 58:475–76
21. Chailakhyan, M. Kh. 1979. Genetic and hormonal regulation of growth flowering and sex expression in plants. *Am. J. Bot.* 66:717–36
22. Chailakhyan, M. Kh., Khryanin, V. N. 1979. Hormonal regulation of sex expression in plants. In *Plant Growth Substances 1979.* Int. Conf. Plant Growth Subst., 10th, Madison, Wis., pp. 331–34
23. Coe, E. H. Jr., 1978. The aleurone tissue of maize as a genetic tool. In *Maize Breeding and Genetics,* ed., D. B. Walden, pp. 447–59. New York: Wiley. 794 pp.
24. Coe, E. H. Jr., 1979. Specification of the

anthocyanin biosynthetic function by B and R in maize. *Maydica* 24:49–58

25. Coe, E. H. Jr., McCormick, S. M., Modena, S. A. 1981. White pollen in maize. *J. Hered.* 72:318–20

26. Coe, E. H. Jr., Neuffer, M. G. 1977. The genetics of corn. In *Corn and Corn Improvement*, ed. G. F. Sprague, pp. 111–223. Madison: Am. Soc. Agron. 774 pp.

27. Coe, E. H. Jr., Neuffer, M. G. 1978. Embryo cells and their destinies in the corn plant. In *The Clonal Basis of Development*, ed. S. Subtelny, I. M. Sussex, pp. 113–29. New York: Academic. 261 pp.

28. Coyne, D. P. 1978. Genetics of flowering in dry beans (*Phaseolus vulgaris* L.) *J. Am. Soc. Hortic. Sci.* 103:606–8

29. Di Fonzo, N., Fornasari, E., Salamini, F., Reggiani, R., Soave, C. 1980. Interaction of maize mutants floury-2 and opaque-7 with opaque-2 in the synthesis of endosperm proteins. *J. Hered.* 71: 397–402

30. Dilday, R. H., Kohel, R. J., Richmond, T. R. 1975. Genetic analysis of leaf differentiation mutants in upland cotton. *Crop Sci.* 15:393–97

31. Donkin, M. E., Travis, A. J., Martin, E. S. 1982. Stomatal responses to light and CO_2 in the *Argenteum* mutant of *Pisum sativum*. *Z. Pflanzenphysiol.* 107:201–9

32. Duke, S. O., Fox, S. B., Naylor, A. W. 1976. Photosynthetic independence of light induced anthocyanin formation in *Zea mays* seedlings. *Plant Physiol.* 57:192–96

33. Favret, E. A., Favret, G. C., Malvarez, E. M. 1975. Genetic regulatory mechanisms for seedling growth in barley. In *Barley Genetics III. Proc. 3rd Int. Barley Genet. Symp., Garching, W. Germany,* pp. 37–42

34. Ferguson, J. E., Rhodes, A. M., Dickinson, D. B. 1978. The genetics of sugary enhancer (*se*), an independent modifier of sweet corn (*su*). *J. Hered.* 69:377–80

35. Festing, M. F. 1980. *Inbred Strains in Biomedical Research.* New York: Oxford Univ. Press. 483 pp.

36. Fincham, J. R., Sastry, G. R. 1974. Controlling elements in maize. *Ann. Rev. Genet.* 8:15–50

37. Foster, C. A. 1977. Slender: an accelerated extension growth mutant of barley. *Barley Genet. Newsl.* 7:24–27

38. Frankel, R., Galun, E. 1977. *Pollination Mechanisms, Reproduction and Plant Breeding.* Berlin: Springer. 281 pp.

39. Fritz, J. O., Cantrell, R. P., Lechtenberg, V. L., Axtell, J. D., Hertel, J. M. 1981. Brown midrib mutants in sudangrass and grain sorghum. *Crop Sci.* 21:706–9

40. Gale, M. D. 1979. Plant hormones and plant breeding. In *Genetic Variation in Hormone Systems,* ed. J. G. M. Shire, 2:123–49. Boca Raton: CRC. 192 pp.

41. Garwood, D. L., Creech, R. G. 1972. Kernel phenotypes of *Zea mays* L. genotypes possessing one to four mutated genes. *Crop Sci.* 12:119–21

42. Garwood, D. L., Vanderslice, S. F. 1982. Carbohydrate composition of alleles of the sugary locus in maize. *Crop Sci.* 22:367–71

43. Gianfagna, T. J., Davies, P. J. 1981. The relationship between fruit growth and apical senescence in the G2 line of peas. *Planta* 152:356–64

44. Gottschalk, W. 1970. Possibilities of leaf evolution through mutation and recombination. A model for the evolution and further evolution of leguminous leaves. *Z. Pflanzenphysiol.* 63:44–54

45. Gottschalk, W. 1979. A *Pisum* gene preventing transition from the vegetative to the reproductive stage. *Pisum Newsl.* 11:10

46. Graebe, J. E., Ropers, H. J. 1978. Gibberellins. In *Phytohormones and Related Compounds—A Comprehensive Treatise,* ed. D. S. Letham, P. B. Goodwin, R. J. Higgins, pp. 107–204. Amsterdam: Elsevier. 648 pp.

47. Green, M. C., ed. 1981. *Genetic Variants and Strains of the Laboratory Mouse.* New York: Fischer. 476 pp.

48. Greyson, R. I., Waldon, D. B., Cheng, P. C. 1980. Light microscopic transmission electron microscopic and scanning electron microscopic observations of anther development in the genic male sterile MS9 mutant of corn *Zea-mays*. *Can. J. Genet. Cytol.* 22:153–66

49. Grierson, D., Tucker, G., Robertson, N. 1981. The regulation of gene expression during the ripening of tomato fruits. In *Quality in Stored and Processed Vegetables and Fruit,* ed. P. W. Goodenough, R. K. Atkin, pp. 179–91. London: Academic. 398 pp.

50. Hanson, W. D., West, D. R. 1982. Source-sink relationships in soybeans. 1. Effects of source manipulation during vegetative growth on dry matter distribution. *Crop Sci.* 22:372–76

51. Harrison, B. J., Stickland, R. G. 1974. Precursors and genetic control of pigmentation. 2. Genotype analysis of pigment controlling genes in acyanic phenotypes in *Antirrhinum majus*. *Heredity* 33:112–15

52. Harrison, B. J., Stickland, R. G. 1978.

Precursors and genetic control of pigmentation. 4. Hydroxylation and methoxylation stages in anthocyanidin synthesis. *Heredity* 40:127–32

53. Harte, C. 1974. *Antirrhinum majus* L. See Ref. 13, pp. 315–31

54. Hartley, R. D., Jones, E. C. 1978. Phenolic components and degradability of the cell walls of the brown midrib mutant BM-3 of *Zea mays*. *J. Sci. Food Agric.* 29:777–82

55. Harvey, D. M. 1978. The photosynthetic and respiratory potential of the fruit in relation to seed yield of leafless and semileafless mutants of *Pisum sativum* L. *Ann. Bot.* 42:331–36

56. Harvey, D. M., Goodwin, J. 1978. The photosynthetic net carbon dioxide exchange potential in conventional and leafless phenotypes of *Pisum sativum* L. in relation to foliage area, dry matter production and seed yield. *Ann. Bot.* 42: 1091–98

57. Hedley, C. L., Ambrose, M. J. 1979. The effect of shading on the yield components of six leafless pea genotypes. *Ann. Bot.* 44:469–78

58. Hedley, C. L., Ambrose, M. J. 1980. An analysis of seed development in *Pisum sativum*. *Ann. Bot.* 46:89–106

59. Hedley, C. L., Ambrose, M. J. 1981. Designing "leafless" plants for improving yields of the dried pea crop. *Adv. Agron.* 34:225–77

60. Heslop-Harrison, J. 1957. The experimental modification of sex expression in flowering plants. *Biol. Rev.* 32:38–90

61. Hoch, H. C., Pratt, C., Marx, G. A. 1980. Subepidermal air spaces: basis for the phenotypic expression of the *Argenteum* mutant of *Pisum*. *Am. J. Bot.* 67:905–11

62. Hockett, E. A., Reid, D. A. 1981. Spring and winter genetic male-sterile barley stocks. *Crop Sci.* 21:655–59

63. Hole, C. C. 1977. Environmental control of flower in multiflowered cultivars of *Pisum sativum* L. *Ann. Bot.* 41:1217–23

64. Hole, C. C., Hardwick, R. C. 1976. Development and control of the number of flowers per node in *Pisum sativum* L. *Ann. Bot.* 40:707–22

65. Holloway, P. J., Hunt, G. M., Baker, E. A., Macey, M. J. K. 1977. Chemical composition and ultrastructure of the epicuticular wax in 4 mutants of *Pisum sativum* L. *Chem. Phys. Lipids* 20:141–55

66. Holm, E. 1969. A gene on chromosome 3 in barley leading to brachytic growth. *Hereditas* 62:419–21

67. Hopp, H. E., Favret, E. A. 1980. Metabolism of (^{3}H) gibberellin A$_1$ in a *gigas* mutant of barley. *Plant Physiol.* 65S:Abstr. 521

68. Horrucks, R. D., Kerby, T. A., Buxton, D. R. 1978. Carbon source for developing bolls in normal and superokra leaf cotton. *New Phytol.* 80:335–40

69. Hrazdina, G. 1982. Anthocyanins. In *The Flavonoids: Advances in Research 1975–1981*, ed. J. B. Harborne, T. J. Mabry. London: Chapman & Hall. 174 pp.

70. Hrazdina, G., Marx, G. A., Hoch, H. C. 1982. Distribution of secondary plant metabolites and their biosynthetic enzymes in pea *Pisum sativum* L. leaves: anthocyanins and flavonol glycosides. *Plant Physiol.* 70:745–48

71. Ingram, T. J., Browning, G. 1979. Influence of photoperiod on seed development in the genetic line of peas G2 and its relation to changes in endogenous gibberellins measured by combined gas chromatography-mass spectrometry. *Planta* 146:423–32

72. Jackson Laboratory. 1978–79. *50th Annual Report 1978–1979*. Bar Harbor, Me: Jackson Lab.

73. Jende-Strid, B. 1978. Mutations affecting flavonoid synthesis in barley. *Carlsberg Res. Commun.* 43:265–73

74. Jende-Strid, B., Moller, B. L. 1981. Analysis of proanthocyanidins in wild-type and mutant barley (*Hordeum vulgare* L.). *Carlsberg Res. Commun.* 46:53–64

75. Jewer, P. C., Incoll, L. D., Shaw, J. 1982. Stomatal responses of *Argenteum*—a mutant of *Pisum sativum* L. with readily detachable leaf epidermis. *Planta* 155:146–53

76. Johri, M. M., Coe, E. H. 1980. Clonal analysis of development in corn. *Maize Gen. Coop. Newsl.* 54:30–33

77. Johri, M. M., Coe, E. H. 1981. Clonal analysis of development in corn. *Maize Gen. Coop. Newsl.* 55:31–38

78. Jones, J. V., Burgess, J. 1977. Physiological studies on a genetic tumor of *Pisum sativum*. *Ann. Bot.* 41:219–25

79. Jones, R. A. 1978. Effects of floury-2 locus on zein accumulation and RIN metabolism during maize endosperm development. *Biochem. Genet.* 16:27–38

80. Karami, E., Krieg, D. R., Quisenberry, J. E. 1980. Water relations and carbon-14 assimilation of cotton with different leaf morphology. *Crop Sci.* 20:421–26

81. Kerby, T. A., Buxton, D. R., Matsuda, K. 1980. Carbon source-sink rela-

tionships within narrow-row cotton cano-
pies. *Crop Sci.* 20:208–13
82. Khudaira, A. K. 1972. The ripening of
tomatoes. *Am. Sci.* 60:696–707
83. Khush, G. S. 1974. Rice. See Ref. 13,
pp. 31–58
84. Kolattukudy, P. E., Croteau, R., Buck-
ner, J. S. 1976. Biochemistry of plant
waxes. In *Chemistry and Biochemistry of
Natural Waxes*, ed. P. E. Kolattukudy,
pp. 289–347. Amsterdam: Elsevier. 459
pp.
85. Kooistra, E. 1962. On the differences
between smooth and three types of wrinkled peas. *Euphytica* 11:357–73
86. Kubicki, B. 1964. Evolution of sex in
Cucumis sativus. Genet. Pol. 5:14
87. Kubicki, B. 1965. New possibilities of
applying different sex types in cucumber
breeding. *Genet. Pol.* 6:241–50
88. Lamprecht, H. 1948. The inheritance of
the slender-type of *Phaseolus vulgaris*
and some other results. *Agri Hort. Genet.*
5:72–84
89. Lin, B. Y. 1978. Structural modifications of the female gametophyte associated with the indeterminate gametophyte
mutant in maize. *Can. J. Genet. Cytol.*
20:249–58
90. Linde-Laursen, I. 1977. Barley mutant
with few roots. *Barley Genet. Newsl.*
7:43–45
91. Lundqvist, U., Wettstein-Knowles, P.
von. 1982. Dominant mutations at *Cer-
yy* change barley spike wax into leaf
blade wax. *Carlsberg Res. Commun.*
47:29–43
92. McClintock, B. 1956. Controlling elements and the gene. *Cold Spring Harbor
Symp. Quant. Biol.* 21:197–216
93. McComb, A. J. 1977. Control of root and
shoot development. In *The Physiology of
the Garden Pea*, ed. J. F. Sutcliffe, J. S.
Pate, pp. 235–63. London: Academic.
500 pp.
94. MacKill, D. J., Rutger, J. N. 1979. The
inheritance of induced mutant semi-
dwarfing genes in rice. *J. Hered.*
70:335–41
95. Marshall, H. G., Murphy, C. F. 1981.
Inheritance of dwarfness in three oat
crosses and relationship of height to
panicle and culm length. *Crop Sci.* 21:
335–38
96. Marx, G. A. 1973. Pollen morphology of
the *wex* mutant. *Pisum Newsl.* 5:31–32
97. Marx, G. A. 1977. A genetic syndrome
affecting leaf development in *Pisum. Am.
J. Bot.* 64:273–77
98. Marx, G. A. 1977. Classification genetics and breeding. See Ref. 93, pp. 21–43
99. Marx, G. A. 1981. The role of genetics in

breeding for quality of vegetables. See
Ref. 49, pp. 67–81
100. Marx, G. A. 1982. Influence of nonallelic interactions on the phenotype of
na. Pisum Newsl. 14:47–49
101. Marx, G. A. 1982. The Argenteum (*Arg*)
mutant of *Pisum:* genetic control and
breeding behavior. *J. Hered.* 73:413–20
102. Marx, G. A., Duczmal, K. 1972. The *Pl*
gene: its use as an index of seed maturity.
Pisum Newsl. 4:32–33
103. Masaya, P. N. 1978. *Genetic and environmental control of flowering in
Phaseolus vulgaris L.* PhD thesis. Cornell Univ., Ithaca, NY
104. Mathan, D. S., Cole, R. D. 1964. Comparative biochemical study of two allelic
forms of a gene affecting leaf-shape in the
tomato. *Am. J. Bot.* 51:560–66
105. Meicheimer, R. D., Muehlbauer, F. J.,
Hindman, J. L., Gritton, E. T. 1982.
Meristem characteristics of genetically
modified pea (*Pisum sativum* L.) leaf primordia. *Can. J. Bot.* In press
106. Meinke, D. W., Sussex, I. M. 1979.
Embryo lethal mutants of *Arabidopsis
thaliana*. A model system for genetic
analysis of plant embryo development.
Dev. Biol. 12:50–61
107. Meinke, D. W., Sussex, I. M. 1979.
Isolation and characterization of embryo
lethal mutants of *Arabidopsis thaliana*.
Dev. Biol. 12:62–72
108. Mikkelsen, J. D. 1978. The effects of
inhibitors on the biosynthesis of the long
chain lipids with even carbon numbers in
barley spike epicuticular wax. *Carlsberg
Res. Commun.* 43:15–35
109. Murfet, I. C. 1977. Environmental interaction and the genetics of flowering.
Ann. Rev. Plant Physiol. 28:253–78
110. Murfet, I. C. 1977. The physiological
genetics of flowering. See Ref. 93, pp.
385–430
111. Murfet, I. C. 1978. Interaction of the
internode length genes *Le-le* and *Na-na*.
Pisum Newsl. 10:54–55
112. Murfet, I. C. 1982. Flowering in the garden pea: Expression of gene Sn in the
field and use of multiple characters to
detect segregation. *Crop Sci.* 22:
923–26
113. Nandgaonkar, A. K., Baker, L. R. 1981.
Inheritance of multi-pistillate flowering
habit in gynoecious pickling cucumber.
J. Am. Soc. Hortic. Sci. 106:755–57
114. Nelson, O. E. Jr., Burr, B. 1973.
Biochemical genetics of higher plants.
Ann. Rev. Plant Physiol. 24:493–518
115. Neuffer, M. G., Coe, E. H. Jr., 1974.
Corn (Maize). See Ref. 13, pp. 3–30
116. Neuffer, M. G. Jones, L., Zuber, M. S.

1968. *The Mutants of Maize*. Madison: Crop Sci. Soc. Am. 74 pp.

117. Nilan, R. A. 1974. Barley (*Hordeum vulgare*). See Ref. 13, pp. 93–110

118. Nissly, C. R., Bernard, R. L., Hittle, C. N. 1981. Variation in photoperiod sensitivity for time of flowering and maturity among soybean strains of Maturity Group III. *Crop Sci.* 21:833–36

119. Noodén, L. D., Leopold, A. C. 1978. Phytohormones and the endogenous regulation of senescence and abscission. In *Phytohormones and Related Compounds—A Comprehensive Treatise*, ed. D. S. Letham, P. B. Goodwin, T. J. V. Higgins, 2:329–69. Amsterdam: Elsevier. 648 pp.

120. Ojehomon, O. O., Rathjen, A. S., Morgan, D. G. 1968. Effects of daylength on the morphology and flowering of five determinate varieties of *Phaseolus vulgaris* L. *J. Agric. Sci.* 71:209–14

121. Olsen, O-A., Krekling, T. 1980. Grain development in normal and high lysine barley. *Hereditas* 93:147–60

122. Padda, D. S., Munger, H. M. 1969. Photoperiod temperature and genotype interactions affecting time of flowering in beans, *Phaseolus vulgaris* L. *J. Am. Hortic. Sci.* 94:157–60

123. Padma, A., Reddy, G. M. 1977. Genetic behavior of five induced dwarf mutants in an indica rice cultivar. *Crop Sci.* 17:860–63

124. Pegelow, E. J. Jr., Buxton, D. R., Briggs, R. E., Muramoto, H., Gensler, W. G. 1977. Canopy photosynthesis and transpiration of cotton as affected by leaf type. *Crop Sci.* 17:1–4

125. Peterson, G. C., Suksayretrup, K., Weibel, D. E. 1982. Inheritance of some bloomless and sparse-bloom mutants in sorghum. *Crop Sci.* 22:63–67

126. Phillips, L. L. 1974. Cotton (*Gossypium*). See Ref. 13, pp. 111–33

127. Phinney, B. O. 1956. Growth response of single-gene dwarf mutants in maize to gibberellic acid. *Genetics* 42:185–89

128. Phinney, B. O. 1961. Dwarfing genes in *Zea mays* and their relation to the gibberellins. In *Plant Growth Regulation. Int. Conf. Plant Growth Regul., 4th, Yonkers*, pp. 489–501

128a. Phinney, B. O., Spray, C. 1982. Chemical genetics and the gibberellin pathway in *Zea mays* L. In *Plant Growth Substances*, ed. P. F. Wareing, pp. 101–10. New York: Academic

129. Phinney, B. O., West, C. A. 1960. Gibberellins and the growth of flowering plants. *Symp. Soc. Study Dev. Growth* 18:71–92

130. Pike, L. M., Peterson, C. E. 1969. Inheritance of parthenocarpy in the cucumber (*Cucumis sativus* L.). *Euphytica* 18:101–5

131. Ponti, O. M. B. de, Garretsen, F. 1976. Inheritance of parthenocarpy in pickling cucumbers (*Cucumis sativus* L.) and linkage with other characters. *Euphytica* 25:633–43

132. Porter, K. S., Axtell, J. D., Lechtenberg, V. L., Colenbrander, V. F. 1978. Phenotype fiber composition and *in-vitro* dry matter disappearance of chemically induced brown midrib mutants of sorghum. *Crop Sci.* 18:205–8

133. Potts, W. C., Ingram, T. J., Reid, J. B., Murfet, I. C. 1982. Internode length in pea genotypes and the involvement of gibberellins. *Plant Physiol.* 69S:Abstr. 133

134. Potts, W. C., Reid, J. B., Murfet, I. C. 1982. Internode length in *Pisum*. I. The effect of the *Le/le* gene difference on endogenous gibberrellin-like substances. *Physiol. Plant.* 55:323–28

135. Prakken, R. 1970. Inheritance of colour in *Phaseolus vulgaris* L. II. A critical review. *Meded. Landbouwhogesch. Wageningen* 70-23:1–38

136. Prakken, R. 1972. Inheritance of colours in *Phaseolus vulgaris* L. III. On genes for red seedcoat colour and a general synthesis. *Meded. Landbouwhogesch. Wageningen* 72-29:1–82

137. Prakken, R. 1974. Inheritance of colours in *Phaseolous vulgaris* L. IV. Recombination within the complex locus C. *Meded. Landbouwhogesch. Wageningen* 74-24:1–36

138. Proebsting, W. M., Davies, P. J., Marx, G. A. 1976. Photoperiod control of apical senescence in a genetic line of peas. *Plant Physiol.* 58:800–2

139. Proebsting, W. M., Davies, P. J., Marx, G. A. 1978. Photoperiodic-induced changes in gibberellin metabolism in relation to apical growth and senescence in genetic lines of peas (*Pisum sativum* L.) *Planta* 141:231–38

140. Pyke, K. A., Hedley, C. L. 1982. Comparative studies within and between seed and seedling populations of leafed, 'leafless' and 'semi-leafless' genotypes of *Pisum sativum* L. *Euphytica*. In press

141. Rédei, G. P. 1974. *Arabidopsis thaliana*. See Ref. 13, pp. 151–80

142. Reid, J. B. 1979. Flowering in *Pisum*: the effect of age on the gene *Sn* and the site of action of gene *Hr*. *Ann. Bot.* 44:163–73

143. Reid, J. B. 1979. Red-far red reversibility of flower development and apical

senescence in *Pisum*. *Z. Pflanzenphysiol.* 93:297–301

144. Reid, J. B. 1982. Flowering in peas: effect of gene *Hr* on spectral sensitivity. *Crop Sci.* 22:266–68

145. Reid, J. B., Murfet, I. C. 1983. Flowering in *Pisum:* a fifth locus, *veg,* controlling flowering. *Physiol. Plant.* In press

146. Reid, J. B., Murfet, I. C., Potts, W. C. 1983. Internode length in Pisum. II. Additional information on the relationship and action of loci *Le, La, Cry, Na,* and *Lm. J. Exp. Bot.* In press

147. Rice, T. B., Carlson, P. S. 1975. Genetic analysis and plant improvement. *Ann. Rev. Plant Physiol.* 26:279–308

148. Rick, C. M. 1974. The tomato. See Ref. 13, pp. 247–80

149. Roberts, M. H. 1982. List of genes— *Phaseolus vulgaris* L. *Ann. Rep. Bean Improvement Coop.* 25: 109

150. Roberts, R. H., Struckmeyer, B. E. 1948. Anatomical and histological changes in relation to vernalization and photoperiodism. In *Vernalization and Photoperiodism,* ed. A. E. Murneek, R. O. Whyte, pp. 91–100. Waltham: Chronica Botanical Co. 196 pp.

151. Robinson, R. W., Cantliffe, D. J., Shannon, S. 1971. Morphactin-induced parthenocarpy in the cucumber. *Science* 171:1251–52

152. Robinson, R. W., McCreight, J. D., Ryder, E. J. 1983. The genes of lettuce and closely related species. In *Plant Breeding Reviews,* 1: 267–93, Westport, Conn: AVI. 397 pp.

153. Robinson, R. W., Shail, J. W. 1981. A cucumber mutant with increased hypocotyl and internode length. *Cucurbit Genet. Coop.* 4:19–20

154. Robinson, R. W., Whitaker, T. W. 1974. See Ref. 13, pp. 145–50

155. Rutger, J. N., Carnahan, H. L. 1981. A fourth genetic element to facilitate hybrid cereal production—a recessive tall in rice. *Crop Sci.* 21:373–76

156. Sawhney, V. K., Greyson, R. I. 1973. Morphogenesis of the stamenless-2 mutant in tomato. II. Modifications of sex organs in the mutant and normal flowers by plant hormones. *Can. J. Bot.* 51:2473–79

157. Sawhney, V. K., Greyson, R. I. 1979. Interpretations of determination and canalisation of stamen development in a tomato *Lycopersicon esculentum* mutant. *Can. J. Bot.* 57:2471–77

158. Schwabe, W. W. 1957. The study of plant development in controlled environments. In *Control of the Plant Environ-* ment, ed. J. P. Hudson, pp. 16–35. New York: Academic. 240 pp.

159. Sears, E. R. 1974. The wheats and their relatives. See Ref. 13, pp. 59–91

160. Sen, K., Mehta, S. L. 1980. Protein and enzyme levels during development in a high lysine barley *Hordeum vulgare* mutant. *Phytochemistry* 19:1323–28

161. Sheridan, W. F., Neuffer, M. G. 1982. Maize developmental mutants. *J. Hered.* 73:318–29

162. Shewry, P. R., Pratt, H. M., Leggatt, M. M., Miflin, B. J. 1979. Protein metabolism in developing endosperms of high lysine and normal barley. *Cereal Chem.* 56:110–17

163. Shii, C. T., Mok, M. C., Temple, S. R., Mok, D. W. S. 1980. Expression of developmental abnormalities in hybirds of *Phaseolus vulgaris* L. *J. Hered.* 71:218–22

164. Silvers, W. K. 1979. *The Coat Colors of Mice.* Berlin: Springer. 379 pp.

165. Singh, B. B., Jha, A. N. 1978. Abnormal differentiation of floral parts in a mutant strain of soybean. *J. Hered* 69:143–44

166. Sinnott, E. W. 1960. *Plant Morphogenesis.* New York: McGraw-Hill. 550 pp.

167. Smith, H. H. 1974. *Nicotiana.* See Ref. 13, pp. 281–314

168. Snoad, B. 1969. Neoplastic pods. *Pisum Newsl.* 1:19–20

169. Snoad, B. 1981. Plant form, growth rate, and relative growth rate compared in conventional, semi-leafless and leafless. *Sci. Horic.* 14:9–18

170. Soberalske, R. M., Andrew, R. H. 1980. Gene effects on water soluble polysaccharides and starch of near-isogenic lines of sweet corn. *Crop Sci.* 20:201–4

171. Sorrells, M. E., Myers, O. Jr. 1982. Duration of developmental stages of 10 milo maturity genotypes. *Crop Sci.* 22:310–14

172. Spray, C., Phinney, B. O. 1982. The growth response of dwarf-5 and dwarf-1 mutants to 38-substituted derivatives of the biologically active maize gibberellin, GA_1. *Plant Physiol.* 69S:Abstr. 132

173. Statham, C. M., Crowden, R. K. 1972. Biochemical genetics of pigmentation in *Pisum sativum. Phytochemistry* 11:1083–88

174. Statham, C. M., Crowden, R. K. 1974. Anthocyanin biosynthesis in *Pisum:* sequence studies in pigment production. *Phytochemistry* 13:1835–40

175. Steffensen, D. M. 1968. A reconstruction of cell development in the shoot apex of maize. *Am. J. Bot.* 55:354–69

176. Steiner, E. 1974. *Oenothera.* See Ref. 13, pp. 223–45

177. Stelly, D. M., Palmer, R. G. 1982. Variable development in anthers of partially male-sterile soybeans. *J. Hered.* 73: 101–8

178. Steward, A. D., Hunt, D. M. 1982. *The Genetic Basis of Development.* New York: Wiley. 221 pp.

179. Styles, E. D., Ceska, O. 1977. The genetic control of flavonoid synthesis in maize. *Can. J. Genet. Cytol.* 19:289–302

180. Styles, E. D., Ceska, O., Seah, K. T. 1973. Developmental differences in action of R and B alleles in maize. *Can. J. Genet. Cytol.* 15:59–72

181. Teas, H. J. 1957. Physiological genetics. *Ann. Rev. Plant Physiol.* 8:393–412

182. Tigchelaar, E. C., McGlassom, W. B., Buescher, R. W. 1978. Genetic regulation of tomato fruit ripening. *HortScience* 13:508–13

183. Tsuchiya, T. 1969. Characteristics and inheritance of radiation-induced mutations in barley. *IAEA Bull.* SM 121/ 36:573–90

184. Tullloch, A. P. 1976. Chemistry of waxes of higher plants. See Ref. 84, pp. 235–87

185. Ullrich, S. E., Eslick, R. F. 1978. Inheritance of the associated kernel characters, high lysine and shrunken endosperm, of the barley mutant Bomi, Riso 1508. *Crop Sci.* 18:828–31

186. Ullrich, S. E., Eslick, R. F. 1978. Lysine and protein characterization of spontaneous shrunken endosperm mutants of barley. *Crop Sci.* 18:809–12

187. Ullrich, S. E., Eslick, R. F. 1978. Lysine and protein characterization of induced shrunken endosperm mutants of barley. *Crop Sci.* 18:963–66

188. Waddington, C. H. 1957. *The Strategy of the Genes.* London: Allen & Unwin. 262 pp.

189. Walbot, V., Thompson, D., Coe, E. H. Jr., Johri, M. M. 1979. Meristem function during maize development. *34th Ann. Corn Sorghum Res. Conf.* 92–103

190. Wallace, D. H., Enriquez, G. A. 1980. Daylength and temperature effects on days to flowering of early and late maturing beans (*Phaseolus vulgaris* L.) *J. Am. Soc. Hortic. Sci.* 105:583–91

191. Wehner, T. C., Gritton, E. T. 1981. Horticultural evaluation of eight foliage types of peas near-isogenic for the genes *af, tl,* and *st. J. Am. Soc. Hortic. Sci.* 106:272–78

192. Weibel, D. E., Sotomayer-Rios, A., Pava, H. M., McNew, R. W. 1982. Relationship of black layer to sorghum kernel moisture content and maximum kernel weight in the tropics. *Crop Sci.* 22:219–23

193. Wellensiek, S. J. 1969. The physiological effects of flower forming genes in pea. *Z. Pflanzenphysiol.* 5:388–402

194. Wettstein-Knowles, P. von. 1968. Mutations affecting anthocyanin synthesis in the tomato. I. Genetics, histology, and biochemistry. *Hereditas* 60:317–46

195. Wettstein-Knowles, P. von. 1969. Mutations affecting anthocyanin synthesis in the tomato. II. Physiology. *Hereditas* 61:255–75

196. Wettstein-Knowles, P. von. 1979. Genetics and biosynthesis of plant epicuticular waxes. In *Advances in the Biochemistry and Physiology of Plant Lipids,* ed. L. A. Appelqvist, C. Liljenberg, pp. 1–26. Amsterdam: Elsevier. 461 pp.

197. Wettstein-Knowles, P. von. 1982. Biosynthesis of epicuticular lipids as analyzed with the aid of gene mutations in barley. In *Biochemistry and Metabolism of Plant Lipids,* ed K. F. Wintermans, P. J. C. Kuiper. Amsterdam: Elsevier. In press

198. Whitaker, T. W. 1974. *Cucurbita.* See Ref. 13, pp. 135–44

199. Wilkinson, R. E., Cummins, D. G. 1981. Epicuticular fatty acid, fatty alcohol, and alkane contents of bloom and bloomless sorghum "Redbine 60" leaves. *Crop Sci.* 21:397–400

200. Zeiger, E., Stebbins, G. L. 1972. Developmental genetics in barley. A mutant for stomatal development. *Am. J. Bot.* 59:143–48

Ann. Rev. Plant Physiol. 1983. 34:419–40

CONCEPT OF APICAL CELLS IN BRYOPHYTES AND PTERIDOPHYTES

E. M. Gifford, Jr.

Department of Botany, University of California, Davis, California 95616

CONTENTS

INTRODUCTION

The purpose of this review is to examine and give examples of bryophytes (liverworts and mosses) and certain lower vascular plants (pteridophytes) that demonstrate the presence of a definite apical cell. Most of the recent investigations on lower vascular plants pertain to ferns and *Equisetum* (horsetails, scouring rushes). These studies have led to conflicting concepts as to the function and importance of an apical cell in root and shoot growth. No attempt will be made to review apical organization in *Lycopodium, Selaginella,* or *Isoetes* because of space limitations. The reader is referred to reviews that incorporate these genera (7, 97), and especially to recent studies of *Isoetes* (75, 76).

The establishment of apical growth and dominance were important events in the evolution of green land plants. The role of auxin in apical dominance is

419

fairly well understood, but its interaction with other hormones and hormonal control of apical organization is less well understood. Many bryophytes and lower vascular plants have a conspicuous apical cell in their apical meristems, and the question might be asked: what role does an apical cell play in development? Is it an initial cell and hence the ultimate source of all cells of roots and shoots? Does it have only a regulatory role in development? Does it play more than one role? For the moss gametophyte the answer is yes for the first question and possibly for the second. For ferns there are as yet no definitive answers (especially for shoot apices).

Some investigators consider the apical cell to be mitotically quiescent, or nearly so. Their concepts will be discussed at appropriate places in the text, and the results from recent studies of histogenesis, cell kinetics, and histochemistry will be presented for evaluation.

Throughout this review the terms "segments" and "merophytes" (31, 44) will be used interchangeably in referring to immediate, unicellular derivatives of an apical cell and to the multicellular structural units derived from them (Figures 1, 2, 4). The boundaries of a multicellular merophyte or segment are often recognizable some distance from the site of origin, especially in roots (Figure 2).

BRYOPHYTES

Descriptions of apical organization of bryophytes will be concerned mainly with the gametophytic (haploid) phase of the life cycle of liverworts and mosses. Some attention will be given to early sporophyte development in mosses.

Liverworts

A thalloid liverwort gametophyte is usually a prostrate, green, dorsiventral, dichotomously branched structure with a more or less conspicuous midrib. The apical meristems are situated in a notch or crease at the tip of each branch. There are varying reports as to whether there is a row of apical cells or whether one can distinguish a specialized apical cell within the group.

Most of our knowledge on apical organization in thalloid liverworts is based upon research from the end of the last century and early part of this century. *Riccia* is considered to be a primitive form. At the surface of the apical meristem, Campbell (14) reported that there is a row of three to five or more apical cells. Segments are cut off mainly toward the dorsal side and soon divide periclinally (new walls parallel to the surface). Subsequent divisions give rise to the bulk of the parenchymatous tissue. Ventral segments may give rise to some parenchyma, rhizoids, and scales. Dichotomous branching results from longitudinal divisions, separating two groups of apical cells. Apical growth in

Marchantia is very similar to that of *Riccia,* but the presence of a definite apical cell was described for *Asterella* (51) and *Targionia* (30). It is obvious that much could be learned from additional investigations of these plants with the use of modern techniques.

In the Jungermanniales the gametophyte may be either a simple foliaceous type or is differentiated into leaves and stem. Although phylogenetically the leaves of leafy liverworts and mosses are not homologous with those of vascular plants, "leaf" and "stem" will be used in a descriptive sense in this review. In all cases there is a single apical cell, and the form of the plant is the result of the process of development of segments cut off by the apical cell. In the foliaceous subgroup the apical cell is two-sided (dolabrate) and development is very precise, as in some species of *Fossombronia* (50) and in Pallavicinia (49). In those species in which there is a definite distinction between stem and leaves, the leaves lie in three vertical series, corresponding to the three cutting faces of the pyramidal apical cell. The apical cell lies with one cutting face toward the ventral side and two toward the dorsal side. Leaves are formed from the dorsal segments. Segments cut off on the ventral side may function in the same manner or they may give rise only to stem tissue. Branching is reported never to be truly dichotomous, i.e. the apical cell never divides into two equal apical cells as in the alga *Dictyota.*

Basile (5a and references therein) demonstrated the developmental effects of exogenously applied hydroxy-L-proline and 2,2'-dipyridyl on leafy liverworts. *Plagiochila artica* normally has dorsiventral symmetry with two rows of leaves arising from lateral derivatives of the apical cell. Ventral leaf formation, however, is suppressed soon after the first division of a merophyte. In contrast, in the presence of these antagonists of hydroxyproline-containing proteins, growth and development of ventral merophytes are not suppressed, resulting in the formation of ventral leaves and radial symmetry of the gametophyte. Basile concluded that hydroxyproline-containing proteins, through interaction with other components of cell walls, regulate several aspects of morphogenesis in leafy liverworts. Additional interesting experiments are anticipated.

Mosses

While most of the literature on liverworts is quite old, interest in apical structure of mosses remains active. In mosses there are two phases in the development of a gametophyte. A spore germinates to form a thalloid structure (e.g. *Sphagnum*) or a filamentous protonema from which the upright leafy branches (gametophores) develop. Sex organs (antheridia and archegonia) are produced on the gametophores.

Phyllotaxy in mosses can be distichous (two rows of leaves), tristichous (three distinct rows of leaves), or helical in which phyllotactic ratios comparable to higher plants are evident (15). In those mosses (e.g. *Fissidens*) with distichous phyllotaxy, the apical cell has two cutting faces, which accounts for this type of leaf arrangement. An example of tristichous arrangement is *Fontinalis,* in which the three ranks of leaves are on line with the three cutting faces of the tetrahedral apical cell. In the helical types of phyllotaxy, the leaves are initiated from the three cutting faces of the apical cell, but in a slightly different manner. The new wall (separating the apical cell and newly formed segment) is not parallel to a lateral (cutting face) wall of the apical cell. The new wall forms an angle of about 20° to the lateral wall of the apical cell, for example, in *Anomodon viticulosus* (10) and *Polytrichum formosum* (46). The newly formed segment is not bilaterally symmetrical; in transverse section, it is wedge-shaped in anticipation of helical phyllotaxy. Earlier it was thought that torsions developing in the stem resulted in helical patterns, but later work has shown that the pattern is established initially at the shoot apex (10, 46, 47, 101).

The apical cell in mosses is quite large, as in *Anomodon* (10, 11), where dimensions may be 50 μm long and 16.5 μm wide, and leaf and axis development can be related to activity of the apical cell. However, the apical cell has rarely been seen in mitosis. This prompted Ho (57) to propose that the apical cell is not active during vegetative growth, because it was never observed in division. It would become active at the time of reproduction in the formation, directly or indirectly, of antheridia and archegonia, somewhat analogous to the activation of the "méristème d'attente" in angiosperms at the time of flowering. However, one must infer that divisions do occur when one considers the plane of division of an immediate segment (see later). Others have commented on the low frequency of mitosis, but nevertheless some apical cells were found in division (6, 11, 52, 53, 68). In a histochemical, autoradiographic, and cytophotometric study of *Polytricum formosum,* Hallet (47) reported that the apical cell divides once in 360 hr, in 72 hr for cells of leaf primordia, 96 hr for immediate derivatives of the apical cell, and 120 hr for cells of the leaf bases. The duration of mitosis was 3.6 hr so that the probability of finding a division figure is $\frac{3.6}{360}$ = 1%, which matched the mitotic index (percentage of cells in mitosis in a given cell population) of 1.0% for the apical cell. Also, the apical cell has a rather unique cytology: large nucleus and nucleolus, cytoplasm rich in RNA, like a meristematic cell of higher plants. At the same time it has an extensive vacuolar system and small plastids with starch—characteristic of differentiated cells. From cytophotometric measurements of DNA, the apical cell, immediate derivatives, leaf primordia, and cells of the leaf bases were C-2C, with no evidence of endopolyploidy. On the other hand, the leptoids (phloem-like tissue) and hydroids (xylem-like tissue) undergo endopolyploidy, sometimes becoming octoploid.

Hébant (52) reported that the apical cell of *Polytrichum commune* divided on the average of two times every 3 days during an active growth period. He also noted a difference in polarity in the apical cell during quiescence and active growth periods. During quiescence, the nucleus occupies the upper part of the cell, along with a zone of dense cytoplasm. When mitotically active, the nucleus occurs in a more central position, whereas the dense cytoplasm moves laterally. After division, the dense cytoplasm is found in the apical cell, and the segment inherits the more vacuolate portion of cytoplasm (53). Perhaps the numerous microtubules observed in the cytoplasm of the apical cell during interphase could be involved in these protoplasmic movements.

The first division of a segment is periclinal (with reference to the shoot apex) and the new wall is strictly perpendicular to a cutting face of the apical cell (Figure 1), or slightly oblique (6, 46, 52). For *Polytrichum commune* the outer cell (toward the surface) divides, giving rise to two cells: the upper cell forms a projection that may partially cover the free apex of the shoot apical cell and which will give rise to a bilateral apical cell of the leaf primordium. The lower cell contributes, by successive divisions, to the formation of the leaf base tissue and cortex. The original inner cell of the first segment division is the origin of a cylinder of central hydroids, leptoids, and parenchyma-like cells (52).

Despite the low frequency of mitosis in the apical cell of mosses, the apical cell is essential in giving rise to segments which are then responsible for the development of leaves and the gametophytic axis. The first division of a segment is perpendicular or oblique to a lateral face of the apical cell (Figure 1).

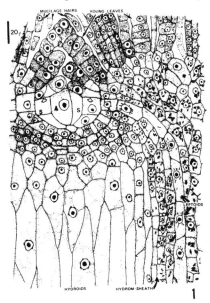

Figure 1. Longitudinal section, shoot tip of *Polytrichum commune*. A, apical cell; S, segment (merophyte); open arrow, cell wall resulting from first division of a segment. (Modified from *The Conducting Tissues of Bryophytes*, A. R. Gantner Verlag.)

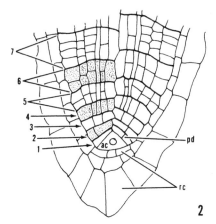

2

Figure 2. Longitudinal section, *Azolla* root. ac, apical cell; pd, results of first cell division (periclinal) in a merophyte; rc, root cap; 1,2,3,..., unicellular merophyte to multicellular merophytes [from (80)].

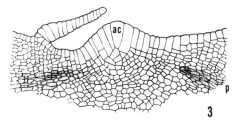

3

Figure 3. Longitudinal section, shoot apex of *Polypodium;* ac, apical cell; p, procambium [modified from (99)].

If a segment divided so that the new wall were parallel to an apical cell lateral face, one could imagine that the apical cell would not be essential for providing new units to the shoot. However, this is not the case. Most investigators agree that the apical cell is an initial in that all units of the shoot can be traced back to it (6, 46, 47, 101). A comparable analysis of the growth of shoots of most lower vascular plants with apical cells becomes much more difficult because of apical size and the complex nature of organogenesis.

Discussion of bryophyte apical growth has pertained mainly to the haploid phase (gametophyte). In most bryophytes the sporophyte (sporogonium) consists of a foot embedded in the gametophyte, a seta or stalk, and a spore capsule. It is only in the more advanced mosses that a definite apical cell is established early in embryogeny at the future capsule-end and at the foot-end of the sporophyte. The apical cell involved in histogenesis of the capsule and part of the seta has two cutting faces and may be active for varying lengths of time (19).

FERNS AND EQUISETUM

Root Apex

HISTOGENESIS   The majority of ferns and all investigated species of *Equisetum* display a conspicuous tetrahedral apical cell in the root that is the focal point of the network of cell walls (Figure 2). The histogenic role played by the apical cell has been the subject of interest and discussion for over 100 years. Does the cell have special inherent qualities as a continuing initial or is it no more than a gap in a system of confocal intersecting cell walls? Before describing the results of recent investigations on the apical cell, a few comments will be made on eusporangiate ferns (Marattiaceae and Ophioglossaceae) and the Osmundaceae; the latter family is generally included with the better known leptosporangiate ferns. The roots of established *Marattia* and *Angiopteris* are said to have two or even four equal apical cells (97), although Schüepp (86) has presented an analysis of the *Angiopteris* root meristem in which a cubical apical cell could account for described cell networks and the apparent lack of a typical, single tetrahedral cell. For *Ophioglossum*, however, the presence of a single apical cell has been well-documented, as well as its relationship to root-borne buds (32, 66, 81, 84). The Osmundaceae are considered by some botanists to occupy an intermediate position (on the basis of several attributes) between the Marattiaceae and leptosporangiate ferns; thin roots of *Osmunda* are reported to have a single apical cell and thick roots may have two or more equal apical cells (97). Root apical organization in the Marattiaceae and Osmundaceae needs to be reinvestigated.

As mentioned previously, the root apical cell of most ferns typically is a pyramidal tetrahedron with the base directed toward the root cap (Figure 2). Merophytes are produced from the three lateral (proximal) faces of the apical cell (if one acknowledges that it divides regularly) and constitute the basic building units of the root. From the distal face, merophytes are initiated that contribute to the root cap. In some ferns (e.g. *Azolla*) the apical cell does not contribute to the root cap, because early in root initiation a separate meristem, the calyptrogen, is formed which is responsible for continued root cap renewal. The first division of a laterally produced merophyte is usually periclinal (with reference to the surface of the root), resulting in the formation of two cells, often of unequal size (Figure 2, 6A, B). Radial longitudinal divisions then occur in one or both of the cells, which result in the formation of two "sextant sectors." Similar divisions occur in all newly initiated merophytes. Continued cell divisions in the sextants establish the basic tissue regions of the root, and the six sectors or "sextants" often can be easily identified, even at considerable distances from the root tip. A root cap merophyte divides longitudinally, followed by a longitudinal division in each of the two cells. Additional cell divisions in older merophytes give rise to the multicellular root cap.

In understanding root histogenesis, the first division in a merophyte is important. Whether the plane is periclinal or radially longitudinal (Figure 6A, B) means that all new merophytes must originate from the apical cell (96). If the initial division were parallel (Figure 6B) to the lateral face of the apical cell, one could envision that the apical cell is not essential as an initiating cell; new merophytes could be added by cells adjacent to the apical cell. However, this is not the case in any investigated fern or for *Equisetum scirpoides* (35). In the older literature, the plane of the first division in a merophyte was stated to be radially longitudinal, but Conard (23) and Bartoo (4, 5) reported that the first division is periclinal in both *Dennstaedtia* and *Schizaea*. Similar results were obtained recently for *Ceratopteris thalictroides* (17) and two species of *Azolla* (44, 80).

EXPERIMENTAL AND QUANTITATIVE STUDIES   Despite the well-established mode of root histogenesis, some investigators believe that the apical cell is mitotically quiescent, somewhat comparable to the quiescent center of investigated angiosperms. Others believe that quiescence spreads to adjacent cells so that a multicellular quiescent center develops in a fashion even more comparable to investigated angiosperm roots.

The following discussion is a chronology of research that led to the concept of mitotic quiescence and the contradictory evidence presented recently. The first modern study of relative mitotic activity of the apical cell and of other regions of the root was made by Buvat & Roger-Liard (13) for *Equisetum arvense*. By plotting the sites of mitoses for many apices, they concluded that there were essentially no differences in mitotic activity between the apical cell and surrounding cells. On the basis of incoropration of labeled precursors, Clowes (20, 21) reported that the apical cell of *Azolla* synthesized DNA and protein at the same rate as adjacent cells. He concluded that there is no quiescent center in the *Azolla* roots. Gifford (33) reported the localization of ^{3}H-thymidine in the apical cell nucleus of *Ceratopteris thalictroides*, which indicated that DNA synthesis does occur. Gifford (33) stated that if the synthesis of DNA does not lead to endopolyploidy, then the results confirmed the concept that the apical cell does divide. D'Amato & Avanzi (27, 28) pointed out [as Gifford (33) had implied] that one cannot determine whether synthesis of DNA heralds mitosis or may result in endopolyploidy. D'Amato & Avanzi (27) and Avanzi & D'Amato (1, 2) concluded, on the basis of autoradiography and cytophotometry, that the apical cell of roots of lower vascular plants is quiescent and blocked at the post-DNA synthetic phase (G$_2$). According to them, the apical cell of *Marsilea strigosa* "divides rarely, if at all," even though they observed mitoses in 2 of 13 apical cells of main roots and in 5 of 62 cells of lateral roots (27). They reported that the apical cell often was not labeled with ^{3}H-thymidine, while labeling was present in adjacent cells. The results of cytophotometric measurements of DNA revealed that the apical cell generally contained a 4C amount (the G_2 phase of a diploid mitotic cycle), some were 8C (the results of endoreduplication or endomitosis), or even 16C. Similar results were obtained for other ferns: two species of *Blechnum, Polypodium aureum,* and *Ceratopteris thalictroides* (1, 2, 26). The apical cells and surrounding cells incorporated ^{3}H-uridine and ^{3}H-lysine, but to a lesser extent in larger primorida and established roots. Localization of nuclear histones revealed that the nuclei of apical cells and immediately surrounding cells stained much more lightly than other cells. No quantitative data were obtained. However, these results prompted Avanzi and D'Amato to reevaluate their previous data as follows: they suggested that endoreduplication of DNA in the apical cell may signify that it is active in the amplification of ribosomal genes. The light staining of nuclear histones could signify a reduction of histones that would allow DNA-dependent RNA synthesis to occur (2). Sossountzov (88) also provided evidence for mitotic quiescence from a study of *Marsilea drummondii;* only 17% of the apical cells were labeled with ^{3}H-thymidine during a 24 hr period. She concluded that the apical cell cannot be considered as an initial with special histogenic properties.

In summary, D'Amato and Avanzi propose that the root apical cell of *Marsilea* (and possibly of all other lower vascular plants with a single apical cell) is active mitotically only in the early initiation and organization of a root primordium. They further propose that in older roots the apical cell, as well as adjacent cells, undergo endopolyploidy and/or gene amplification and are engaged in macromolecular syntheses rather than cell division. The concept that a mitotically inactive apical cell (including a few surrounding cells) may be analogous to the quiescent center of angiosperm roots is appealing and has been endorsed. For example, Clowes (22) reversed his earlier opinion, and Torrey & Feldman (95) called attention to this inviting comparison. However, Clowes (22a) has returned to his original position that there is no quiescent center in roots with a tetrahedral apical cell—the geometry of the root apex "no longer demands a quiescent center."

The concept of mitotic quiescence has been challenged recently by several investigators on the basis of histogenesis, determination of the mitotic index (percentage of mitoses in a given cell population), determination of the duration of the cell cycle and of mitosis, and measurements of DNA content. From data on the initiation and growth of lateral roots over a given period of time, combined with the results of cell lineage studies and planes of cell divisions, the apical cell of *Ceratopteris thalictroides* was reported to divide as frequently as its derivatives, although mitotic activity declined and the duration of the cell cycle increased as roots became older (17).

Two species of the floating water fern *Azolla* have been investigated in great detail. Unlike many other ferns, the roots of *Azolla* become determinate very early and may reach only a few centimeters in length. Merophytes are produced from the three lateral faces of the apical cell, and a very precise number of divisions in them precedes any appreciable cell enlargement. Merophyte boundaries can be identified at some distance from the root tip (Figure 2). By careful analysis of root histogenesis, Gunning et al (44) reported that the apical cell of A. pinnata undergoes a total of about 55 divisions by the time a root is only 1 mm in length. Subsequent growth is the result of cell division and cell enlargement within the merophytes. For A. filiculoides, the mitotic index (MI) and duration of the cell cycle were determined for selected cell populations of 1-mm and 5-mm roots. The apical cell MI was highest in 1-mm roots but declined markedly in 5-mm roots. There was a similar decline, but not as dramatic, in immediate derivatives and cells of the remaining meristem. The cell cycle duration of the apical cell was 23 hr in 1-mm roots (determined by the colchicine-induced metaphase accumulation technique) but increased dramatically in 5-mm roots (193 hr). These data correlated well with the decrease in mitotic index of 5-mm roots (39, 80).

Chiang (16) investigated the effects of prolonged treatment of Ceratopteris pteridoides roots with colchicine. For a 24-hr treatment with 0.05% of the drug, 50% of the apical cells of roots 3-5 mm in length were multinucleate. An apical cell with two nuclei was considered to be the result of one mitosis; three nuclei, the result of two mitoses; four or more, the result of three or more mitoses. The presence of two or more nuclei in the same cell is good evidence that nuclear divisions have occurred, although Polito (82) has suggested that the supernumerary nuclei may be the result of fragmentation of a restitution nucleus. Similar results were obtained for roots of Ophioglossum petiolatum, Marsilea crenata, and Ceratopteris thalictroides; no cells adjacent to the apical cell possessed more nuclei than the apical cell of the same root (18).

Kuligowski-Andrès (62) concluded that the root apical cell of Marsilea vestita remains active throughout the growth phase, becoming only somewhat quiescent late in root development. The apical cell of young root primordia incorporated ^{3}H-thymidine in 40% of the roots. In older roots fewer apical cells were labeled, but labeling was never reduced to zero. By careful analysis of the production of segments and subsequent divisions in them, Vallade & Bugnon (96) concluded that the mitotic rate (authors used the term rhythm) of the apical cell of M. diffusa was somewhat more than three times greater than that of other cells of the meristem, and they consider the apical cell to be an initial in the classical sense.

In an intensive study of Marsilea vestita, Kurth (64) has provided compelling evidence for the initiating role of the apical cell. Roots of five length classes, ranging from 1–120 mm, were analyzed from collections made at 2-hr intervals for 24 hr to determine the MI, duration of the cell cycle, and ploidy levels. Data were obtained for the apical cell, five regions of the root proper, and two regions in the root cap. The mitotic index of the apical cell tended to be above the overall mean mitotic index for the entire meristem. The cell-cycle duration of the apical cell, determined by the metaphase accumulation technique, ranged from 12.1–25.2 hr, that of the other regions from 16.1–41.5 hr. The DNA content of the apical cell of other cell populations was determined by measuring the amount of fluorescence of nuclei stained with pararosaniline. The apical cell of all classes of root lengths never contained more than the 4C amount of DNA. A few nuclei of other selected cell populations contained DNA contents well above the 4C level. From the total data, one would have to conclude that the apical cell performs as a normal cycling cell and does not undergo endopolyploidy. Comparable results were obtained for *Equisetum scirpoides* (35).

CONCLUDING REMARKS   For roots it is apparent that there is no uniformity in the results obtained on the description of apical cells, even for species of the same genus. Evidence for the concept of quiescence relies heavily upon reported endopolyploidy and the large nuclear size of the apical cell especially, and of surrounding cells as a root grows in length. Evidence for the opposing view (that the apical cell is mitotically active) is derived from histogenesis, mitotic index data, determination of the cell-cycle duration, and cytophotometric measurements of DNA content. The disparity in DNA measurements is difficult to explain, although different methods of measurement were used by the two groups. Also, Feulgen staining (use of basic fuchsin) or the use of pararosaniline (one of the constituents of basic fuchsin) can result in quite different staining intensities and patterns, even after the same fixative is used (64).

The choice of material may also affect the results and conclusions drawn from any investigation. The main, rhizome-borne roots of some ferns may become very long. Lateral roots may be produced but often remain quite short. In others (e.g. *Azolla*), the rhizome-borne roots remain unbranched and reach a final length of only a few centimeters. In the latter example, the apical cell probably produces all of the merophyte initials by the time a root is one to a few millimeters in length (44, 80). Gunning (42) has suggested that there is a programmed senescence for the determinate *Azolla pinnata* root in that beginning at about the thirty-fifth apical cell division there is a progressive diminution in frequency of plasmodesmata between the apical cell and lateral derivatives. The result is that the apex of the root, particularly the apical cell, becomes more and more isolated symplastically, which might explain its limited life span. Determinate lateral roots on main roots of other taxa may behave in similar fashion. The apical cell may become quiescent and even undergo endopolyploidy, but probably only after forming a more or less definite number of merophytes. This may be the situation for the lateral roots of *Marsilea strigosa,* for which Cremonini (24) described mitotic quiescence of the apical cell. If, in a lower vascular plant, an inhibited lateral root assumed dominance upon damage to the main root, and if the apical cell had previously undergone endopolyploidy, all cells of the new portion of the lateral root conceivably could be polyploid. An experimental study of reactivated lateral roots would be worthwhile.

Finally, some comments of Barlow (3) are relevant to root organization. The roots of angiosperms have a conspicuous quiescent center (QC), the cells of which progress through the mitotic cycle slowly, have a long G_1 phase, and an indeterminate reproductive (cell division) life-span. Barlow refers to these cells as "founder" cells. Surrounding the QC are initial cells and their derivatives, both of which progress through the mitotic cycle quickly, and the derivative cells have a determinate reproductive life-span. If an apical cell of a pteridophyte root divides as freqently as its surrounding derivatives, for example, as reported for *Marsilea* (64) and *Equisetum* (35), then it could be considered both an initial cell and a founder cell. If it divides only rarely during active root growth, it would be a founder cell. I favor the former view.

Shoot Apex

HISTOGENESIS   In many ferns it is often difficult to relate the origin of leaves and buds directly to derivatives of the shoot apical cell in conventionally prepared longitudinal sections. Exceptions are ferns with small and/or elongate shoot tips. Moss leaves and cells of merophyte sectors of fern roots, in general, can more readily be traced back directly to derivatives in contact with the apical cell. However, Bierhorst (8) maintained that this also can be done for large fern shoot tips if the surface of the apex is examined. Ferns occupied an important place in the formulation and elaboration of the Apical Cell Theory by nineteenth century botanists, e.g. Nägeli and Hofmeister. Subsequent research showed that the concept could not be applied universally to vascular plants but could be applied to shoot apices of the majority of ferns, *Equisetum,* certain species of *Selaginella,* and *Psilotum.* In shoots, the apical cell is commonly tetrahedral and comprised of three initiating faces (Figures 3, 4, 7), or is three-sided with two "cutting" faces. The latter type is reported to be associated with certain rhizomatous ferns. Developing fern leaves also have a conspicuous apical cell that may be tetrahedral initially, but which, after 1 or 2 divisions, usually becomes bilateral (8, 54).

Support for presence of a single apical cell and its reputed role in shoot growth can be found in representative publications that appeared in the first half of the present century (5, 23, 85, 99). These studies were of a descriptive nature, but nevertheless the authors concluded that the apical cell is the focal point of growth and the ultimate source of new building units (segments or merophytes) even though many cell divisions may occur in them before the initiation of leaves, buds, roots, and the differentiation of stem tissues. For the small, fingerlike apex of *Ceratopteris*, Hébant-Mauri (56) was able to relate leaf initiation to specific segments derived from the apical cell [as well as for *Trichomanes* (54)], as Bonnet (9) did for *Salvinia*, and Gifford & Polito (37, 38) for *Azolla*. In the latter example, a detailed study was not made of leaf initiation, but mitotic indices and rates of cell division leave very little doubt about the mitotic activity of the apical cell. Hébant-Mauri (55) concluded that the apical cell of the large apex of *Dicksonia* is mitotically active in that the boundaries of merophytes derived from the apical cell are very sharp and can be traced back to the apical cell.

The strongest proponent of the histogenic importance of the apical cell is Bierhorst (8). He examined the surface of partially cleared slabs of the shoot apices of more than 50 filicalean ferns. He maintains that an examination of longitudinal sections does not provide suitable evidence for localization of the apical cell. Examination of the surface reveals cell packets (merophytes) that can be traced back to the apical cell and which can be correlated with phyllotaxy (Figure 4). By counting the number of cells in surface view and their relationship to the number of cells seen in longitudinal section for the same merophyte, he was able to estimate the frequency of cell divisions. In general, merophyte cells had a higher frequency, but in one example the apical cell appeared to divide as frequently as cells in young merophytes. Bierhorst cites another interesting fact: no phenotypic sectorial chimeras have ever been described for the 7500–8000 species of living ferns, which might not be the case if the apex were a generalized meristem.

The effects of mutagenic agents (river pollutants) on ferns and their relationship to the presumed presence of mitotically active shoot and leaf apical cells have been described recently (58–60).

The dormant shoot tip of *Onoclea*, when viewed in polarized light, shows distinct multicellular domains with respect to surface wall microfibrils which can be related to successive segmentations of the apical cell. There are three sectors for the shoot, matching the three lateral initiating faces of the tetrahedral apical cell and two sectors for the leaf tip that can be related to the lenticular or bilateral leaf apical cell (65). Additional studies on the biophysics of cell-pattern development in ferns would be desirable, to include cellulose microfibril orientation (41) and position of prophase microtubule bands (43). The first division in a newly initiated merophyte could occur in three different planes (Figure 7): 1. in a plane parallel to a lateral face of the apical cell (Figure 7C); 2. transversely [periclinal, with reference to the surface of the shoot apex (Figure 7D)]; 3. radially longitudinal (Figure 7E). There are many examples of type 2 in ferns, but not of type 1. In *Equisetum*, however, the first division is of type 1, followed by radial longitudinal divisions in the two cells. Types 2 and 3

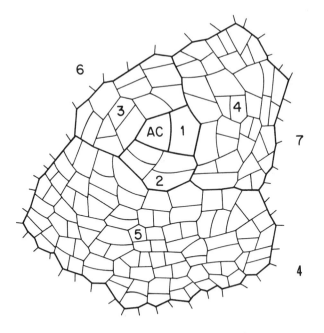

Figure 4.   *Osmunda regalis*, surface view of shoot apex showing merophyte sectors, 1,2,3,...;
AC, apical cell [based on (8)].

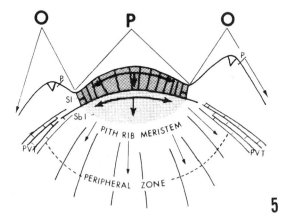

Figure 5.   Schematic representation of apical zonation in the shoot tip of a fern. O, organogenic
zone, to include the provascular tissue (PVT); p, promeristem; p, leaf primordia; SbI, subsurface
initials; SI, surface initials [from (67)].

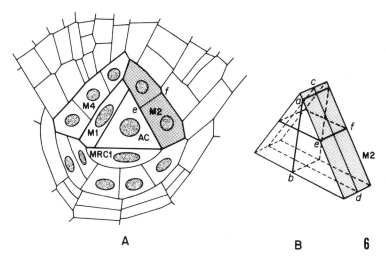

A   B   6

Figure 6. Schematic representation of longitudinal section of a fern root tip (A) and possible planes of first division in a merophyte (B). (A), merophytes M1, M2, and M4 contributed proximally to root proper (M3 is out of plane of section); MRC1, youngest root cap merophyte; AC, apical cell; e–f, cell wall resulting from periclinal division (in reference to root surface in M2. (B), a–b, radial longitudinal cell wall; c–d, wall parallel to a lateral face of the apical cell; e–f, transverse wall, periclinal with reference to root surface, and comparable to e–f in (A).

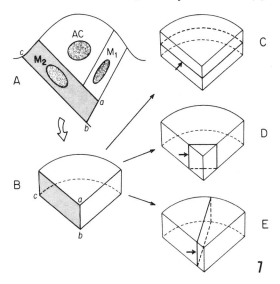

Figure 7. Schematic representation of longitudinal section of shoot tip (A), merophyte 2 (B), and possible planes of division in M2 (C,D,E). In (C) the new cell wall (arrow) would be parallel to a lateral face of the apical cell, AC; D, transverse division (or periclinal with reference to surface of shoot tip); E, radial longitudinal division.

especially lend credence to the concept that the apical cell does divide because it is difficult to understand how a two-celled merophyte could give rise to a younger, unicellular merophyte that is immediately adjacent to the apical cell.

APICAL ZONATION   Some authors see merit in describing fern apices on a zonate basis (29, 61, 72, 87, 88, 92), similar to descriptions of seed plants (34). McAlpin & White (67) adopted this point of view in a study of 22 genera. Their descriptions were based on longitudinal and transverse sections. They favor the term promeristem that includes two cell layers: surface and subsurface initials. The apical cell, when identifiable, would be included in the layer of surface initials. The surface initials divide anticlinally in the perpetuation of cells at the surface, and periclinally in adding cells to the subsurface group which, in turn, continue to add cells to the future procambium and ground tissue (Figure 5). The protoderm gradually becomes individualized from the surface layer; leaf primordia arise from a cell complex that consists of both surface and subsurface cells. A somewhat similar zonate pattern was reported for *Botrychium* and *Dennstaedtia* in which localization of cell constituents supported the concept (93, 94). The above authors, and also Hagemann (45), have cited similarities in organization of the zonate apices in ferns and seed plants, especially gymnosperms. The two concepts—emphasis on the importance of the apical cell or upon zonation—do not seem to be mutually exclusive. Bierhorst's (8) results are compelling in showing that an apical cell can be identified if the apical surface is examined, but in terms of subsequent planes of cell division, degrees of cell differentiation, etc, the concept of zonation can be meaningful.

EXPERIMENTAL AND QUANTITATIVE STUDIES   Wardlaw (98), in recent times, was the first to perform extensive experiments on the fern apex. If the apical cell was punctured, leaves continued to arise in their normal phyllotactic pattern until all the available space was used. He concluded that the apical cell does not determine leaf position, nor is it responsible for leaf initiation. If the apical cell and adjacent cells were separated from young leaf primordia by a tangential incision that partially severed tissue connections between them, a potential leaf primordium developed as a bud rather than a leaf. Reviews of subsequent refinements of these early studies are available in the literature (25, 48, 100).

    In addition to surgical and in vitro growth studies, emphasis has been placed on determining the mitotic index, rates of mitosis, and DNA content of the apical cell and of other selected cell populations. Buvat & Liard (12) recorded the distribution of mitoses in the shoot apical meristem of *Equisetum arvense*. They concluded that cells subjacent to the apical cell were more active mitotically than the apical cell, although they reported that the apical cell was found in division in 7 of 65 shoot tips examined, which results in a mitotic index of

10.8%. For *E. scirpoides* (36) the average mitotic index of the apical cell of shoot tips collected every 2 hr for 24 hr was 3.9%. However, on the basis of a detailed histogenic study it was concluded that the apical cell must be a continuing initial, that is, it is responsible for adding new growth units to the shoot.

Sossountzov (88) cultured *Marsilea drummondii* aseptically in a nutrient solution to which ^{3}H-thymidine and ^{3}H-uridine were added for varying lengths of time. Only 3% of the apical cells incorporated thymidine, whereas the majority incorporated uridine. A higher percentage of thymidine-labeled cells occurred in subjacent regions, which led her to recognize three zones: an apical zone (apical cell and a few adjacent cells), a subapical zone (responsible for regeneration of leaf initiation sites), and a pith meristem. She called attention to the similarity of the *Marsilea* apical zone to that of seed plants in which a group of mitotically inactive cells constitutes the "méristème d'attente." Sossountzov (89–91) stated that the apical cell is the most differentiated cell of the apex, but it does show, at the same time, characteristics of a meristematic cell. She concluded, however, that the apical cell of an adult plant cannot be considered an initial or an "organizer" with a regulatory function as stated by Michaux (72). While the data are interesting and informative, no histogenic study was made such as that of Schneider (85), who related the production of new building units to the apical cell. *Marsilea* exhibits heterophylly during ontogeny, and Kuligowski-Andrès (63) found that the apical cell was mitotically active during early stages of growth. The mitotic index was reduced to zero during the later phases of heterophyllous development, although she acknowledged that only 20 nuclei per stage were counted and that perhaps the cell cycle is long, which reduces the possibility of observing mitosis. In labeling experiments, however, the apical cell always became labeled with ^{3}H-thymidine. She concluded that the apical cell is probably mitotically active even in the adult plant, and that the apical cell must be engaged in syntheses important to plant growth.

An intensive study of cell kinetics, histochemistry, and relative DNA content was made by Michaux (69–74, 77) for the zonate shoot apex of *Pteris cretica,* in which a definite apical cell can be identified. She found that the apical cell and immediate surrounding cells divide regularly in young plants, but very soon they acquire a more differentiated state. The apical cell of young plants has a diploid cycle, 2C and 4C amounts of DNA. In older plants some nuclei had a DNA content higher than 4C—up to 16C, which was correlated with the accumulation of starch and lipids. The apical cell loses its histogenic role which is then assumed by subjacent cells (lateral zone) somewhat analogous to the "anneau initial" or subapical zone discussed by Sossountzov (88). A quantitative study of RNA metabolism (77) revealed that the apical cell throughout three growth phases contained the greatest amount of total RNA, but the highest concentrations per cytoplasmic unit and rate of synthesis were in cells of the lateral zone. These results, in addition to the presence of large numbers of plasmodesmata in the lateral walls of the apical cell, are cited in support for a regulatory function of the apical cell. Comparable results of cell kinetic studies were obtained for *Polypodium vulgare* (78). At the beginning of the adult phase, cells of the apical zone (apical cell and surrounding cells) had a low mitotic index and a cell cycle duration twice as long as cells of the lateral zone (subapical zone).

Equisetum has long been cited as the classical example of a lower vascular plant with a definite shoot apical cell. Of interest is the fact that *Sphenophyllum*, an extinct Paleozoic genus in the same major group as *Equisetum*, was shown to have had a distinct apical cell (40). D'Amato & Avanzi (28) reported that a large proportion of the apical cells of *Equisetum arvense* were polyploid. Very few apical cells contained only the 2C amount of DNA. Most of them had a 4C amount, corresponding to the DNA postsynthesis phase (G_2) of the diploid level, or G_1 of tetraploid cells. Such nuclei could be held at the 4C level or they may undergo further endopolyploidy—up to 8–16C. They regarded this as an explanation of the very rare occurrence of cell division. Quite opposite results were obtained by Gifford & Kurth (36) for *E. scirpoides*. Although the mitotic index was not especially high, an analysis of cell lineages indicated that the apical cell is responsible for initiating new merophytes. Measurements of DNA content of the apical cell, subjacent cells, and the remaining cells distal to the site of leaf initiation revealed no clear evidence of endopolyploidy. Differences in the results of the two studies may reside in the selection of research materials (36).

Gifford & Polito (37, 38) reported that the apical cell of *Azolla filiculoides* is as active mitotically as subjacent cells. This shoot tip, however, is atypical of the majority of ferns in that it is slender and fingerlike. The form of the apex is similar in *Ceratopteris thalictroides;* the apical cell is mitotically active and cycles at a slightly greater frequency than subjacent cell populations (82). There was no evidence that endopolyploidy occurs in the apical cell of either taxon (37, 38, 83). Also, all cells of the promeristem of the leafless stolons of *Nephrolepis multiflora* had a diploid DNA cycle (67).

CONCLUDING REMARKS   Just as for pteridophyte roots with an apical cell, the definition of what constitutes an "initial" is important to any discussion of shoot apices (see *Root Apex*—CONCLUDING REMARKS). Almost all investigators would consider the shoot apical cell of a moss gametophyte to be an initial because each leaf (and a portion of the shoot axis) can be related directly to a derivative of the apical cell by structural analysis. In many ferns the initiation of a derivative and the eventual origin of a leaf are separated considerably in time and space. As mentioned previously, there may be a difference in the mitotic activity of the apical cell in small, fingerlike shoot apices and the larger, flatter ones. In *Ceratopteris* and *Azolla,* representatives of the former type, the apical cell appears to be as active mitotically as derivatives (37, 38, 82). For large apices it is difficult to deny that segments or merophyte sectors exist if the apex is observed in surface view—even of older or "adult" plants (8). The apical cell of young sporophytes, in general, may be more active mitotically than in adult plants and the apical cell of an adult plant could be considered as a "founder" cell if it divides only occasionally (3). The apical cell would not have to divide very frequently to establish new merophytes in which the rate of cell division could be quite high. Endopolyploidy and the reputed mitotic inactivity of the apical cell are difficult to assess. These conditions may exist in the determinate shoot apex of *Equisetum,* for example, after all the growth units have been initiated. In ferns these conditions might signal loss of a regulatory role and lead to necrosis of the leading apex; a dormant bud could then be activated and become the dominant axis. On the other hand, one should take into consideration the growing body of literature on differential DNA replication, e.g. amplification and underreduplication in somatic cells of plants and animals, and its significance in morphogenesis (79).

ACKNOWLEDGMENTS

The author wishes to thank Dr. Eva Kurth for reading the review and making helpful suggestions, and also Susan Larson for her assistance in preparing the illustrations.

Literature Cited

1. Avanzi, S., D'Amato, F. 1967. New evidence on the organization of the root apex in leptosporangiate ferns. *Caryologia* 20: 257–64
2. Avanzi, S., D'Amato, F. 1970. Cytochemical and autoradiographic analyses on root primordia and root apices of *Marsilea strigosa*. (A new interpretation of the apical structure in cryptogams). *Caryologia* 23:335–45
3. Barlow, P. W. 1976. Towards an understanding of the behaviour of root meristems. *J. Theor. Biol.* 57:433–51
4. Bartoo, D. R. 1929. Origin and development of tissues in root of *Schizaea rupestris*. *Bot. Gaz.* 87:642–52
5. Bartoo, D. R. 1930. Origin of tissues of *Schizaea pusilla*. *Bot. Gaz.* 89:137–53
5a. Basile, D. V. 1979. Hydroxyproline-induced changes in form, apical development, and cell wall protein in the liverwort *Plagiochila arctica*. *Am. J. Bot.* 66:776–83
6. Berthier, J. 1972. Recherches sur la structure et le développement de l'apex du gamétophyte feuillé des mousses. *Rev. Bryol. Lichénol.* 38:421–551
7. Bhambie, S. 1972. Meristems in pteridophytes. *J. Indian Bot. Soc.* 51:77–92
8. Bierhorst, D. W. 1977. On the stem apex, leaf initiation and early leaf ontogeny in filicalean ferns. *Am. J. Bot.* 64: 125–52
9. Bonnet, A.-L.-M. 1955. Contribution à l'étude des Hydroptéridées. Recherches sur *Salvinia auriculata* Aubl. *Ann. Sci. Nat. Bot. Sér.* 11, 16:529–600
10. Bonnot, E.-J. 1967. Sur la structure de l'apex du gamétophyte feuillé de la mousse *Anomodon viticulosus* (L.) Hook. et Tayl. *Bull. Soc. Bot. Fr.* 114:4–11
11. Bonnot, E.-J. 1968. Sur la structure et les propriétés de la cellule apicale du gamétophyte feuillé des Bryales. *Bull. Soc. Bot. Fr. Mém.* 115:208–22
12. Buvat, R., Liard, O. 1953. Interprétation nouvelle du fonctionnement de l'apex

d'*Equisetum arvense* L. *C. R. Acad. Sci.* 237:88–90

13. Buvat, R., Roger-Liard, O. 1954. La prolifération cellulaire dans le méristème radiculaire d'*Equisetum arvense* L. *C. R. Acad. Sci.* 238:1257–58

14. Campbell, D. H. 1918. *The Structure and Development of Mosses and Ferns.* London: Macmillan. 708 pp.

15. Chadefaud, M. 1960. Les vegetaux non vasculaires. Cryptogamie. In *Traité de Botanique Systématique,* ed. M. Chadefaud, L. Emberger, Vol. I. Paris: Masson et Cie

16. Chiang, S.-H. T. 1972. The time of mitosis in the root apical cell of *Ceratopteris pteridoides. Taiwania* 17:14–26

17. Chiang, S.-H., Gifford, E. M. Jr. 1971. Development of the root of *Ceratopteris thalictroides* with special reference to apical segmentation. *J. Indian Bot. Soc.* 50A:96–106

18. Chiang, S.-H. T., Lu, C.-Y. 1972. A supplementary study on the cell division of root apical cells in some pteridophytes. *Taiwania* 17:229–38

19. Chopra, R. S., Sharma, P. D. 1958. Cytomorphology of the genus *Pogonatum palis. Phytomorphology* 8:41–60

20. Clowes, F. A. L. 1956. Localization of nucleic acid synthesis in root meristems. *J. Exp. Bot.* 7:307–12

21. Clowes, F. A. L. 1958. Protein synthesis in root meristems. *J. Exp. Bot.* 9:229–38

22. Clowes, F. A. L. 1968. Anatomical aspects of structure and development. In *Root Growth,* ed. W. J. Whittington, pp. 3–19. London: Butterworth

22a. Clowes, F. A. L. 1982. The growth fraction of the quiescent centre. *New Phytol.* 91:129–35

23. Conard, H. S. 1908. *The Structure and Life-History of the Hayscented Fern.* Carnegie Inst. Washington Publ. 94

24. Cremonini, R. 1974. Frequenza e localizzazione delle mitosi in primordi di radici laterali di *Marsilea strigosa. G. Bot. Ital.* 108:155–59

25. Cutter, E. G. 1965. Recent experimental studies of the shoot apex and shoot morphogenesis. *Bot. Rev.* 31:7–113

26. D'Amato, F. 1975. Recent findings on the organization of apical meristems with single apical cell. *G. Bot. Ital.* 109: 321–34

27. D'Amato, F., Avanzi, S. 1965. DNA content, DNA synthesis and mitosis in the root apical cell of *Marsilea strigosa. Caryologia* 18:383–94

28. D'Amato, F., Avanzi, S. 1968. The shoot apical cell of *Equisetum arvense,* a quiescent cell. *Caryologia* 21:83–89

29. de Albertis, J., Paolillo, D. J. Jr. 1972. The concept of incipient vascular tissue in fern apices. *Am. J. Bot.* 59:78–82

30. Deutsch, H. 1912. A study of *Targionia hypophylla. Bot. Gaz.* 53:492–503

31. Douin, Ch. 1923. Recherches sur le gamétophyte des Marchantiées. III. Le thalle stérile des Marchantiées. Développement basilaire des feuilles et autres organes lateraux chez les Muscinées. *Rev. Gén. Bot.* 35:487–508

32. Gewirtz, M., Fahn, A. 1960. The anatomy of the sporophyte and gametophyte of *Ophioglossum lusitanicum* L. ssp. *lusitanicum. Phytomorphology* 10:342–51

33. Gifford, E. M. Jr. 1960. Incorporation of H^3-thymidine into shoot and root apices of *Ceratopteris thalictroides. Am. J. Bot.* 47:834–37

34. Gifford, E. M. Jr., Corson, G. E. Jr. 1971. The shoot apex in seed plants. *Bot. Rev.* 37:143–229

35. Gifford, E. M. Jr., Kurth, E. 1982. Quantitative studies of the root apical meristem of *Equisetum scirpoides. Am. J. Bot.* 69:464–73

36. Gifford, E. M. Jr., Kurth, E. 1983. Quantitative studies of the vegetative shoot apex of *Equisetum scirpoides. Am. J. Bot.* 70:74–79

37. Gifford, E. M. Jr., Polito, V. S. 1981. Growth of *Azolla filiculoides. BioScience* 31:526–27

38. Gifford, E. M. Jr., Polito, V. S. 1981. Mitotic activity at the shoot apex of *Azolla filiculoides. Am. J. Bot.* 68:1050–55

39. Gifford, E. M. Jr., Polito, V. S., Nitayangkura, S. 1979. The apical cell in shoots and roots of certain ferns: a reevaluation of its functional role in histogenesis. *Plant Sci. Lett.* 15:305–11

40. Good, C. W., Taylor, T. N. 1972. The ontogeny of Carboniferous articulates: The apex of *Sphenophyllum. Am. J. Bot.* 59:617–26

41. Green, P. B. 1982. Organogenesis—A biophysical view. *Ann. Rev. Plant Physiol.* 31:51–82

42. Gunning, B. E. S. 1978. Age-related and origin-related control of the numbers of plasmodesmata in cell walls of developing *Azolla* roots. *Planta* 143:181–90

43. Gunning, B. E. S., Hardham, A. R., Hughes, J. E. 1978. Pre-prophase bands of microtubules in all categories of formative and proliferative cell division in *Azolla* roots. *Planta* 143:145–60

44. Gunning, B. E. S., Hughes, J. E., Hardham, A. R. 1978. Formative and proliferative cell divisions, cell differentiation, and developmental changes in the meris-

tem of *Azolla* roots. *Planta* 143:121–44

45. Hagemann, W. 1964. Vergleichende Untersuchungen zur Entwicklungsgeschichte des Farnsprosses. I. Morphogenese und Histogenese am Sprosscheitel leptosporangiater Farne. *Beitr. Biol. Pflanz.* 40:27–64

46. Hallet, J.-N. 1969. Remarques sur la structure et sur le fonctionnement de l'apex du *Polytrichum formosum* Hedw. *C. R. Acad. Sci. Sér. D* 268:916–19

47. Hallet, J.-N. 1972. Morphogenèse du gamétophyte feuillé du *Polytrichum formosum* Hedw. I. Étude histochimique, histoautoradiographique et cytophotométrique du point végétatif. *Ann. Sci. Nat. Bot. Sér.* 12, 13:19–118

48. Halperin, W. 1978. Organogenesis at the shoot apex. *Ann. Rev. Plant Physiol.* 29:239–62

49. Haupt, A. W. 1918. A morphological study of *Pallavicinia lyellii*. *Bot. Gaz.* 66:524–33

50. Haupt, A. W. 1920. Life history of *Fossombronia cristula*. *Bot. Gaz.* 69:318–31

51. Haupt, A. W. 1929. Studies in Californian Hepaticae. I. *Asterella californica*. *Bot. Gaz.* 87:302–18

52. Hébant, C. 1973. Studies on the development of the conducting tissue-system in the gametophytes of some Polytrichales. I. Miscellaneous notes on apical segmentation, growth of gametophytes, and diversity in histo-anatomical structures. *J. Hattori Bot. Lab.* 37:211–27

53. Hébant, C., Hébant-Mauri, R., Barthonnet, J. 1978. Evidence for division and polarity in apical cells of bryophytes and pteridophytes. *Planta* 138:49–52

54. Hébant-Mauri, R. 1973. Fonctionnement apical et ramification chez quelques fougères du genre *Trichomanes* L. (Hyménophyllacées). *Adansonia Sér.* 2, 13:495–526

55. Hébant-Mauri, R. 1975. Apical segmentation and leaf initiation in the tree fern *Dicksonia squarrosa*. *Can. J. Bot.* 53:764–72

56. Hébant-Mauri, R. 1977. Segmentation apicale et initiation foliaire chez *Ceratopteris thalictroides* (Fougère leptosporangiée). *Can. J. Bot.* 55:1820–28

57. Ho, P. H. 1956. Étude de la mitose et spécialement du mégachromocentre chez quelques muscinées. *Rev. Gén. Bot.* 63:273–80

58. Klekowski, E. J. Jr. 1976. Mutational load in a fern population growing in a polluted environment. *Am. J. Bot.* 63:1024–30

59. Klekowski, E. J. Jr., Berger, B. B. 1976.

Chromosome mutations in a fern population growing in a polluted environment: A bioassay for mutagens in aquatic environments. *Am. J. Bot.* 63:239–46

60. Klekowski, E. J. Jr., Klekowski, E. 1982. Mutation in ferns growing in an environment contaminated with polychlorinated biphenyls. *Am. J. Bot.* 69:721–27

61. Kuehnert, C. C., Miksche, J. P. 1964. Application of the 22.5 Mev deutron microbeam to the study of morphogenetic problems within the shoot apex of *Osmunda claytoniana*. *Am. J. Bot.* 51:743–47

62. Kuligowski-Andrès, J. 1977. Étude l'organogenèse radiculaire chez le *Marsilea vestita*. *Flora* 166:333–56

63. Kuligowski-Andrès, J. 1978. Contribution à l'étude d'une fougère, *Marsilea vestita* (Marsileacées), du stade embryon au stade sporophyte adulte. III. Le méristème apical du sporophyte juvénile: Étude histochimique et histoautoradiographique. *Ann. Sci. Nat. Bot. Ser.* 12, 19:219–48

64. Kurth, E. 1981. Mitotic activity in the root apex of the water fern *Marsilea vestita* Hook. and Grev. *Am. J. Bot.* 68:881–96

65. Lintilhac, P. M., Green, P. B. 1976. Patterns of microfibrillar order in a dormant fern apex. *Am. J. Bot.* 63:726–28

66. Maheshwari, P., Singh, H. 1934. The morphology of *Ophioglossum fibrosum* Schum. *J. Indian Bot. Soc.* 13:103–23

67. McAlpin, B. W., White, R. A. 1974. Shoot organization in the Filicales: The promeristem. *Am. J. Bot.* 61:562–79

68. Merl, E. M. 1917. Scheitelzellsegmentierung und Blattstellung der Laubmoose. *Flora* 109:189–212

69. Michaux, N. 1968. Étude cytologique du méristème apical du *Pteris cretica* (L.). *C. R. Acad. Sci. Sér. D* 267:1442–44

70. Michaux, N. 1970. Détermination, par cytophotométrie, de la quantité d'ADN contenue dans le noyau de la cellule apicale des méristèmes jeunes et adultes du *Pteris cretica* L. *C. R. Acad. Sci. Sér. D* 271:656–59

71. Michaux, N. 1971. Durée du cycle mitotique dans le méristème apical du jeune sporophyte du *Pteris cretica* L. *C. R. Acad. Sci. Sér. D* 273:336–39

72. Michaux, N. 1971. Structure et fonctionnnement du méristème apical du *Pteris cretica* L. I. Étude cytologique, histochimique et histoautoradiographique. *Ann. Sci. Nat. Bot. Sér.* 12, 12:17–125

73. Michaux, N. 1971. Structure et fonc-

tionnement du méristème apical du *Pteris cretica* L. II. Étude cytophotométrique. *Ann. Sci. Nat. Bot. Sér.* 12, 12:147–88

74. Michaux-Ferrière, N. 1973. Culture et comportement in vitro du méristème apical adulte du *Pteris cretica* L. *C. R. Acad. Sci. Sér.* D 277:2149–52

75. Michaux-Ferrière, N. 1976. Étude comparative, histologique, histochimique et cytologique du bourgeon apical de l'*Isoetes setacea* durant les periodes de croissance active et de resistance à la sécheresse. *Phytomorphology* 26:210–19

76. Michaux-Ferrière, N. 1978. Les proteines nuclèaires basiques dans le méristème caulinaire de l'*Isoetes setacea* au cours de son cycle annuel. *Phytomorphology* 28:58–73

77. Michaux-Ferrière, N. 1981. Quantitative study of RNA in the shoot apex of *Pteris cretica* L. during its development. *Z. Pflanzenphysiol.* 101:233–47

78. Michaux-Ferrière, N. 1981. Variation de la durée des cycles cellulaires au cours du passage de l'état jeune à l'état adulte dans le méristème caulinaire du *Polypodium vulgare. Can. J. Bot.* 59:1811–16

79. Nagl, W. 1982. Cell growth and nuclear DNA increase by endoredupliction and differential DNA replication. In *Cell Growth,* ed. C. Nicolini, pp. 619–51. New York: Plenum

80. Nitayangkura, S., Gifford, E. M. Jr., Rost, T. L. 1980. Mitotic activity in the root apical meristem of *Azolla filiculoides* Lam., with special reference to the apical cell. *Am. J. Bot.* 67:1484–92

81. Peterson, R. L. 1970. Bud development at the root apex of *Ophioglossum petiolatum. Phytomorphology* 20:183–90

82. Polito, V. S. 1979. Cell division kinetics in the shoot apical meristem of *Ceratopteris thalictroides* Brongn. with special reference to the apical cell. *Am. J. Bot.* 66:485–93

83. Polito, V. S. 1980. DNA microspectrophotometry of shoot apical meristem cell populations in *Ceratopteris thalictroides* (Filicales). *Am. J. Bot.* 67:274–77

84. Rostowzew, S. 1891. Récherches sur l'*Ophioglossum vulgatum* L. *Overs. K. Danske Vidensk. Selsk. Forh.* 2:54–82

85. Schneider, F. 1913. Beiträge zur Entwicklungsgeschichte der Marsiliaceen. *Flora* 105:347–60

86. Schüepp, O. 1966. *Meristeme.* Basel: Birkhäuser

87. Soma, K. 1966. On the shoot apices of *Dicranopteris dichotoma and Diplop-*

terygium glaucum. Bot. Mag. 79:457–66

88. Sossountzov, L. 1969. Incorporation des précurseurs tritiés des acides nucléiques dans les méristèmes apicaux du sporophyte de la fougère aquatique, *Marsilea drummondii* A. Br. *Rev. Gén. Bot.* 76:109–56

89. Sossountzov, L. 1970. Étude au microscope électronique de l'apex du bourgeon terminal, chez la Fougère aquatique *Marsilea drummondii* A. Br. *Rev. Gén. Bot.* 77:451–85

90. Sossountzov, L. 1972. Structure et fonctionnement du méristème apical des Pteridophytes: présent et avenir. *Bull. Soc. Bot. Fr.* 119:341–52

91. Sossountzov, L. 1976. Infrastructure comparée de l'apex de bourgeons en activité et de bourgeons au repos chez une fougère, *Marsilea drummondii* A. Br. *Cellule* 71:273–308

92. Steeves, T. A. 1963. Morphogenetic studies on *Osmunda cinnamomea* L. The shoot apex. *J. Indian Bot. Soc.* 42:225–36

93. Stevenson, D. W. 1976. Shoot apex organization and origin of the rhizomeborne roots and their associated gaps in *Dennstaedtia cicutaria. Am. J. Bot.* 63:673–78

94. Stevenson, D. W. 1976. The cytohistological and cytochemical zonation of the shoot apex of *Botrychium multifidum. Am. J. Bot.* 63:852–56

95. Torrey, J. G., Feldman, L. J. 1977. The organization and function of the root apex. *Am. Sci.* 65:334–44

96. Vallade, J., Bugnon, F. 1979. Le rôle de l'apicale dans la croissance de la racine du *Marsilea diffusa. Rev. Cytol. Biol. Vég.* 2:293–308

97. von Guttenberg, H. 1966. Histogenese der Pteridophyten. *Handbuch der Pflanzenanatomie,* Band VII, Teil 2. Berlin: Gebrüder Borntraeger

98. Wardlaw, C. W. 1952. *Phylogeny and Morphogenesis.* London: Macmillan. 536 pp.

99. Wetter, R., Wetter, C. 1954. Studien über das Erstarkungswachstum und das primäre Dickenwachstum bei leptosporangiaten Farnen. *Flora* 141:598–631

100. White, R. A. 1974. Experimental investigations of fern sporophyte development. In *The Experimental Biology of Ferns,* ed. A. F. Dyer, pp. 505–49. London: Academic. 656 pp.

101. Wigglesworth, G. 1956. Further notes on *Polytrichum commune* L. *Trans. Br. Bryol. Soc.* 3:115–20

Ann. Rev. Plant Physiol. 1983. 34:441–75

THE BIOLOGY OF STOMATAL GUARD CELLS[1]

Eduardo Zeiger

Department of Biological Sciences, Stanford University, Stanford, California 94305

CONTENTS

INTRODUCTION

Stomata provide higher plants with a primary means to adjust to their continuously changing environment (10). Like the membrane of a cell, stomata function at the organismal level to preserve biological integrity while at the same time allowing active physical exchange between the aerial parts of the plant and its habitat. Since the most conspicuous role of stomata is the regulation of water vapor loss and CO_2 uptake, the study of stomatal function has centered around transpiration and photosynthesis. However, an overview

[1]Dedicated to Professor Alberto Soriano, Facultad de Agronomia, Universidad de Buenos Aires, who taught me plant physiology and human wisdom.

441

0066-4294/83/0601-0441$02.00

of stomatal function reveals a broader picture, pointing to a central role of stomata in the maintenance of plant homeostasis. Homeostasis is a fundamental biological property of living organisms, which need to regulate their internal milieu while interacting with their environment (13). Adaptive homeostatic ranges are significantly broader in plants than in animals (81, 120), but the overall conservation of homeostasis is equally crucial in both kingdoms.

Whether stomata are opening to facilitate leaf cooling (39, 116), closing to prevent SO_2 toxicity after a volcanic eruption (176), or changing pore apertures in the coupling of conductance with prevailing photosynthetic rates (178), a common consequence of all these stomatal responses is the maintenance of plant homeostasis. Most of the current explanations for the role of stomata have homeostatic implications. Thus, for instance, the optimization theory (10, 11, 20) which proposes that stomata maximize the rate of carbon gain while minimizing water loss, describes a particular case of homeostasis. As an explanation of stomatal function, however, it is limited to stomatal responses that are coupled to photosynthesis. Other stomatal responses, such as stomatal movements at night (73) or those of nonphotosynthetic, senescing leaves (190), are excluded. Within an homeostatic concept, on the other hand, a coupling between stomatal conductance and photosynthesis is only one of the many complex stomatal responses to internal or external changes. Within this perspective, an emerging theme for understanding the biology of stomatal function is then not its connection with photosynthesis, but rather its role in the maintenance of the homeostasis of the plant. Some of the homeostatic aspects of stomatal movements, like stomatal closure under water stress, (52, 66) are obvious, while many others are poorly understood. This is hardly surprising, though, when we consider the bewildering complexity of living organisms, a complexity which is pervasive in stomatal physiology itself.

Stomata respond to a large number of internal and external signals. The recognition of the maintenance of biological homeostasis as the basis of stomatal functioning gives a sharp biological context to the analysis of the reception and transduction of these signals. The focus of this paper will thus be on the biological properties of stomata, which are dominated by the guard cells, the main source of biological control in the stomatal pathway.

The last review on stomatal movements in this series was published in 1975 (114), and I shall cover the main developments in the field since that time. The impressive progress in the intervening years has provided us with a broader understanding of the metabolic properties of guard cells and the complex physiology of stomata. In reviewing these developments, I shall propose that the control of active ion transport in guard cells is a key integrating mechanism which transduces internal and external signals into the appropriate stomatal movement.

A classical botanical subject, stomata have received attention for more than a century and are the subject of thousands of papers. A survey of this literature reveals a number of hypotheses dealing with stomatal movements. Many of them retain validity because of their explanation of an important aspect of stomatal function, but none provides us with a comprehensive interpretation of the properties of stomata. Because of this prolific literature, almost everything that will appear in this article has already been, to some extent, stated before. No attempt to cover the literature nor to do justice to many of the existing hypotheses will be made. I do hope, however, that a focus on the homeostatic nature of stomatal function and the relationship between active ion transport and sensory transduction in guard cells will provide us with a better understanding of stomatal behavior and its integration in the physiology of the plant.

METHODOLOGY

Recent methodological advances in experimental stomatal physiology have had a significant impact on stomatal research. These developments have not only opened experimental avenues, but they have also set new standards which should be kept in mind when designing further work. In this section I shall review some of these advances and their implications, with the important exception of microhistochemistry of isolated guard cells which has been reviewed elsewhere (90).

Epidermal Peels

A classical technique in stomatal research, epidermal peels allow the analysis of stomatal responses in the absence of mesophyll and have contributed to important breakthroughs in stomatal research (71). The epidermal peel technique has been critically evaluated (16, 71, 171). The angle of peeling, which can cause mechanical damage to the guard cells and affect the proportion of living epidermal cells, the relative humidity of the environment in which the peels are held, and the temperature of incubation during the experimental treatment are experimental variables that may strongly affect the results. Another aspect of work with peels is the effect of the concentration of the osmoticum (usually KCl) on stomatal aperture. Because of the mechanical advantage of the epidermal cells (16, 72), turgor losses caused by high osmotica lead to stomatal opening. These passive turgor changes increase baseline apertures and decrease the hydroactive component of the final aperture values. In addition, supraoptimal KCl concentrations have been reported to alter some physiological properties of guard cells (161, 169, 175). It is therefore desirable not to exceed optimal KCl levels in the incubation medium, which for *Vicia faba* is as low as 1 mM (34). Optimal KCl concentrations for *Commelina communis* have been reevaluated (161), showing that the commonly used 100 mM KCl should be reduced to 50 mM; we (139; A. Schwartz and E. Zeiger, unpublished) have obtained maximal apertures (15 μm) and low baseline levels (2–3 μm) in the same species using 20 to 30 mM KCl.

Mesophyll contamination is another critical problem with epidermal peels. Recent studies have shown that, because of the higher activity of some enzymes in mesophyll cells, as compared with guard cells (91), levels of contamination previously considered tolerable are unacceptable for biochemical studies. Mesophyll chloroplasts adhering to the peels are also a source of contamination (181, 182). Mesophyll chloroplasts not enclosed in intact cells are hard to detect under bright field microscopy, but can be readily seen under fluorescence microscopy (181).

Sonication (88) and low pH treatments (61, 62, 156) have been used to eliminate contaminants from peels. Judged by aperture measurements (61, 71), guard cells remain unaffected but, as discussed below, aperture measurements alone cannot reveal if their whole metabolic machinery is intact.

It is hoped that future work with peels will incorporate available knowledge on the experimental strengths and weaknesses of this methodology, thus minimizing the amount of contradictory data which has been so common in previous years.

Guard Cell Protoplasts

Protoplasts provide a unique means of studying biochemical and physiological properties of guard cells (23, 181). In contrast with their mesophyll counterparts, guard cell protoplasts are severely damaged if subjected to a steep plasmolysis at the beginning of the enzymatic treatment, because of the rupture of yet to be characterized plasmalemma-wall attachments. This difficulty is obviated by a progressive increase in the osmoticum in the presence of a cellulolytic enzyme, which facilitates the detachment of the plasmalemma from the cell wall and yields viable guard cell protoplasts (186).

Because of their thickened walls, guard cells are very resistant to cellulolytic enzymes and long digestion times—10 to 12 hr at room temperature or 5 to 8 hr at 30°C (98, 134, 146a, 186)—are required to release the protoplasts. Younger leaves require shorter digestion times and provide higher yields. The differential wall thickness of the three classes of leaf cells, on the other hand, provides a useful way of separating guard cell protoplasts from those of mesophyll and epidermal cells, which are released first. The use of microchambers is useful for the characterization of protoplast isolation conditions in untested species, and for some physiological studies (43a, 181, 186). Methods yielding a higher number of protoplasts, as required for biochemical analyses, are also available (98, 134, 146a).

Isolation of guard cell protoplasts from *Allium cepa, Commelina communis, Nicotiana tabacum, Paphiopedilum harrisianum, Pisum sativum,* and *Vicia faba* has been reported (43a, 98, 129, 146a, 181, 186). Studies with guard cell protoplasts have provided an unambiguous demonstration of a direct response of guard cells to blue and white light, abscisic acid (ABA), and fusicoccin (FC) (7, 43a, 134, 146a, 187). Protoplasts are also useful for the isolation of guard cell organelles. Vacuoles can be released from bursting protoplasts and restabilized by further manipulation of the medium (181). Guard cell protoplasts are also a source of isolated chloroplasts (98, 135), although the problem of contaminating mesophyll remains to be completely overcome.

A special problem with chloroplasts isolated from guard cell protoplasts is potential damage by the long digestion procedure (146a). Digestion damage to PSII can explain the conflicting results regarding the reported functional competence of guard cell chloroplasts isolated from protoplasts (98, 135).

Gas Exchange

Measurement of leaf transpiration is a common technique for the study of stomatal movements in the intact leaf, for the characterization of mesophyll-epidermis interactions, and for the analysis of stomatal function under field conditions. Stomatal conductance to water vapor can be calculated from transpiration measurements (20, 39).

Most papers dealing with stomatal conductance report results obtained with transient, diffusion porometers (45), instruments having important methodological limitations (71). These limitations have been largely overcome by the development of a steady state, null balance porometer (3, 4), which is now commercially available.

Improvements have also been made in the design and applications of gas exchange systems that simultaneously measure stomatal conductance and net photosynthesis in attached leaves at constant temperature and VPD (5, 39). The simultaneous measurement of rates of photosynthesis and stomatal conductance and concomitant calculations of prevailing intercellular CO_2 concentrations has generated a new understanding of the physiology of the intact leaf and of the functional coupling between photosynthesis and stomatal conductance (11, 17, 178). Recent advances include the development of a portable gas exchange system which uses the null balance approach (21). The validity of the calculations of intercellular CO_2 concentrations obtained from net photosynthesis and stomatal conductance measurements was recently corroborated in an elegant study comparing calculated values with simultaneous, direct determinations (142).

TURGOR REGULATION IN GUARD CELLS

Stomatal opening and closing ensues because of relative turgor changes in the guard cells with respect to the surrounding epidermal tissue (32, 73, 116). Because of the mechanics of the guard cell walls (116, 179), increasing turgor increases the aperture of the pore; guard cell deflation reverses the process and the pore closes. Aperture changes hence depend on turgor differentials in the epidermis, which are related to the water potential of the cells. Lasting turgor differences can only be maintained along osmotic gradients, which generate water fluxes and turgor changes.

Two distinct processes are connected with stomatal movements (73, 116). One, usually termed hydroactive, includes changes in cellular ionic and organic contents, is markedly dependent on prevailing physiological and metabolic conditions, and results in the modulation of osmotic gradients. The other, called hydropassive, involves subsequent water fluxes, changes in turgor relationships, and mechanical responses of the guard cell walls resulting in physical changes in the pore dimensions. Although of undeniable importance, the later series of events is assumed to be unavailable for major cellular control and will not be discussed in detail here; interested readers will find extensive analysis of the subject elsewhere (9, 30, 72, 116, 146, 179).

Ionic Gradients in the Epidermis

The major importance of osmotic changes for stomatal movements was recognized early in stomatal physiology, but the involvement of K^+, the main cation transported in and out of guard cells, was only unraveled many decades later (32). The dominant participation of K^+ in stomatal movements has been firmly established in many species (32, 116). Guard cells, in fact, are not exceptional in their use of K^+ as a major osmoticum; K^+ is widely utilized for turgor build-up in the majority of plant cells (29, 76).

Convincing evidence from work with several species (32, 61, 62, 107, 116) has shown that in fully open stomata, the K^+ content of the guard cells is several-fold higher than that of the surrounding tissue, indicating the operation of a concentration mechanism against diffusion gradients. A selective uptake of K^+ in the presence of other monovalent ions has been demonstrated (34); selectivity is enhanced by Ca^{2+} (100) which is known to preserve membrane integrity and ion transport mechanisms. Commelina stomata open widely in the presence of NaCl (67), but reponses to light, ABA, and CO_2 are correlated with the presence of K^+ and diminish with increasing concentration of Na^+ (40, 67). If the influx of cations is driven by a primary proton gradient, one would expect that in the absence of K^+, other permeable monovalent cations would be taken up down the prevailing electrochemical gradient, as observed with Na^+. The preponderance of K^+ in guard cells from most species and its enhancement of stomatal responses to light, ABA, and CO_2 (67) warrant the conclusion that normal ion uptake mechanisms during stomatal opening are geared for the use of K^+ as the dominant cation.

The correlation between K^+ content and the degree of aperture is usually high (58, 116) and consistent with the notion that K^+ contributes about half of the expected increase in osmotic potentials needed for the build-up of turgor. Recent findings in *Commelina* could indicate that another, hitherto undefined osmoticum is used in the early opening phase (61, 62). These observations, however, were made in acid-treated peels incubated for long periods of time, conditions that could alter the normal physiology of the guard cells, as recognized by the authors.

The picture regarding anions is more complex (2). In *Allium cepa,* a species lacking starch in its guard cell chloroplasts, there is a high correlation between the K^+ and Cl^- content of the guard cells (136, 137); furthermore, stomata from onion fail to open when Cl^- is omitted from the incubation medium (136). As far as we know, all species utilize some Cl^- as a counterion for K^+ (2, 36, 109, 164), but, in species having starch in the guard cell chloroplast, organic acids, mainly malate, are also involved (2, 91, 116). The relative proportions of Cl^- and $malate^{2-}$ found in open stomata vary widely with species and experimental conditions. In studies with epidermal peels, malate levels increase as the concentration of Cl^- in the incubation medium is lowered (116); in peels from *Vicia* incubated in a medium without Cl^-, the levels of malate measured after 4 hr at pH 5.6 compensated about 91% of the K^+ without apparent effects on stomatal apertures (118). From the limited information available (1, 2, 92, 93), variable ratios of $Cl^-/malate^{2-}$ are also to be expected in open stomata of intact leaves. Besides malate, citrate and possibly glutamate and aspartate can also serve as counterions of K^+ (2, 91).

Available evidence indicates that the increase in organic acid content found in guard cells upon stomatal opening can be attributed to de novo synthesis in the guard cells proper. The postulated biochemical pathway utilizes PEP from starch degradation (94). The cytoplasmic enzyme PEP carboxylase (95) catalyzes CO_2 fixation into PEP, generating oxaloacetic acid; oxaloacetic acid is then reduced to malate by malic dehydrogenase (2, 91, 116). The synthesis of malic acid generates two protons (183); the implications of this proton source are discussed below. Like the utilization of K^+ for turgor build-up, the metabolism of organic acids and their use as counterions for K^+ are not unique to guard cells (89).

The uptake of ions against concentration gradients has specific electrical correlates which can be analyzed with electrophysiological measurements (56–58, 147, 155). Guard cells, however, are difficult to impale with glass electrodes because of the thickness of their walls; they also have a tendency to expel the electrode when the protoplasm moves because of turgor changes (82). Because of these obstacles, advances in this aspect of stomatal physiology have fallen behind expectations.

There is conclusive evidence for membrane potential differences (MPD), inside negative, in guard cells (26, 82, 99, 129, 189). These findings agree with the widespread occurrence of negative MPD in most plant cells (57, 155) and are consistent with the high concentrations of electrolytes found in open stomata (62, 93, 108, 109, 116).

A negative MPD could be attributed to passive, Nernstian diffusion potentials resulting from differential ionic concentrations across a semipermeable membrane, or it could be generated by metabolic activity in the cell (57, 76, 155). In the first case, the MPD is a consequence of existing ionic gradients maintained by nonelectrogenic mechanisms. In the second instance, the MPD is a specific component of the activity of an electrogenic pump which drives ion uptake. A conclusive distinction between these alternatives is crucial for our understanding of the bioenergetics of guard cells.

In their study on open and closed *Commelina* guard cells, Saftner & Raschke (129) recently claimed that guard cell MPDs are solely dependent on the passive diffusion of ions, and they concluded that guard cells behave as leaky compartments where ions redistribute according to unspecific electrochemical gradients. *Commelina* stomata, however, have an elaborate group of subsidiary cells, which probably constitute a series of ionic compartments (59), dampening electrical events recorded between the guard cells and the bathing medium. An anomalous lack of sensitivity to the inhibitor CCCP was also reported (35, 101, 112). Interpretation of these results is further complicated by the omission of Ca^{2+} from the bathing medium and by a reported lack of selectivity for monovalent cations (34, 100). In addition, there are discrepancies between the reported MPD and those predicted by direct measurements of K^+ and Cl^- in the same species (108, 109).

Evidence for a light-sensitive electrogenic pump in guard cells was first provided by a study of *Tradescantia*, showing light-induced hyperpolarizations (26). We extended these observations with *Allium* stomata subjected to light-dark cycles (82, 189). The observed hyperpolarizations in the light and depolarizations in the dark (up to 50 mV in either direction) supported the notion of light-dependent electrogenic proton efflux in the guard cells. The rapidity of the responses, with lag times usually ranging between 1 and 45 sec, makes it unlikely that the electrical changes were Nernstian potentials resulting from diffusion gradients. The observed light-induced hyperpolarizations were also inconsistent with a mechanism actively pumping K^+ into the cells, which would cause electrical changes of the opposite polarity. A conclusive demonstration of a light-dependent electrogenic pump in guard cells, however, will require concurrent, dual electrode measurements of membrane conductance (155).

Additional data on the electrophysiological properties of guard cells from several species will be helpful for the understanding of the basis of active ion transport in stomatal movements. Although admittedly difficult (8, 129), this experimental approach should prove highly rewarding. From our experience (82), chances of success are enhanced by the use of beveled electrodes, with resistances measured after beveling, of around 60 MΩ. As in the isolation of guard cell protoplasts, the use of thin-walled guard cells from young leaves is also helpful. Lowering of guard cell turgor by addition of an osmoticum to the incubation medium might be useful to uncouple electrical events associated with early ion fluxes from swelling of the cell protoplasm, which tend to perturb the recordings.

A Chemiosmotic Mechanism of Stomatal Function

Uptake of both cations and anions against diffusion gradients has been demonstrated in many plant systems (57). It can be explained by (a) the operation of electroneutral devices, (b) the active uptake of a single ion species into the cell, or (c) the generation of a primary electrochemical gradient across the membrane. Chemiosmotic theory models the metabolic and electrical events associated with ion transport, depending on a primary gradient (80, 110).

We have claimed that most if not all of the known properties of the guard cells point to a chemiosmotic mechanism, dependent on a primary proton gradient, as the driving force for all the ion fluxes associated with stomatal movements (145, 183, 187, 189). A role of proton fluxes in stomatal movements has been proposed (50, 102, 114, 160), although the broad implications of chemiosmosis have been largely overlooked. The importance of proton fluxes in stomates has been recognized increasingly in the last few years (14, 24, 62, 162).

Guard cells extrude protons under a variety of conditions conducive to stomatal opening (24, 117, 183), and proton secretion is blocked by vanadate, an inhibitor of plasmalemma ATPases (24). The fungal toxin fusicoccin (FC), which promotes proton extrusion in a variety of tissues, including guard cells (68, 115), causes stomata to open widely at micromolar concentrations [Figure 1 and references (115, 132, 163)]. Reported membrane hyperpolarization upon exposure to FC suggests that FC responses in guard cells are mediated by an electrogenic proton pump (155). Stomatal opening is also sensitive to CCCP, an effective proton gradient dissipator (35, 101, 112).

A chemiosmotic postulate yields some straightforward predictions about ion transport during stomatal movements (145, 183). Active proton efflux would generate an electrical and concentration (pH) gradient across the guard cell plasmalemma and K^+ would be taken up passively, down its electrochemical gradient. A balanced H^+/K^+ exchange would dissipate the electrical gradient; the pH gradient would remain available for the uptake of Cl^- through a Cl^-/OH^- antiport. Either because of low permeability or concentration, Cl^- uptake would become limiting in the early stages of stomatal opening, triggering biosynthesis of malic acid which would then be the dominant counterion for K^+.

The magnitude of the prevailing proton motive force (pmf) can be calculated from pmf = $\Delta\psi$ - $z\Delta pH$, where $\Delta\psi$ is the magnitude of the MPD, ΔpH is the difference between the intracellular and extracellular pH, and z is a constant = 2.3 RT/F; the units are millivolts. Conventional electrical equivalents of this equation have been presented (110).

Some implications of the operation of an electrogenic pump on the guard cells are of marked interest. If we take -70 mV (82) as the cell resting potential and assume that, as in most plant cells, the permeability of the guard cell plasmalemma to K^+ is the dominant one (76), we can calculate that $[K^+]_o \cong 15$ $[K^+]_i$ (155). The observed light-induced hyperpolarization of up to 50 mV (82) would further displace equilibrium conditions toward K^+ uptake. It is thus clear that K^+ accumulation is energetically inexpensive for guard cells maintaining a negative MPD.

Because of its negative charge, the reverse conditions apply to Cl^- which, in cells with an MPD of -70 mV, would have an equilibrium potential markedly favoring efflux. As compared with K^+, the energy cost of Cl^- uptake would be further increased by the lower permeability and concentration of this anion. This high cost of anion uptake, well characterized in roots and marine algae (57), is more extreme in guard cells because of their use of ions in repeated cycles of opening and closing, as opposed to conditions prevailing in long-term turgor build-up.

This high cost of Cl^- uptake might also explain why Cl^- is not balancing all the K^+ uptake in guard cells; the required energy output could exceed the maximum capacity of the electrogenic pump. That Cl^- uptake is limiting during stomatal opening is supported by evidence that in the presence of increasing levels of Cl^- in the medium, which would shift equilibrium conditions toward higher intracellular Cl^- concentrations, the amount of malate found in opened stomata declines (118, 164).

The activity of an electrogenic, primary proton gradient driving ion fluxes in the guard cells also has important regulatory consequences. Any signal affecting stomatal movements could modulate stomatal apertures if transduced into a defined magnitude of the pmf. Light provides the best example; if quantum fluxes are proportionally transduced into a pmf value, any given light intensity will result, at steady-state conditions, in a specified stomatal aperture.

This paradigm also underscores a key regulatory role of ion transport in the metabolic control of stomatal movements. Any factor causing changes in stomatal behavior, whether a physical parameter like temperature, a small metabolite like CO_2, or a hormone like ABA, will interact with the stomatal apparatus through a metabolic process leading to turgor changes which must involve the modulation of ion content in the guard cells. In that sense, the metabolic control of stomatal movements is functionally homologous to the regulation of ion transport; the mode of action of all these dissimilar factors affecting stomatal responses could have the modulation of ion fluxes as a common metabolic step.

An Energy Pool in Guard Cells

Guard cell bioenergetics has been a controversial aspect of stomatal physiology (32, 114, 116). Chemiosmotic theory, on the other hand, underscores some specific sources of energy sustaining proton efflux in biological systems, mainly membrane-bound ATPases and electron transport chains (28). Thus, both oxidative phosphorylation and photophosphorylation are potential energy sources driving stomatal movements (32, 75, 182, 183, 189, 191). In addition, the activity of a blue light photoreceptor within a membrane-bound, electron transport chain could also participate in the bioenergetics of the guard cells (39, 54, 149, 162, 180, 187).

I recently obtained evidence for the operation of an energy pool during stomatal opening in experiments with epidermal peels of *Commelina* kept in microchambers (181, 186) continuously perfused with 20 mM KCl. Comparisons of opening rates in preparations treated with 10 μM FC in the dark or 2000 μmol m^{-2} s^{-1} of white light showed that both treatments cause stomata to open widely at comparable rates (Figure 1A; E. Zeiger, unpublished experiments). The FC-induced opening was inhibited by 50 μM KCN at pH 6.5 (Figure 1B), in agreement with the notion that proton extrusion in response to FC depends on oxidative phosphorylation (68). Light, on the other hand, reversed the KCN inhibition (Figure 1B), indicating the operation of a light-dependent energy source other than oxidative phosphorylation.

Fig 1A-1B

This notion of two distinct energy sources driving ion transport in guard cells

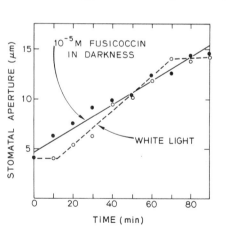

Figure 1A   The kinetics of stomatal opening in response to light (o) and fusicoccin (FC) (●). Epidermal peels of *Commelina communis* were mounted in a microchamber and stomatal apertures monitored in a microscope (181). Continuous perfusion with 20 mM KCl started 15 min prior to the experimental treatment. At time zero peels were exposed to light from an halogen lamp (ca 2000 μE m^{-2} s^{-1}) or to 10 μM FC in darkness. Each data point is the average of three experiments, 90 measurements per point. Measurements in the FC treatment were obtained under low-intensity green light.

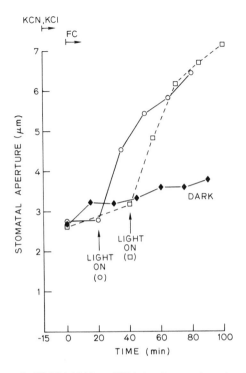

Figure 1B   Light reversal of KCN inhibition of FC-induced stomatal opening. KCl (20 mM) and KCN (50 μM, pH 6.5) were included in the perfusion medium at the onset of perfusion. FC (10 μM) and light were added as indicated by arrows. Other experimental conditions as in Figure 1A.

has had strong advocates in the stomatal literature (32, 191), and it is consistent with the concept of an energy pool for ion transport in other plant systems (41, 55, 57).

OXIDATIVE PHOSPHORYLATION AND THE GUARD CELL MITOCHONDRIA   That mitochondrial respiration can drive stomatal opening is indicated by studies with respiratory inhibitors or under anoxia (32, 54, 101, 112, 163, 166, 173). Mitochondria are abundant in guard cells (32, 73, 114), and oxidative phosphorylation in these mitochondria is the most likely energy source of ATP for stomatal opening in darkness under CO_2-free air (67, 113).

PHOTOPHOSPHORYLATION AND GUARD CELL CHLOROPLASTS
Several lines of evidence point to a role of photophosphorylation by guard cell chloroplasts in stomatal movement. In the leaf epidermis of higher plants, chloroplasts usually are found only in guard cells

(73).[2] Chloroplasts are highly conserved in guard cells of all tissues, including coleoptile, albino portions of variegated leaves and flowers (73, 131, 182). Stomatal responses from etiolated tissue with achlorophyllous guard cells are sluggish (53, 85).

Studies on the wavelength dependence of stomatal opening in isolated peels show that at moderate to high light intensities the spectral sensitivity of opening closely resembles that of photosynthesis (33, 48). These structural and functional correlates argue for an important yet elusive role of guard cell chloroplasts in stomatal movements (32, 50, 73, 98, 103, 182, 191).

New experimental approaches are providing more information on the role of guard cell chloroplasts. Microhistochemistry with a few guard cell pairs showed that guard cell chloroplasts lack key enzymes of the photosynthetic carbon reduction (Calvin) pathway (96). This conclusion was verified by both functional and immunological criteria, and it is consistent with findings with epidermal peels and guard cell protoplasts (91, 133). A recent report on the detection of RuBP carboxylase in several species using indirect immunofluorescence (63) is at variance with that trend but needs confirmation. The bulk of available evidence thus favors the view that guard cell chloroplasts seem unable to fix CO_2 through the Calvin cycle.

Fluorescence spectroscopy provided another means to study guard cell chloroplasts (75, 87, 98, 146a, 182). Analysis of fluorescence emission spectra from guard cell chloroplasts of albino segments from variegated leaves of *Chlorophytum* at 77°K demonstrated the presence of light-harvesting pigments of both PSI and PSII (182). Other studies with purified guard cell chloroplasts of *Vicia* reported identical results (98, 146a). Recently, PSII activity has also been found histochemically in guard cell chloroplasts from barley (K. Vaughn and W. H. Outlaw, unpublished).

In *Chlorophytum,* we also characterized chlorophyll *a* fluorescence transients at room temperature. These transients provide a diagnostic tool for the functional properties of PSII (75, 182). "Fast" measurements (time courses of a few seconds at low light intensity) yielded DCMU-sensitive, variable fluorescence transients with typical kinetics associated with the reduction of a pool of electron acceptors by PSII (49); restoration kinetics of the transients by far-red light showed that PSI was functioning as an electron sink for PSII. These results indicate that guard cell chloroplasts have functional PSI and PSII. Electron transport in guard cell chloroplasts from *Vicia* has also been demonstrated (87, 146a).

Negative evidence for the activity of PSII in guard cell chloroplasts was reported in a study comparing oxygen exchange rates from chloroplasts isolated

---

[2]Subsidiary cells from the stomata of *Maranta leuconeura* var. *massangeana* have chloroplasts, constituting an exceptional case which might justify future work with this species.

from guard cell protoplasts of *Vicia* with their mesophyll counterparts (135). This discrepancy can be attributed to damage of the PSII by the digestion procedure required to isolate guard cell protoplasts (146a). Digestion damage to chloroplasts of guard cell protoplasts without any concomitant inhibition of the blue light-dependent photosystem could also explain why guard cell protoplasts of onion only respond to blue light while lacking any red light response (187).

Analysis of "slow" fluorescent transients (time courses of a few minutes at relatively high light intensities) provided additional information on the properties of guard cell chloroplasts from *Chlorophytum* (75), and *Vicia* (87, 146a). These slow fluorescence transients have been correlated with redox reactions and photophosphorylation in chloroplasts (49). Like their mesophyll counterparts, guard cell chloroplasts showed the slow transients, supporting the notion of their capacity to conduct electron transport and to photophosphorylate. Unlike mesophyll chloroplasts, guard cell chloroplasts lacked one of the kinetic transitions (75), indicating some metabolic differences between the two types of organelles.

Significant differences were observed in the magnitude of transients from *Vicia* guard cell chloroplasts in epidermal peels as compared with those from isolated protoplasts (87, 146a). The reduced magnitude seen with peels could be restored upon addition of ions, especially K^+ and phosphate, whereas ABA had an antagonistic effect (87). These observations were interpreted as evidence for an effect of ion transport in the guard cells on the energy state of their chloroplasts (87).

Recent findings on the properties of guard cell chloroplasts in senescing leaves (190) led to the characterization of another experimental system yielding fluorescence transients of guard cell chloroplasts in intact leaves. We found that guard cell chloroplasts from senescing leaves of several perennial and annual species continued to fluoresce for several days after mesophyll chloroplasts had lost most or all of their fluorescence. Guard cells containing the fluorescing chloroplasts were shown to be functional both in the intact leaf and in epidermal peels (190). In yellow leaves of *Ginkgo biloba* lacking mesophyll fluorescence, we characterized slow fluorescence transients from guard cell chloroplasts and found that, as in *Chlorophytum* and *Vicia,* the transients in *Ginkgo* were devoid of the SM transition characteristic of mesophyll chloroplasts while showing the fluorescence quenching associated with photophosphorylation (190).

Unexpectedly, the fluorescence transients of guard cell chloroplasts from *Chlorophytum* were sensitive to CO_2 (75). The response was restricted to the late fluorescence quenching associated with photophosphorylation; the early transients connected with electron transport were not affected. Although CO_2 was supplied as bicarbonate, the pH dependence of the response clearly indicated that CO_2 was the active ionic species (75). A most intriguing aspect of the CO_2 effect on guard cell chloroplasts is its apparent correlation with physiological responses of stomata to CO_2. The quantitative inhibition of the fluorescence quenching and a concomitant reduction of the photophosphorylating capacity of the chloroplasts would reduce stomatal opening as the intercellular CO_2 concentration increased. In mesophyll chloroplasts, the quenching of the chlorophyll fluorescence is interpreted as a dissipation of the high energy state because of CO_2 fixation (47); in guard cell chloroplasts that explanation is inconsistent with their reported lack of Calvin-cycle enzymes (91). Biochemical analysis should reveal if this response of guard cell chloroplasts to CO_2 underlies a specific reaction of these organelles or if it is a more indirect effect mediated by metabolic reactions outside the chloroplast proper.

THE BLUE LIGHT PHOTOSYSTEM IN GUARD CELLS: A THIRD COMPONENT OF THE ENERGY POOL?   Unequivocal evidence documents a specific blue light response of stomata. Stomatal opening is consistently higher in the blue than in the red, indicating that, in addition to the photoresponses predicted from the absorption properties of chlorophyll (33, 48), a blue light photosystem has been activated (39, 54, 86, 105, 144, 162, 184). The swelling of onion guard cell protoplasts in blue but not in red light, the wavelength dependence of malate biosynthesis in guard cells from epidermal peels of *Vicia,* and the observed increases in transpiration induced by a blue light step in the Gramineae (6, 88, 148, 187) provide experimental evidence for a blue light effect independent of the photosynthetic active radiation (PAR)-dependent system of guard cells (184).

Both in isolated peels and in intact leaves, stomatal responses to blue light show an extremely low threshold; red light, on the other hand, is ineffective below significantly higher threshold intensities which vary with species and environmental conditions (39, 70, 86, 162, 184). These data indicate that below the threshold for red light, the PAR-dependent system is inactive; therefore, stomatal responses under low levels of blue light should reveal the bioenergetics of the blue light-dependent system.

In intact leaves irradiated with blue light intensities below the threshold of the PAR-dependent system, measured increases in stomatal conductances are about 20 to 50% of the saturated conductance levels under white light (39, 70, 86, 162, 184). In similar experiments with epidermal peels under CO_2-free air, stomata opened widely (54, 105, 139, 162), making a more dramatic case for the need of an energy source. Two distinct sources of energy could operate under those conditions: oxidative phosphorylation or a direct photochemical supply from the blue light photosystem. Like the FC-induced opening, stomatal responses to low-intensity blue light are generally both anoxia- and KCN-sensitive (54, 139), thus pointing to the involvement of respiration. A demonstrated enhancement of respiration and of PEP-carboxylase activity by blue light (46) has been invoked as a mechanism for the stomatal response to blue light (54, 116). However, the mode of action of blue light on respiration remains obscure; it could involve primary modulations of membrane activity (46). The central role of ion fluxes on stomatal responses makes this a very likely possibility. Thus, the KCN-sensitive, oxygen-requiring, blue light response could involve a membrane-bound flavin operating within an electron transport chain which uses O_2 as a final acceptor and has a KCN-sensitive electron carrier. Such a mechanism has been postulated in the blue light-dependent responses of *Neurospora* and corn coleoptile (140, 141) and could be widespread in plant cells, including guard cells (180, 187), although KCN sensitivity is lacking in the in vitro systems studied to date (W. R. Briggs, personal communication).

In their analysis of stomatal opening in *Commelina* under low-intensity blue light, Travis & Mansfield (162) recently calculated that available photon fluxes absorbed by the guard cells could supply sufficient energy to take up the required K^+. Although admittedly tentative, this study underscores the need for further work on the bioenergetics of the blue light photosystem of the guard cells.

IMPLICATIONS OF AN ENERGY POOL IN GUARD CELLS   The availability of at least two different energy sources, each capable of supporting saturating levels of opening (Figure 1), is intriguing. Each energy source could be connected with specific stomatal responses, such as the blue light photosystem with the blue light-dependent opening at dawn (185), oxidative phosphorylation with VPD responses (51) and the possible energy requirements of stomatal closing (32), and photophosphorylation with stomatal opening at moderate to high light intensities (144, 182).

The operation of an energy pool in guard cells has other implications. Stomatal opening in the presence of an inhibitor could be caused by the activity of energy sources other than the one being blocked. For instance, opening in white light in the presence of DCMU (35) does not prove that guard cells are incapable of photophosphorylation; opening could be sustained by the blue light photosystem or oxidative phosphorylation, in spite of a fully inhibited PSII. On the other hand, stomatal opening under red light, in which the energy pool would be expected to depend solely on photophosphorylation, was drastically inhibited by DCMU both in *Vicia* and *Commelina* (139; A. Schwartz and E. Zeiger, unpublished). In general, the evaluation of any component of the energy pool is best undertaken under conditions ensuring that other sources are inactive. The preceding discussion also indicates that large or even saturating levels of stomatal opening are not per se an unambiguous indication that all the components of the energy pool are operational.

Genotypic, phenotypic, and ontogenetic variations in the makeup of the energy pool are likely to be found. Genetic and developmental variations in the photosynthetic mesophyll are well known (74); they could also be found in guard cell chloroplasts. This plasticity could change the photosynthetic output of the PAR-system of guard cells in adaxial and abaxial surfaces of the leaf (104–106) and could alter the guard cell chloroplasts concomitantly with leaf adaptations to shade (151). Much of the variability seen in stomata as a result of different growing conditions could be caused by changes in energy pool components.

STOMATA FROM *PAPHIOPEDILUM*   Stomata from the orchid *Paphiopedilum* lack guard cell chloroplasts (83, 128, 181), yet they respond to both blue and red light and to VPD, ABA, and CO_2, (69, 83, 84). This apparent normalcy has been interpreted as evidence that guard cell chloroplasts generally are not required during stomatal movements (67, 83, 98, 116). On the other hand, the concept of an energy pool in the guard cells indicates that this conclusion could be misleading; stomata from *Paphiopedilum* could be devoid of the PAR-dependent system yet operate by relying on alternative energy sources. If the latter interpretation is correct, stomatal responses in *Paphiopedilum* should differ from those exhibited by species having a PAR-dependent system; recent findings in our laboratory indicate that this is the case.

Measurements of stomatal conductance and net photosynthesis in *P. insigne* showed that saturating rates in this species are very low, even for shade plants (172). With those characteristics, the conditions in which photophosphorylation by guard cell chloroplasts seem critical—large conductance levels at moderate to high light intensities (33, 144, 184)—would not be realized in *Paphiopedilum*.

In addition, gas exchange analysis of attached leaves of *P. insigne* under blue and red light showed that, at light saturation, stomatal conductance in red light was one half or less of that observed in the blue, with corresponding intercellular CO_2 values markedly lower under red than under white or blue light (C. Grivet and E. Zeiger, unpublished). Reported conductance increases under red light (83) thus seem to be caused by an indirect effect, such as a decrease in intercellular CO_2 concentrations, and not by red light. Finally, recent measurements of stomatal opening in epidermal peels of *P. harrisianum* incubated in 25 mM MES, pH 6.5, and 60 mM KCl showed an effect of blue but not of red light (H. Meidner, S. Assmann, and E. Zeiger, unpublished).[3]

---

[3]An apparent lack of K^+ requirement in *Paphiopedilum* guard cells has been postulated to be related to their lack of chloroplasts (97). However, there is no obvious connection between the lack of one of the components of the energy pool and the specific nature of the osmoticum used for turgor build-up.

Stomata of *Paphiopedilum* seem therefore to constitute an exceptional case in which the genetic changes leading to the loss of guard cell chloroplasts were not deleterious because of the prevailing low levels of conductance in this genus. Recent work indicates that these stomata lack the PAR-dependent responses and rely on the blue light photosystem for their responses to illumination and on oxidative phosphorylation for light-independent, stomatal opening. Low stomatal conductances and slow growth rates are correlated with this special adaptation.

Intracellular pH Regulation in Guard Cells

The regulation of intracellular pH, an important homeostatic requirement for all cells (150), is particularly important in guard cells. Within a chemiosmotic scheme, the magnitude of ΔpH would be significantly different if the cytoplasm becomes alkaline as protons are extruded or if the intracellular pH is maintained nearly constant.

Evidence from work with dyes and with pH electrodes (108, 130) indicates that the intracellular pH of the guard cells rises when stomata open. On the other hand, recent determinations of the cytoplasmic pH of root tip cells using ^{31}P NMR (121, 122) showed that the pH remained constant under several conditions in which rates of proton extrusion were high; CO_2 fixation into malate was the presumed buffering mechanism (122). Rates of proton extrusion in guard cells, however, are higher than in roots (36, 122), and thus malate biosynthesis might not be sufficiently high or rapid (7) to prevent some cytoplasmic alkalinization.

A supply of protons from malic acid biosynthesis in the guard cell cytoplasm has been invoked as a prerequisite for the operation of the proton pump (14, 132), under the assumption that the proton concentration in the cytoplasm is limiting for proton pumping. However, there is no evidence that a proton source other than H_2O is required for chemiosmosis, as demonstrated by observations with artificial vesicles of bacteriorhodopsin, the best characterized, light-sensitive proton pump. These vesicles translocate protons at high rates from an aqueous solution of KCl or NaCl (18). If that is also the case with guard cells, the generation of protons from malate biosynthesis would not affect the rate of proton pumping but, more importantly, would dissipate the pH gradient. A collapse of ΔpH would inhibit Cl^- uptake, resulting in the replacement of Cl^- by malate as the counterion for K^+ and in the stabilization of the intracellular pH in the absense of Cl^-/OH^- exchange (183).

Some insight about the intracellular pH in guard cells is also emerging from studies on their intrinsic vacuolar fluorescence (181). This green fluorescence was first discovered in guard cell vacuoles in peels of the genus *Allium*, mounted in H_2O and irradiated with blue light (188). Some data suggest that the chromophore could be tonoplast-bound (180, 181).

Guard cell vacuoles from genera other than *Allium* do not fluoresce. In *Vicia*, however, the fluorescence can be induced by exposure to ammonium hydroxide, ammonium chloride at pH >8, 50 mM HEPES at pH >6.6, CCCP and nigericin (180). In the presence of NH_3, fluorescence intensity increased with concentration, indicating that intracellular alkalinization is correlated with fluorescence induction (180). The effect of CCCP, nigericin, and the impermeable buffers could be associated with changes in energy gradients, although indirect effects on the intracellular pH cannot be ruled out. NH_3 and HEPES also induced the vacuolar fluorescence in epidermal cells (180), while the proton translocators affected only guard cell fluorescence. Whether onion guard cells, which show the fluorescence without any pretreatment, have a higher intracellular pH remains to be established.

The presumed alkalinization by NH_3 led to the prediction that FC treatment, which causes high rates of proton extrusion (68), would also induce the fluorescence. That was confirmed with guard cells of *Commelina*, *Malva parviflora*, *Vicia*, and *Xanthium strumarium* (181; E. Zeiger, unpublished). Fluorescence intensity increased with FC concentration, saturating at 5 μM in *Vicia* and 10 μM in *Commelina* (E. Zeiger, unpublished).

Both the identity and function of the vacuolar chromophore remain unknown. Because of its excitation by blue light and its emission spectrum (peak at 525 nm), the chromophore could be a flavin (188). The fast decay of the vacuolar fluorescence in *Allium* suggested to us that the compound was not a flavonoid, which, while having some spectroscopic characteristics in common with flavins, has a much slower fluorescent decay (188).

Vierstra et al (165) isolated a flavonoid, kaempferol, from extracts of epidermal peels of *Vicia*. The compound, present in millimolar concentrations, had a fluorescence emission maximum at 494 nm. This study provides convincing evidence that kaempferol is a major, green fluorescing compound in epidermal peels of *Vicia* but does not prove that the flavonoid is the chromophore observed in the guard cell vacuoles under microfluorospectrophotometry. Because of the relatively small proportion of guard cells in peels, a less abundant flavin could have gone undetected in the chromatographs. Furthermore, epidermal flavonoids vary widely with species (158), in contrast with the striking similarity of the autofluorescence in guard cell vacuoles of *Allium* and that induced by FC in the other species. The presence of several blue-absorbing compounds in the epidermis of *Commelina* and *Allium* has been recently reported (15).

The fact that flavins are usually found in very low concentrations in plant membranes (119, 154), in contrast to the abundance of the chromophore indicated by the relatively high fluorescence intensity of the guard cell vacuoles, has been raised as an argument against the likelihood of the latter being a flavin (119, 165). However, most of the data on flavin concentrations originate from measurements in phototropic systems, in which the response to blue light does not require a high energy expenditure. Guard cells, on the other hand, are highly specialized in ion transport, and they could have evolved an unusually high flavin content.

A correlation between the time courses of fluorescence induction by FC and stomatal opening in both *Commelina* and *Vicia* indicates that the fluorescent chromophore responds to metabolic changes associated with stomatal movements (Figure 2). The transient rise in fluorescence intensity could reflect an initial alkalinization of the guard cells caused by high rates of proton extrusion induced by FC; the subsequent decrease could be attributed to restabilization of the intracellular pH caused by the onset of malic acid biosynthesis. Whether the fluorescing compound is responding passively to fluctuations in the pmf or is an integral part of an energy-transducing mechanism remains to be established.

THE BIOLOGY OF STOMATAL RESPONSES

Light

A distinction between stomatal reponses to light and CO_2 has been a classical problem in stomatal physiology. Both light and CO_2 elicit marked stomatal responses; in the intact leaf the two effects are tightly coupled because increasing irradiation simultaneously opens stomata and stimulates mesophyll photosynthesis which depletes intercellular CO_2 (39). One of the most dramatic recent advances in stomatal physiology, however, has been the unequivocal demonstration of a specific light response of stomata.

Both in isolation and in the intact leaf, guard cells respond to light quality and intensity. The swelling of guard cell protoplasts in response to light demonstrates that photoreception is within the guard cells proper (43a, 187). Intensity-dependent increases in stomatal apertures in epidermal peels exposed to light in CO_2-free air characterize the photoresponses of guard cells in the intact epidermis (22, 162). In the whole leaf, the classical work of Heath & Russell (31) showing an effect of light independent of CO_2 has been recently substantiated by several groups. Gas exchange analysis with *Eucalyptus pauciflora* (177), *Xanthium strumarium, Gossypium hirsutum, Phaseolus vulgaris, Zea mays,* and *Perilla frutescens* (143) characterized changes in stomatal conductance as a function of white light irradiation and intercellular CO_2 concentration; the light and CO_2-dependent components of the stomatal responses were analyzed with the method of Farquhar et al (19). In all species, the light component was found to be the dominant one.

In a different approach, specific light effects on stomatal conductance were demonstrated in experiments with blue or red light (39, 86) and in dual beam experiments with attached leaves of *Malva parviflora* (184). In the latter study,

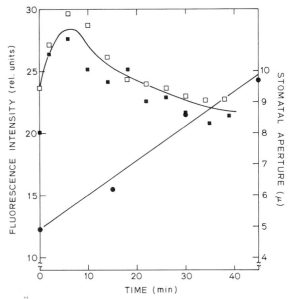

Figure 2 FC-dependent stomatal opening and vacuolar fluorescence in guard cells from *Commelina communis*. Peels in microchambers on a microscope stage were perfused with 10 μM FC in 20 mM KCl. Stomatal opening (●) was measured with an ocular micrometer. Data points are the average of two experiments, 50 measurements per point. Fluorescence intensity at 525 nm of single guard cells was measured for 2 sec with a microspectrophotometer (open and solid squares, two different experiments, 10 measurements per point).

we used lasers as a source of monochromatic light and added low-intensity blue or far-red light to a leaf irradiated with high-intensity red light. Addition of blue light caused a marked increase in conductance with a concomitant increase in intercellular CO_2 concentration without significant changes in photosynthesis. Addition of far-red light increased photosynthesis without affecting stomatal conductance, and the intercellular CO_2 concentration decreased. Stomatal and mesophyll responses to light are thus not obligatorily coupled (39, 184); furthermore, stomatal responses to light quality can override the CO_2 responses.

A direct stimulation of the pmf appears as the most plausible mode of action of light in stomatal responses. Light stimulates proton extrusion in guard cells (24, 36), induces photophosphorylation in guard cell chloroplasts (75, 87) and, in the blue, sustains wide stomatal apertures at very low quantum fluxes (54, 105, 162). Light-stimulated electrogenic pumps have been demonstrated in several plant systems (56, 155). In nature, a large portion of the daily increases in conductance is attained in the early morning as light intensity rises (185). A light-dependent activation of the pmf would drive stomatal conductance to near saturation, with other environmental and metabolic factors providing fine adjustments in apertures through subsequent modulations of the energy gradients. Prevailing conductance values at any time would thus depend on a dominant light component and an integrated, superimposed modulation of other stomatal signals on the pmf. Low VPD values common in the early morning (138) would minimize interactions with the humidity response in the opening phase. Later in the day, with higher temperatures and lower humidities, VPD modulations would exert a major impact on the time courses of stomatal closing, which can vary from early afternoon to dusk (77, 138).

Stomatal responses to light quality largely depend on the sensitivity of the two photoreceptor systems of the guard cells, chlorophyll in the chloroplasts and a blue light photoreceptor, presumably a flavin (184). Their combined effect results in a wavelength dependence of stomatal opening showing major peaks in the blue and in the red, with the blue peak enhanced by the contribution of the blue light photosystem (39, 86, 144, 162, 180, 184). As discussed previously, light intensity has a marked effect on the observed ratios between the blue and red peaks because of the differential threshold intensities of the PAR and blue-light photosystems.

A recently reported action spectrum for stomatal conductance in *Xanthium* (144) showed a major peak in the blue with a minor shoulder in the red. Because action spectra are in principle independent of light intensity, these observations would indicate that the reported blue/red ratio reflects the general relationship between the PAR and blue light photosystems in guard cells. However, this spectrum was calculated from the intercept of fluence curves corresponding to conductance values of about 50% of the saturating levels observed under white light (143). The apparent dominant blue response is thus comparable to the preponderance of the blue light photosystem usually observed at low conductance values (33, 184). The fluence curves reported (144) indicate that progressively higher red peaks would have been observed in spectra normalized for larger conductance levels.

The specific roles of the PAR and blue light photosystems of the guard cells in stomatal function are poorly understood. A calculation of apparent photon efficiencies of blue and red light on stomatal conductance of attached *Malva* leaves showed that at moderate to high irradiances, the efficiencies of blue and red light were very similar while blue light had a higher efficiency at low irradiances (184). This indicates that the PAR-dependent system of the guard cells is the energy source driving stomatal conductance at moderate to high irradiances. This PAR response of guard cells appears mediated by CO_2-sensitive photophosphorylation in their chloroplasts (75).

Guard cells relying solely on a PAR-dependent photosystem, however, would have two major limitations. First, they would be unable to respond to light intensities below the activation threshold of the PAR-dependent system.

Second, the CO_2 coupling would make them prone to oscillations (9, 39) because of rapid responses of photosynthesis to variations in light intensity. The operation of the blue light photosystem in guard cells appears to eliminate these limitations. Because of its apparent saturation at low light intensities (184), the high gain output of the blue light photosystem would be saturated at moderate to high irradiances and provide guard cells with a light sensor independent of the PAR responses. The high sensitivity of the blue light photosystem would facilitate stomatal responses at low irradiances and control stomatal opening at dawn (185).

A possible adaptive advantage of the blue light photosystem which would provide the leaf with higher carbon gains because of the enhanced stomatal opening in the early morning has been investigated (185). Under nonlimiting water supply, the calculated additional carbon gain was small, but it could be substantially larger under moderate or high water stress, conditions in which stomata close early in the day (52, 138).

A participation of phytochrome as a third photosystem in guard cells (27) reemerged from recent work on *Commelina* (126, 127) showing that far-red light alone or simultaneously supplied with red light inhibits stomatal opening. Phytochrome could play a role in the control of membrane permeability during stomatal movements (127) and the entrainment of circadian rhythms.

Carbon Dioxide

In both light and darkness, stomatal apertures are inversely correlated with CO_2 concentrations (67, 116). CO_2 fixation into PEP by PEP carboxylase is the only well-characterized metabolic reaction of CO_2 in guard cells (116). This reaction is certain to affect guard cell metabolism via its modulation of malic acid biosynthesis, but it is not easily interpreted as a primary step in CO_2-induced stomatal closure. Closing responses to CO_2 can be observed within seconds (113), whereas increases in malate synthesis are significantly slower (1, 2, 88). Furthermore, if CO_2-limited malate synthesis were regulating stomatal movements, stomatal apertures would be expected to increase in high CO_2 concentrations; the opposite is observed (32). CO_2 fixation into malate could perhaps be unrelated to the observed CO_2 effects on stomatal apertures and instead could be homologous to the widespread metabolic regulation of intracellular pH by organic acid biosynthesis (122, 150). The specific, rapid effect of CO_2 in stomatal movements could be mediated by changes in membrane permeability, an amplified acidification of the guard cell cytoplasm, and a concomitant decrease in the pmf, or the use of HCO_3^- as a counterion for K^+ (24, 32, 116, 183).

Recent observations of the modulation of photophosphorylation in guard cell chloroplasts of *Chlorophytum* by CO_2 (75) prompted us to reconsider the classical notion of a CO_2 effect on stomatal movements mediated by guard cell chloroplasts (32, 73, 103, 191). This hypothesis had been ruled out because of the lack of Calvin cycle enzymes in these organelles (91), but CO_2 modulation of photophosphorylation is not incompatible with an absence of photosynthetic CO_2 fixation.

An hypothesis of CO_2 modulation of photophosphorylation in guard cells predicts that CO_2 concentrations, inhibition of photophosphorylation, and stomatal closure would be positively correlated. In addition, at any given value of intercellular CO_2 concentration, the magnitude of the inhibition would decrease with light intensity. Some studies on the kinetics of stomatal conductance as a function of intercellular CO_2 concentrations and irradiance levels are strikingly consistent with this prediction [see for instance Figure 4 in (177)]. CO_2 responses, however, can vary substantially with species (39, 143, 177) and water stress (116), and the photophosphorylating capacity of guard cell chloroplasts could change with growth conditions (74). Clearly, direct tests of this hypothesis are needed.

Using notions developed for control theory, effects of CO_2 on stomatal conductance in the light have been analyzed in terms of the gain of a CO_2 feedback loop (19, 116). The magnitude of this loop would be large if the CO_2 control of stomatal conductance were significant. Empirical determinations, however, have shown small values for the gain of the CO_2 feedback loop (20, 116, 143). That is consistent with a large effect of light on stomatal opening and a CO_2 modulation of the output of the PAR-dependent system of the guard cells. Large, parallel responses of photosynthesis and stomatal conductance to irradiance (39, 184) could be coupled effectively through small, CO_2-dependent adjustments of stomatal apertures. More precise determinations should establish whether these combined light and CO_2 effects are sufficient to maintain the observed constant, intercellular CO_2 concentrations in the leaf (178) or if another mesophyll messenger needs to be invoked (20).

A CO_2 effect on guard cell chloroplasts is unlikely to mediate stomatal responses to CO_2 in darkness; hence two classes of physiological responses to CO_2 can be distinguished. Responses in the dark would include CO_2-dependent opening in epidermal peels (67, 73), nocturnal opening in CAM plants (44), and CO_2 responses in chloroplast-less guard cells from *Paphiopedilum* (83; C. Grivet and E. Zeiger, unpublished), and would utilize energy from oxidative phosphorylation. On the other hand, the CO_2 reponses in the light would involve CO_2 modulation of photophosphorlyation in guard cell chloroplasts (75).

ABA and Other Phytohormones

There is unquestionable evidence that ABA can cause stomatal closure and convincing data showing a relationship between endogenous ABA levels and stomatal responses to water stress (66, 79, 111, 167). Unresolved questions pertaining to the mode of action of ABA on stomatal closure are the time courses of ABA biosynthesis, transport, and degradation and their correlation with stomatal movements; the compartmentalization of ABA in the leaf; and the specific mechanism whereby ABA causes stomatal closure (66, 78, 79, 168).

Inconsistencies between measured levels of endogenous ABA and the rapidity of stomatal closure upon the application of stress made it difficult to explain the fast stomatal responses in terms of ABA action. Improved measurements of ABA content in the epidermis, however, have recently shown that ABA increases in epidermal peels somewhat faster than in the mesophyll (168), suggesting that some of the early effects of ABA could be directly dependent on guard cell metabolism rather than on apoplastic transport.

Fast, transient K^+ and Cl^- efflux from guard cells treated with ABA (60, 170) was observed without concomitant changes in the rates of K^+ uptake (60). In addition, epidermal cells seem important in the ABA-induced redistribution of osmoticum (37, 38). These findings point to ABA effects on membrane permeability rather than a direct modulation of the pmf. The low concentrations of ABA required to cause stomatal closure—10^{-9} M in *Kalanchoe* (43)—rule out a direct effect of ABA on ΔpH, but amplification processes similar to those discussed for CO_2 should be kept in mind. Synergistic effects of ABA and CO_2 on stomatal closure are well documented (116, 152, 174). On the other hand, the dissociation constant of ABA is in the physiological range of several cell compartments, and a regulatory mechanism depending on its distribution along prevailing pH gradients within and between the leaf cells has been proposed (12). An inhibition by ABA of fluorescence transients associated with photophosphorylation in *Vicia* guard cell chloroplasts has been reported recently (87).

Exogenously supplied ABA accelerates leaf senescence and the endogenous levels of ABA are highest when senescence is most rapid; these effects are antagonized by kinetin (25); in turn, both phytohormones have parallel, antagonistic effects on stomatal movements in senescing leaves (159), suggesting that stomatal behavior is a component of leaf senescence. The extended longevity of guard cells and their chloroplasts in senescing leaves (190) could have a bearing on this phenomenon.

Several natural and synthetic cytokinins have been shown to promote stomatal opening in epidermal peels (42), but their role in the stomatal physiology of nonsenescing leaves is not known. Recent work on the gas exchange of tomato plants with flooded roots characterized changes in stomatal conductance preceding changes in leaf water potential (K. J. Bradford, unpublished). Evidence for an effect of root cytokinins on the levels of stomatal conductance in the leaf was documented, suggesting that these phytohormones could play a role in the control of stomatal conductance.

Hormonal treatment of epidermal peels also provided information on the differential responses of adaxial and abaxial stomata to stimuli. An analysis of the reactivity of abaxial and adaxial stomata in peels showed that their differential responses were intrinsic rather than a result of the environment of each leaf surface (104). Treatment with IAA and FC, however, increased the usually reduced responses of adaxial stomata to the levels observed in the abaxial ones (104, 106). The IAA effects have been further investigated in abaxial peels of *Commelina* (152, 153), in which IAA-ABA antagonistic interaction with CO_2 was demonstrated, indicating that an in vivo interaction between the two phytohormones could modulate stomatal movements.

Humidity and Temperature

In experiments with attached leaves kept in gas exchange cuvettes, decreases in stomatal conductance as a function of increasing VPD have been reported in many species (39, 51, 138). An adaptive advantage of the VPD response has been convincingly demonstrated (51, 138, 146).

Stomata of epidermal peels also show a humidity response independent of temperature (51). Available evidence indicates that peristomatic transpiration plays a role in the early stages of the stomatal responses to humidity (51); it is also clear that, at steady state conditions, metabolic adjustments in the guard cells must accompany initial hydropassive responses. The observed metabolic changes in guard cells responding to humidity lagged behind the changes in aperture (51). This apparent reversal of the paradigm explaining the regulation of stomatal movements by modulated ionic changes preceding water fluxes and turgor build-up poses a paradox. In an extreme view of the problem, Maier-Maercker (64, 65) proposed that hydropassive events mediated by subsidiary (and epidermal) cells and subsequent turgor changes in the epidermis actually control the responses of the guard cells. This hypothesis appears untenable in view of the definitive evidence showing that guard cell protoplasts or guard cells in peels in which other epidermal cells have been destroyed exhibit a wide range of responses similar to those observed in intact leaves. Thus the question of turgor regulation in the guard cells responding to humidity and its transduction into metabolic adjustments required for the establishment of steady-state conditions is unresolved.

Stomatal responses to temperature are hard to separate from those to humidity because VPD usually increases with temperature. Specific stomatal responses to temperature, however, are well documented (39, 73, 116). Interesting discrepancies between temperature effects in the light and in darkness were studied recently (123, 124). As reported by other investigators (51, 157), aperture maxima in the light were observed at 35°C, with aperture values decreasing at higher and lower temperatures. In darkness, the largest apertures were found at the highest temperature tested, 45°C. A subsequent study showed that, in the light, closing at high temperature correlated with damage to the guard cell chloroplasts (125). These findings are consistent with the hypothesis that guard cell chloroplasts provide a driving force for stomatal opening in the light, while opening in darkness depends on oxidative phosphorylation. The temperature responses in each condition would thus reflect the thermal responses of chloroplasts and mitochondria, respectively. Whether the guard cells have a specific temperature sensor in addition to these temperature-dependent metabolic responses remains to be elucidated.

CONCLUDING REMARKS

Advances in recent years provided several remarkable "firsts" in stomatal physiology, including direct determination of K^+ and organic acid contents in isolated guard cells; isolation of guard cell protoplasts; characterization of light-induced hyperpolarization in single guard cells; localization of light-harvesting pigments in guard cell chloroplasts; discovery of their lack of Calvin-cycle enzymes; demonstration of the capacity of these organelles to conduct electron transport, to photophosphorylate, and to respond to CO_2; and elucidation of a stomatal response to blue light under field conditions. These findings emphasize the notorious biological complexity and functional versatility of stomata and point to their crucial importance in the physiology of higher plants.

The concept of a central role of stomata in the maintenance of plant homeostasis provides us with a unifying theme for understanding stomatal responses. Guard cells integrate internal and external stimuli and modulate stomatal apertures to balance the opposite needs of preserving biological integrity while allowing an active physical exchange between the plant and its environment.

Metabolically dependent ion transport appears as a dominant regulatory process in the build-up of turgor in the guard cells. A chemiosmotic mechanism ensuing from a primary, electrogenic proton gradient in guard cells integrates all the ion fluxes connected with stomatal movements. A modulation of the pmf by stimuli affecting stomatal responses would provide a common metabolic step to their mode of action in guard cells. A conclusive elucidation of this mechanism and its quantitative changes under different conditions is likely to constitute one of the most exciting aspects of future stomatal research. A chemiosmotic hypothesis of stomatal movements also underscores the operation of an energy pool driving proton extrusion in the guard cells. Components of this pool include oxidative and photophosphorylation, and perhaps the blue light photosystem. This aspect of guard cell physiology offers a remarkable potential for enhancing our understanding of the biology of plant cells.

The unique role of light in plant biology and the impact of stomata on plant adaptation and productivity give particular relevance to the study of the photobiological properties of stomata. Recent progress on the characterization of stomatal responses to light quality and intensity provides us with a new perspective on stomatal function and its impact on plant adaptation. Further advances in this field could open the way for the use of the emerging knowledge in general agricultural practices and for the cultivation of marginal lands.

Only a few years ago, I fancied that stomatal physiology was a largely "underdeveloped" area of plant physiology in which many unanswered questions were ripe for straightforward solutions. Perhaps because of the sobering wisdom from writing this review, or because of the impressive progress of recent years, that notion is now replaced by one of nearly overwhelming complexity. I am certain, though, that this complexity will continue to provide its students with knowledge and excitement.

ACKNOWLEDGMENTS

I thank Professor Hans Meidner, Drs. A. Bloom, D. Cosgrove, S. Davis, G. Farquhar, A. Schwartz, A. Schafer, P. Sharpe, G. Tallman, and S. Assmann for valuable discussions. My thanks also to S. Sakaguchi for software development, Eleanor Crump for editing the manuscript, Judy Nishimoto, Sally Moran, and Scott Grayburn for invaluable logistic support, and Stephanie Williams for artwork. This work was supported by grants from the National Science Foundation PCM 8012060 and Department of Energy 81ER10924.

Literature Cited

1. Allaway, W. G. 1973. Accumulation of malate in guard cells of *Vicia faba* during stomatal opening. *Planta* 110:63–70
2. Allaway, W. G. 1981. Anions in stomatal operation. In *Stomatal Physiology*, ed. P. G. Jarvis, T. A. Mansfield. *Soc. Exp. Biol.* (SS) 8:71–85. Cambridge: Cambridge Univ. Press. 295 pp.
3. Beardsell, M. F., Jarvis, P. G., Davidson, B. 1972. A null-balance diffusion porometer suitable for use with leaves of many shapes. *J. Appl. Ecol.* 9:677–90
4. Bingham, G. E., Coyne, P. I. 1977. A portable, temperature-controlled, steady-state porometer for field measurements of transpiration and photosynthesis. *Photosynthesis* 11:148–60
5. Bloom, A. J., Mooney, H. A., Björkman, O., Berry, J. 1980. Materials and methods for carbon dioxide and water exchange analysis. *Plant Cell Environ.* 3:371–76
6. Brogårdh, T., Johnsson, A. 1975. Regulation of transpiration in *Avena*. Responses to white light steps. *Physiol. Plant.* 35:115–25

7. Brown, P. H., Outlaw, W. H. 1982. Effect of fusicoccin on dark $^{14}CO_2$ fixation by *Vicia faba* L. guard cell protoplasts. *Plant Physiol.* 70:1700–3
8. Cheeseman, J. M., Edwards, M., Meidner, H. 1982. Cell potentials and turgor pressures in epidermal cells of *Tradescantia* and *Commelina*. *J. Exp. Bot.* 33:761–70
9. Cowan, I. R. 1977. Stomatal behavior and environment. *Adv. Bot. Res.* 4:117–228
10. Cowan, I. R. 1982. Regulation of water use in relation to carbon gain in higher plants. In *Physiological Plant Ecology*, ed. O. L. Lange, P. S. Nobel, C. B. Osmond, H. Ziegler. *Encycl. Plant Physiol.* (NS) 12B:589–613. Heidelberg: Springer
11. Cowan, I. R., Farquhar, G. D. 1977. Stomatal function in relation to leaf metabolism and environment. *Symp. Soc. Exp. Biol.* 31:471–505
12. Cowan, I. R., Raven, J. A., Hartung, W., Farquhar, G. D. 1982. A possible role for abscisic acid in coupling stomatal

conductance and photosynthetic carbon metabolism in leaves. *Aust. J. Plant Physiol.* 9:489–98

13. Curtis, H. 1979. *Biology.* New York: Worth. 1043 pp.

14. Dittrich, P., Mayer, M., Meusel, M. 1979. Proton-stimulated opening of stomata in relation to chloride uptake by guard cells. *Planta* 144:305–9

15a. Donkin, M. E., Martin, E. S. 1981. Blue light absorption by guard cells of *Commelina communis* and *Allium cepa.* *Z. Pflanzenphysiol.* 102:345–52

16. Edwards, M., Meidner, H., Sheriff, D. W. 1976. Direct measurements of turgor pressure potentials of guard cells. II. The mechanical advantage of subsidiary cells, the *Spannungsphase,* and the optimum leaf water deficit. *J. Exp. Bot.* 27:163–71

17. Ehleringer, J., Björkman, O. 1977. Quantum yields for CO_2 uptake in C_3 and C_4 plants. Dependence on temperature, CO_2, and O_2 concentration. *Plant Physiol.* 59:86–90

18. Eisenbach, M., Caplan, S. R. 1979. The light-driven proton pump of *Halobacterium halobium:* mechanism and function. *Curr. Top. Membr. Transp.* 12:165–248

19. Farquhar, G. D., Dubbe, D. R., Raschke, K. 1978. Gain of the feedback loop involving carbon dioxide and stomata, theory and measurement. *Plant Physiol.* 62:406–12

20. Farquhar, G. D., Sharkey, T. D. 1982. Stomatal conductance and photosynthesis. *Ann. Rev. Plant Physiol.* 33:317–45

21. Field, C., Berry, J. A., Mooney, H. A. 1982. A portable system for measuring carbon dioxide and water vapour exchange of leaves. *Plant Cell Environ.* 5:179–86

22. Fischer, R. A. 1968. Stomatal opening in isolated epidermal strips of *Vicia faba.* I. Response to light and to CO_2-free air. *Plant Physiol.* 43:1947–52

23. Galun, E. 1981. Plant protoplasts as physiological tools. *Ann. Rev. Plant Physiol.* 32:237–66

24. Gepstein, S., Jacobs, M., Taiz, L. 1982. Inhibition of stomatal opening in *Vicia faba* epidermal tissue by vanadate and abscisic acid. *Plant Sci. Lett.* 28:63–72

25. Gepstein, S., Thimann, K. V. 1980. Changes in the abscisic acid content of oat leaves during senescence. *Proc. Natl. Acad. Sci. USA* 77:2050–53

26. Gunar, I. I., Zlotnikova, I. F., Panichkin, L. A. 1975. Electrophysiological investigation of cells of the stomate complex in spiderwort. *Sov. Plant. Physiol.* 22:704–7

27. Habermann, H. M. 1973. Evidence for two photoreactions and possible involvement of phytochrome in light-dependent stomatal opening. *Plant Physiol.* 51:543–48

28. Harold, F. M. 1977. Membranes and energy transduction in bacteria. *Curr. Top. Bioenerg.* 6:83–149

29. Hastings, D. F., Gutknecht, J. 1978. Potassium and turgor pressure in plants. *J. Theor. Biol.* 73:363–66

30. Heath, O. V. S., Meidner, H. 1981. Feedback processes in the opening of leaf stomata in light. *Proc. R. Soc. London Ser. B* 213:161–70

31. Heath, O. V. S., Russell, J. 1954. Studies in stomatal behavior. VI. An investigation of the light responses of wheat stomata with the attempted elimination of control by the mesophyll. Part 2: Interactions with external carbon dioxide, and general discussion. *J. Exp. Bot.* 5:269–92

32. Hsiao, T. C. 1976. Stomatal ion transport. In *Transport in Plants II,* ed. U. Lüttge, M. G. Pitman. *Encycl. Plant Physiol.* (NS) 2B:195–221. Berlin: Springer

33. Hsiao, T. C., Allaway, W. E., Evans, L. T. 1973. Action spectra for guard cell Rb^+ uptake and stomatal opening in *Vicia faba. Plant Physiol.* 51:82–88

34. Humble, G. D., Hsiao, T. C. 1969. Specific requirement of potassium for light-activated opening of stomata in epidermal strips. *Plant Physiol.* 44:230–34

35. Humble, G. D., Hsiao, T. C. 1970. Light-dependent influx and efflux of potassium of guard cells during stomatal opening and closing. *Plant Physiol.* 46:483–87

36. Humble, G. D., Raschke, K. 1971. Stomatal opening quantitatively related to potassium transport; evidence from electron probe analysis. *Plant Physiol.* 48:447–53

37. Itai, C., Meidner, H. 1978. Functional epidermal cells are necessary for abscisic acid effects on guard cells. *J. Exp. Bot.* 29:765–70

38. Itai, C., Meidner, H. 1978. Effect of abscisic acid on solute transport in epidermal tissue. *Nature* 271:653–54

39. Jarvis, P. G., Morison, J. I. L. 1981. The control of transpiration and photosynthesis by the stomata. See Ref. 2, pp. 247–79

40. Jarvis, R. G., Mansfield, T. A. 1980.

Reduced stomatal responses to light, carbon dioxide and abscisic acid in the presence of sodium ions. *Plant Cell Environ.* 3:279–83

41. Jeschke, W. D. 1976. Ionic relations of leaf cells. See Ref. 32, pp. 160–94

42. Jewer, P. C., Incoll, L. D. 1980. Promotion of stomatal opening in the grass *Anthephora pubescens* Nees by a range of natural and synthetic cytokinins. *Planta* 150:218–21

43. Jewer, P. C., Incoll, L. D., Howarth, G. L. 1981. Stomatal responses in isolated epidermis of the crassulacean acid metabolism plant *Kalanchoe diagremontiana* Hamet et Perr. *Planta* 153:238–45

43a. Jewer, P. C., Incoll, L. D., Shaw, J. 1982. Stomatal responses of *Argenteum*—a mutant of *Pisum sativum* L. with readily detachable leaf epidermis. *Planta* 155:146–53

44. Kluge, M., Ting, I. P. 1978. *Crassulacean Acid Metabolism.* New York: Springer. 209 pp.

45. Körner, Ch., Scheel, J. A., Bauer, H. 1979. Maximum leaf diffusive conductance in vascular plants. *Photosynthesis* 13:45–82

46. Kowallik, W. 1982. Blue light effects on respiration. *Ann. Rev. Plant Physiol.* 33:51–72

47. Krause, G. H. 1973. The high-energy state of the thylakoid system as indicated by chlorophyll fluorescence and chloroplast shrinkage. *Biochim. Biophys. Acta* 292:715–28

48. Kuiper, P. C. J. 1964. Dependence upon wavelength of stomatal movement in epidermal tissue of *Senecio odoris*. *Plant Physiol.* 39:952–55

49. Lavorel, J., Etienne, A. -L. 1977. *In vivo* chlorophyll fluorescence. In *Primary Processes of Photosynthesis,* ed. J. Barber, pp. 203–68. *Topics in Photosynthesis,* Vol. 2. Oxford: North-Holland. 516 pp.

50. Levitt, J. 1974. The mechanism of stomatal movement—once more. *Protoplasma* 82:1–17

51. Lösch, R., Tenhunen, J. D. 1981. Stomatal responses to humidity—phenomenon and mechanism. See Ref. 2, pp. 137–61

52. Ludlow, M. M. 1980. Adaptive significance of stomatal responses to water stress. In *Adaptation of Plants to Water and High Temperature Stress,* ed. N. C. Turner, P. J. Kramer, pp. 123–38. New York: Wiley. 482 pp.

53. Lurie, S. 1977. Stomatal development in etiolated *Vicia faba:* relationship between structure and function. *Aust. J. Plant Physiol.* 4:61–68

54. Lurie, S. 1978. The effect of wavelength of light on stomatal opening. *Planta* 140:245–49

55. Lüttge, U., Pitman, M. G. 1976. Transport and Energy. In *Transport in Plants II,* ed. U. Lüttge, M. G. Pitman. *Encycl. Plant Physiol.* (NS) 2A:251–59. Berlin: Springer

56. MacRobbie, E. A. C. 1965. The nature of the coupling between light energy and active ion transport in *Nitella translucens*. *Biochim. Biophys. Acta* 94:64–73

57. MacRobbie, E. A. C. 1970. The active transport of ions in plant cells. *Q. Rev. Biophys.* 3:251–94

58. MacRobbie, E. A. C. 1977. Functions of ion transport in plant cells and tissues. *Int. Rev. Biochem. II* 13:211–47

59. MacRobbie, E. A. C. 1981. Ion fluxes in 'isolated' guard cells of *Commelina communis*. *J. Exp. Bot.*32: 545–62

60. MacRobbie, E. A. C. 1981. Effects of ABA in 'isolated' guard cells of *Commelina communis* L. *J. Exp. Bot.* 32:563–72

61. MacRobbie, E. A. C., Lettau, J. 1980. Ion content and aperture in "isolated" guard cells of *Commelina communis* L. *J. Membr. Biol.* 53:199–205

62. MacRobbie, E. A. C., Lettau, J. 1980. Potassium content and aperture in "intact" stomatal and epidermal cells of *Commelina communis* L. *J. Membr. Biol.* 56:249–56

63. Madhavan, S., Smith, B. N. 1982. Localization of ribulose bisphosphate carboxylase in the guard cells by an indirect, immunofluorescence technique. *Plant Physiol.* 69:273–77

64. Maier-Maercker, U. 1979. "Peristomatal transpiration" and stomatal movement: a controversial view. I. Additional proof of peristomatal transpiration by hygrophotography and a comprehensive discussion in the light of recent results. *Z. Pflanzenphysiol.* 91:25–43

65. Maier-Maercker, U. 1979. "Peristomatal transpiration" and stomatal movement: a controversial view. II. Observation of stomatal movements under different conditions of water supply and demand. *Z. Pflanzenphysiol.* 91:157–72

66. Mansfield, T. A., Davies, W. J. 1981. Stomata and stomatal mechanisms. In *The Physiology and Biochemistry of Drought Resistance in Plants,* ed. L. G. Paleg, D. Aspinall, pp. 314–46. San Francisco: Academic. 492 pp.

67. Mansfield, T. A., Travis, A. J., Jarvis, R. G. 1981. Responses to light and carbon dioxide. See Ref. 2, pp. 119–35

68. Marrè, E. 1979. Fusicoccin: a tool in plant physiology. *Ann Rev. Plant Physiol.* 30:273–88

69. Mayo, J. M., Ehret, D. 1980. The effects of abscisic acid and vapor pressure deficit on leaf resistance of *Paphiopedilum leeanum*. *Can. J. Bot.* 58:1202–4

70. Meidner, H. 1968. The comparative effects of blue and red light on the stomata of *Allium cepa* L. and *Xanthium pennsylvanicum*. *J. Exp. Bot.* 19:146–51

71. Meidner, H. 1981. Measurements of stomatal aperture and responses to stimuli. See Ref. 2, pp. 25–49

72. Meidner, H., Bannister, P. 1979. Pressure and solute potentials in stomatal cells of *Tradescantia virginiana*. *J. Exp. Bot.* 30:255–65

73. Meidner, H., Mansfield, T. A. 1968. *Physiology of Stomata.* London: Mc-Graw-Hill. 179 pp.

74. Melis, A., Harvey, G. W. 1981. Regulation of photosystem stoichiometry, chlorophyll *a* and chlorophyll *b* content and relation to chloroplast ultrastructure. *Biochim. Biophys. Acta* 637:138–45

75. Melis, A., Zeiger, E. 1982. Chlorophyll *a* fluorescence transients in mesophyll and guard cells. Modulation of guard cell photophosphorylation by CO_2. *Plant Physiol.* 69:642–47

76. Mengel, K., Kirkby, E. A. 1980. Potassium in crop production. *Adv. Agron.* 33:59–110

77. Meyer, W. S., Green, G. C. 1981. Comparison of stomatal action of orange, soybean and wheat under field conditions. *Aust. J. Plant Physiol.* 8:65–76

78. Milborrow, B. V. 1980. A distinction between the fast and slow responses to abscisic acid. *Aust. J. Plant Physiol.* 7:749–54

79. Milborrow, B. V. 1981. Abscisic acid and other hormones. See Ref. 66, pp. 347–88

80. Mitchell, P. 1979. Compartmentation and communication in living systems. Ligand conduction: a general catalytic principle in chemical, osmotic and chemiosmotic reaction systems. *Eur. J. Biochem.* 95:1–20

81. Mohr, H. 1972. *Lectures on Photomorphogenesis.* New York: Springer. 237 pp.

82. Moody, W., Zeiger, E. 1978. Electrophysiological properties of onion guard cells. *Planta* 139:159–65

83. Nelson, S. D., Mayo, J. M. 1975. The occurrence of functional non-chlorophyllous guard cells in *Paphiopedilum* spp. *Can. J. Bot.* 53:1–7

84. Nelson, S. D., Mayo, J. M. 1977. Low K^+ in *Paphiopedilum leeanum* leaf

epidermis: implications for stomatal functioning. *Can. J. Bot.* 55:489–95

85. Ogawa, T. 1979. Stomatal responses to light and CO_2 in greening wheat leaves. *Plant Cell Physiol.* 20:445–52

86. Ogawa, T. 1981. Blue light response of stomata with starch-containing (*Vicia faba*) and starch-deficient (*Allium cepa*) guard cells under background illumination with red light. *Plant Sci. Lett.* 22:103–8

87. Ogawa, T., Grantz, D., Boyer, J., Govindjee. 1982. Effects of cations and abscisic acid on chlorophyll *a* fluorescence in guard cells of *Vicia faba*. *Plant Physiol.* 69:1140–44

88. Ogawa, T., Ishikawa, H., Shimada, K., Shibata, K. 1978. Synergistic action of red and blue light and action spectra for malate formation in guard cells of *Vicia faba* L. *Planta* 142:61–65

89. O'Leary, M. H. 1982. Phosphoenolpyruvate carboxylase: An enzymologist's view. *Ann. Rev. Plant Physiol.* 33:297–315

90. Outlaw, W. H. Jr. 1980. A descriptive evaluation of quantitative histochemical methods based on pyridine nucleotides. *Ann. Rev. Plant Physiol.* 31:299–311

91. Outlaw, W. H. Jr. 1982. Carbon metabolism in guard cells. In *Cellular and Subcellular Localization in Plant Metabolism,* ed. L. L. Creasy, G. Hrazdina, pp. 185–222. New York: Plenum

92. Outlaw, W. H. Jr., Kennedy, J. 1978. Enzymic and substrate basis for the anaplerotic step in guard cells. *Plant Physiol.* 62:648–52

93. Outlaw, W. H. Jr., Lowry, O. H. 1977. Organic acid and potassium accumulation in guard cells during stomatal opening. *Proc. Natl. Acad. Sci. USA* 74:4434–38

94. Outlaw, W. H. Jr., Manchester, J. 1979. Guard cell starch concentration quantitatively related to stomatal aperture. *Plant Physiol.* 64:79–82

95. Outlaw, W. H. Jr., Manchester, J., DiCamelli, C. A. 1979. Histochemical approach to properties of *Vicia faba* guard cell phosphoenolpyruvate carboxylase. *Plant Physiol.* 64:269–72

96. Outlaw, W. H. Jr., Manchester, J., DiCamelli, C. A., Randall, D. D., Rapp, B., Veith, G. M. 1979. Photosynthetic carbon reduction pathway is absent in chloroplasts of *Vicia faba* guard cells. *Proc. Natl. Acad. Sci. USA* 76:6371–75

97. Outlaw, W. H. Jr., Manchester, J., Zenger, V. E. 1982. Potassium involvement not demonstrated in stomatal movements of *Paphiopedilum*. Qualified con-

firmation of the Nelson-Mayo report. *Can. J. Bot.* 60:240–44

98. Outlaw, W. H. Jr., Mayne, B. C., Zenger, V. E., Manchester, J. 1981. Presence of both photosystems in guard cells of *Vicia faba* L., implications for environmental signal processing. *Plant Physiol.* 67:12–16

99. Pallaghy, C. K. 1968. Electrophysiological studies in guard cells of tobacco. *Planta* 80:147–53

100. Pallaghy, C. K. 1970. The effect of Ca^{++} on the ion specificity of stomatal opening in epidermal strips of *Vicia faba*. *Z. Pflanzenphysiol.* 62:58–62

101. Pallaghy, C. K., Fischer, R. A. 1974. Metabolic aspects of stomatal opening and ion accumulation by guard cells in *Vicia faba*. *Z. Pflanzenphysiol.* 71:332–44

102. Pallas, J. E. Jr. 1969. Stomatal operation. *What's New in Plant Physiology* 1:1–6

103. Pallas, J. E. Jr., Dilley, R. A. 1972. Photophosphorylation can provide sufficient adenosine 5'-triphosphate to drive K^+ movements during stomatal opening. *Plant Physiol.* 49:649–50

104. Pemadasa, M. A. 1981. Abaxial and adaxial stomatal behaviour and responses to fusicoccin on isolated epidermis of *Commelina communis* L. *New Phytol.* 89:373–84

105. Pemadasa, M. A. 1982. Abaxial and adaxial stomatal responses to light of different wavelengths and to phenylacetic acid on isolated epidermes of *Commelina communis* L. *J. Exp. Bot.* 33:92–99

106. Pemadasa, M. A. 1982. Differential abaxial and adaxial stomatal responses to indole-3-acetic acid in *Commelina communis* L. *New Phytol.* 90:209–19

107. Penny, M. G., Bowling, D. J. F. 1974. A study of potassium gradients in the epidermis of intact leaves of *Commelina communis* L. in relation to stomatal opening. *Planta* 119:17–25

108. Penny, M. G., Bowling, D. J. F. 1975. Direct determination of pH in the stomatal complex of *Commelina*. *Planta* 122:209–12

109. Penny, M. G., Kelday, L. S., Bowling, D. J. F. 1976. Active chloride transport in the leaf epidermis of *Commelina communis* in relation to stomatal activity. *Planta* 130:291–94

110. Poole, R. J. 1978. Energy coupling for membrane transport. *Ann. Rev. Plant Physiol.* 29:437–60

111. Radin, J. W., Ackerson, R. C. 1982. Does abscisic acid control stomatal closure during water stress? *What's New in Plant Physiology* 13:9–12

112. Raghavendra, A. S. 1981. Energy supply for stomatal opening in epidermal strips of *Commelina benghalensis*. *Plant Physiol.* 67:385–87

113. Raschke, K. 1972. Saturation kinetics of the velocity of stomatal closing in response to CO_2. *Plant Physiol.* 49:229–34

114. Raschke, K. 1975. Stomatal action. *Ann. Rev. Plant Physiol.* 26:309–40

115. Raschke, K. 1977. The stomatal turgor mechanism and its responses to CO_2 and abscisic acid: observations and a hypothesis. In *Regulation of Cell Membrane Activities in Plants*, ed. E. Marrè, O. Ciferri, pp. 173–83. Amsterdam: North-Holland. 332 pp.

116. Raschke, K. 1979. Movements of stomata. In *Physiology of Movements*, ed. W. Haupt, M. E. Feinleib. *Encycl. Plant Physiol.* (NS) 7:383–441. Berlin: Springer

117. Raschke, K., Humble, G. D. 1973. No uptake of anions required by opening stomata of *Vicia faba*: guard cells release hydrogen ions. *Planta* 115:47–57

118. Raschke, K., Schnabl, H. 1978. Availability of chloride affects the balance between potassium chloride and potassium malate in guard cells of *Vicia faba* L. *Plant Physiol.* 62:84–87

119. Raven, J. A. 1981. Light quality and solute transport. In *Plants and the Daylight Spectrum*, ed. H. Smith, pp. 375–90. New York: Academic. 508 pp.

120. Raven, J. A. 1981. Introduction to metabolic control. In *Mathematics and Plant Physiology*, ed. D. A. Rose, D. A Charles-Edwards, pp. 3–27. New York Academic. 320 pp.

121. Roberts, J. K. M., Ray, P. M., Wade-Jardetzky, N., Jardetzky, O. 1980. Estimation of cytoplasmic and vacuolar pH in higher plant cells by ^{31}P NMR. *Nature* 283:870–72

122. Roberts, J. K. M., Ray, P. M., Wade-Jardetzky, N., Jardetzky, O. 1981. Extent of intracellular pH changes during H^+ extrusion by maize root-tip cells. *Planta* 152:74–78

123. Rogers, C. A., Powell, R. D., Sharpe, P. J. H. 1979. Relationship of temperature to stomatal aperture and potassium accumulation in guard cells of *Vicia faba*. *Plant Physiol.* 63:388–91

124. Rogers, C. A., Sharpe, P. J. H., Powell, R. D. 1980. Dark opening of stomates of *Vicia faba* in CO_2-free air. Effect of temperature on stomatal aperture and potassium accumulation. *Plant Physiol.* 65:1036–38

125. Rogers, C. A., Sharpe, P. J. H., Powell, R. D., Spence, R. D. 1981. High-temperature disruption of guard cells of

Vicia faba. Effect on stomatal aperture. *Plant Physiol.* 67:193–96

126. Roth-Bejerano, N., Itai, C. 1981. Involvement of phytochrome in stomatal movement: effect of blue and red light. *Physiol. Plant.* 52:201–6

127. Roth-Bejerano, N., Nejidat, A., Itai, C. 1982. Phytochrome-membrane interactions as a factor in stomatal opening. *Physiol. Plant.* 56:80–83

128. Rutter, J. C., Willmer, C. M. 1979. A light and electron microscopy study of the epidermis of *Paphiopedilum* spp., with emphasis on stomatal ultrastructure. *Plant Cell Environ.* 2:211–19

129. Saftner, R. A., Raschke, K. 1981. Electrical potentials in stomatal complexes. *Plant Physiol.* 67:1124–32

130. Scarth, G. W. 1932. Mechanism of the action of light and other factors on stomatal movement. *Plant Physiol.* 7:481–504

131. Schittengruber, B. 1953. Chromoplasten fehlen den Schliesszellen. *Protoplasma* 42:102–4

132. Schnabl, H. 1978. The effect of Cl- upon the sensitivity of starch-containing and starch-deficient stomata and guard cell protoplasts towards potassium ions, fusicoccin and abscisic acid. *Planta* 144:95–100

133. Schnabl, H. 1981. The compartmentation of carboxylating and decarboxylating enzymes in guard cell protoplasts. *Planta* 152:307–13

134. Schnabl, H., Bornman, C. H., Ziegler, H. 1978. Studies on isolated starch-containing (*Vicia faba*) and starch-deficient (*Allium cepa*) guard cell protoplasts. *Planta* 143:33–39

135. Schnabl, H., Hampp, R. 1980. *Vicia* guard cell protoplasts lack photosystem II activity. *Naturwissenschaften* 67:465–66

136. Schnabl, H., Raschke, K. 1980. Potassium chloride as stomatal osmoticum in *Allium cepa* L., a species devoid of starch in guard cells. *Plant Physiol.* 65:88–93

137. Schnabl, H., Ziegler, H. 1977. The mechanism of stomatal movement in *Allium cepa* L. *Planta* 136:37–43

138. Schulze, E. D., Lange, O. L., Evenari, M., Kappen, L., Buschbom, U. 1980. Long-term effects of drought on wild and cultivated plants in the Negev Desert. II. Diurnal patterns of net photosynthesis and daily carbon gain. *Oecologia* 45:19–25

139. Schwartz, A., Zeiger, E. 1982. Bioenergetics of stomatal opening. *Plant Physiol.* 69:83 (Abstr.)

140. Senger, H. 1982. The effect of blue light on plants and microorganisms. *Photochem. Photobiol.* 35:911–20

141. Senger, H., Briggs, W. R. 1981. The blue light receptor(s): primary reactions and subsequent metabolic changes. *Photochem. Photobiol.* 6:1–38

142. Sharkey, T. D., Imai, K., Farquhar, G. D., Cowan, I. R. 1982. A direct confirmation of the standard method of estimating intercellular partial pressure of CO_2. *Plant Physiol.* 69:657–59

143. Sharkey, T. D., Raschke, K. 1981. Separation and measurement of direct and indirect effects of light on stomata. *Plant Physiol.* 68:33–40

144. Sharkey, T. D., Raschke, K. 1981. Effect of light quality on stomatal opening in leaves of *Xanthium strumarium* L. *Plant Physiol.* 68:1170–74

145. Sharpe, P. J. H., Zeiger, E. 1981. Chemiosmotic hypothesis of ion transport in guard cells. In *Integrated View of Guard Cells*, ed. C. Rogers, pp. 47–64. Symp. South. Sect. Soc. Plant Physiol. 64 pp.

146. Sheriff, D. W. 1979. Stomatal aperture and the sensing of the environment by guard cells. *Plant Cell Environ.* 2:15–22

146a. Shimazaki, K., Gotow, K., Kondo, N. 1982. Photosynthetic properties of guard cell protoplasts from *Vicia faba* L. *Plant Cell Physiol.* 23:871–79

147. Simons, P. J. 1981. The role of electricity in plant movements. *New Phytol.* 87:11–37

148. Skaar, H., Johnsson, A. 1978. Rapid, blue light induced transpiration in *Avena*. *Physiol. Plant.* 43:390–96

149. Skaar, H., Johnsson, A. 1980. Light induced transpiration in a chlorophyll deficient mutant of *Hordeum vulgare*. *Physiol. Plant.* 49:210–14

150. Smith, F. A., Raven, J. A. 1979. Intracellular pH and its regulation. *Ann. Rev. Plant Physiol.* 30:289–311

151. Smith, H. 1982. Light quality, photoperception, and plant strategy. *Ann. Rev. Plant Physiol.* 33:481–518

152. Snaith, P. J., Mansfield, T. A. 1982. Control of the CO_2 responses of stomata by indol-3ylacetic acid and abscisic acid. *J. Exp. Bot.* 33:360–65

153. Snaith, P. J., Mansfield, T. A. 1982. Stomatal sensitivity to abscisic acid: Can it be defined? *Plant Cell Environ.* 5:309–11

154. Song, P. -S., Fugate, R. D., Briggs, W. R. 1980. Flavin as a photoreceptor for phototropic transduction: fluorescence studies of model and corn coleoptile systems. In *Flavins and Flavoproteins*, ed. K. Yagi, T. Yamano, pp. 443–53. Tokyo: Japan Sci. Soc. 740 pp.

155. Spanswick, R. M. 1981. Electrogenic ion pumps. *Ann. Rev. Plant Physiol.* 32:267–89

156. Squire, G. R., Mansfield, T. A. 1972. A simple method of isolating stomata on detached epidermis by low pH treatment: observations of the importance of the subsidiary cells. *New Phytol.* 71:1033–43

157. Stålfelt, M. G. 1962. The effect of temperature on opening of the stomatal cells. *Physiol. Plant.* 15:772–79

158. Swain, T. 1976. Nature and properties of flavonoids. In *Chemistry and Biochemistry of Plant Pigments,* ed. T. W. Goodwin, 1:425–63. London: Academic. 870 pp. 2nd ed.

159. Thimann, K. V., ed. 1980. *Senescence in Plants.* Boca Raton, Florida: CRC. 276 pp.

160. Thomas, D. A. 1975. Stomata. In *Ion Transport in Plant Cells and Tissues,* ed. D. A. Baker, J. L. Hall, pp. 377–412. London: North-Holland. 437 pp.

161. Travis, A. J., Mansfield, T. A. 1979. Stomatal responses to light and CO_2 are dependent on KCl concentration. *Plant Cell Environ.* 2:319–23

162. Travis, A. J., Mansfield, T. A. 1981. Light saturation of stomatal opening on the adaxial and abaxial epidermis of *Commelina communis. J. Exp. Bot.* 32:1169–79

163. Turner, N. C. 1973. Action of fusicoccin on the potassium balance of guard cells of *Phaseolus vulgaris. Am. J. Bot.* 60:717–25

164. Van Kirk, C. A., Raschke, K. 1978. Presence of chloride reduces malate production in epidermis during stomatal opening. *Plant Physiol.* 61:361–64

165. Vierstra, R. D., John, T. R., Poff, K. L. 1982. Kaempferol 3-O-Galactoside, 7-O-Rhamnoside is the major green fluorescing compound in the epidermis of *Vicia faba. Plant Physiol.* 69:522–25

166. Walker, D. A., Zelitch, I. 1963. Some effects of metabolic inhibitors, temperature and anaerobic conditions on stomatal movement. *Plant Physiol.* 38:390–96

167. Walton, D. C. 1980. Biochemistry and physiology of abscisic acid. *Ann. Rev. Plant Physiol.* 31:453–89

168. Weiler, E. W., Schnabl, H., Hornberg, C. 1982. Stress-related levels of abscisic acid in guard cell protoplasts of *Vicia faba* L. *Planta* 154:24–28

169. Weyers, J. D. B., Hillman, J. R. 1979. Uptake and distribution of ABA in *Commelina* leaf epidermis. *Planta* 144:167–72

170. Weyers, J. D. B., Hillman, J. R. 1980. Effects of abscisic acid on $^{86}Rb^+$ fluxes in *Commelina communis* L. leaf epidermis. *J. Exp. Bot.* 31:711–20

171. Weyers, J. D. B., Travis, A. J. 1981. Selection and preparation of leaf epidermis for experiments on stomatal physiology. *J. Exp. Bot.* 32:837–50

172. Williams, W. E., Grivet, C., Zeiger, E. 1982. Photosynthesis and stomatal conductance in leaves of *Paphiopedilum:* low rates, typical of shade plants. *Plant Physiol.* 69:84 (Abstr.)

173. Willmer, C. M., Mansfield, T. A. 1970. Effects of some metabolic inhibitors and temperature on ion-stimulated stomatal opening in detached epidermis. *New Phytol.* 69:983–92

174. Wilson, J. A. 1981. Stomatal responses to applied ABA and CO_2 in epidermis detached from well-watered and water-stressed plants of *Commelina communis* L. *J. Exp. Bot.* 32:261–69

175. Wilson, J. A., Ogunkanmi, A. B., Mansfield, T. A. 1978. Effects of external potassium supply on stomatal closure induced by abscisic acid. *Plant Cell Environ.* 1:199–201

176. Winner, W. E., Mooney, H. A. 1980. Responses of Hawaiian plants to volcanic sulfur dioxide: stomatal behavior and foliar injury. *Science* 210:789–91

177. Wong, S. C., Cowan, I. R., Farquhar, G. D. 1978. Leaf conductance in relation to assimilation in *Eucalyptus pauciflora* Sieb. ex Spreng. Influence of irradiance and partial pressure of carbon dioxide. *Plant Physiol.* 62:670–74

178. Wong, S. C., Cowan, I. R., Farquhar, G. D. 1979. Stomatal conductance correlates with photosynthetic capacity. *Nature* 282:424–26

179. Wu, H.-I., Sharpe, P. J. H. 1979. Stomatal mechanics. II. Material properties of guard cell walls. *Plant Cell Environ.* 2:325–44

180. Zeiger, E. 1980. The blue light response of stomata and the green vacuolar fluorescence of guard cells. In *The Blue Light Syndrome,* ed. H. Senger, pp. 629–36. Berlin: Springer. 665 pp.

181. Zeiger, E. 1981. Novel approaches to the biology of stomatal guard cells: protoplast and fluorescence studies. See Ref. 2, pp. 103–17

182. Zeiger, E., Armond, P., Melis, A. 1981. Fluorescence properties of guard cell chloroplasts. *Plant Physiol.* 67:17–20

183. Zeiger, E., Bloom, A. J., Hepler, P. K. 1978. Ion transport in stomatal guard cells: A chemiosmotic hypothesis. *What's New in Plant Physiology* 9:29–32

184. Zeiger, E., Field, C. 1982. Photocontrol of the functional coupling between photosynthesis and stomatal conductance in the intact leaf. *Plant Physiol.* 70:370–75

185. Zeiger, E., Field, C., Mooney, H. A. 1981. Stomatal opening at dawn: possible roles of the blue light response in nature. See Ref. 119, pp. 391–407

186. Zeiger, E., Hepler, P. K. 1976. Production of guard cell protoplasts from onion and tobacco. *Plant Physiol.* 58:492–98

187. Zeiger, E., Hepler, P. K. 1977. Light and stomatal function: blue light stimulates swelling of guard cell protoplasts. *Science* 196:887–89

188. Zeiger, E., Hepler, P. K. 1979. Blue light-induced, intrinsic vacuolar fluorescence in onion guard cells. *J. Cell Sci.* 37:1–10

189. Zeiger, E., Moody, W., Hepler, P., Varela, F. 1977. Light-sensitive membrane potentials in onion guard cells. *Nature* 270:270–71

190. Zeiger, E., Schwartz, A. 1982. Longevity of guard cell chloroplasts in falling leaves: implication for stomatal function and cellular aging. *Science* 218:680–82

191. Zelitch, I. 1965. Environmental and biochemical control of stomatal movement in leaves. *Biol. Rev.* 40:463–82

SUBJECT INDEX

477

CUMULATIVE INDEXES

CONTRIBUTING AUTHORS, VOLUMES 28–34

A

Amesz, J., 29:47–66
Amrhein, N., 28:123–32
Andersen, K., 29:263–76
Archer, R. R., 28:23–43

B

Bahr, J. T., 28:379–400
Bandurski, R. S., 33:403–30
Barber, J., 33:261–95
Bauer, W. D., 32:407–49
Beale, S. I., 29:95–120; 34:241–78
Bedbrook, J. R., 30:593–620
Beevers, H., 30:159–93
Bell, A. A., 32:21–81
Benedict, C. R., 29:67–93
Bernstam, V. A., 29:25–46
Berry, J., 31:491–543
Bewley, J. D., 30:195–238
Biale, J. B., 29:1–23
Björkman, O., 31:491–543
Boardman, N. K., 28:355–77
Bottomley, W., 34:279–310
Brenner, M. L., 32:511–38
Buchanan, B. B., 31:341–74
Bünning, E., 28:1–22
Butler, W. L., 29:345–78

C

Castelfranco, P. A., 34:241–78
Chaney, R. L., 29:511–66
Chapman, D. J., 31:639–78
Clarke, A. E., 34:47–70
Clarkson, D. T., 31:239–98
Cogdell, R. J., 34:21–45
Cohen, J. D., 33:403–30
Conn, E. E., 31:433–51
Craigie, J. S., 30:41–53
Cramer, W. A., 28:133–72
Cronshaw, J., 32:465–84

D

Davies, D. D., 30:131–58
Dennis, D. T., 33:27–50

Diener, T. O., 32:313–25
Digby, J., 31:131–48

E

Eisbrenner, G., 34:105–36
Eisinger, W., 34:225–40
Elbein, A. D., 30:239–72
Ellis, R. J., 32:111–37
Elstner, E. F., 33:73–96
Evans, H. J., 34:105–36
Evans, L. T., 32:485–509
Evert, R. F., 28:199–222

F

Farquhar, G. D., 33:317–45
Feldman, J. F., 33:583–608
Fincher, G. B., 34:47–70
Firn, R. D., 31:131–48
Fischer, R. A., 29:277–317
Flavell, R., 31:569–96
Flowers, T. J., 28:89–121
Foy, C. D., 29:511–66
French, C. S., 30:1–26

G

Galston, A. W., 32:83–110
Galun, E., 32:237–66
Gantt, E., 32:327–47
Giaquinta, R. T., 34:347–87
Gifford, E. M. Jr., 34:419–40
Gifford, R. M., 32:485–509
Glass, A. D. M., 34:311–26
Goldsmith, M. H. M., 28:439–78
Goodwin, T. W., 30:369–404
Graham, D., 33:347–72
Gray, M. W., 33:373–402
Green, P. B., 31:51–82
Greenway, H., 31:149–90
Grisebach, H., 30:105–30
Guerrero, M. G., 32:169–204
Gunning, B. E. S., 33:651–98

H

Hahlbrock, K., 30:105–30
Halperin, W., 29:239–62

Hanson, A. D., 33:163–203
Hanson, J. B., 31:239–98
Hardham, A. R., 33:651–98
Harding, R. W., 31:217–38
Haselkorn, R., 29:319–44
Haupt, W., 33:205–33
Heath, R. L., 31:395–431
Heber, U., 32:139–68
Hedden, P., 29:149–92
Heldt, H. W., 32:139–68
Hendricks, S. B., 28:331–54
Hitz, W. D., 33:163–203
Hotta, Y., 29:415–36
Howell, S. H., 33:609–50

J

Jacobsen, J. V., 28:537–64
Jensen, R. G., 28:379–400
Junge, W., 28:503–36

K

Kamiya, N., 32:205–36
Kochert, G., 29:461–86
Kolattukudy, P. E., 32:539–67
Kolodner, R. 30:593–620
Kowallik, W., 33:51–72
Kung, S. D., 28:401–37

L

Labavitch, J. M., 32:385–406
Lang, A., 31:1–28
Laties, G. G., 33:519–55
Lea, P. J., 28:299–329
Leaver, C. J., 33:373–402
Letham, D. S., 34:163–97
Lieberman, M., 30:533–91
Loewus, F. A., 34:137–61
Loewus, M. W., 34:137–61
Loomis, R. S., 30:339–67
Lorimer, G. H., 32:349–83
Losada, M., 32:169–204
Lucas, W. J., 34:71–104

M

MacMillan, J., 29:149–92
Malkin, R., 33:455–79

CHAPTER TITLES, VOLUMES 28–34